SEVENTH CANADIAN EDITION

ELECTRICAL WIRING: RESIDENTIAL

BASED ON THE 2015 CANADIAN ELECTRICAL CODE

RAY C. MULLIN • TONY BRANCH • SANDY F. GEROLIMON • CRAIG TRINEER • BILL TODD • PHIL SIMMONS

Electrical Wiring: Residential, Seventh Canadian Edition

by Ray C. Mullin, Tony Branch, Sandy F. Gerolimon, Craig Trineer, Bill Todd, and Phil Simmons

Vice President, Editorial Higher Education:
Anne Williams

Publisher:
Jackie Wood

Marketing Manager:
Alexis Hood

Technical Reviewer:
Tony Poirier

Developmental Editor:
Katherine Goodes

Photo Researcher and Permissions Coordinator:
Julie Pratt

Production Project Manager:
Jennifer Hare

Production Service:
MPS Limited

Copy Editor:
June Trusty

Proofreader:
MPS Limited

Indexer:
Elizabeth Walker

Design Director:
Ken Phipps

Managing Designer:
Franca Amore

Cover Design:
Courtney Hellam

Cover and Chapter Opener Image:
idea for life/Shutterstock.com

Compositor:
MPS Limited

Printed and bound in the United States of America
1 2 3 4 18 17 16 15

For more information contact Nelson Education Ltd., 1120 Birchmount Road, Toronto, Ontario, M1K 5G4. Or you can visit our Internet site at http://www .nelson.com

Library and Archives Canada Cataloguing in Publication Data

Mullin, Ray C., author
Electrical wiring : residential / Ray C. Mullin, Tony Branch, Sandy F. Gerolimon, Craig Trineer, Bill Todd, Phil Simmons. — Seventh Canadian edition.

Includes index.
Revision of: Electrical wiring : residential. 6th Canadian ed. Toronto : Nelson Education, [2012], ©2013.
Based on the 2015 Canadian Electrical Code.

ISBN 978-0-17-657045-3 (paperback)

1. Electric wiring, Interior. 2. Dwellings—Electric equipment. I. Title.

TK3285.M84 2015 621.319'24
C2015-903084-6

ISBN-13: 978-0-17-657045-3
ISBN-10: 0-17-657045-4

Brief Contents

Contents

UNIT 13

UNIT 26

Introduction

This seventh Canadian edition of *Electrical Wiring: Residential* is based on the 2015 edition of the *Canadian Electrical Code, Part I (CEC)*, the safety standard for electrical installations. The authors welcome readers' suggestions for improvements that may be included in subsequent editions of the text, as all of the authors are, themselves, tradesmen and see the advantage of having a second set of eyes on a problem. Trade knowledge is constantly evolving as new products and new methodologies are incorporated into the community knowledge base, so the authors are constantly looking for new perspectives.

Electrical Wiring: Residential provides an entry-level text that is both comprehensive and readable. It is suitable for colleges, technical institutes, vocational/technical schools, and electrical programs in high schools, and for readers who want to study the subject on their own.

The wiring in the home illustrated in *Electrical Wiring: Residential* incorporates more features than are absolutely necessary, to present as many *Canadian Electrical Code (CEC)* rules as possible so that students are provided with the information they need to complete a safe installation. The text focuses on the technical skills required to perform electrical installations. Topics include calculating conductor sizes, calculating voltage drop, sizing services, connecting electrical appliances, grounding and bonding equipment, and installing various fixtures. These are critical skills that can make the difference between an installation that "meets code" and one that is exceptional. An electrician must understand that the reason for following *CEC* regulations is to achieve an installation that is essentially free from hazard to life and property.

The *CEC* is the basic standard for the layout and construction of all electrical systems in Canada. However, some provincial and local codes may contain specific amendments that must be adhered to in all electrical wiring installations in those jurisdictions. Therefore, regional authority supersedes the *CEC*.

The authors encourage the reader to develop a detailed knowledge of the layout and content of the *CEC*, which must be used in conjunction with a comprehensive study of *Electrical Wiring: Residential* to derive the greatest benefit from the text.

The *CEC* is divided into numbered sections, each covering a main division. These sections are further divided into numbered rules, subrules, paragraphs, and subparagraphs. All references in the text are to the *section* or to the *section rule number*. For example, *Rule 8–200(1)(a)(i)* refers to *Section 8, Rule 8–200, Subrule 1, Paragraph a, Subparagraph i*. This explanation should assist the student in locating *CEC* references in the text.

Thorough explanations are provided throughout, as the text guides the student through the steps necessary to become proficient in the techniques and *CEC* requirements described.

Preface

Electrical Wiring: Residential, Seventh Canadian Edition, will prove a valuable resource to instructors and students alike. It includes 2015 *Canadian Electrical Code, Part I* references and wiring techniques. Each chapter is a complete lesson ending with review questions to summarize the material covered. The chapters are sequenced to introduce the student to basic principles and wiring practices, and progress to more advanced areas of residential electrical wiring.

This text assumes no prior knowledge of residential electrical wiring, but the student will need a reasonable level of mechanical aptitude and skill to be successful in the practical application of the techniques discussed.

The text guides students through the working drawings for a residential electrical installation, the proper wiring of receptacles, and the minimum required number of lighting and power branch circuits. Voltage drop calculations based on the *CEC* are shown. Grounding, bonding, and ground-fault circuit interrupters are discussed, together with the lighting branch circuits for all of the rooms in the house, as well as for the garage, workshop, and exterior. Special-purpose outlets for ranges, dryers, air conditioners, water heaters, and water pumps are explained, as are electric, oil, and gas heating systems; heat detectors, smoke detectors, and smoke alarms; swimming pools, spas, and hot tubs; and low-voltage-signal systems, security systems, and low-voltage remote-control systems, or "smart house wiring."

The text details service entrance equipment and the installation of the electrical service, including the calculations for the sizing of the service entrance conductors, conduit, switch, grounding conductor, and panel.

FEATURES

- New! All text material fully updated to reference the 2015 edition of the *CEC*.
- New! The different types of technical drawings are now discussed in Unit 2.
- New! Green technology content is updated and increased, where applicable.
- New! Referencing of appendixes has been increased where mentioned in the text.
- New! *Canadian Electrical Code* cross-referencing index, which references the *CEC* sections mentioned in the text.
- New! Four-colour art and photos throughout.
- New! Four-colour text design.
- Metrication is in keeping with the 2015 *CEC*.
- Descriptions of Canadian practice and applications.
- Numerous diagrams of Canadian Standards Association (CSA)-approved equipment found in everyday wiring installations.

- Step-by-step explanations of wiring a typical residence.
- Complete set of blueprints to help apply *CEC* theory to real working drawings.

INSTRUCTOR RESOURCES

The **Nelson Education Teaching Advantage (NETA)** program delivers research-based instructor resources that promote student engagement and higher-order thinking to enable the success of Canadian students and educators. Be sure to visit Nelson Education's **Inspired Instruction** website at www.nelson.com/inspired to find out more about NETA.

Accessing Instructor Supplements from SSO Dashboard

1. Go to http://www.nelson.com/instructor and log in using the instructor e-mail address and password.
2. Enter author, title, or ISBN in the **Add a title to your bookshelf** search.
3. Click **Add to my bookshelf** to add instructor resources.
4. At the Product page, click the **Instructor Companion site** link.

The following instructor resources have been created for *Electrical Wiring: Residential*, Seventh Canadian Edition, and can be accessed from the SSO dashboard.

MindTap

Offering personalized paths of dynamic assignments and applications, **MindTap** is a digital learning solution that turns cookie-cutter into cutting-edge, apathy into engagement, and memorizers into higher-level thinkers. MindTap enables students to analyze and apply chapter concepts within relevant assignments, and allows instructors to measure skills and promote better outcomes with ease. A fully online learning solution, MindTap combines all student learning tools—readings, multimedia, activities, and assessments—into a single Learning Path that guides the student through the curriculum. Instructors personalize the experience by customizing the presentation of these learning tools to their students, even seamlessly introducing their own content into the Learning Path.

NETA Test Bank

This resource was written by Tony Fazzari, Mohawk College. It includes over 660 multiple-choice questions written according to NETA guidelines for effective construction and development of higher-order questions. Also included are over 125 short-answer questions. The technical check was performed by Tony Poirier, Durham College.

The NETA Test Bank is available in a new, cloud-based platform. Nelson Testing Powered by Cognero® is a secure online testing system that allows you to author, edit, and manage test bank content from any place you have Internet access. No special installations or downloads are needed, and the desktop-inspired interface, with its drop-down menus and familiar, intuitive tools, allows you to create and manage tests with ease. You can create multiple test versions in an instant, and import or export content into other systems. Tests can be delivered from your learning management system, your classroom, or wherever you want. Nelson Testing Powered by Cognero for *Electrical Wiring: Residential*, Seventh Canadian Edition, can be accessed through http://www.nelson.com/instructor.

NETA PowerPoint

Microsoft® PowerPoint® lecture slides for every chapter have been created by the text's co-author, Sandy Gerolimon, Humber College. There is an average of 29 slides per chapter, many featuring key figures, tables, and photographs from *Electrical*

Wiring: Residential, Seventh Canadian Edition. NETA principles of clear design and engaging content have been incorporated throughout, making it simple for instructors to customize the deck for their courses.

Image Library

This resource consists of digital copies of figures, short tables, and photographs used in the book. Instructors may use these JPEGs to customize the NETA PowerPoint program or create their own PowerPoint presentations.

Blueprints

Revised and updated by Nick Palazzo, Humber College, and technically checked by all of the textbook authors, the set of blueprints for the house used as an example accompanies this textbook.

Instructor's Solutions Manual

The Instructor's Solutions Manual to accompany *Electrical Wiring: Residential,* Seventh Canadian Edition, has been prepared by the text's co-author, Sandy Gerolimon, Humber College. To assist the instructor, questions from the end of each unit have been included with answers/solutions. These may be used as additional review or examination questions.

Day One Slides

Day One—Prof In Class is a PowerPoint presentation that instructors can customize to orient students to the class and their textbook at the beginning of the course.

STUDENT ANCILLARIES

MindTap

Stay organized and efficient with **MindTap**—a single destination with all of the course material and study aids you need to succeed. Built-in apps leverage social media and the latest learning technology. For example:

- ReadSpeaker will read the text to you.
- Flashcards are pre-populated to provide you with a jump-start for review—or you can create your own.
- You can highlight text and make notes in your MindTap Reader. Your notes will flow into Evernote, the electronic notebook app that you can access anywhere when it's time to study for the exam.
- Self-quizzing allows you to assess your understanding.

Visit www.nelson.com/student to start using **MindTap**. Enter the online access code from the card included with your textbook. If a code card is not provided, you can purchase instant access at NELSONbrain.com.

Blueprints

Revised and updated by Nick Palazzo, Humber College, and technically checked by the text authors, the set of blueprints accompanies this textbook.

Acknowledgments

Every effort has been made to be technically correct and to avoid typographical errors. The authors wish to acknowledge the valuable assistance of Bill Wright in the preparation of the original architectural drawings for this book. Thank you to Tony Poirier, Durham College, for a thorough technical check of the text for adherence to the *CEC 2015*. Thank you as well to the people at Nelson Education Ltd. for their encouragement and professional advice.

The authors are also indebted to reviewers of each textbook in the *Electrical Wiring* series for their comments and suggestions:

Sam Agnew, Centennial College
Neil Blackadar, Nova Scotia Community College
Rob Brown, Fanshawe College
Michael Gillis, Nova Scotia Community College
Bruce Kellington, University College of the North
George Locke, Mohawk College
Shawn MaCaulay, Nova Scotia Community College
Ken Malczewski, Red River College
Mike Mankulich, Fanshawe College
Keith Mercer, Centennial College,
Tony Poirier, Durham College
Orest Staneckyj, Cambrian College
Violin Voynov, Durham College

We wish to thank the following for contributing data, illustrations, and technical information:

AFC/A Nortek Co.
AFL Global
Anchor Electric Division, Sola Basic Industries
Appleton Electric Co.
Arrow-Hart, Inc.
Brazos Technologies, Inc.
Brian Cameron
BRK Electronics, A Division of Pittway Corporation
Broan-NuTone Canada
Bussmann Division, Cooper Industries
Carlon
Chromalox
Cisco
Commander
Contrast Lighting M.L. Inc.
Cooper Industries
CSA Group
Eaton Canada
Electri-Flex Co.
Emerson Heating Products/Emerson Electric Company
ERICO International Corporation
General Electric Co.
Gould Shawmut
Halo Lighting Division, Cooper Industries
Heyco Molded Products, Inc.
Honeywell
Hubbell Incorporated, Wiring Devices Division
International Association of Electrical Inspectors
Intermatic

IPEX
J. Wampler
Juno
KEPTEL
Kohler Co.
Kohler Power Systems
Landis+Gyr
Legrand/Pass & Seymour, Inc.
Legrand/Wiremold
Leviton Manufacturing
Lightolier Canada
Linksys/Cisco Systems
Midwest Electric Products, Inc.
Milbank Manufacturing Co.
MM Plastics
Moe Light Division, Thomas Industries
National Fire Protection Association
Noma Division, Danbel Industries Inc.
Northern Cables Inc.
Progress Lighting
Rheem Manufacturing Co.
Robert Nolan/Shutterstock
Schneider Electric
Seatek Co.
Siemens Energy and Automation
Sierra Electric Division, Sola Basic Industries
SMART HOUSE, L.P.
SMC
Southwire Company
Square D Co.
Standards Council of Canada
Superior Electric Co.
THERM-O-DISC, Inc.
Thomas & Betts Incorporated
Underwriters Laboratories of Canada
(ULC is accredited by the Standards Council of Canada as a certification organization, a testing organization, a registration organization, and standards organization under the National Standards System of Canada.)
Vitabath, DM Industries
Wilo
Winegard Company
Wiremold SE
Wolberg Electrical Supply Co., Inc.
Woodhead Industries, Inc.
Gary Gudbranson,
www.cottageontheedge.com

For any further information or to give feedback on this textbook, you can contact the authors by email at

sandy.gerolimon@humber.ca
craig.trineer@humber.ca
tony.branch@humber.ca

UNIT 1

General Information for Electrical Installations

OBJECTIVES

After studying this unit, you should be able to

- identify the purpose of codes and standards related to electrical systems
- describe the importance of the *Canadian Electrical Code (CEC)*
- describe the layout of the *CEC*
- identify approved equipment
- explain the importance of electrical inspection
- identify links to websites relating to the electrical trade
- list the agencies that are responsible for establishing electrical standards and ensuring that materials meet the standards
- discuss systems of measurement used on construction drawings
- begin to refer to the *CEC*

SAFETY

Electrical installations are required to be safe. Because of the ever-present danger of electric shock and/or fire due to the failure of an electrical system, electricians and electrical contractors must use approved materials and methods. Therefore, codes and standards are used to ensure that the exposure of users, installers, and maintenance personnel to the shock and fire hazards associated with electrical systems is minimized.

CODES AND STANDARDS

Codes are standards that deal with life safety issues and are enforceable by law. Residential electrical wiring systems are required to meet building codes and electrical codes. Building codes dictate such things as which rooms and areas (such as stairwells) require luminaires (lighting fixtures) and how they are to be controlled. Electrical codes deal with the actual installation of the luminaires and controls. Municipal, provincial, and federal governments can pass laws regarding the installation of electrical equipment.

The *Canadian Electrical Code, Part I (CEC)* is the standard that governs electrical work across Canada. It is adopted by the provinces, is enacted in legislation, and is the basis of a bill (with provincial amendments) that passes through the provincial legislature, becoming law and enforced by municipalities or provincial electrical safety authorities.

The first edition of the *CEC* was published in 1927. It is revised every three years to take into account changes in equipment, wiring systems, and new technologies. The standards steering committee for the *CEC* solicits comments and proposals from the electrical industry and others interested in electrical safety. Information about the structure and operation of the committee, how to propose an amendment, or how to request an interpretation can be found in *CEC Appendix C*. Copies of the *CEC* can be ordered from the Canadian Standards Association (CSA Group).

The object of the *CEC* is to provide a minimum safety standard for the installation and maintenance of electrical systems to safeguard people from the hazards of fire and electric shock arising from the use of electrical energy. Meeting the requirements of the *CEC* ensures an essentially safe installation.

The *CEC* does not take economics into consideration. For instance, a house built using wood frame construction may be wired with non-metallic-sheathed cable or armoured cable. Both meet the requirement of the *CEC,* but one costs several times as much as the other.

The *CEC* is divided up into evenly numbered sections, each one covering a main division of the work. Each section has a title that describes what will be covered by the section. Ten general sections cover basic installation requirements (see Table 1-1). The rest are amending or supplementary sections, which change the general requirements of the *Code* for specific types of equipment or installation (see Table 1-2). Rules in one amending section cannot be

TABLE 1-1

General sections of the *CEC.*

SECTION	TITLE
0	Object, scope, and definitions
2	General rules
4	Conductors
6	Services and service equipment
8	Circuit loading and demand factors
10	Grounding and bonding
12	Wiring methods
14	Protection and control
16	Class 1 and Class 2 circuits
26	Installation of electrical equipment

Courtesy of CSA Group

TABLE 1-2

Some supplementary or amending sections of the *CEC.*

SECTION	TITLE
28	Motors and generators
30	Installation of lighting equipment
32	Fire alarm systems and fire pumps
62	Fixed electric space and surface heating systems
68	Pools, tubs, and spas
76	Temporary wiring

Courtesy of CSA Group

applied to another amending section unless they specifically state that they may.

The sections are divided into subsections and numbered rules with captions. For example, *Section 12–500* contains non-metallic-sheathed cable rules, where *12* identifies the section, *500* is the rule number in the section. *12–500* is the first rule of the subsection that deals with the installation of non-metallic-sheathed cable. The *12–500* series of rules (*12–500* through *12–526*) describe how non-metallic-sheathed cables are to be installed.

EXAMPLE

Rule 6–112 provides information about support for the attachment of overhead supply or consumer's service conductors (see *Appendix B*).

Rule 6–112 in *Appendix B* outlines the components that may be used for an overhead mast service and methods of installation.

Appendix D is tabulated general information. This textbook will refer to several *CEC* tables, including *Tables D1, D3,* and *D6.*

Tables

At the end of the main body of the *CEC* are a number of tables, preceded by a list of all of the tables. At the top of each table is a list of rules that refer you to the table, which can be helpful in looking up rules. The tables cover such things as the ampacity of wires, number of conductors in raceways, number of conductors in boxes, and depth of cover over conductors installed underground.

Diagrams

After the tables are a number of diagrams, preceded by a list of all of the diagrams. Some of the diagrams included show pin configurations for both locking and non-locking receptacles.

Appendixes

Of the 11 *CEC* appendixes, the two that are important for you are *Appendix B* and *Appendix D*. *Appendix B* is "Notes on Rules." These notes give supplemental information about rules. Whenever a rule refers to *Appendix B,* you must read that information. *Appendix B* uses the same numbering system as the main body of the *CEC*. Simply look up in *Appendix B* the rule number from the main body of the *CEC*.

Table of Contents and Index

The *CEC* provides an index and a table of contents to help you look up information. If you know exactly what you want to look up, use the index at the back of the book. If you have just a general idea of what you are looking for, use the table of contents at the front of the book.

Some *Code* Terminology

- Acceptable: Acceptable to the authority having jurisdiction
- May: Permitted or allowable
- Notwithstanding: In spite of
- Practicable: Feasible or possible
- Shall: Indicates a mandatory requirement
- Shall be: Compulsory, mandatory, a requirement
- Shall have: The same as *shall be*
- Shall not: Not permitted, not allowed, must not be
- Shall be permitted: Is acceptable, is allowed, is permitted

This textbook is designed to be used in conjunction with the *CEC*. Throughout the text, reference will be made to the rules, tables, and appendixes found in the 23rd (2015) edition of the *CEC*. To distinguish these from the tables and appendixes in this text, all references to them are in italics.

Standards

Standards describe minimum performance levels for equipment and systems. The *Canadian Electrical Code, Part II,* found in *Appendix A* consists of safety standards for electrical equipment and is a mandatory part of the *CEC*. Another standard used in the electrical field is TIA-570A, which is used when installing communication cabling.

PERSONNEL SAFETY

When it comes to personnel safety, you have to make it your responsibility.

When you ask most people if they work safely, they answer that they do. The question is: Do you know enough about safe working practices to work safely? See if you can answer the five questions below. If you can't, you require additional safety training.

1. What are the proper procedures for locking out and tagging electrical equipment?
2. When is a worker required to wear a safety belt or safety harness on a construction site?
3. What are five pieces of personal protective equipment?
4. What are the eight classes of products identified by WHMIS symbols?
5. What precautions should be taken when using ladders for electrical installations?

You can obtain information about safe working practices from the provincial government, construction safety associations, and construction unions.

TESTING AND ACCREDITATION

All electrical equipment sold or installed in Canada is required to be approved for its intended use. Approved equipment has met specified safety standards set by federal and provincial governments. Currently several companies are accredited to approve electrical equipment. Equipment that has been approved by an approval agency will have an identifying label such as CSA or ULC.

If you are unsure about an approval label other than CSA or ULC, look for C at the eight o'clock position on the label.

For equipment that has not been approved, arrangements may be made for field approval by contacting the inspection authority in your province. This authority should also be able to provide you with a complete list of companies that can approve electrical equipment in your province.

Standards Council of Canada (SCC)

Courtesy of the Standards Council of Canada

The Standards Council of Canada (SCC) is a federal Crown corporation whose mandate is to promote efficient and effective voluntary standardization as a means of advancing the national economy, and benefiting the health, safety, and well-being of the public.

SCC fulfills its mandate by accrediting organizations engaged in standards development, certification and testing, by approving National Standards of Canada (NSC), by coordinating participation in international forums, such as the International Organization for Standardization (ISO) and the International Electrotechnical Commission (IEC), and by developing standardization solutions for government and industry.

Through these activities, SCC facilitates Canadian and international trade, protects consumers, and promotes international standards cooperation. For more information, visit https://www.scc.ca.

CSA Group

Courtesy of CSA Group

CSA Group is a not-for-profit, member-based organization formed in 1919 to develop standards that address the needs of a broad range of organizations and disciplines. These standards can be used for product testing and certification requirements and are often adopted by provincial and federal government authorities to form the basis for legislation and/or compliance requirements.

Thousands of manufacturers work directly with CSA Group to have their products tested and certified and the CSA mark (shown here) appears on billions of products around the world.

Courtesy of CSA Group

The *Canadian Electrical Code* is one of the many codes and standards that are developed by CSA Group. The *Canadian Electrical Code* has been adopted by provincial authorities to become law in the various provinces. For more information, visit www.csagroup.org.

Underwriters Laboratories of Canada (ULC)

Courtesy of the Underwriters Laboratories of Canada

Underwriters Laboratories of Canada (ULC) is an independent product safety testing, certification, and inspection organization. ULC has tested products for public safety for over 90 years and is accredited by the SCC.

ULC also publishes standards, specifications, and classifications for products having a bearing on fire, accident, or property hazards. For more information, contact one of ULC's offices in Montreal, Ottawa, Toronto, or Vancouver, or visit its website at canada.ul.com.

ELECTRICAL INSPECTION

Most electrical safety regulations require that a permit be obtained for the inspection of all work related to the installation, alteration, or repair of electrical equipment. It is the responsibility of the electrical contractor or person doing the work to obtain the permit. When the installation has passed inspection, the inspection authority will issue a connection permit (also called *authorization for connection* or *current permit)* to the supply authority. The supply authority will then make connection to the installation.

Since local codes vary, the electrician should also check with the local inspection authority. Electrical utility companies can supply additional information on local regulations. The electrician should be aware that any city or province adopting the *CEC* may do so with amendments and may also have additional licensing laws.

UNITS OF MEASURE

Two common systems of measurement are used in Canada: the International System of Units (abbreviated SI from its French name: Système international d'unités), or metric, and the American standard unit (ASU), which uses inches and pounds. Although SI is the official system of measurement in Canada, the ASU system is widely used. Refer to Tables 1-3 to 1-5.

TABLE 1-3

Metric units of measure.

PROPERTY	SI UNIT	SYMBOL
Length	Metre	m
Area	Square metre	m^2
Volume	Cubic metre	m^3
Time	Second	s
Speed (velocity)	Metres/second	m/s
Acceleration	Metres/second/squared	m/s^2
Temperature	Degrees Celsius	°C
Force	Newton	N
Energy	Joule	J
Work	Joule	J
Power	Watt	W
Mass	Kilogram	kg
Weight	Kilogram	kg
Pressure	Pascal	Pa
Flow	Litres/minute	L/m
Electric charge	Coulomb	C
Electric current	Ampere	A
Electric potential difference	Volt	V
Luminous flux	Lumen	lm
Illuminance	Lux	lx

TABLE 1-4

SI prefixes.

PREFIX	VALUE
Mega (M)	1 000 000
Kilo	1000
Hecto	100
Deka (deca)	10
Base unit (metre)	1
Deci	0.1
Centi (c)	0.01
Milli (m)	0.001
Micro (μ)	0.000 001
Nano (n)	0.000 000 001

TABLE 1-5

ASU units.

UNIT	VALUE
Inch	1/36 yard
Foot	12 inches
Yard	3 feet
Rod	5½ yards, 16 feet
Mile	320 rods, 1760 yards, 5280 feet

Length

The basic SI unit of measure for length is the metre. The SI system has a base of 10 (the next larger unit is 10 times the size of the smaller unit) and uses prefixes to identify the different units of length.

Construction drawings normally give SI dimensions in millimetres or metres to prevent confusion.

The ASU commonly uses inches, feet, yards, rods, and miles to measure length.

Construction drawings may use feet (′), inches (″), and fractions of an inch as units of measure. A standard steel tape measure that uses ASU units is divided into feet, inches, and fractions of an inch. The first 12 inches of the tape normally has divisions of 1/32 of an inch. The remainder of the tape is divided into 16ths of an inch. Figure 1-1 shows how to read a tape measure using ASU units. An architect's scale is used for working with construction drawings that use ASU units. Scales are covered in Unit 2. Table 1-6 gives useful conversions from ASU to metric units.

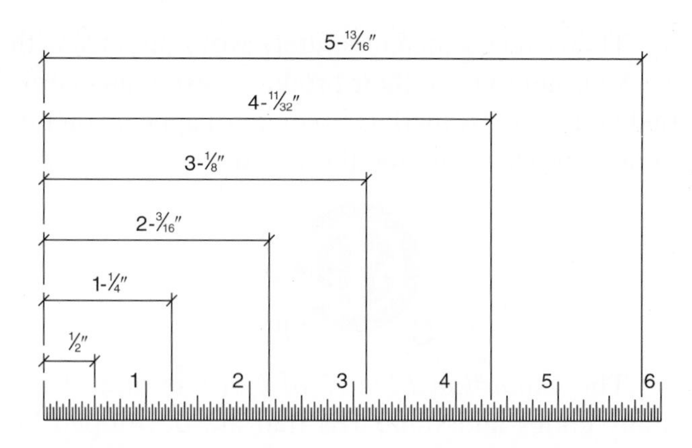

FIGURE 1-1 Reading tape measure divisions.

COMMON CONVERSIONS OF TRADE SIZES

Raceways such as rigid conduit and electrical metallic tubing (EMT) are designated by trade sizes because the actual inside diameter of a conduit or tubing is not its stated size. For example, ½″ (trade size) rigid steel conduit has an inside diameter of 0.632 inches, ½″ EMT has an internal diameter of 0.606 inches, and ½″ ENT (non-metallic tubing) has an internal diameter of 0.574 inches. All are considered to be ½″ (or 16 mm) trade size.

The 2015 *CEC* designates raceways in metric units only: The actual size of the raceway has not changed, only the fact that metric units of measure are used to designate the trade size of the raceway. Table 1-7 lists the equivalent metric and American designator units.

Outlet boxes are still referenced in inches. The volume of the boxes is given in both cubic inches and mL or cm^3.

American wire gauge (AWG) sizes are still used in the 2015 *CEC*. Metric wire sizes are not identical to AWG sizes. Standard metric sizes generally fall between American wire gauge sizes. Table 1-8 gives standard AWG sizes and their circular mil area and metric wire sizes and their equivalent circular mil sizes.

TABLE 1-6

Converting ASU to SI (metric) units.

PROPERTY	ASU UNIT	MULTIPLICATION FACTOR FOR CONVERSION TO SI	SI UNIT
Length	Inch	× 25.4	Millimetre
	Foot	× 304.8	Millimetre
	Yard	× 914.4	Millimetre
	Rod	× 5.0292	Metre
	Furlong	× 201.17	Metre
	Mile	× 1.609 347	Kilometre
Area	Sq. in.	× 645.16	Sq. mm
	Sq. ft.	× 92 903	Sq. mm
	Sq. yd.	× 0.8354	Sq. m
Volume	Cu. in.	× 16.39	Cubic centimetre (cm^3) or millilitre (mL)
	Cu. ft.	× 0.028 32	Cubic metre
	Cu. yd.	× 0.7646	Cubic metre
Liquid volume	Fl. oz.	× 29.6	Millilitre
	Gallon	× 3.78	Litre

TABLE 1-7

Metric and American designators for common trade sizes of raceways.

METRIC DESIGNATOR	AMERICAN DESIGNATOR
12	3⁄8
16	1⁄2
21	3⁄4
27	1
35	1 1⁄4
41	1 1⁄2
53	2
63	2 1⁄2
78	3
91	3 1⁄2
103	4
129	5
155	6

TABLE 1-8

Standard AWG wire sizes and metric (mm^2).

AMERICAN WIRE SIZE (AWG)	CIRCULAR MILS OR EQUIVALENT CIRCULAR MILS	METRIC WIRE SIZE (mM^2)
	937	0.50
20	1020	
18	1620	
	1974	1.0
16	2580	
	2960	1.5
14	4110	
	4934	2.5
12	6530	
	7894	4.0
10	10 380	
	11 840	6.0
8	16 510	
	19 740	10.0
6	26 240	
	31 580	16
4	41 740	
	49 340	25
3	52 620	
2	66 360	
	69 070	
1	83 690	
	98 680	50
1/0	105 600	
2/0	133 100	
	138 100	70
3/0	167 800	
	187 500	95
4/0	211 600	

Courtesy of Thomas & Betts Corporation

REVIEW

Where applicable, responses should be written in complete sentences.

1. What is the difference between a code and a standard?

✓ Standards describe minimum performance levels for equipment and systems. Codes are standards that deal with life safety issues and are enforcable by law.

2. What *Code* sets standards for the installation of electrical equipment?

✓ The Canadian Electrical Code, the (CEC) sets standards for the installation of electrical equipment

3. What authority enforces the standards set by the *CEC?* municipal and/or provincal

The Standards Council of Canada, ~~the CSA and the Underwriters Laboratories of Canada~~ all enforce the standards set by the CEC

4. Does the *CEC* provide minimum or maximum standards?

✓ CEC provides minimum standards.

5. What do the letters CSA signify?

✓ The Canadian Standards Association who test and certify products.

6. Does compliance with the *CEC* always result in an electrical installation that is adequate, safe, and efficient? Why?

✓ No not always. The electrician has to be aware that if their city or province is adopting the CEC may do so with amendments and additional licensing laws. Doesn't take economics into consideration

7. What are the general sections of the *CEC?* Basic installation requirements

Objects, Scope and Definitions, General Rules, Conductors, Services and service equipment, circuit loading and demand factors, grounding and bonding, wiring methods, protection and control, class 1 and class 2 circuits and installation of electrical equipment

8. Is the section of the *CEC* that deals with wiring methods a general section or an amending section?

✓ Wiring methods is a general section of the CEC

9. When is an electrical installation required to be inspected?

✓ When a permit is obtained electrical installation required to be inspected or repair

10. What should you look for when trying to determine if a piece of electrical equipment is approved for use in Canada?

Look for CSA or ULC label

11. If a piece of electrical equipment is not approved for use in Canada, what should you do?

An Application for field approval must be made to the provincal authority

12. When the words “shall be” appear in a code reference, they mean that it (must) (may) (does not have to) be done. (Underline the correct answer.)

Page 3 'Some code terminolgy'

Shall - be: Compulsory, mandatory

UNIT 2

Drawings and Specifications

OBJECTIVES

After studying this unit, you should be able to

- identify the various types of technical drawings
- identify the line types used in construction drawings
- visualize building views
- read common scales
- explain how electrical wiring information is conveyed to the electrician at the construction or installation site
- demonstrate how the specifications are used in estimating cost and in making electrical installations
- identify and explain the application of the various common line types used on drawings
- explain why symbols and notations are used on electrical drawings
- identify symbols used on architectural, mechanical, and electrical drawings

TECHNICAL DRAWINGS

Several types of drawings are used in the technical trades. This book uses three primary types: architectural drawings, schematic drawings, and wiring or layout diagrams (see an example of a schematic drawing and of a wiring diagram in Figure 2-1). Architectural drawings show the floor plans of a building and the relative positions and distances between elements within the building. Schematic drawings show all of the significant components of a system and their relationships to each other, but use simplified, standard icons to represent devices. Wiring diagrams are like schematic diagrams but the pictures used may look like the devices that they are meant to represent and the layout of the pictures may resemble the actual layout of each device relative to each other device.

Technical drawings are used to show the size and shape of an object. The basic elements of a technical drawing are lines, symbols, dimensions, and notations.

Lines

The lines that make up working drawings are sometimes referred to as the *alphabet of lines*, since each one is unique and conveys a special meaning.

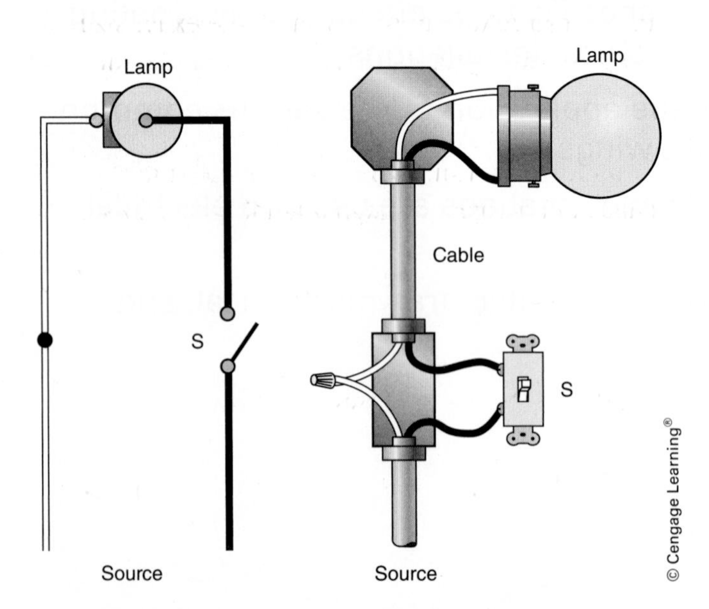

FIGURE 2-1 Example of a schematic drawing (on the left) and of a wiring diagram (on the right).

Border Line	Centre Line
Visible Object Line	Cutting Plane Line
Hidden Line	Section Lines
Construction Line	Extension Lines
Projection Line	Dimension Lines
Phantom Line	Arrowheads
Break Line	Ticks
Contour Line	Leader

FIGURE 2-2 Line types.

Figure 2-2 shows types of lines commonly found on construction drawings.

Visible Lines. Visible lines are continuous lines that show the visible edges of an object. An example of a visible line would be the outside edge of the foundation wall shown on Sheet 3 of the drawings provided with this book.

Hidden Lines. Hidden lines are short dashed lines. They show the edges of an object that are hidden from view by surfaces that are closer to the viewer. Hidden lines are used to show the location and position of the basement windows on Sheet 1 of the drawing set.

Centre Lines. Centre lines show the centre of an arc or circle. They consist of alternating long and short dashes. The exact centre of a circle is normally indicated by the crossing of two short dashes.

Section Lines. A sectional view (also called a section or cross-section) is a view of a building in which we imagine a portion of the building has been sliced off to reveal the internal construction details. Section lines show the solid parts (e.g., walls, floors) of the portion of the building shown in the sectional view. Section lines are also called cross-hatching. Different types of cross-hatching represent different types of materials used in the building. In Section A-A on Sheet 4 of the drawings, different types of cross-hatching indicate concrete used for the basement floor, cement blocks used for the basement walls, and bricks used on the outside walls above grade.

Cutting Plane Lines. Cutting plane lines consist of a long dash followed by two short dashes. They indicate where a building is being sectioned. The basement floor plan, Sheet 4 of the drawings, uses a cutting plane line to show where the building is being sectioned. The arrows at each end of a cutting plane line indicate the direction you are looking when you are viewing the section. Refer to Section A-A on Sheet 4.

Phantom Lines. Phantom lines are short dashed lines (about 1.5 times as long as the dashes for hidden lines) that show an alternative position. Figure 5-11 in Unit 5 shows two single-pole, double-throw (SPDT) switches. Closed switch positions are shown by a phantom line.

— — — — — — — — —

Break Lines. Break lines are used where only a part of the drawing needs to be shown. For example, a break line might be used in a connection diagram where it is necessary to show the termination of both ends of a cable but not the cable in between.

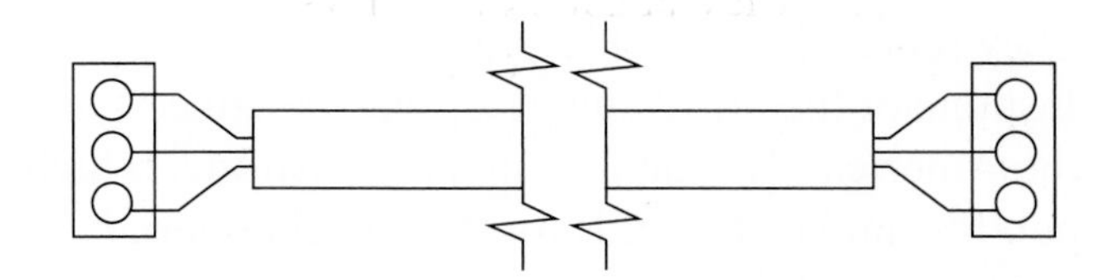

Contour Lines. Contour lines are used on plot plans to show changes in elevation. Dashed contour lines indicate existing grades and solid contour lines show finished grades.

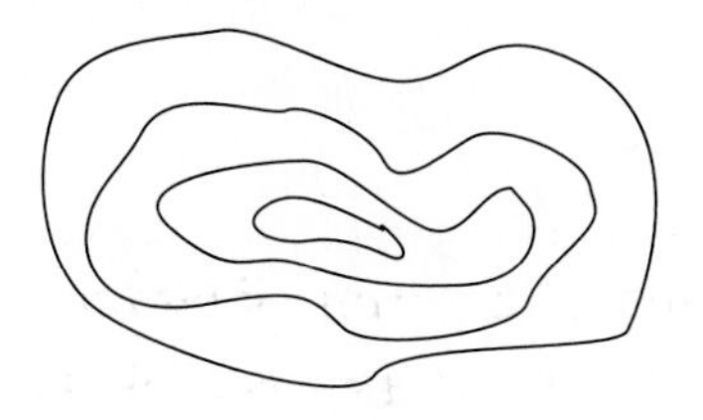

Lineweight refers to the thickness of a line used on a drawing. Object lines have a heavier lineweight (are thicker) than other lines, such as hidden lines or dimension lines. Items of other trades that appear on electrical drawings normally have a lighter weight (are thinner) than the electrical items.

VISUALIZING A BUILDING

Pictorial drawings are three-dimensional drawings that show two or more surfaces of an object in one view. Three types of pictorial drawings are shown in Figure 2-3. Figure 2-4 shows a pictorial drawing of the wiring of a house.

Unfortunately, pictorial drawings distort shapes such as angles, arcs, and circles so that it is difficult to show the actual dimensions. Since it is important that construction drawings show the exact size and shape of an object, most construction drawings are two-dimensional drawings.

Construction drawings are made using orthographic projection, which shows the true size and shape of an object through a number of views. Each

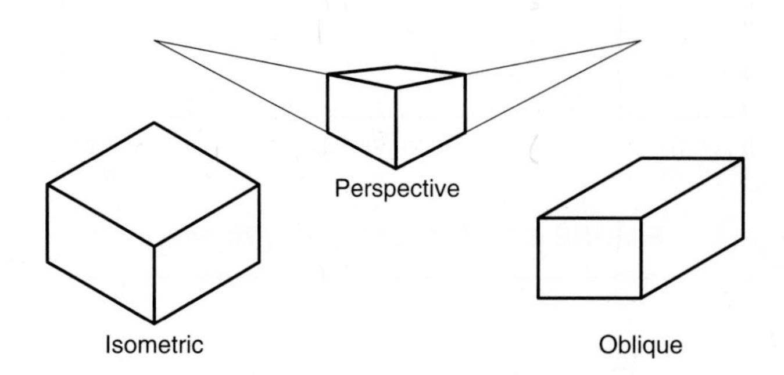

FIGURE 2-3 Pictorial drawings.

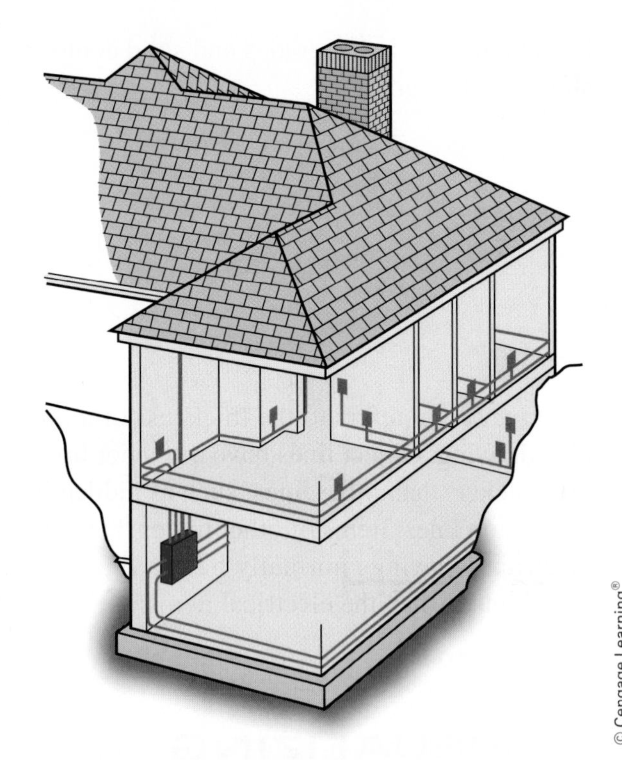

FIGURE 2-4 Three-dimensional view of house wiring.

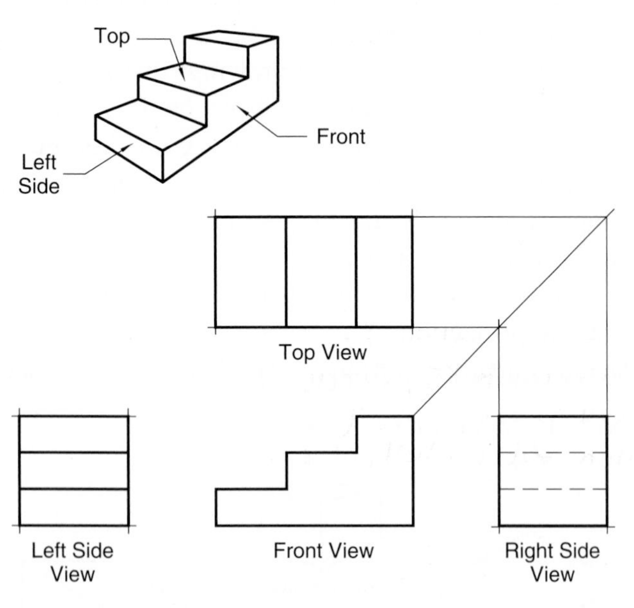

FIGURE 2-5 Orthographic projection.

view shows two dimensions. To visualize an object (such as a house), two or more views will often be required. Figure 2-5 shows an example of orthographic projection.

BUILDING VIEWS

Plan Views

A plan view shows how a building will look when viewed from directly above it. Plan views are used to show lengths and widths. Floor plans and site plans are examples of plan views. Figure 2-6 shows the first-floor plan of the house used as an example in this book.

Elevations

An elevation is a side view of the building. Elevations are used to show heights and widths. Elevations are commonly used to show outside walls, interior walls, and the placement of such things as cupboards and equipment. Figure 2-7 shows the front (south) elevation of the house.

Sections

A sectional view of the building (after a portion of the building has been removed) is used to show the interior construction details of the building. Figure 2-8 shows a portion of Section A-A on Sheet 4.

SYMBOLS AND NOTATIONS

Graphical symbols are used on construction drawings to represent equipment and components. Symbols are used to show the size, location, and ratings of equipment. The Standards Council of Canada (SCC) approves national standards for symbols used on construction drawings and accredits organizations such as the CSA Group that are engaged in standards development. CSA standard Z99.3-1979 (1989) includes symbols for electrical drawings. They relate closely to the symbols published by the Institute of Electrical and Electronics Engineers and the American National Standards Institute, which publish similar standards for the United States.

Figure 2-9 is a portion of a floor plan showing the symbols for two three-way switches controlling an overhead light and two duplex receptacles. Symbols commonly used on construction drawings are shown in Figure 2-10 (A through E).

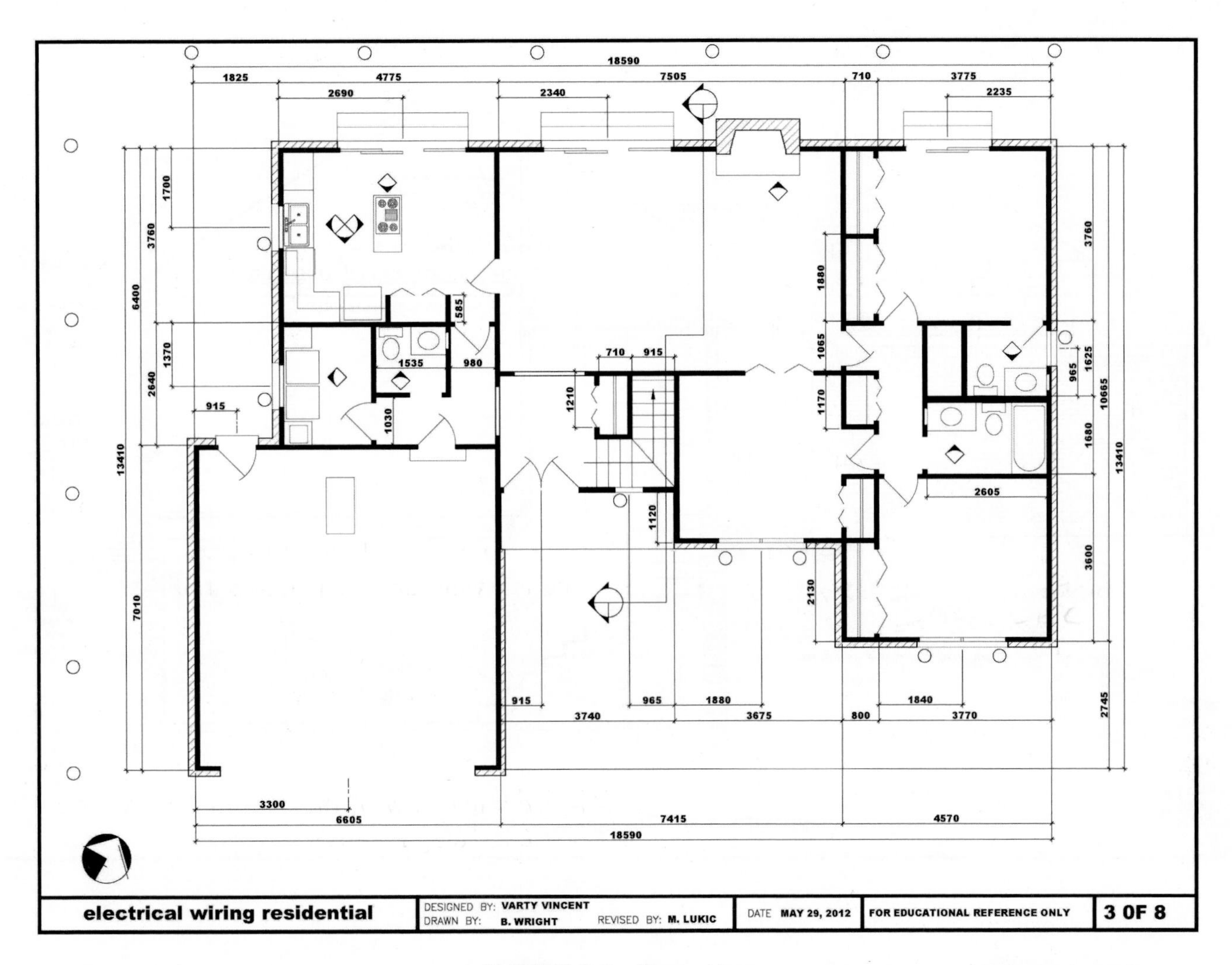

FIGURE 2-6 Floor plan.

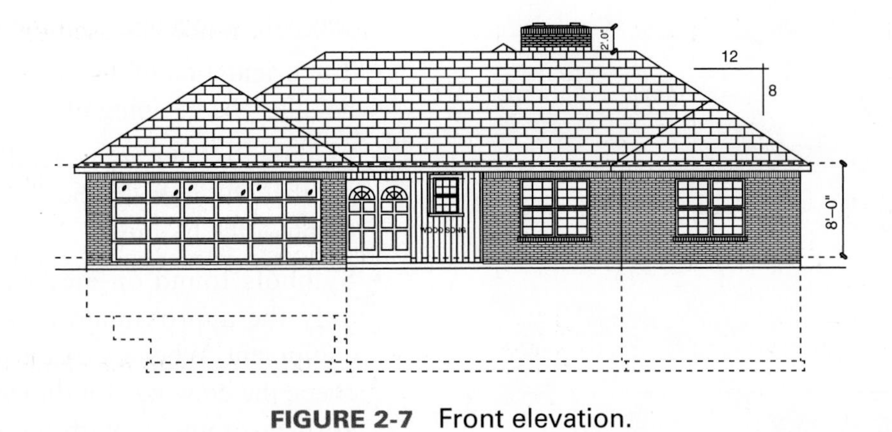

FIGURE 2-7 Front elevation.

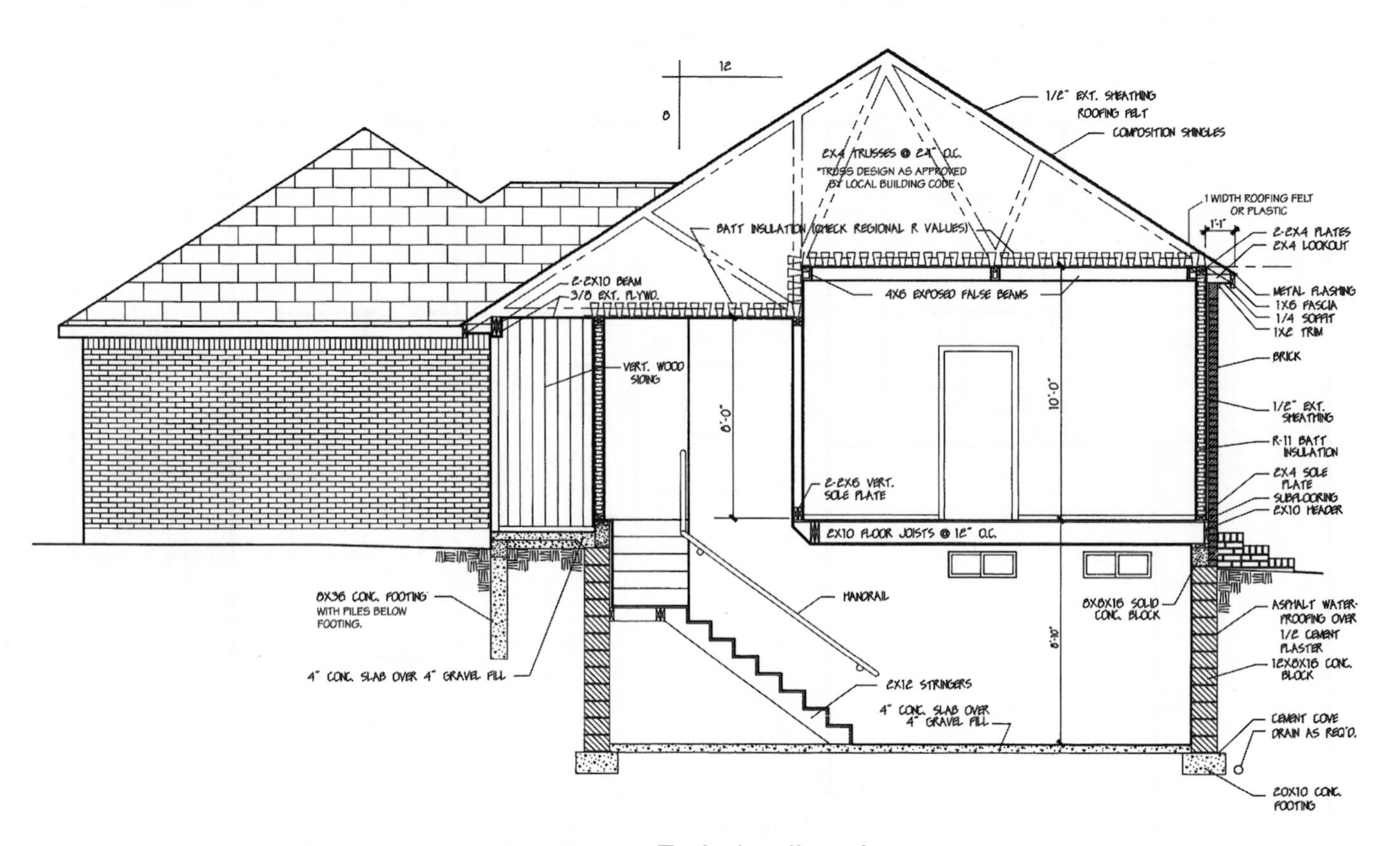

FIGURE 2-8 Typical wall section.

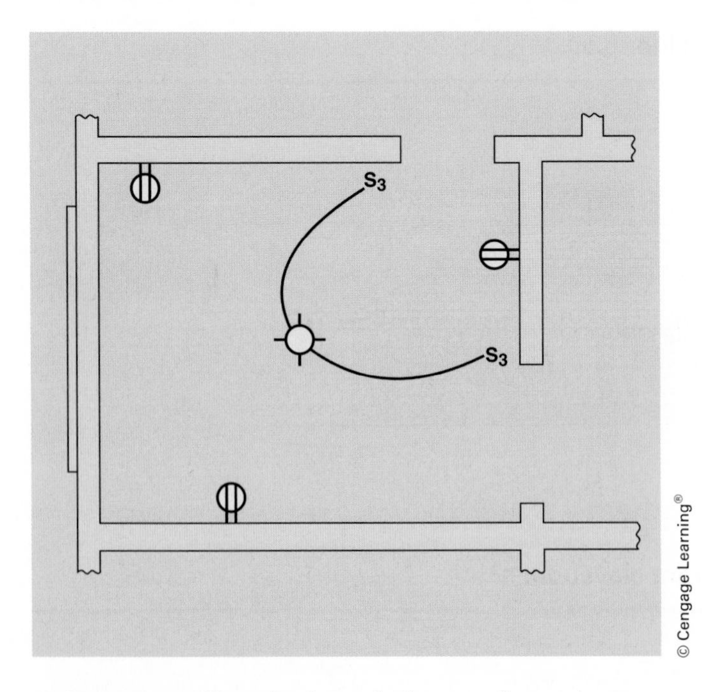

FIGURE 2-9 Electrical symbols on a floor plan.

Notes Explaining Symbols

- When symbols are used to indicate future equipment, they are shown with dashed lines and will have a note to indicate such use.
- The orientation of a symbol on a drawing does not alter the meaning of the symbol.
- Lighting fixtures are generally drawn in a manner that will indicate the size, shape, and orientation of the fixture.
- Symbols found on electrical plans indicate only the approximate location of electrical equipment. When exact locations are required, check the drawings for dimensions and confirm all measurements with the owner or owner's representative.

A notation will generally be found on the plans (blueprints) next to a specific symbol calling attention to a variation, type, size, quantity, or other

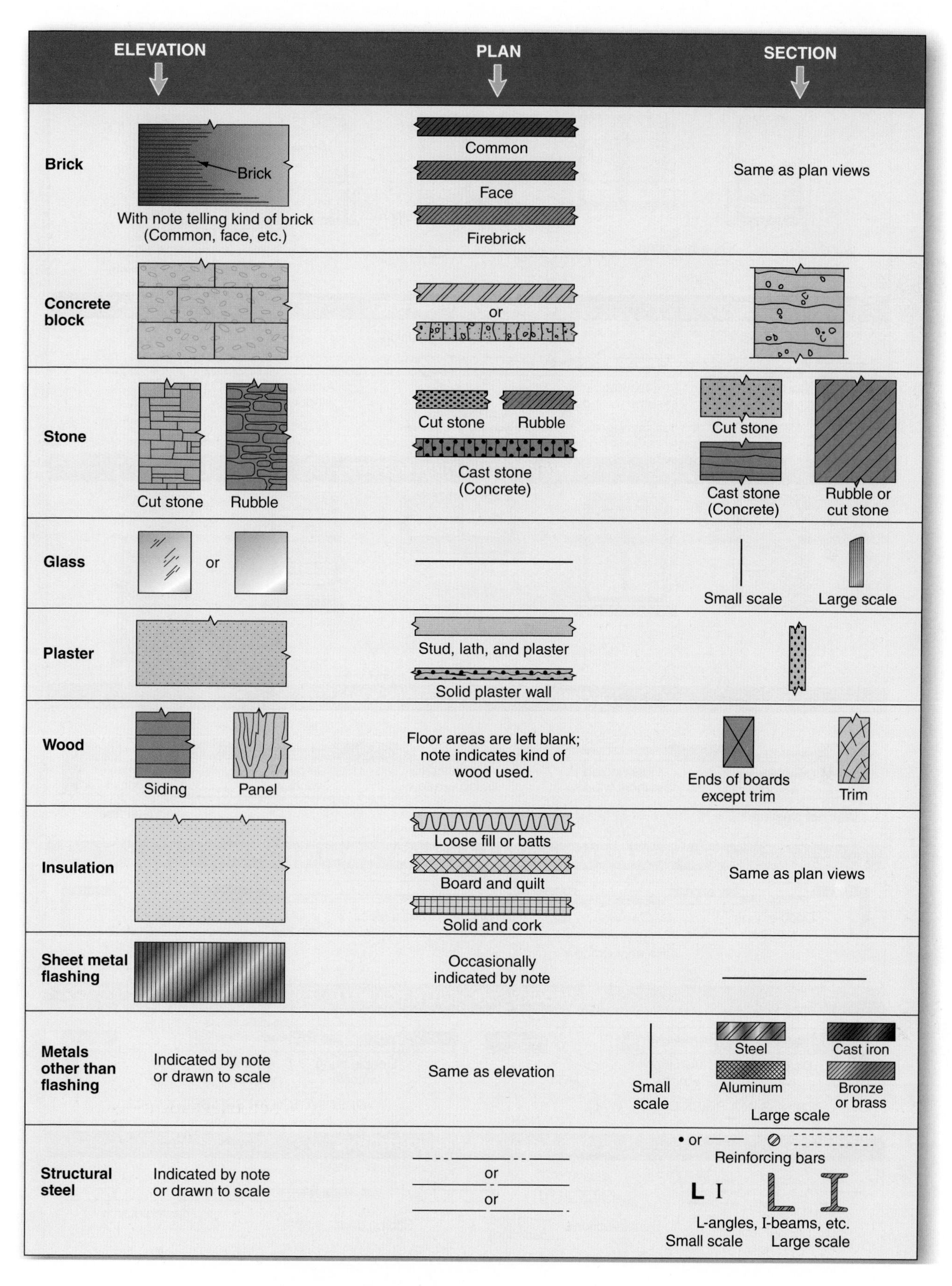

FIGURE 2-10A Architectural drafting symbols.

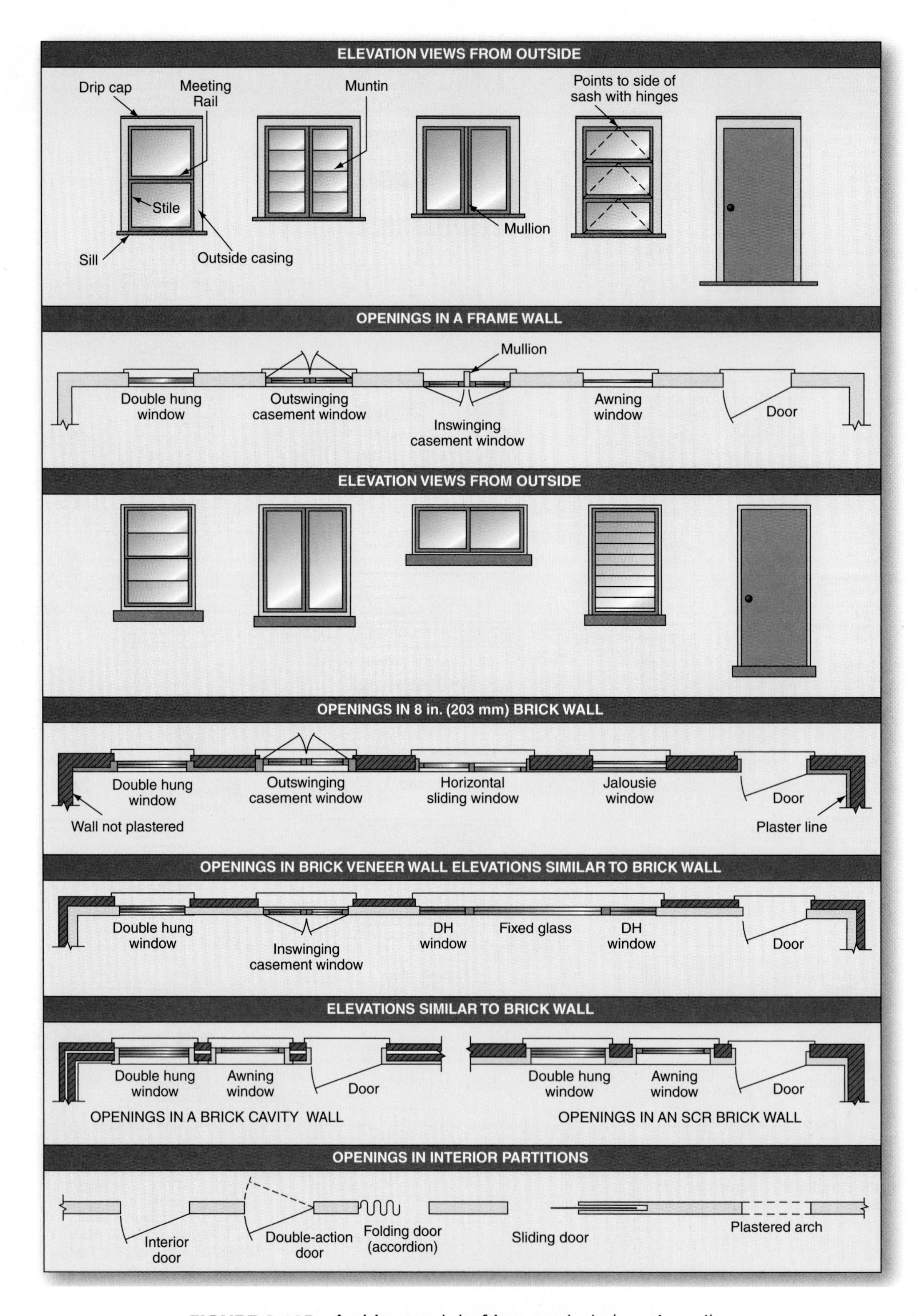

FIGURE 2-10B Architectural drafting symbols (continued).

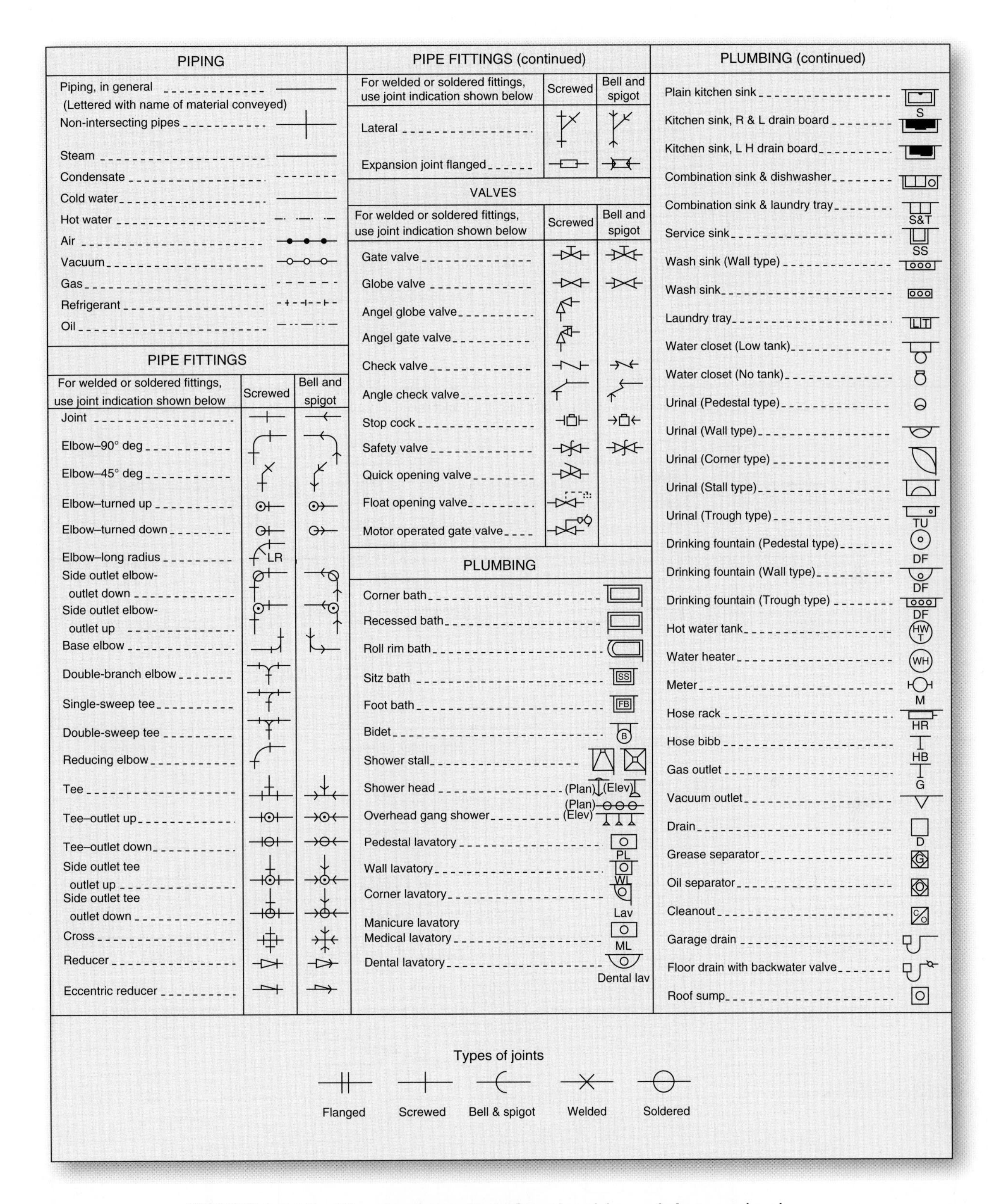

FIGURE 2-10C Standard symbols for plumbing, piping, and valves.

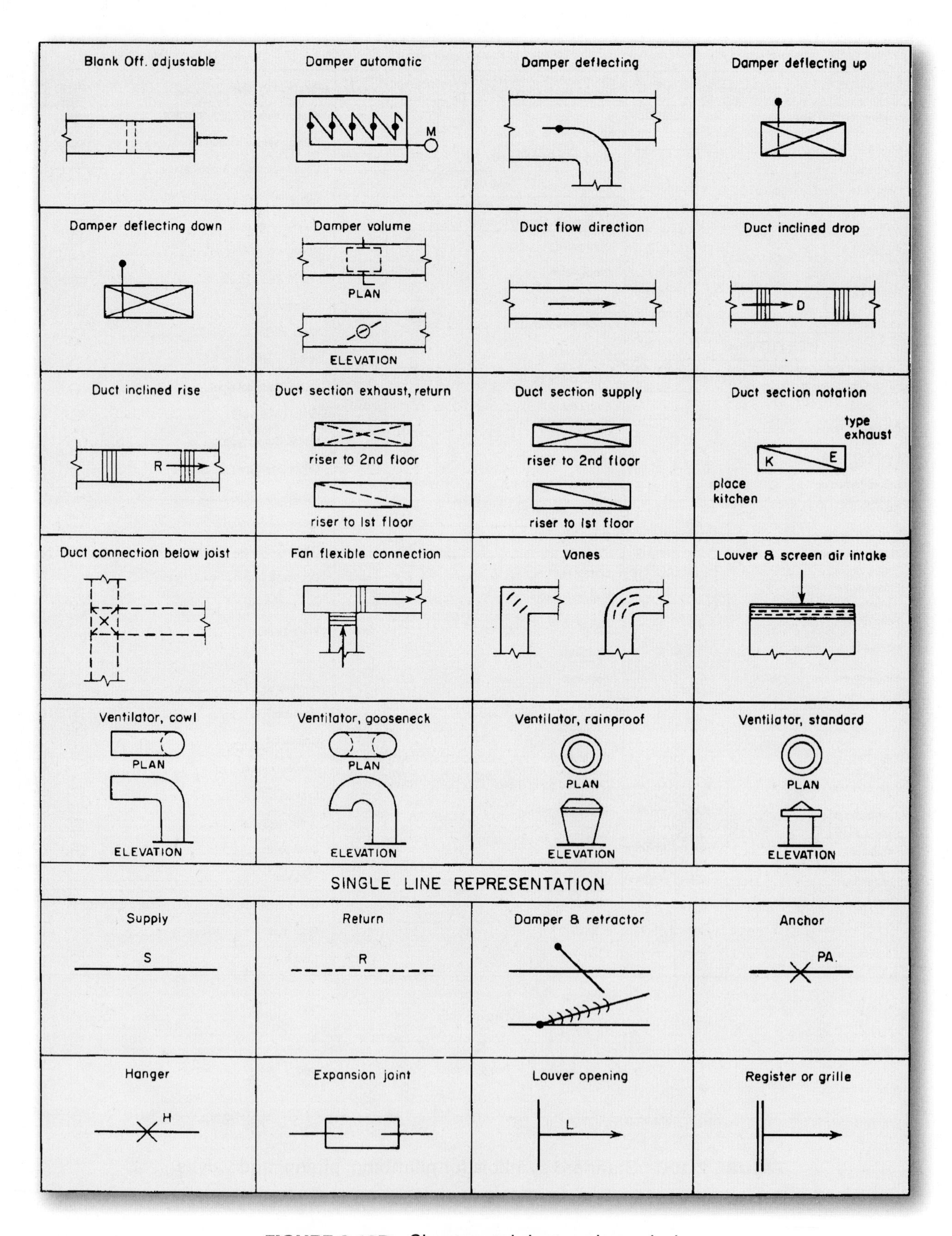

FIGURE 2-10D Sheet metal ductwork symbols.

ELECTRICAL FLOOR PLAN SYMBOLS

$ OR S	Single pole switch
$3 OR S3	3-way switch
$4 OR S4	4-way switch
$D OR SD	Door switch
$DIM OR SDIM	Dimmer switch
$RCM OR SRCM	Remote control master
$RC OR SRC	Remote control switch
$M OR SM	Motor starter switch
$a OR Sa	Lowercase letter designates switching arrangement
	Single receptacle 5-15R
	Duplex receptacle 5-15R
	Duplex rec. 5-15r floor-mounted
	Duplex receptacle 5-20 RA
	Split switched duplex receptacle 5-15R
	3-conductor split receptacle 5-15R
D	Dryer receptacle
R	Range receptacle
A	Special-purpose outlet
	Multioutlet assembly
	Track lighting length and number of fixtures as noted
	Ceiling outlet
	Wall outlet
A d	Fluorescent luminaire Uppercase letter denotes style Lowercase letter denotes switching arrangement
	Recessed fluorescent luminaire
	Fluorescent strip luminaire
C	Clock outlet
X	Exit light
F	Fan outlet
	Ionization type smoke detector
	Photoelectric type smoke detector
T	Thermostat
T	Transformer size and type as noted
	Pushbutton
	Buzzer
	Bell
D	Door opener
CH	Door chime
TV	TV outlet
WH	Water heater
	Data outlet
	Telephone outlet
	Data/telephone outlet
F	Paddlefan
L	Paddlefan with lamp
	Electrical panel flush-mounted
	Electrical panel surface-mounted DP distribution panel LP lighting panel PP power panel
	Branch-circuit wiring. Short lines indicate ungrounded conductors. The long line indicates an identified or neutral conductor. A backward slash indicates a bonding conductor. The arrow indicates a home run.

FIGURE 2-10E Electrical symbols for floor plans.

necessary information. In reality, a symbol might be considered to be a notation because, according to the dictionary, symbols "represent words, phrases, numbers, and quantities."

Another method of using notations to avoid cluttering up a blueprint is to provide a system of symbols that refer to a specific table. For example, the written sentences on plans could be included in a table referred to by a notation. Figure 5-4 in Unit 5 is an example of how this can be done. The special symbols that refer to the table would have been shown on the actual plan.

If the word "typical" appears as a notation, it means that construction details are uniform throughout the building. An example would be "typical two-bedroom suite." This indicates all that the two-bedroom suites are identical.

DIMENSIONS

Dimensions are used to show the size of an object. Metric dimensions are normally expressed in millimetres or metres.

When the American system of units is used on building drawings such as floor plans, measurements are expressed in feet, inches, and fractions of an inch. This is referred to as *architectural practice.* There are 12 inches in 1 foot. A ′ mark after a number indicates a measurement in feet (e.g., 2′). If the number is followed by ″, it represents a dimension in inches (e.g., 2″). Architectural dimensions may be expressed in feet and inches or in inches; for example, a dimension of 3′6″ may be expressed as 42″. The standard on construction drawings is to use feet and inches.

Building dimensions are normally shown on the architectural drawings from which all measurements should be taken. Be sure to confirm dimensions with the owner or architect when exact positioning of equipment is required.

SCALE

A drawing of an object that is the same size as the object is said to be drawn full scale or at a scale of 1:1.

Construction drawings reduce the size of a building down to fit on a piece of paper by drawing everything proportionally smaller than it actually is. This is called *reduced scale.* A drawing that uses a scale of 1:50 indicates that the building is 50 times larger than the drawing. A drawing with a scale of ¼″ = 1 ft has a scale of 1:48.

Enlarged scale would produce a drawing larger than the actual object. Printed circuit boards are drawn using enlarged scale since many of the components used in the manufacture of printed circuit boards are very small. An example of enlarged scale is 2:1, which indicates the drawing is twice as large as the component.

Types of Scales (Measuring Instruments)

Three types of scales are commonly used on construction drawings. When the drawing is in feet and inches, an architectural scale is used. Figure 2-11 shows a portion of an architect's triangular scale showing the ⅛″ and ¼″ scales.

For metric drawings, a metric scale is used. For working with civil drawings (e.g., roads, dams, bridges), an engineering scale is used. Table 2-1 lists common construction drawing scales.

Figure 2-12 gives examples of reading scales.

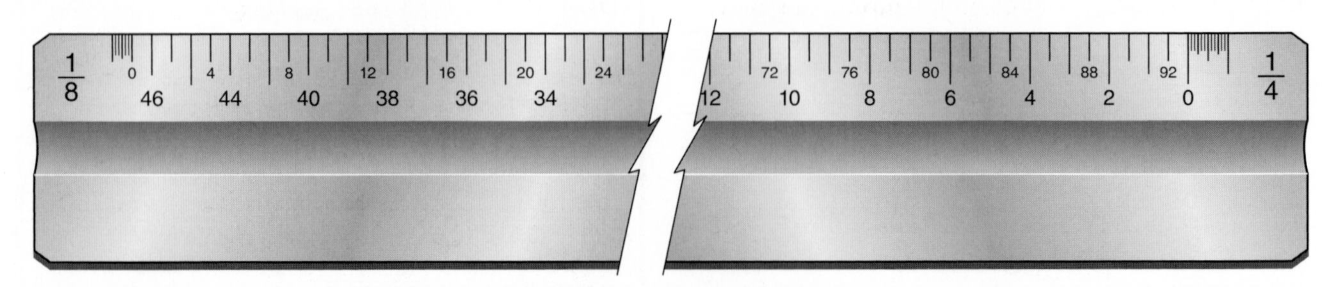

FIGURE 2-11 Architect's scale.

TABLE 2-1

Common construction drawing scales.

METRIC	ARCHITECTURAL	ENGINEERING
1:20	1/2″ = 1′0″	1″ = 10
1:50	1/4″ = 1′0″	1″ = 20
1:100	1/8″ = 1′0″	1″ = 30
1:200	1/16″ = 1′0″	1″ = 40
1:500	1/32″ = 1′0″	1″ = 50

THE WORKING DRAWINGS

The architect or engineer uses a set of working drawings or plans to make the necessary instructions available to the skilled trades that are to build the structure. The plans show sizes, quantities, and locations of the materials required and the construction features of the structural members. Each skilled construction trade—masons, carpenters, electricians, and others—must study and interpret these details of construction before the actual work is started.

The electrician must be able to convert the two-dimensional plans into an actual electrical installation and visualize the many views of the plans and coordinate them into a three-dimensional picture, as shown in Figure 2-4.

The ability to visualize an accurate three-dimensional picture requires a thorough knowledge of blueprint reading. Since all of the skilled trades use a common set of plans, the electrician must be able to interpret the lines and symbols that refer to the electrical installation and also to interpret those used by the other construction trades. The electrician must know the structural makeup of the building and the construction materials to be used.

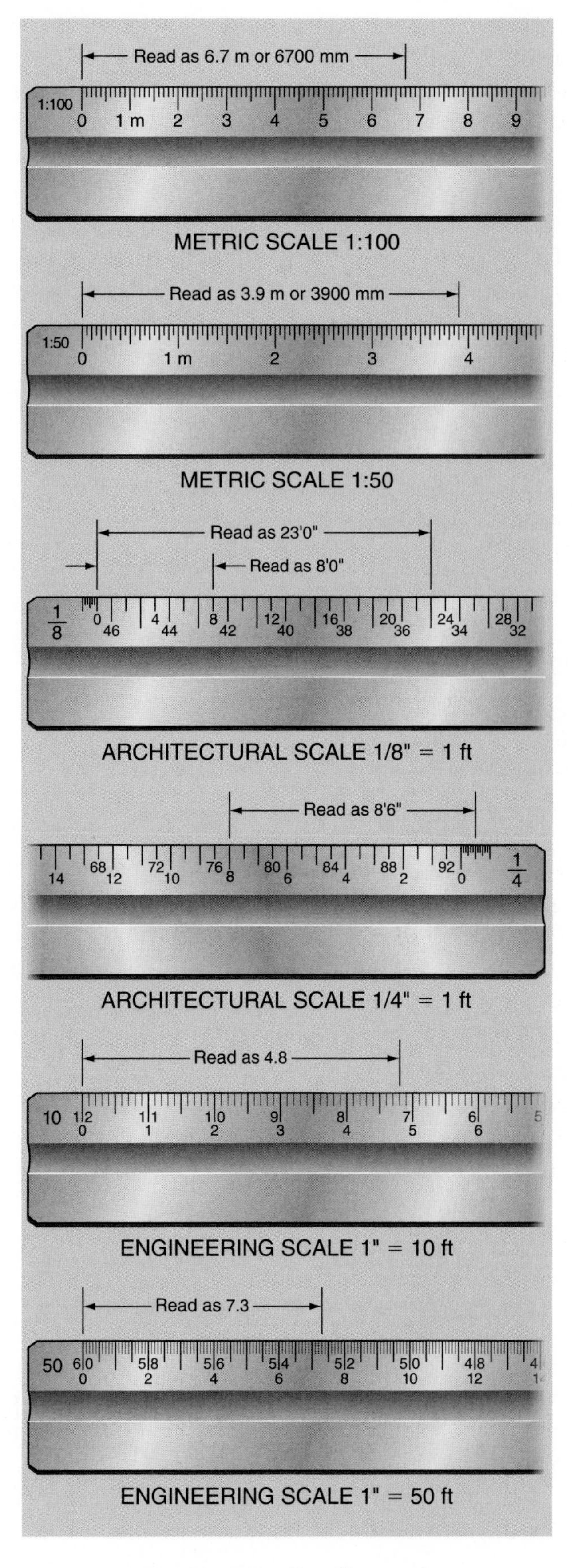

FIGURE 2-12 Reading scales.

The Drawing Set

A set of drawings is made up of several numbered sheets. If there are many sheets in a set, it will be broken down into subsets of architectural, electrical, mechanical, etc. The electrical drawings will be numbered E-1, E-2, E-3, etc.

A typical set of construction drawings includes a site plan, floor plans, elevations, details and sections, symbols legend, schedules, diagrams, and shop drawings.

Site (Plot) Plan

The site plan shows

- the dimensions of the lot
- a north-pointing arrow
- the location of utilities and their connection points
- any outdoor lighting
- easements (rights for use of some aspect of the property by someone other than the owner; e.g., use of a road that crosses the owner's property)
- contour lines to show changes in elevation
- accessory buildings (e.g., detached garages, storage sheds)
- trees to be saved

Floor Plans

Depending on the complexity of the building, a set of drawings for a single-family dwelling may have only a single floor plan for each storey of the building. As the complexity of the building increases there will be a number of floor plans for each storey (one for each trade). A logical breakdown would be

- Architectural
- Electrical
- Mechanical
- Plumbing
- HVAC (heating, ventilating, and air conditioning)
- Structural

On very complex installations, each electrical system such as power, lighting, voice/data, fire alarm, and security/access control may have its own plan for each storey.

Use architectural floor plans to find room measurements, wall thickness, the sizes of doors and windows, and the swing of doors. The floor plans of other trades will show the locations of the equipment associated with that trade.

Elevations

Elevations show the vertical faces of buildings, structures, and equipment. Elevations are used to show both interior and exterior vertical faces. Sheet 7 of the drawings that accompany this textbook shows elevations of the kitchen cupboards. On Sheet 1 you will find the elevations (south, north, east, and west) of the building.

Details and Sections

Details are large-scale drawings that are used when floor plans and elevations cannot provide sufficient detail. Sheet 8 provides details of the requirements for the installation of a swimming pool. Sheet 7 shows details of the bathroom vanities as well as of the kitchen cabinets. Sections show internal details of construction. See Section A-A on Sheet 4.

Legend of Symbols

A legend of symbols provides the contractor or electrician with the information necessary to interpret the symbols on the drawing.

Schedules

Schedules are lists of equipment. The door schedule on Sheet 7 of the plans describes the size and types of doors required for the house. Electrical schedules are made for such equipment as lighting fixtures (luminaires), panelboards, raceways and cables, motors, and mechanical equipment connections.

Diagrams

One-line and riser diagrams are provided to quickly show the flow of power throughout the building. Riser diagrams are used to show vertical details while using a one-line format to show power, telephone, or fire alarm systems.

Shop Drawings

Shop drawings are used by equipment manufacturers to show details of custom-made equipment for a specific job. Installation diagrams that come with equipment could be considered to be shop drawings.

SPECIFICATIONS

Working drawings are usually complex because of the amount of information they must include. To prevent confusing detail, it is standard practice to include with each set of plans a set of detailed written specifications prepared by the architect.

Specifications are written descriptions of materials and methods of construction. They provide general information for all trades involved in the construction of the building, as well as specialized information for the individual trades. Specifications include information on the sizes, type, and desired quality of the parts to be used in the structure.

Typical specifications include a section on *General Clauses and Conditions* that is applicable to all trades involved in the construction. This section is followed by detailed requirements for the various trades—excavating, masonry, carpentry, plumbing, heating, electrical, painting, and others.

The plan drawings for the residence used as an example for this textbook are included at the back of the text. The specifications for the electrical work indicated on the plans are given in this book's Appendix A.

In the electrical specifications, the list of standard electrical parts and supplies frequently includes manufacturers' names and catalogue numbers of the specified items. This ensures that these items will be the correct size, type, and electrical rating, and that the quality will meet a certain standard. To allow for the possibility that the contractor may be unable to obtain the specified item, the words "or equivalent" are usually added after the manufacturer's name and catalogue number.

The specifications are also useful to the electrical contractor as all of the items needed for a specific job are grouped together and the type or size of each item is indicated. This information allows the contractor to prepare an accurate cost estimate without having to find all of the data in the plans.

REVIEW

Note: For these exercises, refer to the blueprints provided with this textbook. Also, refer to the *CEC* when necessary. Where applicable, responses should be written in complete sentences.

PART 1: DRAWING PLANS

1. Identify three line types shown on Sheet 2.

   Dimension Lines

   Centre Lines

   Hidden Lines

2. Determine the length of the lines for the following.

SCALE		LENGTH
1:50	1/4" = 1'0"	
1:75		
1:100		
⅛″ = 1 ft		
¼″ = 1 ft		

3. What is the purpose of specifications?
To provide general information for all trades involved in the construction of the building as well as specialized information for the individual trades

4. In what additional way are the specifications particularly useful to the electrical contractor?
It ensures the items will be the correct size, type and electrical rating, and that the quality will meet a certain standard

5. What is done to prevent a plan from becoming confusing because of too much detail?
A set of detailed specifications is written by the architect

6. Name three requirements contained in the specifications regarding material (Appendix A).
a. All materials shall be new
b. Bear the appropiate label of the CSA
c. Shall be the size and type on the dwg or specs

7. What are the two main hazards that the *CEC* is designed to prevent?
a. ___
b. ___

8. What phrase is used when a substitution is permitted for a specific item in a specification?
"or equivalent"

9. What is the purpose of an electrical symbol?
Electrical Symbols are used to represent equipment and components. Also they are used for approximate locations

10. What is a notation?
Found next to symbols calling attention to a variation type, size, quantity or type or other necessary information

11. List five electrical notations found on the plans for the residence used as an example in this book. Refer to the blueprints provided with this textbook.
a. Center TV Outlets 300mm from finished floor
b. Phone wall box in recreation room to be 150mm to the right of wet bar...
c. All electrical panels to be mounted on 13mm plywood
d. Freezer Receptiaale single type
e. Luminares "A" and Luminares on landing of stairway controlled by 3-way switches.

PART 2: STRUCTURAL FEATURES

1. To what scale is the basement plan drawn?

   1:50

2. What is the size of the footing for the steel support columns in the basement?

   610 x 610 x 305

3. To what kind of material will the front porch lighting bracket fixture be attached?

   brick

4. Give the size, spacing, and direction of the ceiling joists in the workshop.

5. What is the size of the lot on which this residence is located?

   46m x 30m

6. In what compass direction is the front of the house facing?

   South

7. How far is the front garage wall from the curb?

   10 670 mm

8. How far is the side garage wall from the property lot line?

   7620 mm + 1.525 m

9. How many steel support columns are in the basement, and what size are they?

   5, 76mm

10. What is the purpose of the I-beams that rest on top of the steel support columns?

11. Is the entire garage area to have a concrete floor?

    No

12. Where is access to the attic provided?

    garage

13. Give the thickness of the outer basement walls.

    305 conc foundation

14. What material is indicated for the foundation walls?

    concrete

15. Where are the smoke detectors located in the basement?

    bottom of stairs

16. What is the ceiling height in the basement workshop from the bottom of joists to floor?

 2435

17. Give the size and type of the front door.

18. Who is to furnish the range hood?

19. Who is to install the range hood?

UNIT 3

Service Entrance Calculations

OBJECTIVES

After studying this unit, you should be able to

- determine the total calculated load of the residence
- calculate the size of the service entrance conductors, including the size of the neutral conductors
- understand the *CEC* requirements for services, *Rule 8–200*
- fully understand special *CEC* rules for single-family dwelling service entrance conductor sizing
- perform an optional calculation for computing the required size of service entrance conductors for the residence

Before calculating the ampacity of service and service entrance conductors required for the residence, you must take an inventory of all loads in excess of 1500 watts. The calculations are based on the *Canadian Electrical Code (CEC)*. Always consult local electrical codes for any variations in requirements that may take precedence over the *CEC*.

SIZE OF SERVICE ENTRANCE CONDUCTORS AND SERVICE DISCONNECTING MEANS*

Service entrance conductors (*Rule 8–200*) and the disconnecting means shall not be smaller than

1. 60-ampere, three-wire, for a single-family dwelling when the floor area, excluding the basement floor area, is less than 80 m^2, *Rule 8–200(1)(b)(ii)*.
2. 100-ampere, three-wire, for a single-family dwelling when the floor area, excluding the basement floor area, is more than 80 m^2, *Rule 8–200(1)(b)(i)*.

The load on a residential service is a noncontinuous load, *Rule 8–200(3)*. This means that the service conductors, the service switch, and the panel are sized based on the demand amperes. To correctly calculate the demand amperes for the service, you must follow the steps listed in *Rule 8–200(1)(a)*, as follows:

Step 1. Calculate the basic load based on the living area of the house, *Rule 8–200(1)(a)(i)*.

Step 2. Calculate the electric space-heating load and the air-conditioning load. Use the larger of these loads in calculating the demand if you know that they cannot be used at the same time, *Rule 8–106(4)*. If it is possible for the electric heat and the air conditioning to operate at the same time, you must include both loads in calculating the demand. If baseboard heating is used with individual thermostats in each room or area, refer to *Rule 62–118(3)* for demand watts. If the residence has an electric furnace, the load must be included at 100% of its rating, *Rule 62–118(4)*.

Step 3. Calculate the load for the electric range or the combined load for a separate cooktop and oven.

Step 4. Add the kilowatt rating of any swimming pool heaters or hot-tub and spa heaters. In this case, there should be no de-rating, *Rule 8–200(1)(a)(v)*.

Step 5. Electric vehicle-charging equipment is loaded at 100%, *Rule 8–200(1)(a)(vi)*.

Step 6. A demand factor of 25% must be applied to any additional loads in excess of 1500 watts if an electric range is included. If there is no electric range because the residence has a gas range, the first 6-kW loads in excess of 1500 watts are added at 100%. If the load is greater than 6 kW, the remainder should be added at 25%, *Rule 8–200(1)(a)(vii)*.

For clarity, these calculations have been summarized in Tables 3-1 and 3-2. No electric vehicles are called for in the drawings, so there are no references to loading or equipment in the tables.

TABLE 3-1

Demand watts calculation for service and neutral conductor.

SERVICE CALCULATION

	Demand Watts
Load	Line 1, Line 2
Basic	7 000 W
Heat	13 000 W
Range	6 820 W
Other—Dryer	1 425 W
—Water heater	750 W
—Water pump	600 W
	29 595 W
Lines 1, 2; I = P/E	29 595/240 = 123.3 A
Neutral; I = P/E	

*Courtesy of CSA Group

TABLE 3-2

Service requirements.

MINIMUM CODE REQUIREMENTS FOR A SERVICE		
Reference	**Minimum Service Calculation**	**Panel A for Residence**
Minimum service ampacity, *Rule 8–104(1)*	123.3 A	200 A
Minimum service switch	200 A	200 A
Minimum fuse or breaker, *Table 13*	125 A	200 A
Minimum ground conductor, *Rule 10–812*	#6 copper	#4 copper
Minimum bonding jumper, *Table 41*	#6 copper	#4 copper
Minimum AWG L1, L2, *Table 2*	#2 RW75	3/0 RW75
Minimum AWG neutral, *Table 2*		3/0 RW75 (jacketed)
Minimum conduit size, *Tables 8, 9, 10*	35 mm	53 mm
Minimum panel ampacity	125 A	200 A
Minimum number of circuits, *Rule 8–108(1)(c)*	24	30

SERVICE CALCULATIONS FOR A SINGLE-FAMILY DWELLING, *RULE 8–200*: CALCULATING FLOOR AREA

Ground-Floor Area

To estimate the total basic load for a dwelling, you must calculate the occupied floor area of the dwelling. Note in the residence plans that the ground-floor area has an irregular shape. In this case, the simplest method of calculating the occupied floor area is to determine the total floor area using the inside dimensions of the dwelling. Then subtract open porches, garages, and other unfinished or unused spaces from the total area if they are not adaptable for future use, *Rule 8–110.*

Many open porches, terraces, patios, and similar areas are commonly used as recreation and entertainment areas. Therefore, adequate lighting and receptacle outlets must be provided.

To simplify the calculation, the author has chosen to round off dimensions for those areas (garage, porch, and portions of the inset at the front of the house) not to be included in the computation of the general lighting load.

Figure 3-1 shows the procedure for calculating the watts needed for the ground floor of this residence.

For practical purposes, the outside dimensions of this building could be taken as being 17 m and 11 m.

Basement Area

The basement area of the home contains the furnace, the hot water tank, and other service equipment. According to *Rule 8–110,* only 75% of the inside dimensions are to be used in calculating the living area of the basement because 25% houses the above equipment.

For our calculations, we make these assumptions:

1. Basement walls are 305 mm thick.
2. Garage is unexcavated so that additional support for the weight of vehicles is not required.

On this basis, our calculations can be made as detailed in the following:

Ground-Floor Area

$$16.765 - (0.1 + 0.1)\text{ m} = 16.565\text{ m}$$
$$10.670 - (0.1 + 0.1)\text{ m} = 10.470\text{ m}$$
$$16.565\text{ m} \times 10.470\text{ m} = 173.44\text{ m}^2$$

Less area not included:

$$2.135\text{ m} \times 3.675\text{ m} = -\ 7.85\text{ m}^2$$
$$3.255\text{ m} \times 3.740\text{ m} = -12.17\text{ m}^2$$
$$3.255\text{ m} \times 4.775\text{ m} = \underline{-15.54\text{ m}^2}$$

Total ground-floor area $= 137.88\text{ m}^2$

Basement Area

$$17.045 - (0.305 + 0.305)\text{ m} = 16.435\text{ m}$$
$$10.945 - (0.305 + 0.305)\text{ m} = 10.34\text{ m}$$

$$16.435\text{ m} \times 10.34\text{ m} = 169.94\text{ m}^2$$

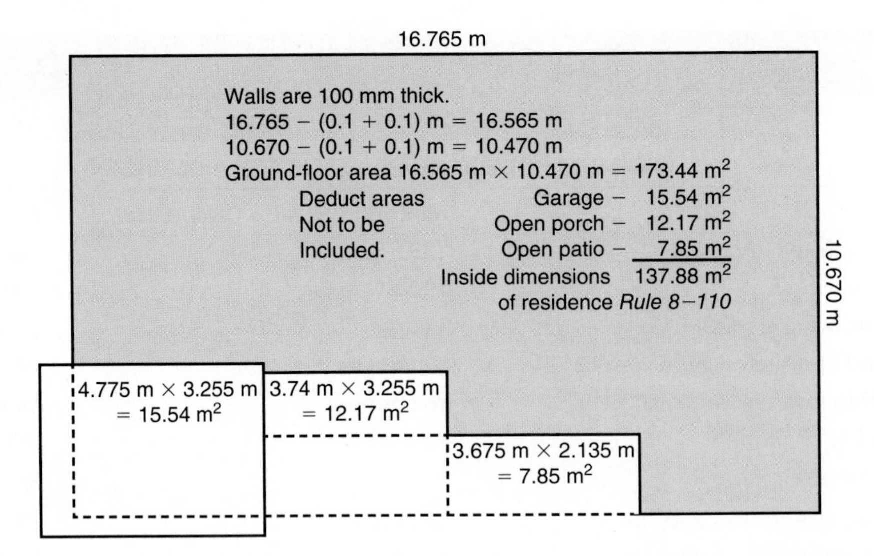

FIGURE 3-1 Determining the ground-floor living area to be used to calculate basic load, *Rule 8–110.*

Less area not included:

$2.1 \text{ m} \times 3.75 \text{ m} = -\ \ 7.88 \text{ m}^2$
$3.45 \text{ m} \times 3.2 \text{ m} = -11.04 \text{ m}^2$
$3.2 \text{ m} \times 4.71 \text{ m} = \underline{-15.07 \text{ m}^2}$

Total basement area $= 135.95 \text{ m}^2$

Basic Load Calculations, *Rule 8–200(1)(a)(i,ii)*

For this calculation, *Rule 8–200(1)(a)(i,ii)* requires that the load be based on 100% of the living area on the ground floor plus 75% of the living area in the basement.

Area to be considered:
100% of ground floor (139.9 m^2) $= 139.9 \text{ m}^2$
75% of the basement area:
75% of 135.95 m^2 $= \underline{101.96 \text{ m}^2}$
Total area $= 241.86 \text{ m}^2$

Total area $= 241.86 \text{ m}^2$
1st $90 \text{ m}^2 = 5000 \text{ W}$

This leaves 151.86 m^2 ($241.86 \text{ m}^2 - 90 \text{ m}^2$).

An additional 1000 W must be added for each 90 m^2 *or portion thereof* in excess of 90 m^2. Since $151.86/90 = 1.7$, use 2 times:

$$2 \times 1000 \text{ W} = 2000 \text{ W}$$

A. Basic load = 5000 W + 2000 W = 7000 W

Electric Furnace and Air Conditioning, *Rule 8–200(1)(a)(iii)*

Since you know that the air conditioning and heating will not be used simultaneously, use whichever is the greater load in calculating the demand, *Rule 8–106(4).* Therefore, if the air conditioner load is less than the heating load, it need not be included in the calculations. Air conditioner load is

$$30 \text{ amperes} \times 240 \text{ volts} = 7200 \text{ W}$$

This is less than the heating load.

The electric furnace for this home is applied at a demand of 100%, based on *Rule 62–118(1).* If this home had been heated with baseboard heaters, we would have been able to de-rate the amount over 10 kW at a demand factor of 75%, *Rule 62–116(3)(a).*

B. Electric furnace = 13 000 W

Electric Range or Wall-Mounted Oven and Counter-Mounted Cooking Unit Combined, *Rule 8–200(1)(a)(iv)*

Wall-mounted oven = 6600 W
Counter-mounted cooking unit = 7450 W
Total = 14 050 W (14.05 kW)
For rating of 12 kW or less load = 6000 W
14.05 kW − 12 kW = 2.05 kW
Add 2.05 kW × 40% = 820 W

C. Range load = 6820 W

D. Dryer = 5700 W × 25% = 1425 W

E. Water heater = 3000 W × 25% = 750 W

F. Pump = 2400 W × 25% = 600 W

A + B + C + D + E + F = Total demand load
= 29 595 W

Summary of Calculations

Step 1: Basic Load
241.86 m^2
−90 m^2 1st 90 m^2 = 5 000 W
151.86 ÷ 90
= 1.7 or 2 × 1 000 = 2 000 W

Demand watts = 7 000 W **7 000 W**

Step 2: Heat 13 000 W **13 000 W**
(Electric furnace @ 100%)

Step 3: Range 1st 12 kW = 6 000 W
2.05 kW × 40% = 820 W

Demand watts = 6 820 W **6 820 W**

Step 4: Other Loads
Dryer × 25% = 1 425 W
Water heater × 25% = 750 W
Pump × 25% = 600 W

Demand watts = 2 775 W **2 775 W**

Total load (demand watts) **29 595 W**

Service Entrance Conductor Size

The ungrounded conductors are sized from *CEC Table 2* for copper wires and *Table 4* for aluminum. It is common to have aluminum conductors installed up to the line side of the meter base and from the load side of the meter base into the panel.

Take care in this type of installation to ensure that the aluminum conductors are installed in lugs rated and approved for the purpose, *Rule 12–118(3).*

An appropriate joint compound, for example, Pentrox™, must be used to penetrate the oxide film and prevent it from re-forming when terminating or splicing all sizes of stranded aluminum conductors, *Rule 12–118(2).*

It is important to follow the requirements of *Rule 2–134.* All conductors that are exposed to direct sunlight must be approved for the purpose. Conductors that are approved for exposure to sunlight will be surface-marked appropriately.

Neutral Conductor

According to *Rule 4–024(1),* the neutral conductor is sized according to the unbalanced load on the service. This wire shall be the larger of the bonding conductor, sized from *CEC* as specified in *Rule 10–204(2),* or the calculated ampacity as determined from the demand watts calculation shown in Table 3-1. In this example, both methods require a minimum #6 AWG conductor.

This conductor may be run bare provided it is made of copper and is run in a raceway, *Rule 6–308(a).* Verify this with your local inspection authority before proceeding.

The neutral conductor is sized by the maximum unbalanced load between line 1, line 2, and the neutral.

This means that 240-volt loads are not included in calculating the demand watts on the neutral. These may include

- central air conditioner
- electric furnace
- water pump
- water heater

Since some utilities do not allow a reduction in the size of the neutral conductor for services less than 200 amperes, check with provincial codes or the local inspection authority before proceeding with a reduced neutral installation.

Consumer's Service Ampacity Rating

The minimum rating of a consumer's service is based on a comparison between the overcurrent device and the ampacity of the line conductors.

Whichever is the lower value determines the ampacity rating of the service. This would also apply for branch circuits or feeders, *Rule 8–104(1).*

Main Service Panel

The main panel for a service is described in the *CEC* as the service box. This could be a separate switch containing fuses or a circuit breaker. In reality, it is more commonly installed as a combination panel comprising the service entrance portion and the distribution portion.

The standard available sizes of disconnecting means are 30 amperes, 60 amperes, 100 amperes, 200 amperes, and 400 amperes.

Optional available sizes are 125-ampere and 225-ampere panels. However, most residential applications are now either 100 amperes or 200 amperes. Using the demand watts calculation in Table 3-1, it is determined that a 125-ampere service would be sufficient for this residence. This would, however, leave little room for future growth, such as an outbuilding for a workshop or a hot tub. For this reason, it has been decided that a 200-ampere service will be installed.

Main Service Fuse or Circuit Breaker (*Table 13*)

The size of the main fuse or circuit breaker is based on the ampacity of the line conductors, as found in *Table 2*. This ampacity is used to determine the size of the fuse or circuit breaker from *Table 13, Rule 14–104(1)(a).*

TABLE 3-3

CEC Table 16A.

CEC TABLE 16A
MINIMUM SIZE OF CONDUCTORS FOR BONDING CONDUCTORS
(*RULES 10–204, 10–626, 10–814, 10–816, 12–1814, 24–104, 24–202, 30–1030, 68–058,* AND *68–406*)

Size of largest ungrounded conductor	Size of bonding conductor	
Copper, AWG or kcmil	**Copper, AWG or kcmil**	**Aluminum, AWG or kcmil**
14 and 12	14	12
10	12	10
8	10	8
6–4	8	6
3–2/0	6	4
3/0–300	4	2
350–500	3	1
600–750	2	1/0
800–1000	1	2/0
1250–1200	1/0	3/0
Aluminum, AWG or kcmil	**Aluminum, AWG or kcmil**	**Copper, AWG or kcmil**
12	12	14
10 and 8	10	12
6	8	10
4–2	6	8
1–4/0	4	6
250–400	2	4
500–700	1	3
750–1000	1/0	2
1250–1500	2/0	1
1750–2000	3/0	1/0

Note: Where multiple ungrounded conductors are used in parallel runs, parallel bonding conductors should be used in accordance with *Rule 10–814* and installed in close proximity to the corresponding ungrounded conductors to minimize increased impedance in the bonding conductor(s).

Main Service Ground (*Table 16A*)

The main service ground is sized from *Table 16A* (see Table 3-3), and is based on the ampacity of the line 1 and line 2 conductors, *Rule 10–204(1).*

Main Service Bonding Conductor (*Table 41*)

The bonding jumper for metallic service raceways is sized from *Table 41* and is based on the ampacity of the largest service conductor, *Rule 10–614(1) (a,b).* This jumper connects the service raceway by means of a ground bushing to the metal enclosure. Accordingly, it provides a low impedance path for adequate bonding to prevent dangerous conditions that may result from equipment failure or lightning strikes.

Main Service Conduit Size (*Tables 8, 9A–J, and 10A–D*)

The minimum size of conduit for the service entrance is based on 40% conduit fill (*Table 8*). The total area of the conductors is calculated using *Table 10A.* Then the 40% column in *Table 9C* is used to find the minimum size of the raceway. If all of the conductors are of the same type and size, *Table 6(A – K)* may be used to size the raceway. For this service, a 53-mm conduit is adequate for the three jacketed 3/0 RW75 XLPE conductors.

Note that the service mast is usually 63-mm rigid metal, but the conduit from the meter base into the panel must be sized using the method described.

Main Service Panel Ampacity and Number of Circuits

The ampacity of the service panel is based on the ampacity of the main service switch, for example, a 100-ampere switch and a 100-ampere panel.

The minimum number of circuits for the panel is listed in *Rule 8–108(1).* In this case, the 200-ampere service requires a minimum of 30 circuits, *Rule 8–108(1)(c)(ii).* If the house had been heated with baseboard heaters, it would have required 40 circuits. The baseboard heaters would have required a number of two-pole breakers (30 amperes maximum), whereas a central furnace could be fed by a 70-ampere two-pole breaker, taking up only two positions.

Rule 8–108(2) states that two additional spaces must be left for future overcurrent devices. This is sometimes interpreted to mean that two circuits must be added to the required number in *Rule 8–108(1),* but for the purposes of this calculation, *Rule 8–108(2)* is satisfied by leaving two spaces available in the 30-circuit panel. These two spaces could be used to feed a future subpanel.

According to *Table 2,* #1 AWG RW75 XLPE (jacketed) will be the minimum size of conductor for the 29 595 watt (123.3 ampere) load.

For the service to be 200 amperes, 3/0 AWG RW90 (200-ampere) will be required.

The 240-volt loads are not to be included in the calculation for neutral conductors.

Calculation for Subpanel B

The calculation for the subpanel is based on the load supplied. *Rule 6–102(1)* allows for only one service to be installed to the residence. We cannot apply *Rule 8–200* to a subpanel because that rule must be applied to the whole house.

We must provide enough power at the second panel to handle the known loads. Based on the circuit schedule for Panel B in Figure 4-13 in Unit 4, a 60-ampere panel would be adequate for the load. However, if the dryer, oven, and range were operated at the same time, the subpanel feeder breaker could be overloaded and trip. Therefore, the recommended feeder to Panel B is 100 amperes.

To install a 100-ampere panel, we need #3 RW75 XLPE (jacketed) (from Table 2). From *Table 6A,* three #3 RW75 wires require a 27-mm conduit.

Checking the specifications and also Figure 4-2 in Unit 4, we find that three #3 RW75 XLPE (jacketed) conductors supply Panel B.

Main Disconnect

The main disconnect in the residence is a 200-ampere combination panel. Many electric utilities state that the conductors feeding this type of disconnect must have the same ampere rating as the disconnect. Some utilities also state that the neutral conductor cannot be reduced in size because it is not always possible to foresee the type of load that may be connected to the panel. As a result, it is sometimes difficult to conform to all the rules of the electric utility as well as the local electrical code. By installing three 3/0 RW90 conductors for the line and neutral conductors, you will satisfy most local and provincial electrical code requirements.

Grounding Electrode Conductor

The grounding electrode conductor connects the main service equipment neutral bar to the grounding electrode. The grounding electrode might be the underground metallic water piping system, a driven ground rod, or a concrete-encased ground. All of this is discussed in Unit 4.

Grounding electrode conductors are sized according to *Table 16A*. For example, a residence is supplied by 3/0 AWG copper service entrance conductors. Checking *CEC Rule 10–812*, we find that the minimum grounding electrode conductor must be a #4 AWG copper conductor.

SERVICE CALCULATIONS FOR APARTMENTS

Often, single-family homes are constructed or renovated to divide the space into two or more units. This type of building is considered to be an apartment building and requires that the services be calculated according to *Rule 8–202*.

For this example, we assume that our building is constructed with a completely independent apartment in the basement. The living area has not changed.

The heat will be changed to electric baseboard heaters with 7.5 kW located in the ground-floor unit and 5.5 kW in the basement. The floor plans will be modified so that all loads previously allocated to the service will be connected to the ground-floor unit service. The basement unit will have a 3-kW water heater and a 12-kW range. Laundry facilities will not be available in the basement unit.

The service calculation for an apartment unit is almost identical to that of a single dwelling. The only difference is in the basic load calculation.

Ground-Floor Unit

Basic load based on a floor area of 139.9 m^2:

First 45 m^2	3 500 watts	
Next 45 m^2	1 500 watts	
Next 90 m^2 or portion	1 000 watts	
Total range	6 000 watts	6 000 watts
First 12 kW	6 000 watts	
Remaining 2.05 kW × 40%	820 watts	
Total	**6 820 watts**	**6 820 watts**
Other loads:		
Dryer × 25%	1 425 watts	
Water heater × 25%	750 watts	
Pump × 25%	600 watts	
Total	**2 775 watts**	**2 775 watts**
Subtotal		**15 595 watts**
Electric heat		7 500 watts
Grand total		**23 095 watts**

Basement Unit

Basic load based on a floor area of 180.98 m^2

First 45 m^2	3 500 watts	
Next 45 m^2	1 500 watts	
Next 90 m^2 or portion	1 000 watts	
Total Range	6 000 watts	6 000 watts
First 12 kW	6 000 watts	6000 watts
Other loads:		
Water heater × 25%	750 watts	750 watts
Subtotal		**12 750 watts**
Electric heat × 100%		5 500 watts
Grand total		**18 250 watts**

To determine the ampacity of the service for the ground floor:

$$\frac{\text{Demand watts}}{\text{System voltage}} = \frac{23\ 095\ \text{watts}}{240\ \text{volts}} = 96.2\ \text{amperes}$$

Therefore, a 100-ampere service would be acceptable. However, the installer may wish to increase the service to 125 amperes to allow for future growth.

To determine the ampacity of the basement unit service:

$$\frac{\text{Demand watts}}{\text{System voltage}} = \frac{18\ 250\ \text{watts}}{240\ \text{volts}} = 76\ \text{amperes}$$

Therefore a 100-ampere service would be installed.

This type of service installation usually is in the form of a single consumer's service, divided into several parts in a multigang meter socket and terminating in separate service boxes. The ampacity of the consumer's service conductors is not the sum of the individual services, but a calculation based on the individual demand loads and *Rule 8–202(3)*. To perform this calculation, we must begin with the demand of each unit, without any electric heat or air-conditioning loads.

The calculation is

100% of the largest unit = 15 595 × 100%
= 15 595 watts

65% of the next largest unit =
12 750 × 65% = 8 287.5 watts

Subtotal **23 882.5 watts**

Add the electric heat as outlined in *Rule 62–118(2)*:

Total installed electric heat = 7.5 kW + 5.5 kW
= 13 kW

100% of the first 10 kW = 10 kW × 100% = 10 000 watts

75% of the remaining 3 kW = 3 kW × 75% = 2 250 watts

Subtotal **12 250 watts**

Grand total **36 132.5 watts**

Minimum Service Ampacity

$$\frac{\text{Calculated demand}}{\text{System voltage}} = \frac{36\ 132.5\ \text{watts}}{240\ \text{volts}} = 150.6\ \text{amperes}$$

The minimum consumer's service ampacity is considerably less than the sum of the individual services. For this application, the installer would normally install a 200-ampere service to allow for future growth.

REVIEW

Note: Refer to the *CEC* or to the blueprints provided with this textbook when necessary. Where applicable, responses should be written in complete sentences.

1. When a service entrance calculation results in a value of 15 kW or more, what is the minimum size service required by the *CEC*? ____________________

____

2. a. What is the demand load per 90 m^2 for the general lighting load of a residence?

____

b. What are the demand factors for the baseboard heating load in dwellings?

____

____

____

____

3. a. What is the ampere rating of the circuits that are provided for the lighting and power loads? ______

   b. What is the maximum number of outlets permitted on a 15-ampere branch circuit by the *CEC*? ______

   c. How many 15-ampere circuits are included in this residence? ______

4. Why is the air-conditioning load for this residence omitted in the service calculations? What is the *CEC* rule number? ______

5. What demand factor may be applied to loads greater than 1500 watts, such as a clothes dryer and water heater, in addition to an electric range, *Rule 8–200(1)(a)(vi).*

6. What load may be used for an electric range rated at not over 12 kW?

7. What is the load for an electric range rated at 16 kW, *Rule 8–200(1)(a)(v)?* Show calculations.

8. What is the computed load when fixed electric heating is used in a residence? (*Rule 62–116*) ______

9. On what basis is the neutral conductor of a service entrance determined? ______

10. Why is it permissible to omit an electric space heater, water heater, and certain other 240-volt equipment when calculating the size of the neutral service entrance conductor for a residence? ______

11. Calculate the minimum size of service entrance conductors required for a residence containing the following: floor area 8 m × 12 m; 12-kW electric range; 5-kW dryer, 120/240-volt; 2200-watt sauna heater, 120-volt; 12 kW of baseboard heaters with individual thermostat control, 240-volts; two 3-kW, 240-volt air conditioners; 3-kW, 240-volt water heater.

    Determine the sizes of the ungrounded conductors and the neutral conductor. Use type RW90 copper conductors.

    Two # ______ RW90 ungrounded conductors

    One # ______ RW90 neutral (or bare neutral if permitted)

    One # ______ AWG system grounding conductor to water meter

STUDENT CALCULATIONS

Minimum service ampacity

Minimum switch

Minimum fuse or circuit breaker

Minimum bonding jumper

Minimum conduit size

Minimum panel ampacity

Minimum number of circuits in panel

UNIT 4

Service Entrance Equipment

OBJECTIVES

After studying this unit, you should be able to

- define electric service, supply service, consumer's service, overhead service, and underground service
- list the various *CEC* rules covering the installation of a mast-type overhead service and an underground service
- discuss the *CEC* requirements for disconnecting the electric service using a main panel and load centres
- discuss the grounding of interior alternating current systems and the bonding of all service entrance equipment
- describe the various types of fuses
- select the proper fuse for a particular installation
- explain the operation of fuses and circuit breakers
- explain the term *interrupting rating*
- determine available short-circuit current using a simple formula
- perform cost-of-energy calculations
- understand how to read a watt-hour meter

An electric service is required for all buildings containing an electrical system and receiving electrical energy from a utility company. The *Canadian Electrical Code (CEC)* describes the term *service* in two ways: *consumer's service* and *supply service.*

A consumer's service is considered to be all of the wiring from the service box to the point where the utility makes connection. A service box is a main switch and fuses or the main breaker in a combination panel.

CEC Section 0 defines a supply service as the single set of conductors run by the utility from its mains to the consumer's service. On an overhead service, this is usually from the transformer on the pole to the point of attachment on the service mast. With an underground service, this may be from the transformer to the line side of the meter base. If the electric utility defines the point of connection as the transformer, the wires that are run from the transformer to the meter base form part of the consumer's service. Always check with your local utility for the specific bylaws and requirements that pertain to services.

In general, watt-hour meters are located on the exterior of a building. Local codes may permit the watt-hour meter to be mounted inside the building. In some cases, all of the service entrance equipment may be mounted outside the building. This includes the watt-hour meter and the disconnecting means, *Rule 6–408.*

The electric utility generally must be contacted to determine the location of the meter, *Rule 6–408.*

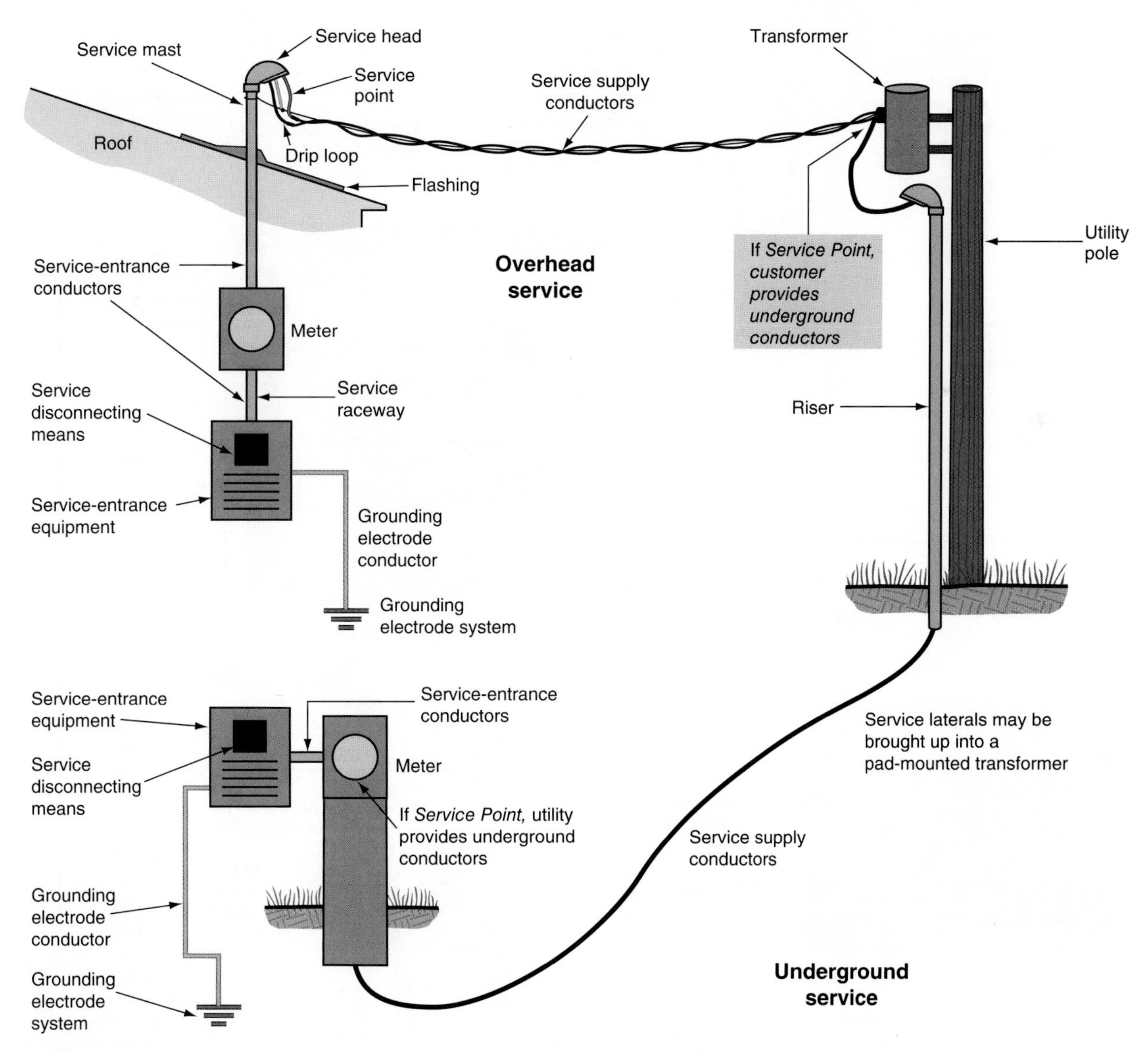

FIGURE 4-1 Service entrance, main panel, subpanel, and grounding for the service in this residence. The metal water pipe from a public water main provides the best ground for a service, *Rule 10–700(1)(c).*

This will usually be within 1 m from the front of the house and 1.8 m above grade to the centre of the socket. The final say on the position of the meter rests with the local supply authority in accordance with *Rule 6–410*.

OVERHEAD SERVICE

An overhead supply service consists of the overhead service conductors (including any splices) that are connected from the mains of the supply authority to the service entrance conductors at the mast for the consumer's service.

A combination panel has two sections, the service box (service entrance portion) and the panelboard (the distribution portion); see Figure 4-1.

In Figure 4-1, there is a dashed line between the two portions of the combination panel. The panelboard (distribution portion of the combination panel) is not considered to be part of the service entrance wiring. Therefore, it is very important that all of the service entrance wiring enters only the service entrance portion of the panel.

Figure 4-1 shows how a pad-mounted transformer feeds the service in the residence, which in turn feeds Panel B. Figure 4-2 illustrates the

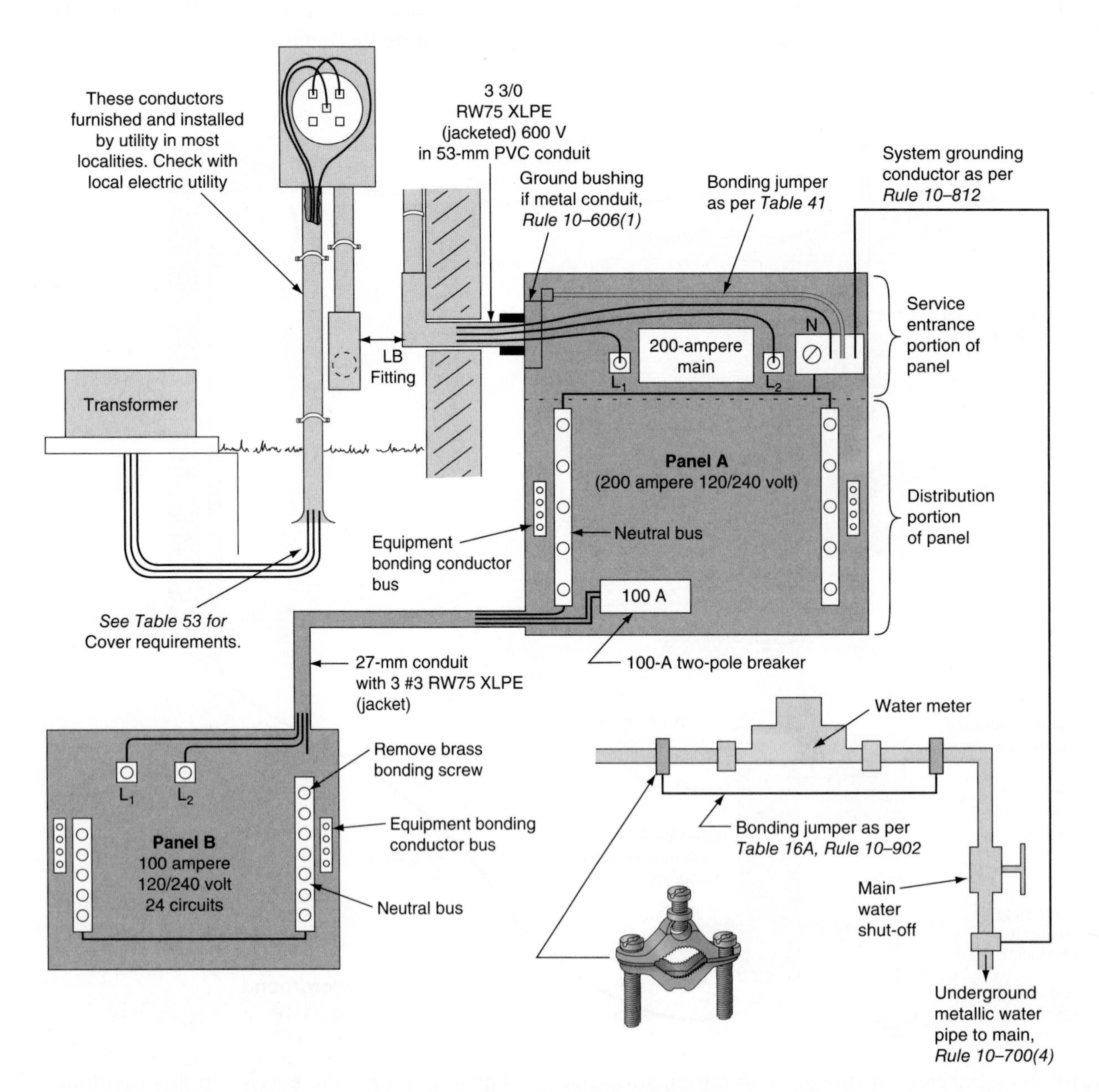

FIGURE 4-2 *CEC* terms for the services. See Figure 4-6 for grounding requirements.

CEC terms for the various components of a service entrance.

MAST-TYPE SERVICE ENTRANCE

The mast service (Figures 4-3 through 4-5) is a commonly used method of installing a service entrance. The mast service is often used on buildings with low roofs, such as ranch-style dwellings, to ensure adequate clearance between the ground and the lowest service conductor.

The mast must be run through the roof, as shown in Figure 4-3, and must be fastened to comply with *CEC* requirements. The specific requirements for installing this type of service are set out in *Rule 6–112* and *CEC Appendix B.*

Clearance Requirements for Mast Installations

When installing an overhead service, you must maintain certain clearances between the conductors and other surfaces such as the ground and nearby parts of the building. The point of attachment of the conductors to the building must not be higher than 9 m above grade, and minimum clearances from the conductors to grade should not be less than

5.5 m for highways, lanes, and alleys
5.0 m for commercial and industrial driveways
4.0 m for residential driveways
3.5 m for sidewalks

The service conductors must not be less than 1 m from a window, door, or porch, *Rule 6–112(3).*

Normally there must be a clearance of 915 mm mm (*Rule 6–112 (4), Appendix B*) between the point of attachment and the roof. However, this may be reduced if a minimum of 600 mm is maintained between the roof and the bottom of the drip loops on the supply conductors (Figure 4-4).

Service Mast as Support (*Rule 6–112*)

The bending force on the conduit increases with an increase in the distance between the roof support and the point where the supply service conductors are attached. The pulling force of service conductors on a mast service conduit increases as the length of the supply service conductors increases. As the length of the service conductors decreases, the pulling force on the mast service conduit decreases.

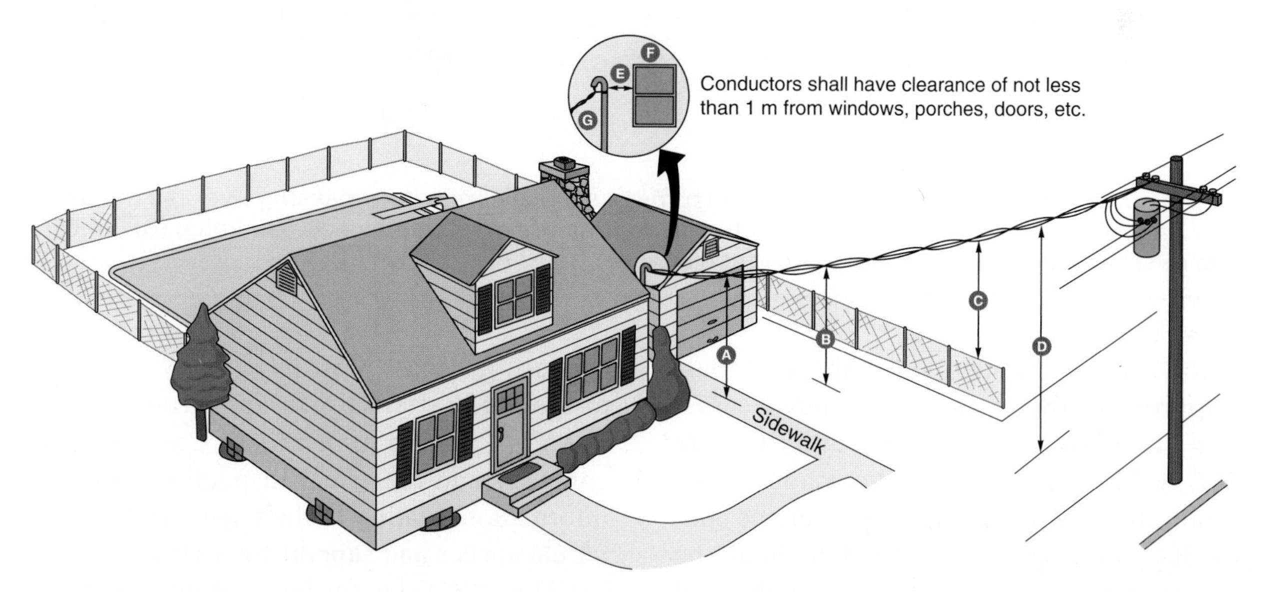

FIGURE 4-3 Clearances for a typical service entrance installation, *Rule 6–112(2).*

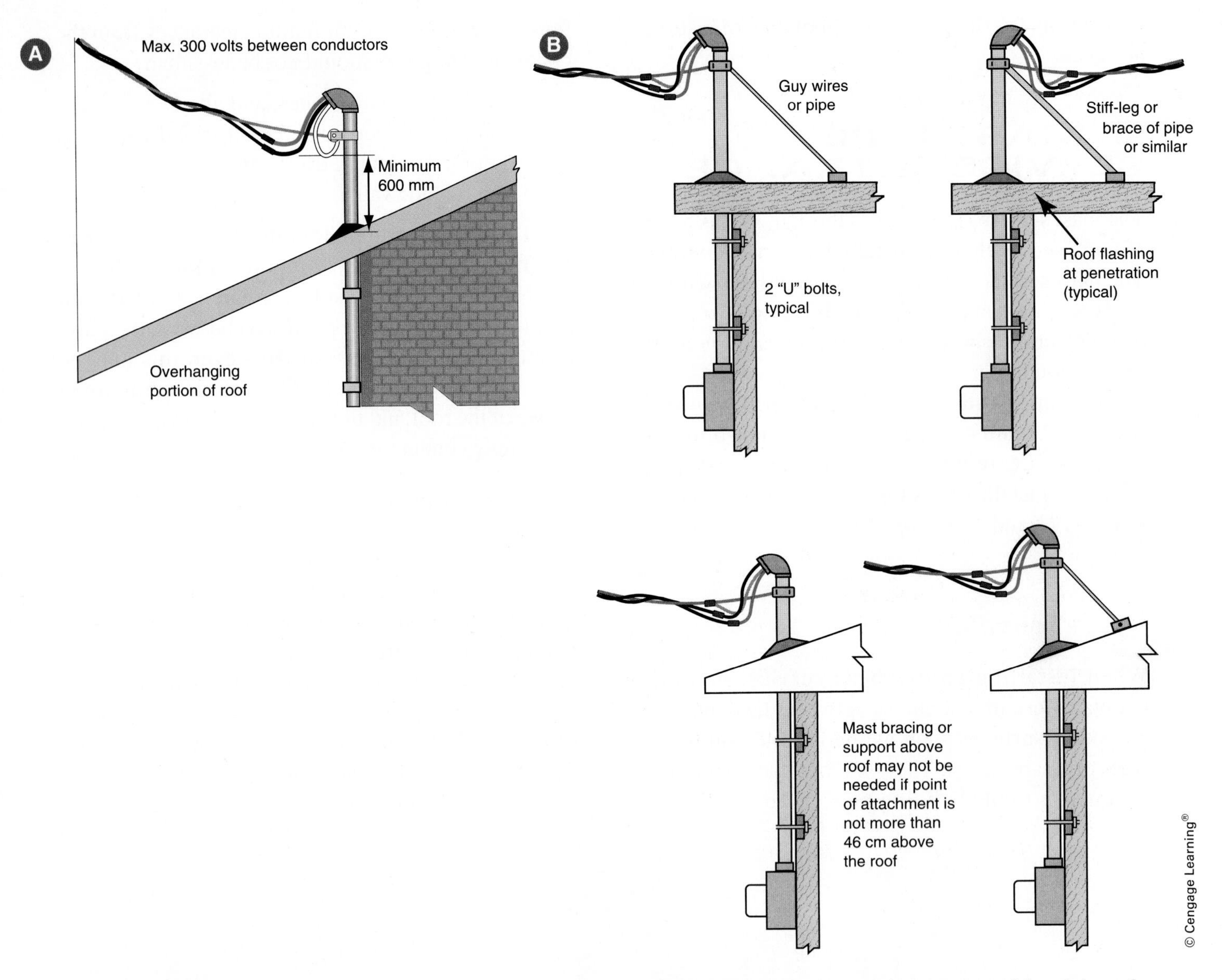

FIGURE 4-4 (A) Clearance requirements for consumer's service conductors passing over residential roofs, *Rule 6–112* and *Appendix B,* where voltage between conductors does not exceed 300 volts. (B) Typical methods of securing and supporting a mast-type service.

If extra support is not provided, the mast must be an approved tubular support member (most common) or 63-mm rigid steel conduit, *Rule 6–112(4,5)*. This size prevents the conduit from bending due to the strain of the supply service conductors. If extra support is provided, it should be in the form of a guy wire that attaches to the roof rafters of the house and the service mast by means of approved fittings. This provides support for long runs of supply service conductors and for the extra weight that ice storms can add to the conductors. If the mast is more than 1.5 m above the uppermost support, the mast must have a guy wire attached to the roof rafters by an approved fitting, *Rule 6–112(4,7,8)* and *Appendix B*. Consult the utility company and electrical inspection authority for information relating to their specific requirements for clearances and support for service masts.

The *CEC* rules for installation and clearances apply to the supply service and consumer's service conductors. For example, the service conductors must be insulated, except where the provisions of *Rule 6–308* are met and approved for exposure to direct sunlight as required by *Rule 2–134*. In

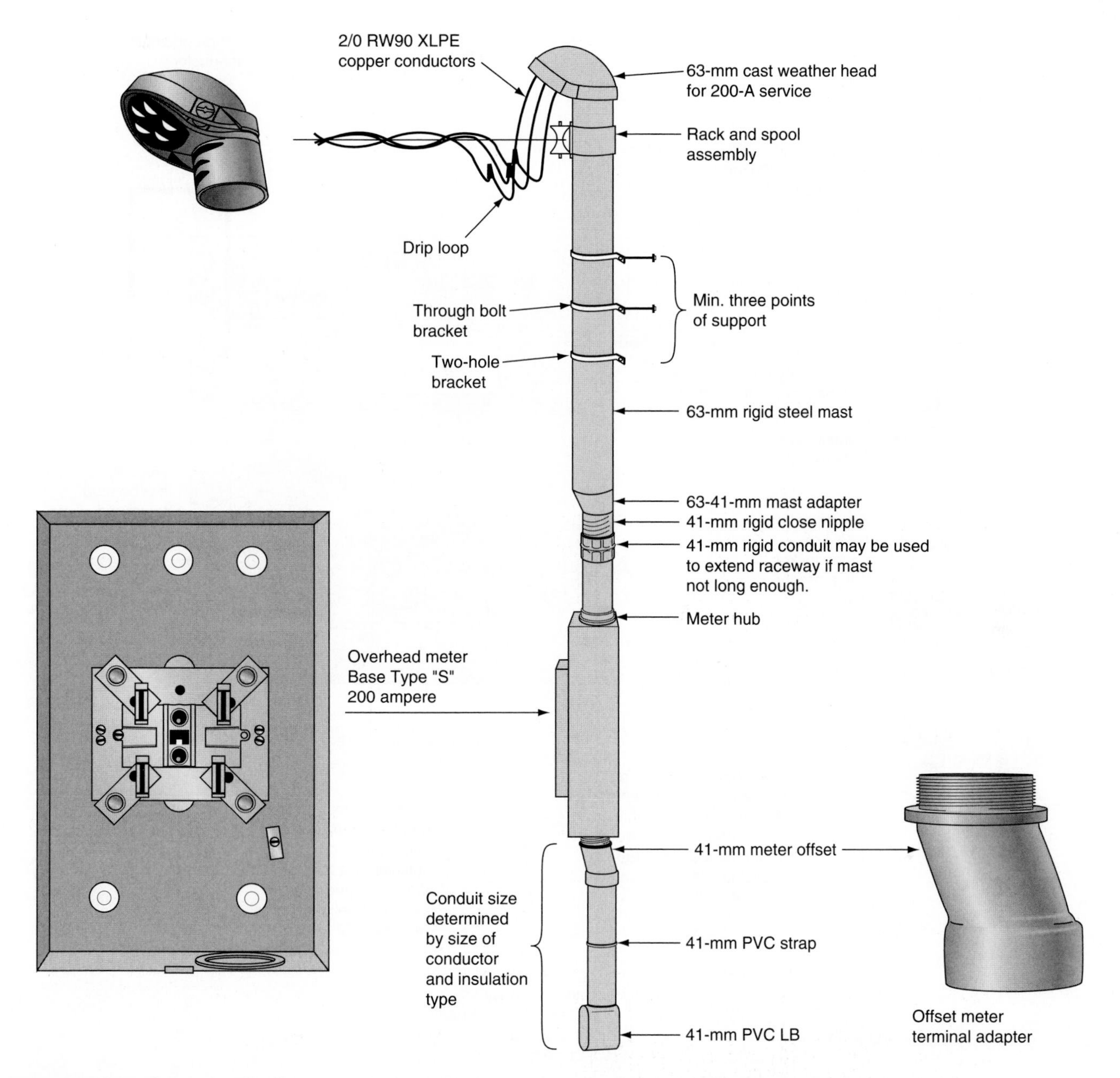

FIGURE 4-5 A method of connecting conduit to meter base on a 200-ampere overhead service entrance installation.

this case, the grounded neutral conductors are not required to have insulation.

Rule 6–112 and *Appendix B* give clearance allowances for the supply service passing over the roof of a dwelling (Figure 4-4).

The installation requirements for a typical service entrance are shown in Figures 4-5 and 4-6. Figure 4-3 shows the required clearances above the ground. The wiring connections, grounding requirements, and *CEC* references are given in Figures 4-5 and 4-6.

UNDERGROUND SERVICE

The underground service is the cable installed underground from the customer's meter socket to the utility's supply. This cable is usually, but not always, supplied by the utility company.

New residential subdivisions often include underground installations of the electrical system. The transformer and primary high-voltage

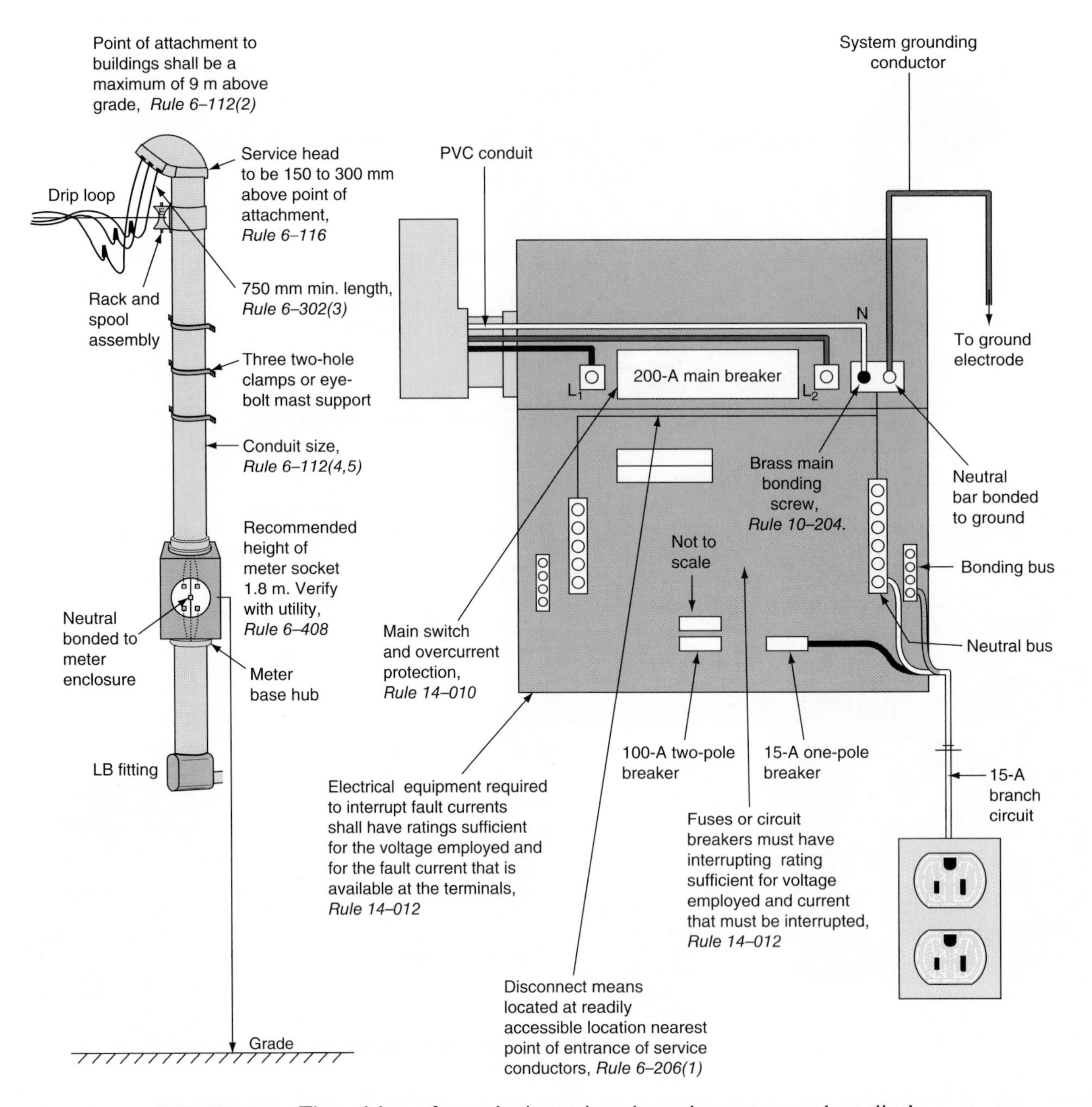

FIGURE 4-6 The wiring of a typical overhead service entrance installation.

conductors are installed by, and are the responsibility of, the utility company.

The conductors installed between the pad-mounted transformers and the meter are called *consumer's service conductors.* The utility runs the services to the lot line, and the electrical contractor completes the installation. Otherwise, the electrical contractor supplies and installs the consumer's service conductors, but the local utility may require a specific type of cable that it may supply to the contractor at a reasonable cost. Always check with the local utility in addition to the inspection authority before commencing work. Figure 4-7 shows a typical underground installation.

The wiring from the external meter to the main service equipment is the same as the wiring for a service connected from overhead lines (Figure 4-6). Some local codes may require conduit to be installed underground from the pole to the service entrance equipment. Requirements for underground services are given in *Rule 6–300.* The

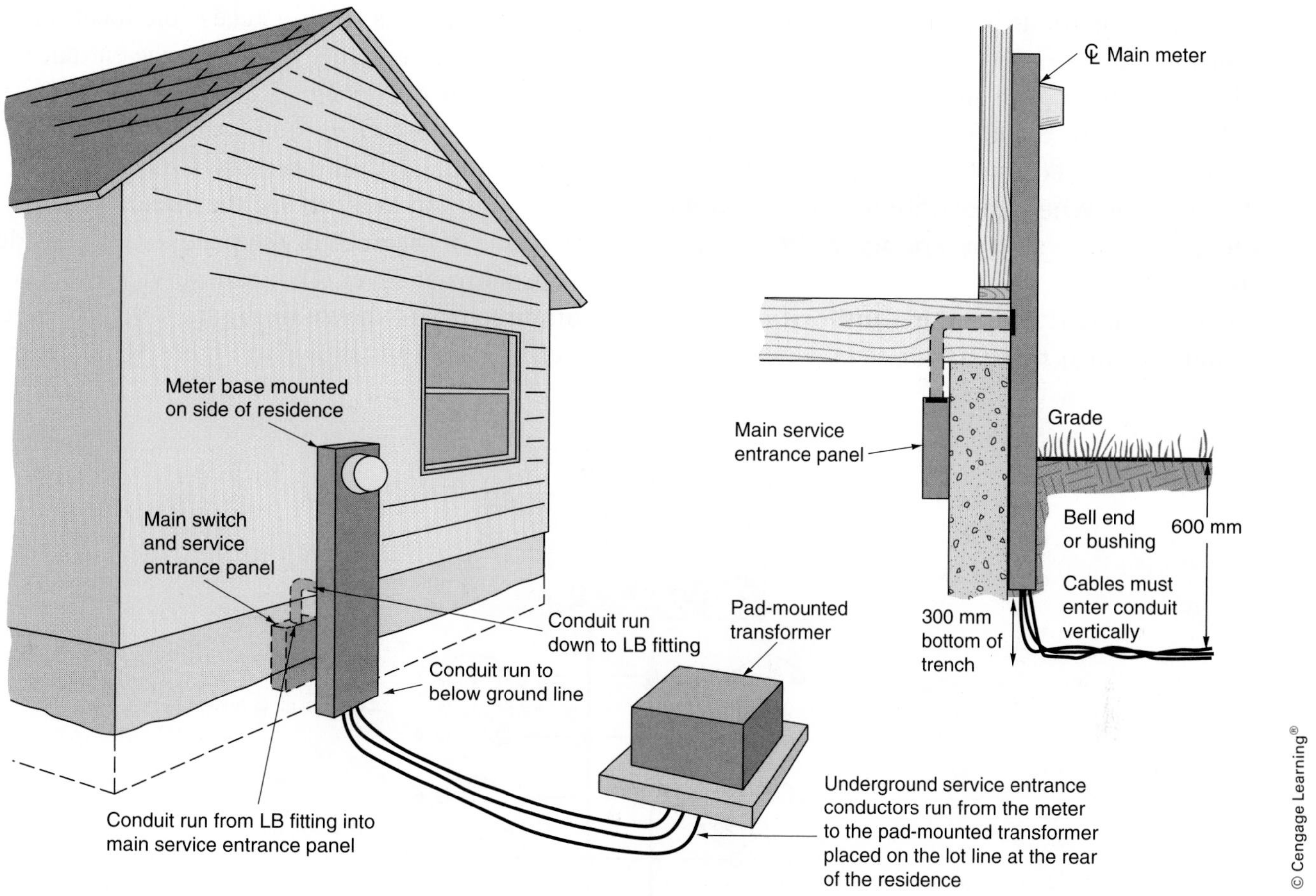

FIGURE 4-7 Underground service.

underground conductors must be suitable for direct burial in the earth.

If the electric utility installs the underground service conductors, the work must comply with the rules established by the utility, *Section 0, Scope.* These rules may not be the same as *Rule 6–300.*

When the underground conductors are installed by the electrician, *Rule 6–300* applies. This rule deals with the protection of conductors against damage and sealing of underground conduits where they enter a building.

MAIN SERVICE DISCONNECT LOCATION

The main service disconnect means shall be installed at a readily accessible location so that the service entrance conductors within the building are as short as possible. See Figure 4-8.

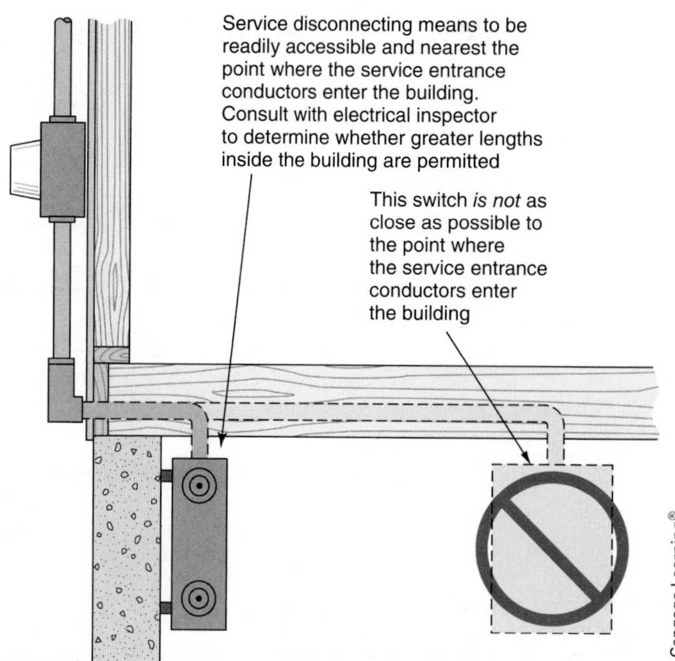

FIGURE 4-8 Main service disconnect location, *Rule 6–206(1)(c).*

The reason for this rule is that the service entrance conductors do not have overcurrent protection other than that provided by the utility's transformer fuses. Should a fault occur on these service entrance conductors within the building (at the bushing where the conduit enters the main switch, for instance), the arcing could result in a fire.

The electrical inspector (authority having jurisdiction) must make a judgment as to what is considered to be a readily accessible location nearest the point of entrance of the service entrance conductors, *Rule 6–206(1)*.

Rule 2–100(3)(a) states that all panels must have a legible circuit directory indicating what the circuits are for. You can see the circuit directory on the inside of the door of the panel or on the side of the panel front cover. The main service Panel A circuit directory is shown in Figure 4-9. The Panel B circuit directory is shown in Figure 4-13.

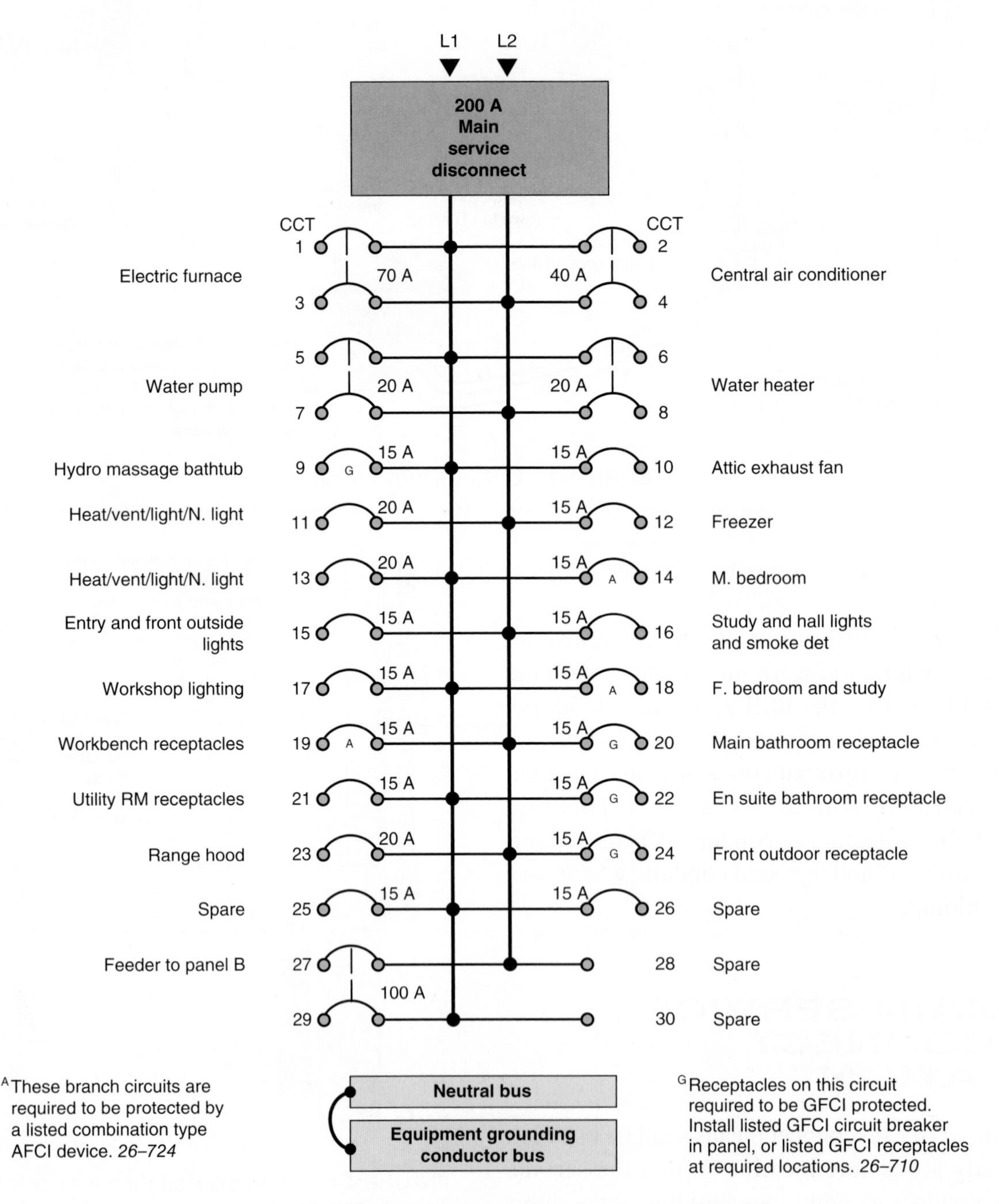

FIGURE 4-9 Circuit schedule of main service Panel A.

Disconnect Means (Panel A)

The requirements for disconnecting the electric services are covered in *Rule 6–200. Rule 6–200(1)* requires that each consumer's service be provided with a single service box.

However, *Subrule (2)* sets out the circumstances in which more than one service box is permitted:

- The subdivision is made at the meter base.
- Maximum rating is 600 amperes and 150 volts to ground.
- The meter base is located outdoors.

A panelboard must have the minimum number of overcurrent devices or positions, as required by *Rule 8–108(1,2).*

The *minimum* rating for a single-family residential service is 100-ampere, three-wire when

- The area excluding the basement is 80 m^2 or more, *Rule 8–200(1)(b)(i)*

or

- The initial computed load is more than 14.4 kW

The minimum rating for a single-family residential service is 60-ampere, three-wire when

- The area is 80 m^2 or less, *Rule 8–200(1)(b)(ii)*

and

- The initial computed load is less than 14.4 kW

For this residence, Panel A provides a 200-ampere circuit breaker for the main disconnect. Panel A also has a number of branch-circuit overcurrent devices to protect the many circuits originating from this panel (Figures 4-9 and 4-10). This type of panel is listed by the Canadian Standards Association (CSA) as a load centre and as service equipment. The panel meets *CEC* requirements and is designed to accommodate full- and half-width breakers. This panel may be surface- or flush-mounted.

Panel A is located in the workshop. The placement of the main disconnect is determined by the location of the meter. The disconnect is mounted at an accessible point as close as possible to the place where the service conductors enter the building. The local utility company generally decides where the supply service, or the consumer's service in the case of an underground service, is to be located, and also provides the electrician with a meter location. The consumer's service is the responsibility of the electrician, and the supply service is the responsibility of the utility.

Courtesy of Siemens Energy and Automation

FIGURE 4-10 Typical main panel with 200-ampere main breaker suitable for use as service entrance equipment, *Rule 6–200(1).*

Rule 10–624(6) states that a grounded circuit conductor shall not be used for bonding non-current-carrying equipment on the load side of the service disconnecting means.

The panelboard must have an approved means for attaching the bonding conductors when cable is used (Figure 4-11).

Bonding. At the main service entrance equipment, the grounded neutral conductor must be bonded to the metal enclosure. For most residential-type panels, this main bonding jumper is a bonding screw furnished with the panel. This bonding screw is inserted through the neutral bar into a threaded hole in the back of the panel itself. The bonding screw will be brass or green in colour and must be clearly visible after it is in place.

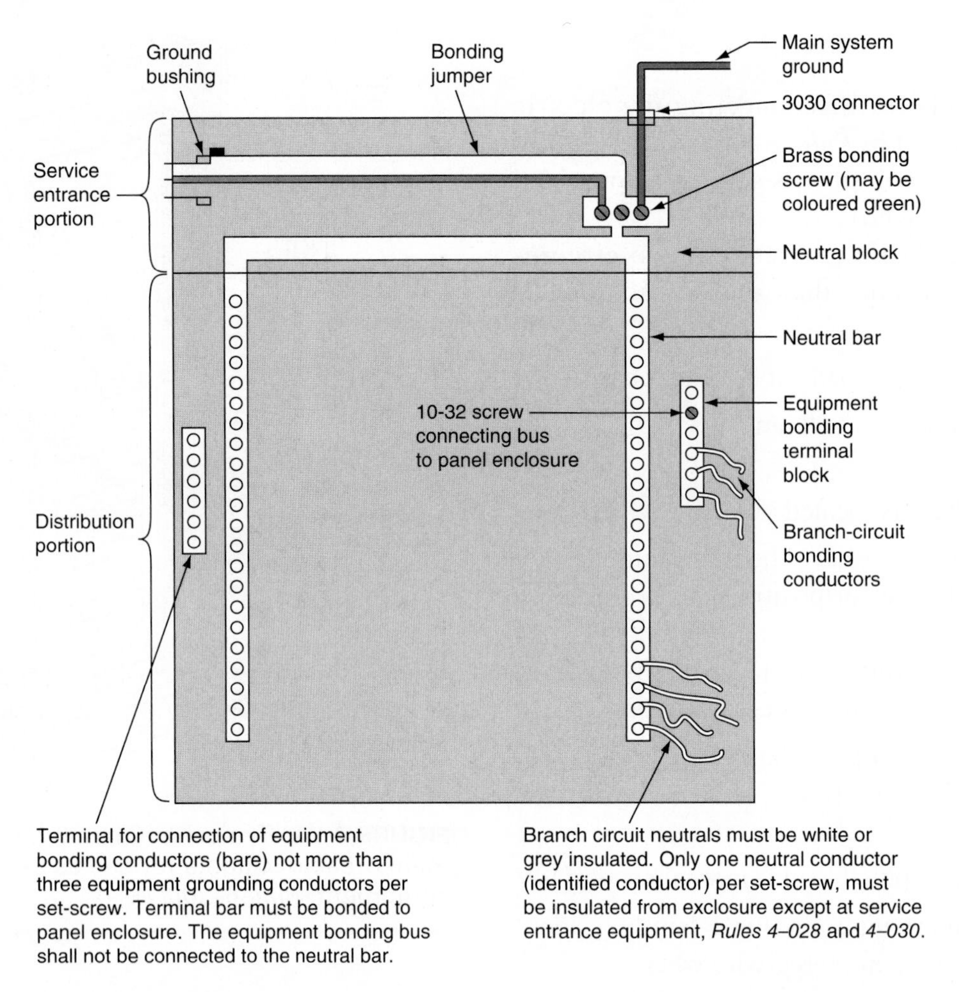

FIGURE 4-11 Connections of service neutral, branch-circuit neutral, and equipment-bonding conductors at main panelboards, *Rule 10–210(1)(b)*.

As previously mentioned, *Rule 10–624(1)* prohibits bonding the grounded neutral of a system to any equipment on the load side of the main service disconnect. Therefore, the brass or green main bonding jumper screw furnished with residential-type panels will be inserted at the main service panel *only*. This screw must be removed at subpanels, such as Panel B in the recreation room of the residence. See Figure 4-11.

Load Centre (Panel B)

When the main service panel is located some distance from areas having many circuits and/or a heavy load concentration such as the kitchen or laundry of this residence, it is recommended that load centres be installed near these concentrations of load. The individual branch-circuit conductors are run to the load centre, not back to the main panel. Thus, the branch-circuit runs are short, and line losses (voltage and wattage) are less than if the circuits had been run all the way back to the main panel.

To determine if there is a cost or benefit and if the lower line losses justify installing an extra load centre, the cost of material and labour to install the extra load centre should be compared to the cost of running many branch circuits back to the main panel.

A typical load centre is shown in Figure 4-12. The circuit schedule for Panel B, the load centre of the residence, is shown in Figure 4-13. Panel B is located in the recreation room. It is fed by three #3 RW75 XLPE (jacketed) conductors run in a 27-mm conduit originating from Panel A. The conductors are protected by a 100-ampere, 240-volt, two-pole overcurrent device in Panel A. The

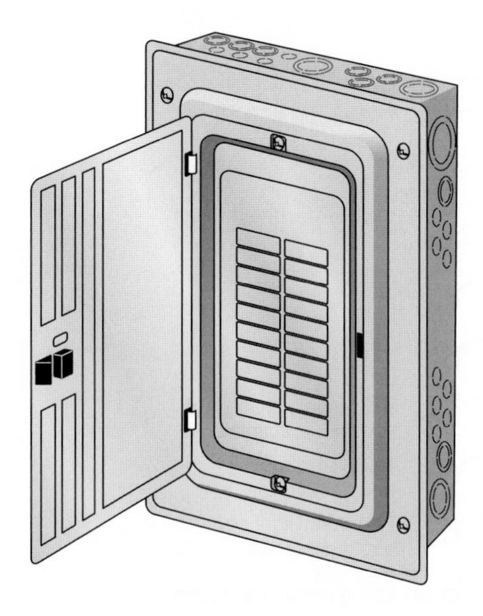
Courtesy of Siemens Energy and Automation

FIGURE 4-12 Typical load centre of the type installed for Panel B in this residence. The bonding screw will be removed from the neutral block, *Rule 10–204(1)(c)*.

overcurrent protection can be fuses or circuit breakers. This is discussed later in this unit.

SERVICE ENTRANCE CONDUIT SIZING

To determine the proper size of conduit, the conduit fill is calculated using the same method as outlined in Unit 3.

Look up the necessary data in *CEC Tables 8, 9(A-J),* and *10(A-D)* (see Table 4-1). If all of the wires in the conduit are the same size, *CEC Table 6(A-K)* may be used for sizing the conduit.

METER

Some electric utility companies offer lower rates for water heaters and electric heating units connected to separate meters. The residence in the plans has only one meter, so all lighting, heating, cooking, and water heater loads are registered on that meter.

For a typical overhead service, a meter socket (Figure 4-14) is mounted at eye level on the outside wall of the house; see Figure 4-7. The service conduit is connected to the socket at a threaded boss (hub) on the top of the socket.

If non-metallic raceways are used for the service conduit, the method of fastening to the top of the meter socket will be different. If a threaded hub is used at the meter socket, a short threaded conduit nipple will be required, along with a non-metallic female adapter. Non-metallic raceways must never be threaded directly into a threaded boss, *Rule 12–1112(2)*. The lower index of expansion of the metal nipple prevents the raceway from breaking inside the threaded boss.

Alternatively, a non-metallic hub may be used in place of the threaded hub, with the raceway fastened with solvent cement.

This conduit would be carried down into the basement, connecting to the main service equipment, as shown in Figures 4-7, 4-8, and 4-15. Proper fittings must be used, and the conduit must be sealed where it penetrates the wall; see Figure 4-15.

This is required by *Rule 6–312(1)* and is achieved by placing duct sealing compound around the conductors in the raceway where they pass through the wall. *Rule 6–312* also requires that the service raceway be suitably drained, either inside or outside the building. However, if the service raceway terminates in the top of the service box, the drain must be outdoors. This is usually accomplished by drilling a small hole in the bottom of the "L" fitting on the outside of the building. The electrician's responsibility is to install the consumer's service conductors from the load side of the meter to the line side of the main service disconnect switch.

Figure 4-16 shows one type of underground meter base. Note the following:

1. The line side is designated. This is important because if the meter is installed with the load at the top it will run backward.
2. These are stud-type connections and require the line conductors to have a crimped-on lug.
3. Line-side insulated jaw shrouds help prevent shocks and faults when installing the cover on the meter base.
4. Insulated support blocks also provide a location for a flat-rate water heater lug, which can be bolted on at the three o'clock position.
5. Load-side lugs are angled for ease of connection without too much cable bending.

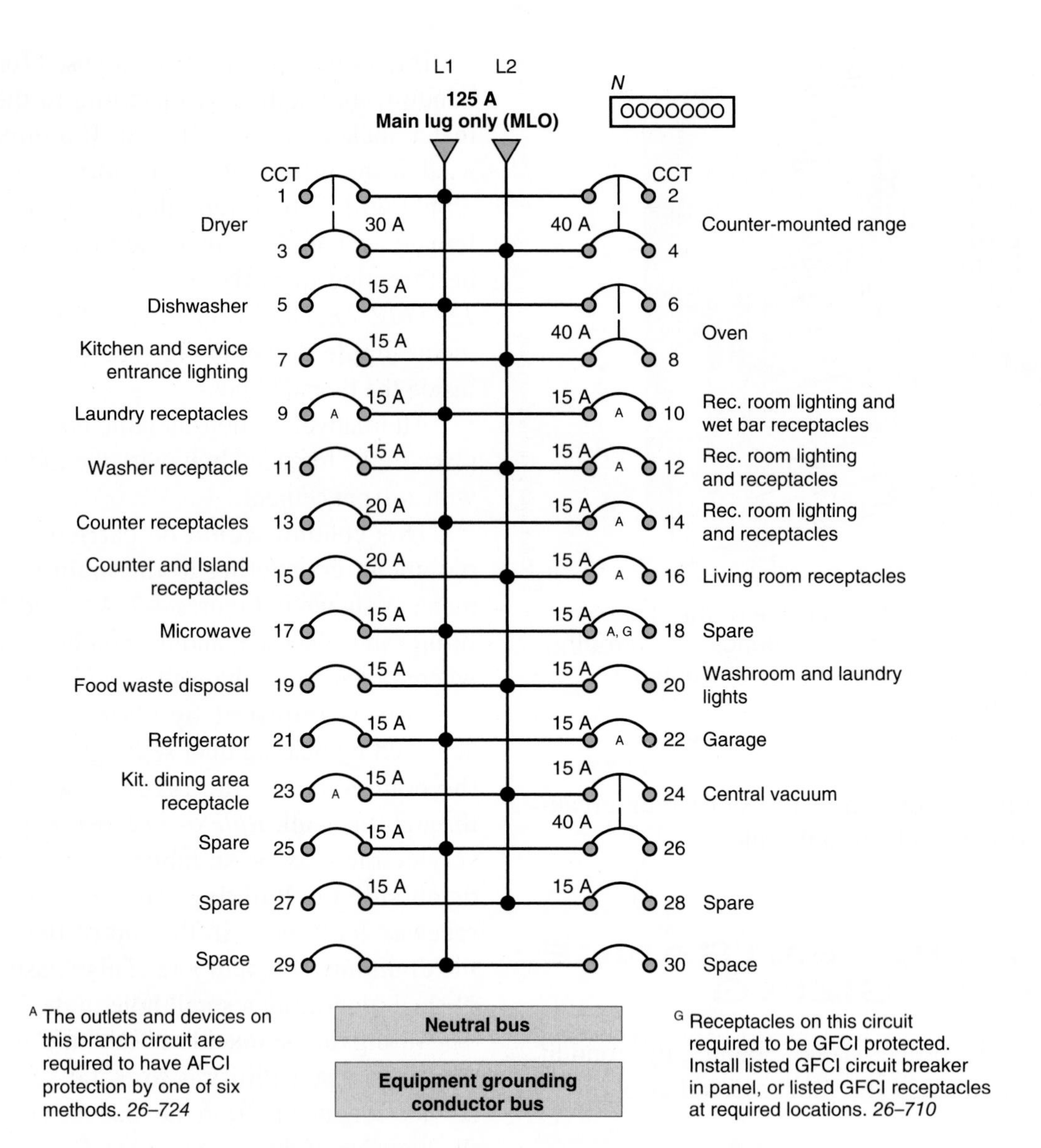

FIGURE 4-13 Circuit schedule of Panel B.

TABLE 4-1

Examples of conduit fill for the service entrance conduit and subpanel feeder conduit in this residence. Obtain conductor area from *Table 10A*. Using *Table 8*, conduit fill is 40%. Use *Table 9I's* 40% column for correct conduit size.

PANEL A CONDUCTOR SIZE BASED ON AMPACITY PER *CEC TABLE 2*		PANEL B CONDUCTOR SIZE BASED ON AMPACITY PER *CEC TABLE 2*	
Three 3/0 RW75 XLPE (jacketed) (600 V)	227.5 $mm^2 \times 3 = 682.5\ mm^2$	Three #3 RW75 XLPE (jacketed) (600 V)	85.01 $mm^2 \times 3 = 255.03\ mm^2$
Conduit size	53 mm	Conduit size	27 mm

Courtesy of Milbank Manufacturing Company

FIGURE 4-14 Typical meter socket.

6. Two-position ground lug. This is not normally used because the meter base is bonded through the grounded service conductor (neutral), *Rule 10–624(1).*
7. The enclosure is 450 × 300 × 112.5 mm, which allows plenty of room for pulling in conductors and terminating the lugs.
8. The knockouts are provided for the conduit from the transformer to the meter base and for the conduit to the main disconnect.

GROUNDING—WHY GROUND?

Electrical systems and their conductors are grounded to minimize voltage spikes when lightning strikes, or when other line surges occur. Grounding stabilizes the normal voltage to ground.

Electrical metallic tubing (EMT) and equipment are grounded so that the equipment voltage to ground is kept to a low value. In this way, the shock hazard is reduced.

Proper grounding means that overcurrent devices can operate faster when responding to ground faults. Effective grounding occurs when a low-impedance (opposition to AC current flow) ground path is provided. A low-impedance ground path means that there is a high value of ground-fault current. As the ground-fault current increases, there is an increase in the speed with which a fuse will open or a circuit breaker will trip. Thus, as the overcurrent device senses and opens the circuit faster, less equipment damage results. This *inverse time characteristic* means that the higher the value of

RULE 6–312(11) requires that when raceways pass through areas having great temperature differences, some means must be provided to prevent passage of air back and forth through the raceway. Note that outside air is drawn in through the conduit whenever a door opens. Cold outside air meeting warm inside air causes condensation, which can cause rusting and corrosion of vital electrical components. Equipment having moving parts, such as circuit breakers, switches, and controllers, are especially affected by moisture. Sluggish action of the moving parts in this equipment is undesirable. Insulation or other type of sealing compound can be inserted as shown to prevent the passage of air. The raceway must also be drained where it enters the building. This is usually accomplished by drilling a small hole in the bottom of the "L" fitting.

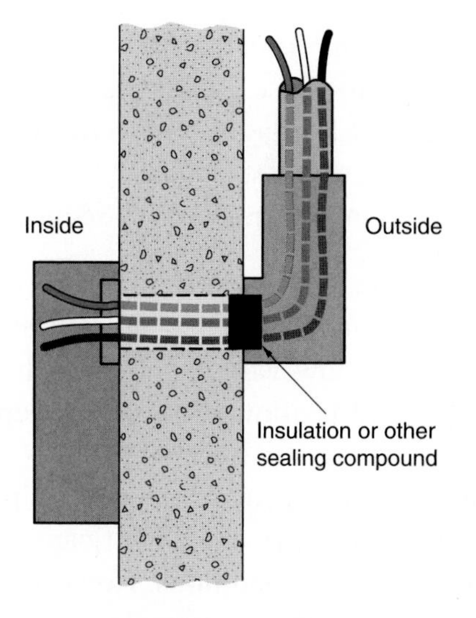

Courtesy of CSA Group

FIGURE 4-15 Installation of conduit through a basement wall.

FIGURE 4-16 An underground type of meter base with studs for utility connection of line conductors.

current, the less time it will take to operate the overcurrent device.

Rule 10–600 and *Appendix B* provide the purpose and requirements for effective grounding in an electrical system. The arcing damage to electrical equipment and conductor insulation is closely related to the value of I^2t, where

I = current flowing from phase to ground, or from phase to phase, in amperes
t = time needed by the overcurrent device to open the circuit, in seconds
I^2t is measured in ampere-squared-seconds

This expression shows that there will be less equipment and/or conductor damage when the fault current is kept to a low value and when the time that the fault current is allowed to flow is kept to a minimum.

Grounding electrode conductors and equipment-grounding conductors carry an insignificant amount of current under normal conditions. However, when a ground fault occurs, these grounding conductors must be capable of carrying whatever value of fault current might flow for the time it takes the overcurrent protective device to clear the fault.

GROUNDING ELECTRODE SYSTEMS

In the grounding electrode system, rather than grounding a single item such as the neutral conductor, the electrician must be concerned with grounding and bonding together an entire system. The term *system* means the service neutral conductor, the grounding electrode, cold water pipes, gas pipes, service entrance equipment, and jumpers installed around meters. If any of these system parts become disconnected or open, the integrity of the grounding system is maintained through other paths. This means that all parts of the system must be tied (bonded) together.

Figure 4-17 and the following steps illustrate what can happen if an entire system is not grounded:

1. A live wire contacts the gas pipe. The bonding jumper (A) is not installed originally.
2. The gas pipe now has 120 volts with respect to ground. The pipe is hot.
3. The insulating joint in the gas pipe results in a poor path to ground; assume the resistance is 8 ohms.
4. The 20-ampere overcurrent device does not open:

$$I = \frac{E}{R} = \frac{120}{8}\text{ amperes} = 15\text{ amperes}$$

5. If a person touches the hot gas pipe and the water pipe at the same time, current flows through the person's body. If the body resistance is 12 000 ohms, the current is

$$I = \frac{E}{R} = \frac{120}{12\,000}\text{ amperes} = 0.01\text{ amperes}$$

   This value of current passing through a human body can cause death.

6. The overcurrent device is "seeing" (15 + 0.01) amperes = 15.01 amperes; however, it still does not open.
7. If the *system grounding* concept had been used, the bonding jumper would have kept the voltage difference between the water pipe and the

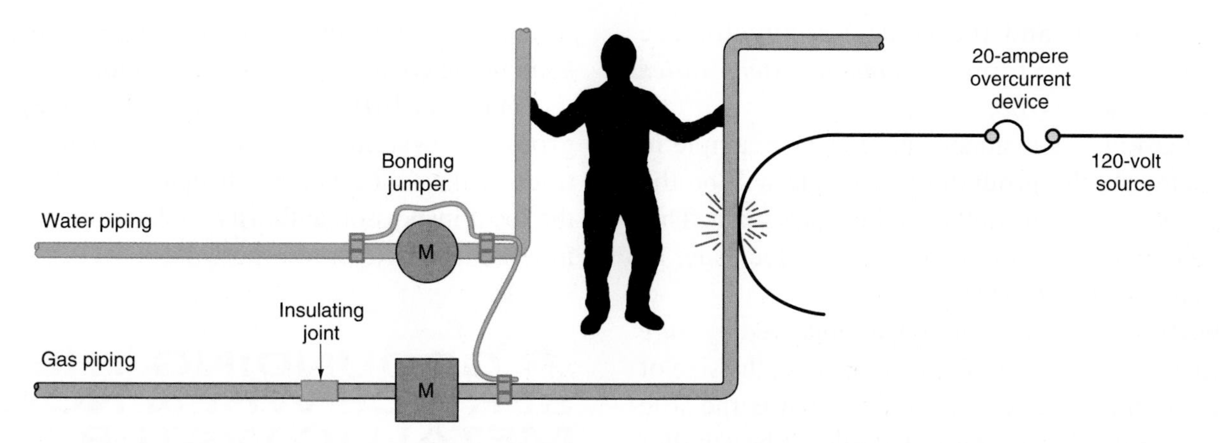

FIGURE 4-17 System grounding.

gas pipe at zero, *Rule 10–406(2)(a)* and (4). Thus, the overcurrent device would open. If 3.05 m of #6 AWG copper wire is used as the jumper, the resistance of the jumper is 0.003 95 ohms. The current is

$$I = \frac{E}{R} = \frac{120}{0.003\ 95} \text{ amperes} = 30\ 380 \text{ amperes}$$

(In an actual system, the impedance of all parts of the circuit is much higher, resulting in a much lower current. The current, however, would be high enough to cause the overcurrent device to open.)

Advantages of System Grounding*

To appreciate the concepts of proper grounding, let us review some important *CEC* rules.

Rule 10–002 sets out the objectives of grounding and bonding, summarized as follows:

- to protect life from the danger of electric shock
- to limit the voltage on a circuit
- to facilitate the operation of electrical apparatus and systems
- to limit voltage on a circuit when exposed to lightning
- to limit AC circuit voltages to ground to 150 volts or less on circuits supplying interior wiring systems

*Courtesy of CSA Group

CEC *Section 0* provides definitions for the terms associated with grounding and bonding electrical systems:

Bonding: A low-impedance path obtained by permanently joining all non-current-carrying metal parts to ensure electrical continuity. Any current likely to be imposed must be conducted safely.

Grounding: A permanent path to earth with sufficient ampacity to carry any fault current liable to be imposed on it.

Bonding Conductor: A conductor that connects the non-current-carrying parts of electrical equipment to the service equipment or grounding conductor.

Grounding Conductor: The conductor used to connect the service equipment to the grounding electrode.

- The potential voltage differences between the parts of the system are minimized, reducing the shock hazard.
- The impedance of the ground path is minimized. This results in a higher current flow in the event of a ground fault: The lower the impedance, the higher the current flow. This means that the overcurrent device will open faster under fault conditions.

Methods of Grounding

Figure 4-19 shows that the metal cold water pipes, the service raceways, the metal enclosures, the service switch, and the neutral conductor are bonded together to form a *grounded system, Rules 10–406(2)* and *10–700.*

Regarding the residence used as an example in this textbook, the grounding electrode will be the service water pipe from the public water main. This is considered an "in-situ" grounding electrode, as outlined in *Rule 10–700(4).*

Where a driven ground rod, as approved by *Rule 10–700(2)(a),* is the grounding electrode, the size of the grounding electrode conductor that is the sole connection to the driven rod shall not be smaller than the requirements of *Rule 10–812.*

If the driven ground rod is not selected as the grounding electrode, any one of these items may be used: at least 6 m of bare copper conductor not smaller than #4 AWG *(Table 43)* or an approved manufactured ground plate; see Figure 4-18. Both of these may be encased in concrete at least 50 mm thick and in direct contact with the earth, such as near the bottom of a foundation or footing.

Either a bare copper conductor or a ground plate may be used when buried directly in the ground to a minimum depth of 600 mm below finished grade. The residence used as an example in this book is being fed from a jacketed 3/0 RW75, the rated ampacity of which is 200 amperes, so our house would require a #3 bonding conductor according to Table 43.

Courtesy of Thomas & Betts Corporation

FIGURE 4-18 Ground plates.

There is little doubt that this concept of a grounded system gives rise to many interpretations of the *CEC*. The electrician must check with the local inspection authority to determine the local interpretation. Refer to Figure 4-19.

GROUNDING THE SERVICE WHEN NON-METALLIC WATER PIPE IS USED

It is quite common to find the main water supply to a residence installed with plastic (PVC) piping. Even interior water piping systems may be non-metallic piping. Some local building codes prohibit the use of non-metallic piping. Regardless, it is important that the electrical system be properly grounded and bonded.

When non-metallic water systems are used, one of the other methods of system grounding as identified in *Rule 10–700* must be used. One of the most common methods is the ground plate.

If the water piping inside the building is non-metallic and only short, isolated sections of metal water piping are used, such as at water heaters, these short sections may not need to be bonded. Check with your local inspection authority to be sure. Sometimes, the water system inside the dwelling is metallic, even though the water service to the street is non-metallic. With this large amount of metal water piping present, it must now be bonded to ground.

SUMMARY—SERVICE ENTRANCE EQUIPMENT GROUNDING

When grounding service entrance equipment (Figure 4-20), you must observe the following *CEC* rules*:

- The system must be grounded when the maximum voltage to ground does not exceed 150 volts, *Rule 10–106(1)(a).*

*Courtesy of CSA Group

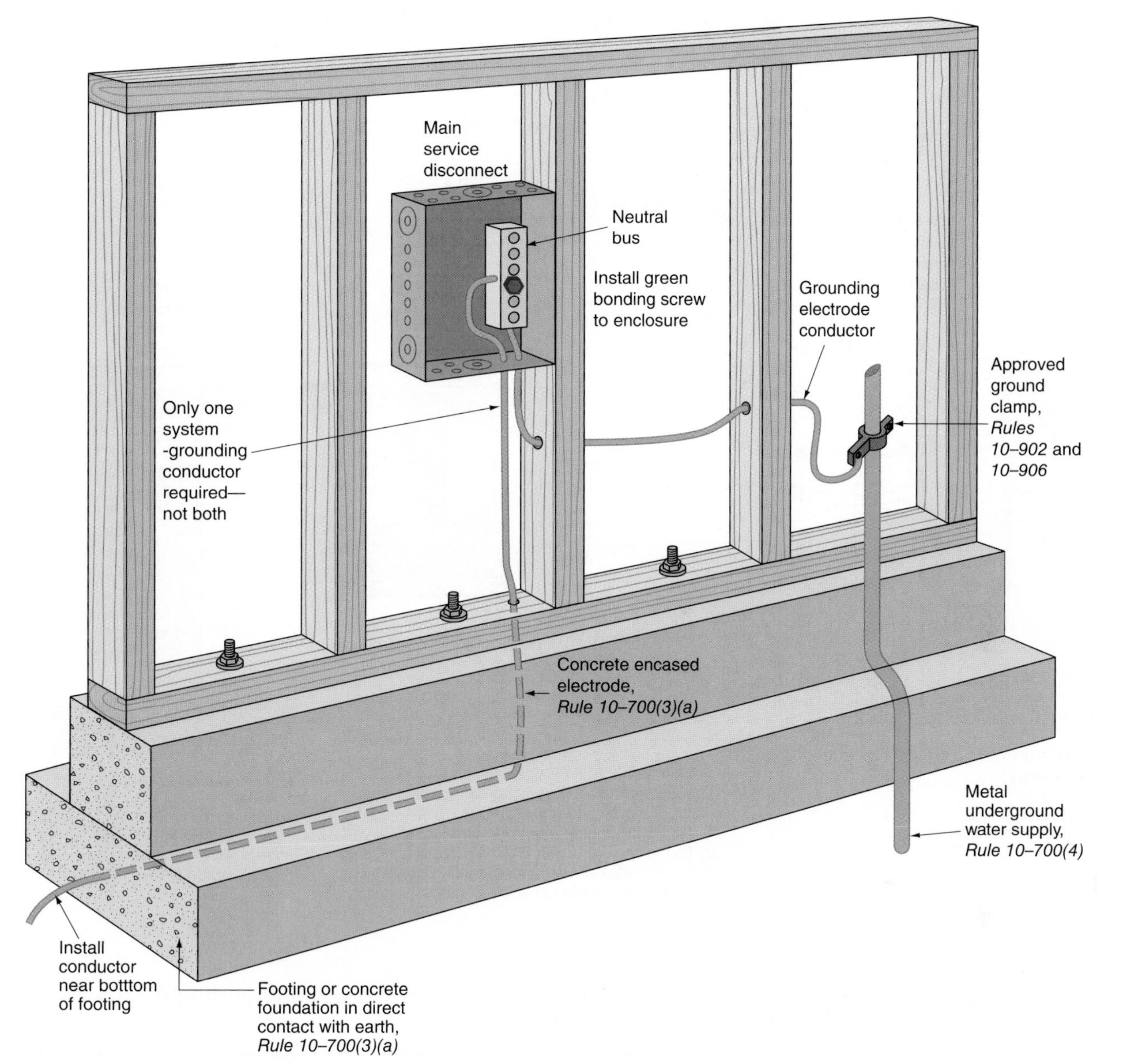

FIGURE 4-19 *Rule 10–700* lists many options for a system-grounding electrode. Here a concrete-encased #3 AWG bare copper conductor is laid near the bottom of the footing, encased by at least 50 mm of concrete. The *minimum* length of the conductor is 6 m. See *Rule 10–700(3)(a)* and *Table 43. Note:* The *CEC* requires that if more than one ground electrode connection exists, they must be bonded together, *Rule 10–702.*

- The system must be grounded if it incorporates a neutral conductor, *Rule 10–106(1)(b).*
- All grounding schemes shall be installed so that no objectionable currents will flow over the grounding conductors and other grounding paths, *Rule 10–200(1,2).*
- The ground electrode conductor must be connected to the supply side of the service-disconnecting means. It must not be connected to any grounded circuit conductor on the load side of the service disconnect, *Rule 10–204(1).*
- The neutral conductor must be grounded, *Rule 10–210(1).*
- All electrical equipment must be bonded together. See *Rule 10–400.*

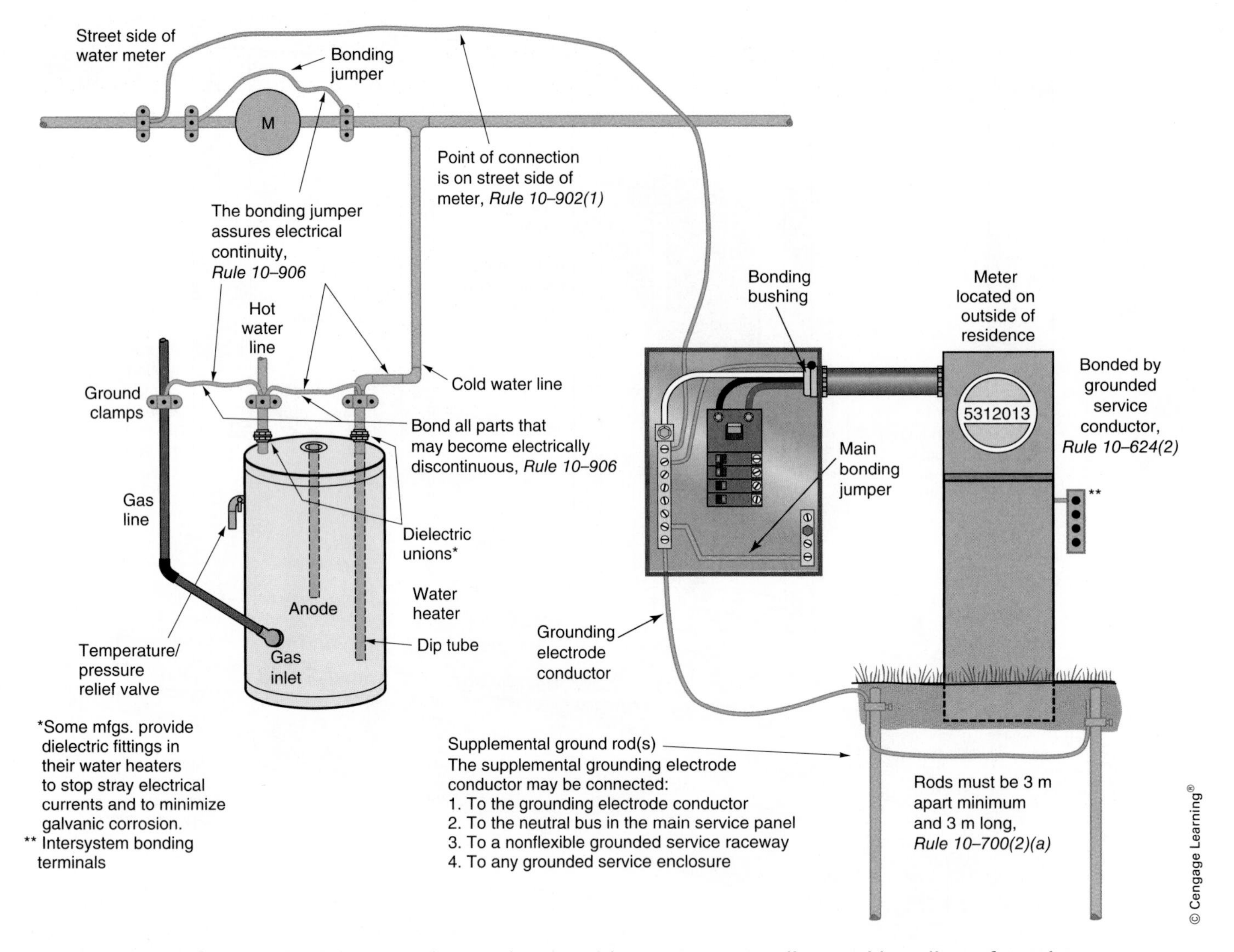

FIGURE 4-20 One method that may be used to provide proper grounding and bonding of service entrance equipment for a typical single-family residence.

- The grounding electrode conductor used to connect the grounded neutral conductor to the grounding electrode must not be spliced, except with thermit welding or compression connectors, *Rule 10–806(1)*.
- The grounding electrode conductor must be sized according to *Rule 10–812*.
- The minimum conductor size for concrete-encased electrodes is found in *Table 43* and *Rule 10–700(3)*.
- The metal water supply system shall be bonded to the grounding conductor of the electrical system, *Rule 10–406(2)* and *Appendix B*.
- When more than one grounding means exist, they must be bonded together, *Rule 10–702*.
- The grounding electrode conductor must be rigidly stapled to the surface of a construction if the run is exposed and without metal covering or protection. This wire must be free from exposure to mechanical injury, *Rule 10–806(2)*.
- The grounding electrode conductor shall be copper, aluminum, or another acceptable material, *Rule 10–802*.
- Bonding shall be provided around all insulating joints or sections of the metal water piping system that may be disconnected, *Rule 10–406(4)*.

- The connection to the grounding electrode must be accessible, *Rule 10–902(2).*
- The grounding conductor must be connected tightly using the proper lugs, connectors, clamps, or other approved means, *Rule 10–906.*

Figure 4-21 illustrates three common types of ground electrodes. These are connected to the neutral in the main service disconnect.

This residence is supplied by three 3/0 RW75 XLPE service entrance conductors. According to *Rule 10–812,* a #6 AWG grounding electrode conductor is required. This conductor may be run in conduit or cable armour, or it may be run exposed if it is not subjected to severe physical damage.

BONDING

Bonding must be done at service entrance equipment. *Rule 10–606* lists the methods approved for bonding this equipment. Grounding bushings

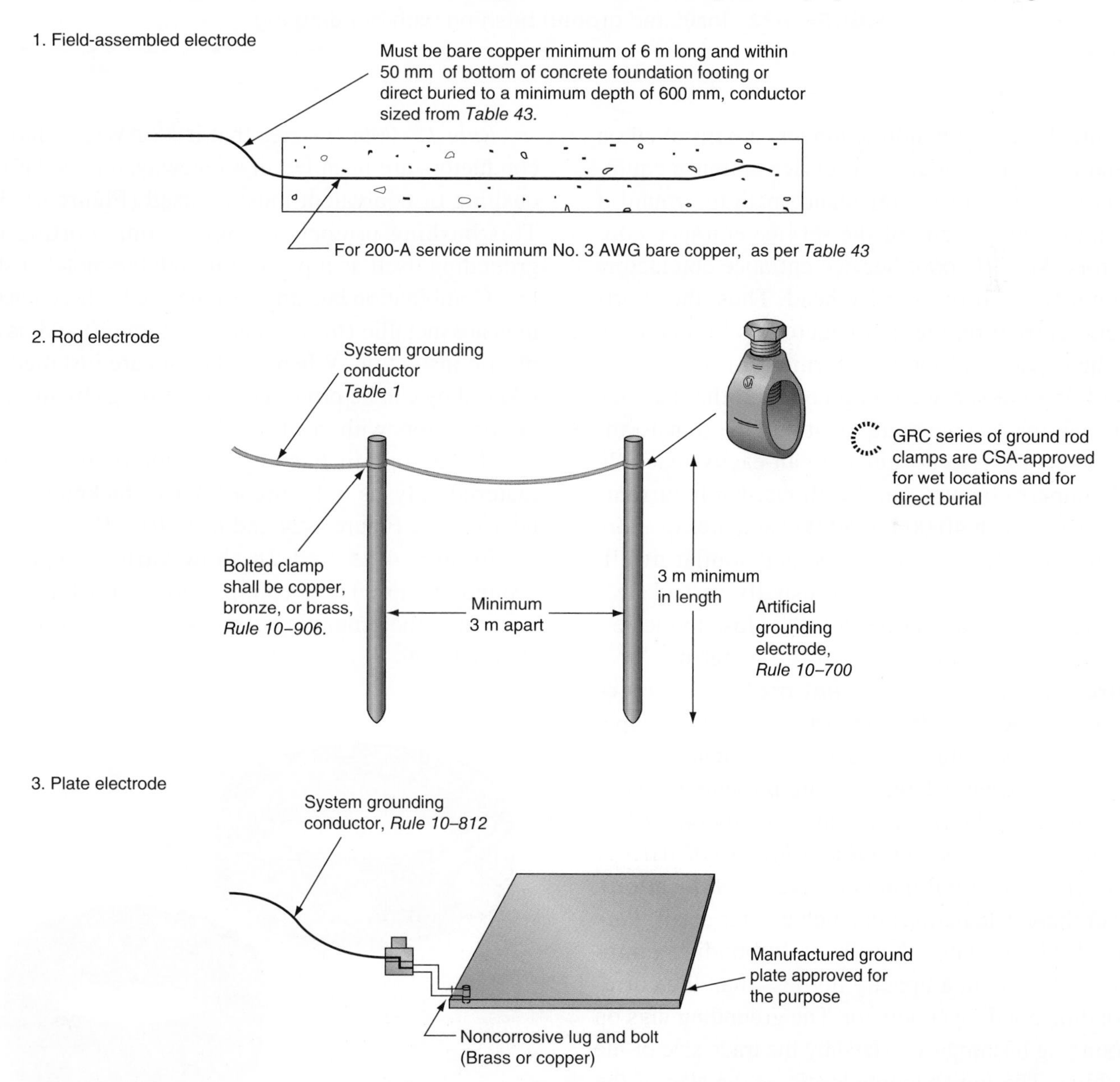

FIGURE 4-21 Three common types of grounding electrodes: (1) a field-assembled grounding electrode, (2) two rod electrodes, and (3) a plate electrode. In many instances, the grounding electrode is better than the service water pipe (in situ) ground electrode, particularly since the advent of non-metallic (PVC) water piping systems.

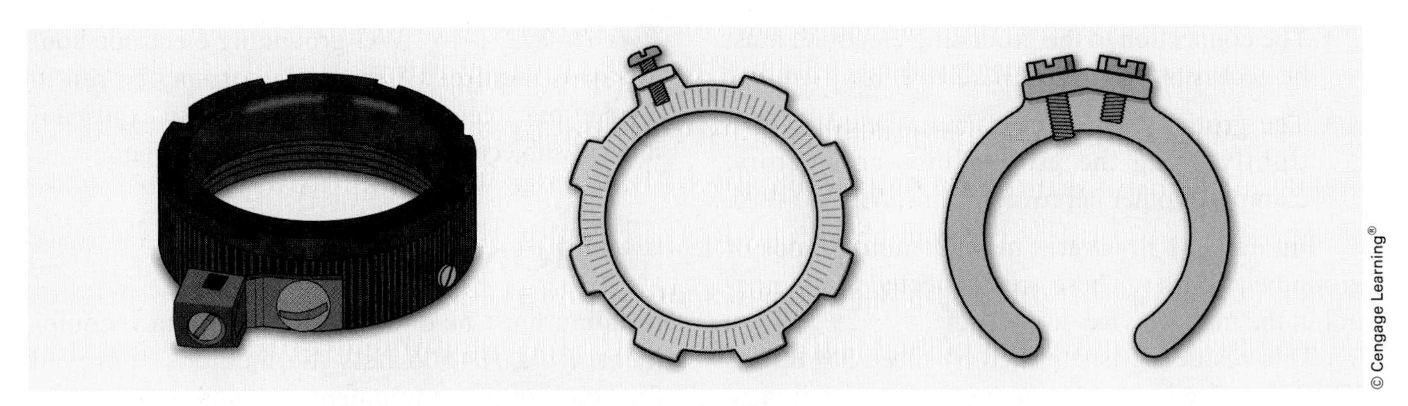

FIGURE 4-22 Insulated ground bushing with bonding lug.

(Figure 4-22) and bonding jumpers are installed on metallic service conduits at service entrance equipment to ensure a low-impedance path to ground if a fault occurs on any of the service entrance conductors, *Rule 10–606*. Service entrance conductors are not fused at the service head. Thus, the short-circuit current on these conductors is limited only by the capacity of the transformer or transformers supplying the service equipment and the distance between the service equipment and the transformers. The short-circuit current can easily reach 20 000 amperes or more in dwellings. Fault currents can easily reach 40 000 to 50 000 amperes or more in apartments, condominiums, and similar dwellings. These installations are usually served by a large-capacity transformer located close to the service entrance equipment and the metering. This extremely high-fault current produces severe arcing, which is a fire hazard. The use of proper bonding reduces this hazard to some extent.

Fault current calculations are presented later in this unit. This book's companion textbooks, *Electrical Wiring: Commercial* and *Electrical Wiring: Industrial* (also published by Nelson Education), cover these calculations in much greater detail. *Rule 10–614(1)(b)* states that the main bonding jumpers must have an ampacity not less than the corresponding bonding conductor. The grounding lugs on grounding bushings are sized by the trade size of the bushing. The lugs become larger as the size of the bushing increases.

The conductor may be sized according to *Table 41* if used in conjunction with two locknuts and a ground bushing.

Rule 12–906(2) states that if #8 AWG or larger conductors are installed in a raceway, an insulating bushing or equivalent must be used (Figure 4-23). This bushing protects the wire from shorting or grounding itself as it passes through the metal bushing. Combination bushings can be used. These bushings are metallic (for mechanical strength) and have plastic insulation. When conductors are installed in EMT, they can be protected at the fittings by the use of connectors with insulated throats.

If the conduit bushing is made of insulating material only, as in Figure 4-23, two locknuts must be used; see Figure 4-24 and *Rule 10–610(b)*.

Figures 4-25 to 4-28 show various types of ground clamps. These clamps and their actual connection to the grounding electrode must conform to *Rule 10–906*.

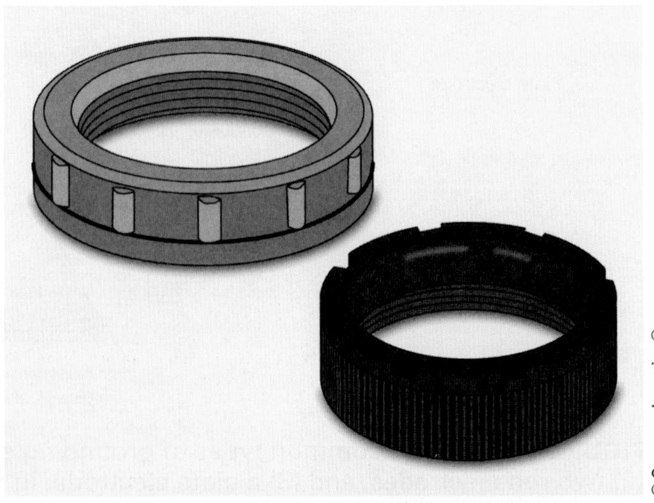

FIGURE 4-23 Insulated plastic bushings.

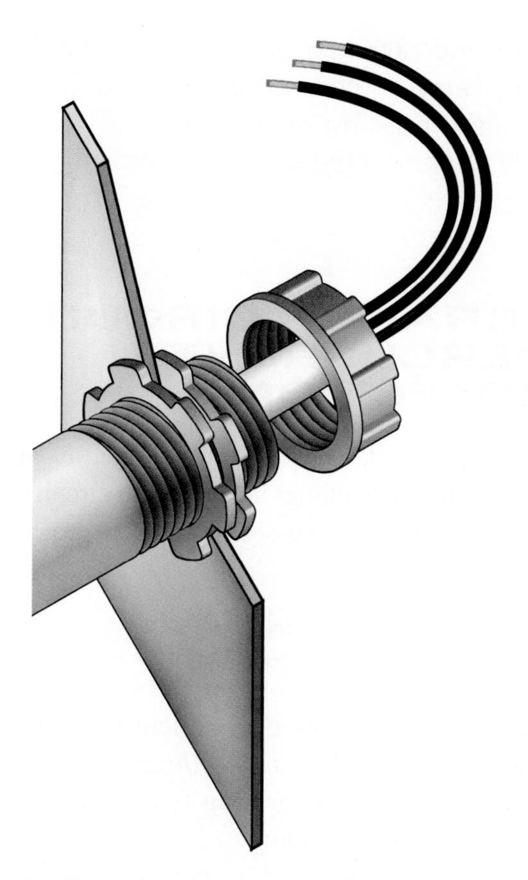

FIGURE 4-24 The use of locknuts. *See Rule 10–610(b).*

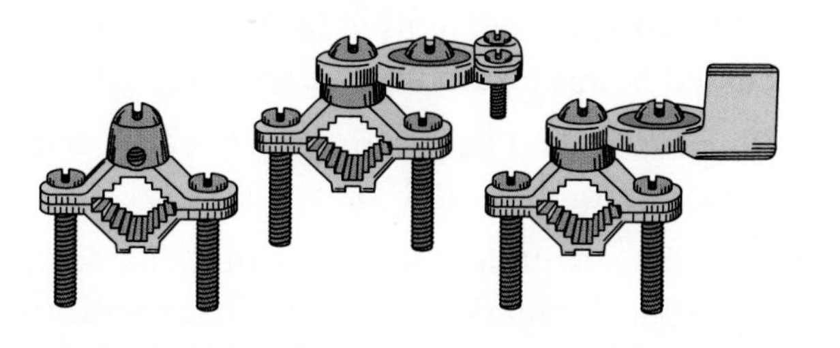

FIGURE 4-25 Typical ground clamps used in residential systems.

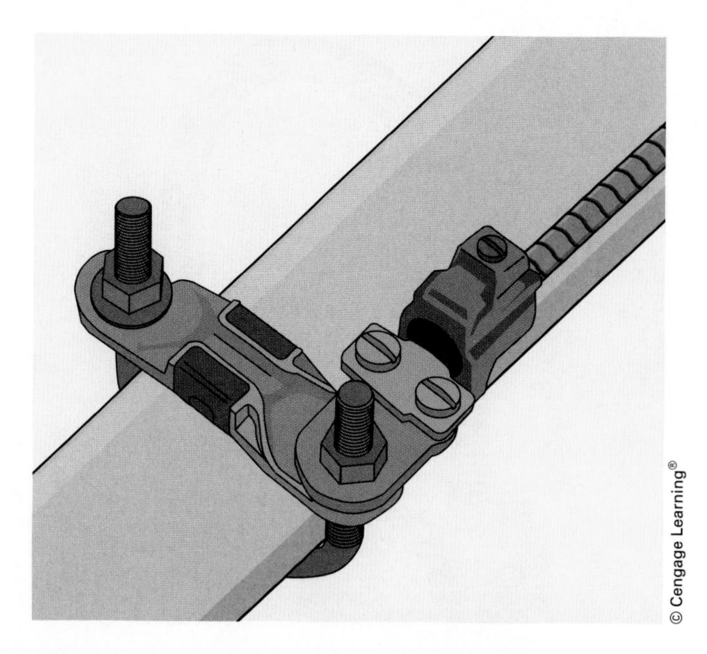

FIGURE 4-26 Armoured grounding conductor connected with ground clamp to water pipe.

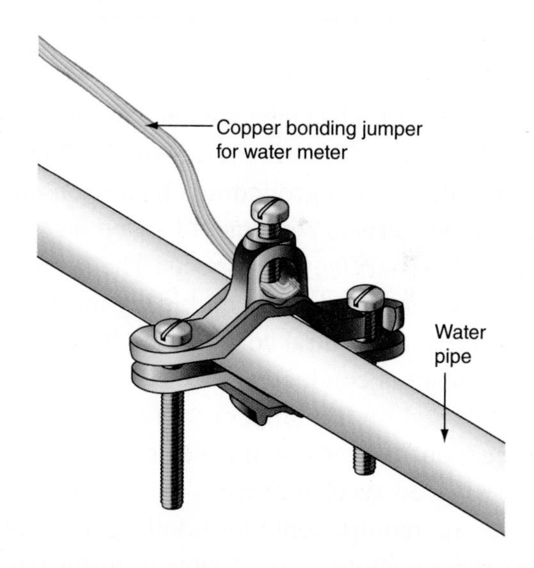

FIGURE 4-27 Ground clamp of the type used to bond (jumper) around water meter.

MULTIPLE METER INSTALLATIONS

If multiple meter sockets are required, as in a multifamily dwelling or a water heater metered at a different rate, a multigang meter socket is used. This is a single enclosure that is capable of holding more than one meter. The meter socket is rated in two ways: the maximum ampacity of the consumer's service conductors feeding the enclosures and the maximum ampacity that may be taken off each meter position. A commonly used two-position meter socket may have a maximum ampacity of 200 amperes on the consumer's service with a maximum ampacity of 100 amperes per meter position.

Since the system grounding conductor is connected in the service box of each service, the grounding of the multiple services is the same as for a single

Courtesy of Thomas & Betts Corporation

FIGURE 4-28 Ground clamp of the type used to attach ground wire to well casings.

service, that is, based on the individual service ampacity and *Rule 10–812*. However, *Rule 10–204(1)(b)* states that the system grounding connection may be made in other service equipment located on the supply side of the service disconnects. The service can therefore be grounded at the multigang meter socket by connecting the neutral conductor to the grounding electrode using a conductor sized according to *Rule 10–812,* based on the ampacity of the consumer's service conductors feeding the meter socket.

If the raceway system from the meter socket is metallic, the requirements for bonding the raceway to the service enclosure, as previously outlined in this section, will apply. If the service raceways are nonmetallic, a separate bonding conductor is required, sized according to *Table 16A,* based on the ampacity of the individual service overcurrent device ratings. Because the neutral conductor is already bonded to the case of the meter enclosure and connected to the grounding electrode from this location, you must apply *Rule 10–204(1)(b)*. This rule requires that the neutral not be connected to the enclosure in the service box of the individual services. Therefore, the brass or green bonding screw must be removed from the neutral bar in the individual services.

The neutral must be bonded to only non-current-carrying components in one location, that location being at the point at which the neutral is connected to the system-grounding electrode.

BRANCH-CIRCUIT OVERCURRENT PROTECTION

The overcurrent devices commonly used to protect branch circuits in dwellings are fuses and circuit breakers. *CEC Section 14* discusses overcurrent protection.

Plug Fuses, Fuseholders, and Fuse Rejecters

Fuses are a reliable and economical form of overcurrent protection. *Rules 14–200* through *14–212* give the requirements for plug fuses, fuseholders, and fuse rejecters, including*

- Plug fuses shall not be used in circuits exceeding 125 volts between conductors. An exception to this rule is for a system having a grounded neutral where no conductor is more than 150 volts to ground. (This is the case for the 120/240-volt system used in the residence discussed in this textbook.) See *Rule 14–202.*
- Plug fuses shall have ratings between 0 and 30 amperes, *Rule 14–208(1).*
- Plug fuses with low melting-point characteristics shall be marked "P," and fuses with time delay shall be marked "D," *Rule 14–200.*
- The screw shell of the fuseholder must be connected to the load side of the circuit.
- Where there is an alteration to a fuse panel, the fuse panel is required to be equipped with fuse rejecters that are designed to limit the size of fuse that can be accommodated by the fuseholder, *Rule 14–204(2).*
- All new installations require fuse rejecters.
- Fuse rejecters are classified at 0 to 15 amperes, 16 to 20 amperes, and 21 to 30 amperes. The

*Courtesy of CSA Group

reason for this classification is given in the following paragraph.

When the electrician installs fusible equipment, the ampere rating of the various circuits must be determined. Based on this rating, a fuse rejecter of the proper size is inserted into the fuseholder. The proper Type P or D fuse is then placed in the fuseholder. Because of the rejecter, the fuse is non-interchangeable, as required by *Rule 14–204(1)*. For example, assume that a 15-ampere rejecter is inserted for the 15-ampere branch circuits in the residence. It is impossible to substitute a fuse with a larger rating without removing the 15-ampere rejecter. If any alterations are made to an existing panel, these rejecters must be installed, *Rule 14–204(2)*. Figure 4-29 shows typical fuse rejecters.

Often the term *Dual-Element* is used when referring to time-delay fuses. Dual-Element is a trade name of fuses made by Bussmann Division, Cooper Industries. The term is used in much the same manner as *Romex* is used for Type NMD 90 cable and *BX* is used for armoured cable.

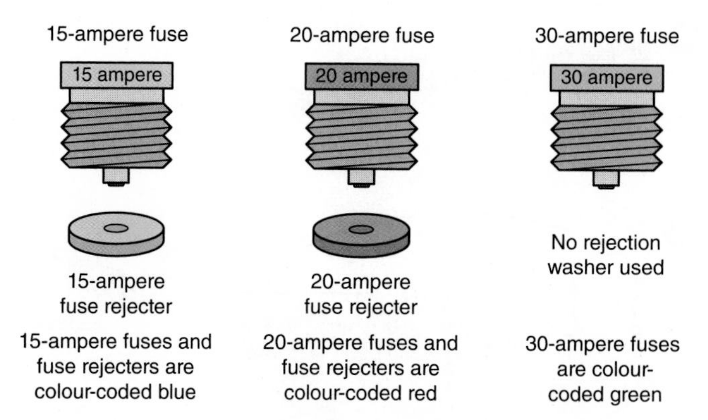

FIGURE 4-29 Plug-type fuses and rejection washers are designed to prevent overfusing in panelboards. Rejection washers can be added to any existing panelboard that uses plug fuses.

CSA has classified fuses according to their size and operating characteristics. Table 4-2 provides information on common low-voltage fuses.

Dual-Element Fuse (Time-Delay)

Figure 4-30 shows the most common type of Dual-Element cartridge fuse. These fuses are available

TABLE 4-2

Characteristics of common low-voltage fuses.

FUSE CLASS	VOLTAGE	CURRENT	COMMENTS
			Plug Fuses
C	125	0–30	Has rejection feature when used with rejection washers; low melting-point plug fuses are Type P nontime-delay and Type D time-delay
S	125	0–30	May be time-delay or nontime-delay
			Cartridge Fuses
H	250 600	0–600	Standard fuse used in circuits where interrupting capacity of 10 000 A or less is required
J	250 600	0–600	UL Class J: 200 000-A interrupting capacity, smaller physical dimensions than Class H fuses
K	250 600	0–600	UL Class K: Will be found on equipment imported from the United States
L	250 600	601–6000	UL Class L: High-temperature fuses
R	250 600	0–600	UL Class R: Rejection fuse may be placed in a standard fuseholder, but standard fuses will not fit in a Class R fuseholder
T	250 600	0–600	UL Class T: Has smaller physical dimensions than either Class H or Class J

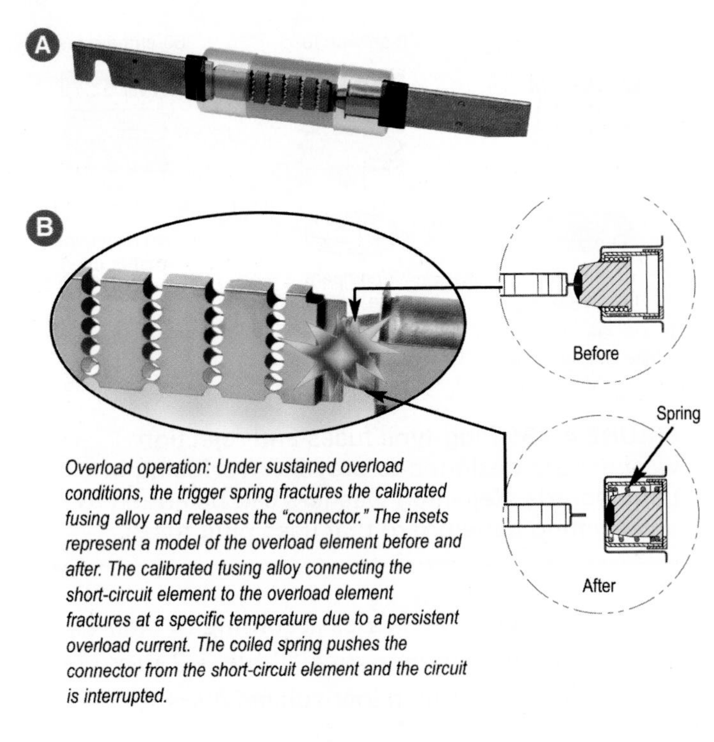

Overload operation: Under sustained overload conditions, the trigger spring fractures the calibrated fusing alloy and releases the "connector." The insets represent a model of the overload element before and after. The calibrated fusing alloy connecting the short-circuit element to the overload element fractures at a specific temperature due to a persistent overload current. The coiled spring pushes the connector from the short-circuit element and the circuit is interrupted.

Short-circuit operation: Modern fuses are designed with minimum metal in the restricted portions which greatly enhance their ability to have excellent current-limiting characteristics – minimizing the short-circuit let-through current. A short-circuit current causes the restricted portions of the short-circuit element to vaporize and arcing commences. The arcs burn back the element at the points of the arcing. Longer arcs result, which assist in reducing the current. Also, the special arc quenching filler material contributes to extinguishing the arcing current. Modern fuses have many restricted portions, which results in many small arclets – all working together to force the current to zero.

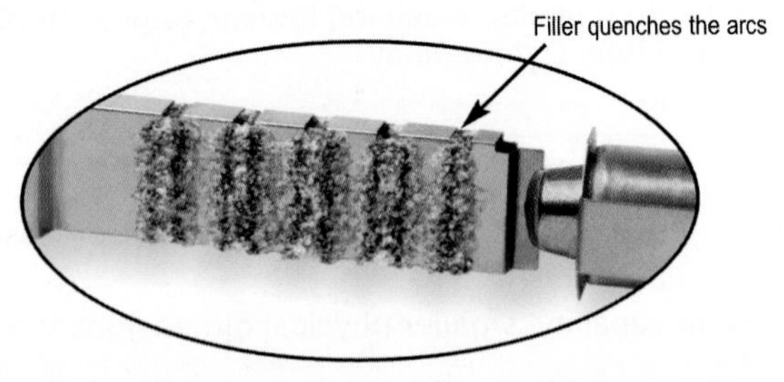

Short-circuit operation: The special small granular, arc-quenching material plays an important part in the interruption process. The filler assists in quenching the arcs; the filler material absorbs the thermal energy of the arcs, fuses together and creates an insulating barrier. This process helps in forcing the current to zero. Modern current-limiting fuses, under short-circuit conditions, can force the current to zero and complete the interruption within a few thousandths of a second.

Courtesy of Eaton

FIGURE 4-30 Cartridge-type Dual-Element fuse (A) is a 250-volt, 100-ampere fuse. The cutaway view in (B) shows the internal parts of the fuse.

in 250-volt and 600-volt sizes with ratings from 0 to 600 amperes, and have an interrupting rating of 200 000 amperes.

A Dual-Element fuse has two fusible elements connected in series: the overload element and the short-circuit element. When an excessive current flows, one of the elements opens. The amount of excess current determines which element opens.

The electrical characteristics of these two elements are very different. Thus, a Dual-Element fuse has a greater range of protection than the single-element fuse. The overload element opens circuits on current in the low overload range (up to 500% of the fuse rating). The short-circuit element handles only short circuits and heavy overload currents (above 500% of the fuse rating).

Overload Element. The overload element opens when excessive heat is developed in the element. This heat may be the result of a loose connection or poor contact in the fuseholder or excessive current. When the temperature reaches 138°C, a fusible alloy melts and the element opens. Any excessive current produces heat in the element. However, the mass of the element absorbs a great deal of heat before the cutout opens. A small excess of current will cause this element to open if it continues for a long period of time. This characteristic gives the thermal cutout element a large time lag on low overloads (up to 10 seconds at a current of 500% of the fuse rating). In addition, it provides very accurate protection for prolonged overloads.

Short-Circuit Element. The capacity of the short-circuit element is high enough to prevent it from opening on low overloads. This element is designed to clear the circuit quickly of short circuits or heavy overloads above 500% of the fuse ratings.

Dual-element fuses are used on motor and appliance circuits where the time-lag characteristic of the fuse is required. Single-element fuses do not have this time lag and blow as soon as an overcurrent condition occurs. The homeowner may be tempted to install a fuse with a higher rating. The fuseholder adapter, however, prevents the use of such a fuse. Thus, the dual-element fuse is recommended for this type of situation.

The standard plug fuse can be used only to replace blown fuses on existing installations. A Type P

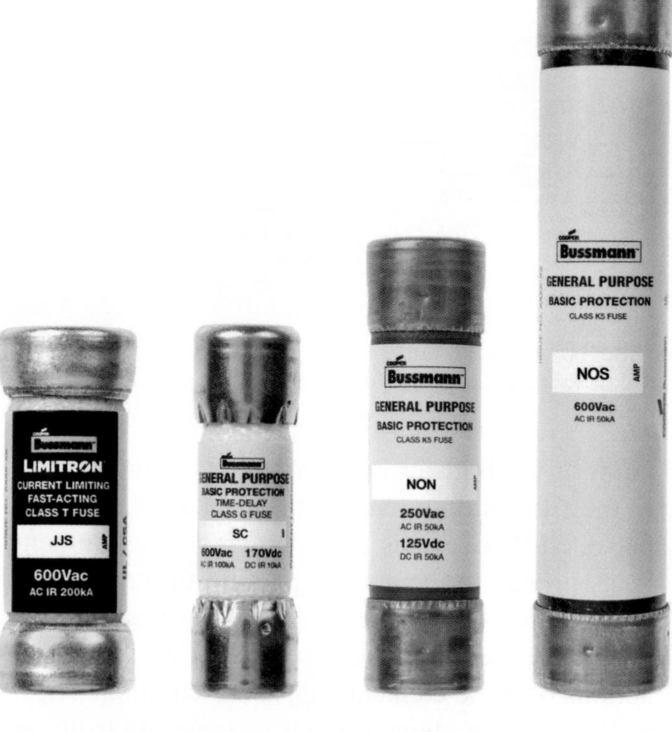

FIGURE 4-31 Cartridge-type fuses.

or D fuse is required on all new installations. The cartridge fuse is a dual-element fuse that is available in both ferrule and knife-blade styles.

Cartridge Fuses

Figure 4-31 shows another style of small-dimension cartridge fuse.

A type of cartridge fuse that is also in use is rated at not over 60 amperes for circuits of 300 volts or less to ground. The physical size of this fuse is smaller than that of standard cartridge fuses. The time-delay characteristics of this fuse mean that it can handle harmless current surges or momentary overloads without blowing. However, these fuses open very rapidly under short-circuit conditions.

These fuses prevent over-fusing since they are size-limiting for their ampere ratings. For example, a fuseholder designed to accept a 15-ampere fuse will not accept a 20-ampere cartridge fuse. In the same manner, a fuseholder designed to accept a 20-ampere fuse will not accept a 30-ampere cartridge fuse.

Class J Fuses

Another type of physically small fuse is the Class J fuse; see Figure 4-32. This has a high-interrupting rating in sizes from 0 to 600 amperes in both 300-volt and 600-volt ratings. Such fuses are used as the main fuses in a panel having circuit-breaker branches; see Figure 4-33. In this case, the Class J fuses protect the low-interrupting capacity breakers against high-level short-circuit currents.

Most manufacturers list service equipment, metering equipment, and disconnect switches that make use of Class J fuses. This allows the safe installation of this equipment where fault currents exceed 10 000 amperes, such as on large services, and any service equipment located close to pad-mounted transformers, such as those found in apartments, condominiums, shopping centres, and similar locations.

The standard ratings for fuses range from 1 ampere to 6000 amperes.

Circuit Breakers

Installations in dwellings normally use thermal-magnetic circuit breakers. On a continuous overload, a bimetallic element in such a breaker moves until it unlatches the inner tripping mechanism of the breaker. Momentary small overloads do not cause the element to trip the breaker. If the overload is heavy or if there is a short circuit, a magnetic coil

FIGURE 4-32 Class J fuses rated at 30, 60, 100, 200, 400, and 600 amperes, 300 volts.

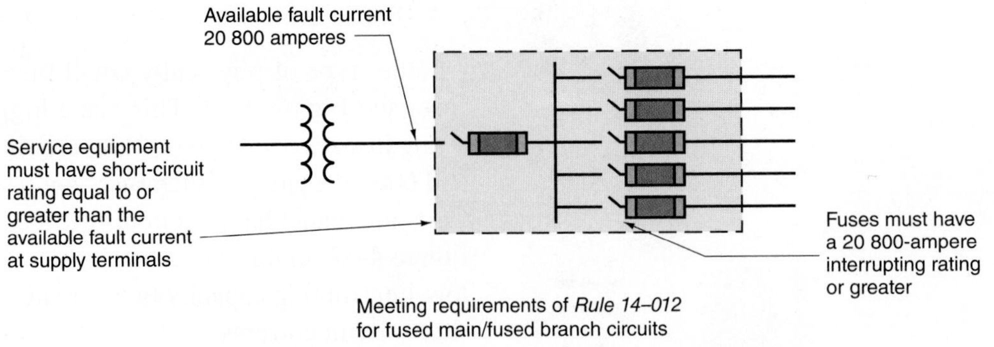

FIGURE 4-33 Fused main/fused branch circuits.

in the breaker causes it to interrupt the branch circuit instantly.

Rules 14–300 to *14–308* give the requirements for circuit breakers, including*

- Circuit breakers shall be trip-free so that the internal mechanism will trip to the OFF position even if the handle is held in the ON position (which can be achieved using a lock-on device), *Rule 14–300(1).*
- Breakers shall indicate clearly whether they are on or off, *Rule 14–300(2).*
- A breaker shall be non-tamperable so that it cannot be readjusted (trip point changed) without dismantling the breaker or breaking the seal, *Rule 14-304.*
- The rating shall be durably marked on the breaker. For small breakers rated at 100 amperes or less and 600 volts or less, the rating must be moulded, stamped, or etched on the handle (or on another part of the breaker that will be visible after the cover of the panel is installed).
- Every breaker with an interrupting rating other than 5000 amperes shall have this rating marked on the breaker. Most have a 10 000-ampere minimum rating.
- Circuit breakers rated at 120 volts and 347 volts and used for fluorescent loads shall not be used as switches unless marked "SWD," *Section 30–710(3)(b).*

Most circuit breakers are ambient temperature compensated. This means that the tripping point of the breaker is not affected by an increase in the surrounding temperature. An ambient-compensated breaker has two elements. One element heats up due to the current passing through it and the heat in the surrounding area. The other element heats up because of the surrounding air only. The actions of these elements oppose each other. Thus, as the tripping element tends to lower its tripping point because of external heat, the second element opposes the tripping element and stabilizes the tripping point. As a result, the current through the tripping element is the only factor that causes the element to open the circuit. It is a good practice to turn the breaker off and on periodically to "exercise" its moving parts.

CEC Table 13 gives the standard ampere ratings of fuses and circuit breakers up to 600 amperes.

*Courtesy of CSA Group

INTERRUPTING RATINGS FOR FUSES AND CIRCUIT BREAKERS*

Rule 14–012(a) states that all fuses, circuit breakers, and other electrical devices that break current shall have an interrupting capacity sufficient for the voltage employed and for the current that must be interrupted.

According to *Rule 14–012(a)* and *Appendix B,* all overcurrent devices, the total circuit impedance, and the withstand capability of all circuit components (wires, contractors, switches) must be selected

*Courtesy of CSA Group

so that minimal damage will result in the event of a fault, either line-to-line or line-to-ground.

The overcurrent protective device must be able to interrupt the current that may flow under any condition (overload or short circuit). Such interruption must be made with complete safety to personnel and without damage to the panel or switch in which the overcurrent device is installed.

Overcurrent devices with inadequate interrupting ratings are, in effect, bombs waiting for a short circuit to trigger them into an explosion. Personal injury may result, and serious damage will be done to the electrical equipment.

Cartridge-type fuses have a maximum interrupting rating of 10 000 amperes root mean square (RMS) symmetrical. Plug fuses can interrupt no more than 10 000 amperes RMS symmetrical. Cartridge dual-element fuses (Figure 4-30) and Class J fuses (Figure 4-32) have interrupting ratings of 200 000 amperes RMS symmetrical. These interrupting ratings are described in CSA Standard C22.2 No. 59.1-M1987. The interrupting rating of a fuse is marked on its label when the rating is other than 10 000 amperes.

Short-Circuit Currents

This text does not cover in detail the methods of calculating short-circuit currents (see this book's companion textbooks, *Electrical Wiring: Industrial* and *Electrical Wiring: Commercial* for more detailed circuit-fault calculations). The ratings required to determine the maximum available short-circuit current delivered by a transformer are the kilovolt-ampere (kVA) and impedance values of the transformer. The size and length of wire installed between the transformer and the overcurrent device must be considered as well.

The transformers used in modern electrical installations are efficient and have very low impedance values. A low-impedance transformer having a given kVA rating delivers more short-circuit current than a transformer with the same kVA rating and a higher impedance. When an electric service is connected to a low-impedance transformer, the problem of available short-circuit current is very serious. The possible situations are shown in Figure 4-33.

The examples given in Figure 4-33 show that the available short-circuit current at the transformer secondary terminal is 20 800 amperes.

Determining Short-Circuit Current

The local power utility and the electrical inspector are good sources of information for determining short-circuit current.

You can use this simplified method to determine the approximate available short-circuit current at the terminals of a transformer.

Step 1. Determine the normal full-load secondary current delivered by the transformer.

For single-phase transformers:

$$I = \frac{\text{kVA} \times 1000}{E}$$

For three-phase transformers:

$$I = \frac{\text{kVA} \times 1000}{E \times 1.73}$$

I = current, in amperes
kVA = kilovolt-amperes (transformer nameplate rating)
E = secondary line-to-line voltage (transformer nameplate rating)

Step 2. Using the impedance value given on the transformer nameplate, find the multiplier to determine the short-circuit current.

$$\text{Multiplier} = \frac{100}{\text{Percent impedance}}$$

Step 3. Short-circuit current = Normal full-load secondary current × Multiplier

EXAMPLE

A transformer is rated at 100 kVA and 120/240 volts. It is a single-phase transformer with an impedance of 1% (from the transformer nameplate). Find the short-circuit current.

SOLUTION For a single phase transformer:

$$I = \frac{kVA \times 1000}{E}$$

$$= \frac{100 \times 1000}{240} \text{ amperes}$$

$$= 417 \text{ amperes, full-load current}$$

For a transformer impedance of 1%:

$$\text{Multiplier} = \frac{100}{\text{Percent impedance}}$$

$$= \frac{100}{1}$$

$$= 100$$

$$\text{Short-circuit current} = I \times \text{Multiplier}$$

$$= 416 \text{ amperes} \times 100$$

$$= 41\,600 \text{ amperes}$$

Thus, the available short-circuit current at the terminals of the transformer is 41 600 amperes.

This value decreases as the distance from the transformer increases. If the transformer impedance is 1.5%,

$$\text{Multiplier} = \frac{100}{1.5} = 66.6$$

$$\text{The short-circuit current} = 416 \times 66.6 \text{ amperes} = 27\,706 \text{ amperes}$$

If the transformer impedance is 2%:

$$\text{Multiplier} = \frac{100}{2} = 50$$

$$\text{The short-circuit current} = 416 \times 50 \text{ amperes} = 20\,800 \text{ amperes}$$

Note: A short-circuit current of 20 800 amperes is used in Figures 4-33 through 4-36.

WARNING: The line-to-neutral short-circuit current at the secondary of a single-phase transformer is about 1.5 times greater than the line-to-line short-circuit current. For the previous example:

Line-to-line short-circuit current
= 20 800 amperes

Line-to-neutral short-circuit current
= 20 800 × 1.5 amperes = 31 200 amperes

Note that CSA Standard C22.2 No. 47 allows the marked impedance on the nameplate of a transformer to vary plus or minus (±) 10% from the actual impedance value of the transformer. Thus, for a transformer marked "2%Z," the actual impedance could be as low as 1.8%Z or as much as 2.2%Z. Therefore, the available short-circuit current at the secondary of a transformer as calculated above should be increased by 10% if you want to be on the safe side when considering the interrupting rating of the fuses and breakers to be installed.

An excellent point-to-point method of calculating fault currents at various distances from the secondary of a transformer that uses different sizes of conductors is covered in detail in the *Electrical Wiring: Commercial* textbook. It is simple and accurate. The point-to-point method can be used to compute both single-phase and three-phase faults.

Calculating fault currents is just as important as calculating load currents. Overloads will cause conductors and equipment to run hot, will shorten their life, and will eventually destroy them. Fault currents that exceed the interrupting rating of the overcurrent protective devices can cause violent electrical explosions of equipment with the potential of serious injury to people standing near it. Fire hazard is also present.

PANELS AND LOAD CENTRES

Fused Main/Fused Branches (Figure 4-33)

- Fused main/fused branch circuits must be listed as suitable for use as service equipment.
- They must be sized to satisfy the required ampacity as determined by service entrance calculations and/or local codes.
- All fuses must have interrupting rating adequate for the available fault current. Panel must have short-circuit rating adequate for the available fault current.
- For overload conditions on a branch, only the branch-circuit fuse(s) will open. All other circuits remain energized. The system is selective.*

*See highlighted note on page 70.

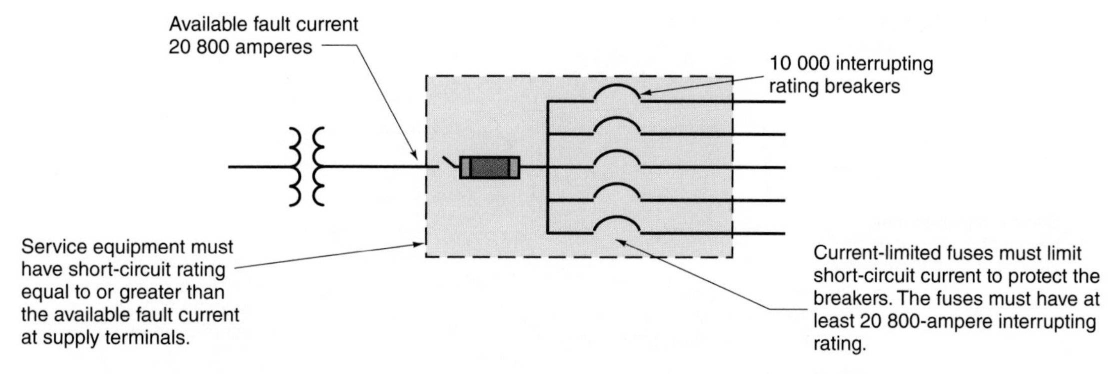

FIGURE 4-34 Fused main/breaker branch circuits.

- For short circuits (line-to-line or line-to-neutral) and for ground faults (line-to-ground), only the fuse protecting the faulted circuit opens. All other circuits remain energized. The system is selective.*

Fused Main/Breaker Branches (Figure 4-34)

- Fused main/breaker branches must be listed as suitable for use as service equipment.
- They must be sized to satisfy the required ampacity as determined by service entrance calculations and/or local codes.
- Main fuses must have interrupting rating adequate for the available fault circuit.
- Panel must have short-circuit rating adequate for the available fault current.
- Main fuses must be current-limiting, such as Class J fuses, so that if a fault exceeding the branch-circuit breaker's interrupting rating (ampere interrupting rating, AIR) occurs, the fuse will limit the let-through fault current to a value less than the maximum interrupting rating of the branch-circuit breaker. Branch-circuit breakers generally have 10 000 amperes interrupting rating.
- For branch-circuit faults (L-N or L-G) exceeding the branch-circuit breaker's interrupting rating, one fuse opens, de-energizing one-half of the panel. The system is nonselective.*
- For branch-circuit faults (L-L) exceeding the branch-circuit breaker's interrupting rating, both main fuses open, de-energizing the entire panel. The system is non-selective.*
- Check time–current curves of fuses and breakers and check the unlatching time of breakers to determine at what point in "time" and "current" the breakers are slower than the current-limiting fuses.
- For normal overloads and low-level faults, this system is selective.

Breaker Main/Breaker Branches (Figure 4-35) (All Standard Plastic-Case Breakers)

- These breaker main/breaker branches must be listed as suitable for use as service equipment.
- They must be sized to satisfy the required ampacity as determined by service entrance calculations and/or local codes.
- Main and branch breakers must have interrupting rating adequate for the available fault current.
- Panel must have short-circuit rating adequate for the available fault current.

*See highlighted note on page 70.

*See highlighted note on page 70.

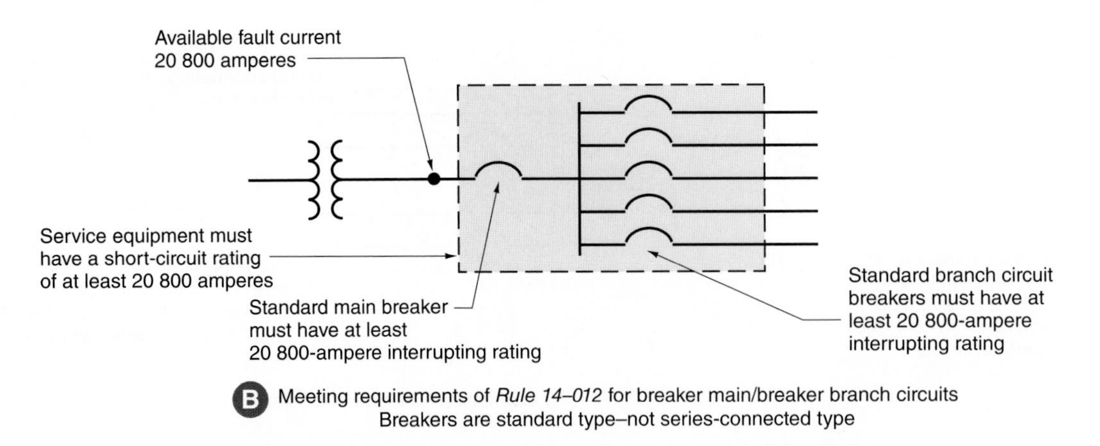

FIGURE 4-35 Breaker main/breaker branch circuits.

- For overload conditions and low-level faults on a branch circuit, only the branch-circuit breaker will trip off. All other circuits remain energized. The system is selective.*
- For branch-circuit faults (short circuits or ground faults) that exceed the instant trip setting of the main breaker (usually five times the breaker's ampere rating), both the branch-circuit breaker and main breaker trip off. For instance, for a branch fault of 700 amperes, a 100-ampere main breaker also trips because the main will trip instantly for faults of 500 (100 × 5) amperes or more. Entire panel is de-energized. System is nonselective.*
- Check manufacturer's data for time–current characteristic curves and unlatching times of the breakers.

Breaker Main/Breaker Branches (Figure 4-36) (Series Connected—Sometimes Called *Series Rated, Rule 14–014*)

- These breaker main/breaker branches must be listed as suitable for use as service equipment.
- They must be sized to satisfy the required ampacity as determined by service entrance calculations and/or local codes.
- They must be listed as series connected.

*See highlighted note on this page.

Note: The terms and theory behind selective and nonselective systems are covered in detail in the *Electrical Wiring: Commercial* textbook. In all instances, you should check the time–current curves of fuses and circuit breakers to be installed. Also check the circuit-breaker manufacturer's unlatching time data, as these will help determine at what values of current the system will be selective or non-selective. Fault current calculations are also covered in the *Electrical Wiring: Commercial* textbook.

- Main breaker will have interrupting rating higher than the branch breaker. For example, main breaker has a 22 000-ampere interrupting rating, whereas branch breaker has a 10 000-ampere interrupting rating.
- Panel must be listed and marked with its maximum fault current rating.
- For branch-circuit faults (L-L, L-N, L-G) above the instant trip factory setting (i.e., 200-ampere main breaker with 5 × setting: 200 × 5 = 1000 amperes), both branch and main breakers will trip. The system is non-selective.
- For overload conditions and low-level faults on a branch, only the branch breaker will trip. The system is selective.
- Obtain time–current curves and unlatching time data from the manufacturer.
- The panel's integral rating is attained because when a heavy fault occurs, both main and branch breakers open. The two arcs in series add impedance; thus, the fault current is reduced to

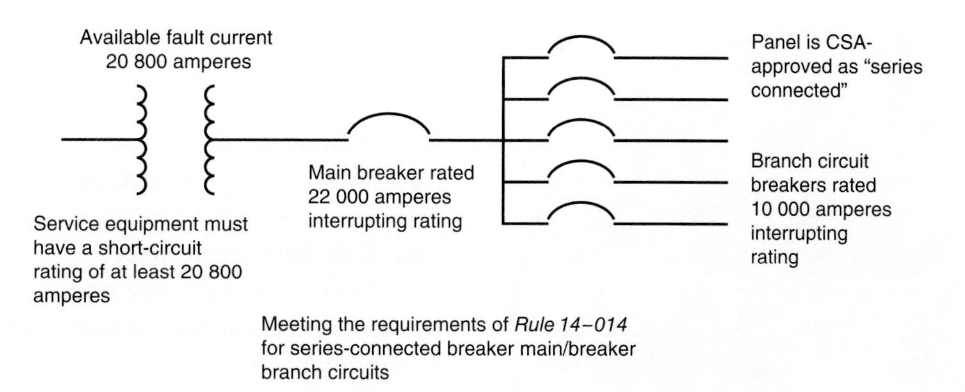

Courtesy of CSA Group

FIGURE 4-36 Breaker main/breaker branch circuits (series-connected plastic-case type).

a level less than the 10 000 AIR rating of the branch-circuit breaker.

- For series-rated panelboards where the available fault current exceeds the interrupting rating of the branch-circuit breakers, the equipment must be marked by the manufacturer as having a series combination interrupting rating at least equal to the maximum available fault current, *Rule 14–014(d).*
- When panels containing low interrupting rating circuit breakers are separated from the equipment in which the higher interrupting rating circuit breakers protecting the panel are installed, the system becomes a field-installed *series-rated system.* This panel must be field-marked (by the installer) and this marking must be readily visible, *Rule 14–014(e).* This marking calls attention to the fact that the actual available fault current exceeds the interrupting rating of the breakers in the panel, and that the panel is protected by properly selected circuit breakers back at the main equipment. The label will appear on panelboards and enclosed panelboards and may appear as follows:

When protected by ___ A maximum HRC ___ fuse or (manufacturer's name and type designation) circuit breaker rated not more than ___ A, this panelboard is suitable for use on a circuit capable of delivering not more than ___ A RMS symmetrical, ___ V maximum.

CAUTION: Closely check the manufacturer's catalogue numbers to make sure that the circuit-breaker combination you are intending to install *in series* is recognized by the CSA Group for that purpose. This is extremely important because many circuit breakers have the same physical size and dimensions, yet have different interrupting and voltage ratings, and have not been tested as *series-rated* devices. To install these as a series combination constitutes a violation of *Rules 14–012* and *14–014,* which could be dangerous.

READING THE METER

Figure 4-37 shows a typical single-phase watt-hour meter with five dials. From left to right, the dials represent tens of thousands, thousands, hundreds, tens, and units of kilowatt-hours.

Starting with the first dial on the left, record the last number the pointer has passed. Continue doing this with each dial until the full reading is obtained. The reading on the five-dial meter in Figure 4-38 is 18 672 kWh.

If the meter reads 18 975 one month later (Figure 4-39), by subtracting the previous reading of 18 672, it is found that 303 kWh were used during the month.

Courtesy of Landis+Gyr

FIGURE 4-37 Typical single-phase watt-hour meter.

COST OF USING ELECTRICAL ENERGY

A watt-hour meter is always connected into some part of the service entrance equipment. In residential metering, a watt-hour meter is normally installed as part of the service entrance.

All electrical energy consumed is metered so that the utility can bill the customer on a monthly or bimonthly basis.

The kilowatt (kW) is a convenient unit of electrical power. One thousand watts (W) is equal to one kilowatt (kW). The watt-hour meter measures both wattage and time. As the dials of the meter turn, the kilowatt-hour (kWh) consumption is continually recorded.

Utility rates are based on so many cents per kilowatt-hour.

The number is multiplied by the rate per kilowatt-hour, and the power company bills the consumer for the energy used. The utility may also add a fuel adjustment charge.

EXAMPLE

One kWh will light a 100-watt light bulb for ten hours. One kWh will operate a 1000-watt electric heater for one hour. Therefore, if the electric rate is 8 cents per kilowatt-hour, the use of a 100-watt bulb for ten hours would

© Cengage Learning®

FIGURE 4-38 The reading on this five-dial meter is 18 672 kWh.

FIGURE 4-39 One month later, the meter reads 18 975 kWh, indicating that 303 kWh were used during the month.

cost 8 cents. A 1000-watt electric heater could be used for one hour at a cost of 8 cents.

$$\text{kWh} = \frac{\text{Watts} \times \text{Hours}}{1000} = \frac{100 \times 10}{1000}\ \text{kWh} = 1\ \text{kWh}$$

The cost of the energy used by an appliance is

$$\text{Cost} = \frac{\text{Watts} \times \text{Hours used} \times \text{Cost per kWh}}{1000}$$

EXAMPLE

Find the cost of operating a colour television for eight hours. The set is rated at 175 watts. The electric rate is 9.6 cents per kWh.

SOLUTION

$$\text{Cost} = \frac{175 \times 8 \times 9.6}{1000}\ \text{cents} = 13.4\ \text{cents}$$

Assume that a meter now reads 18 672 kWh. The previous meter reading was 17 895 kWh. The difference is 777 kWh.

The following is how a typical bill might look.

GENERIC ELECTRIC COMPANY

Days of Service: 33	From 01–28–15	To 03–01–15	Due Date 03–25–15
Present reading			18 672
Last reading			17 895
Kilowatt-hours			777
Rate/kWh			0.08
Amount			$62.16
Total			$62.16 + applicable taxes

Some utilities apply a fuel adjustment charge that may increase or decrease the electric bill. These charges, on a "per kWh" basis, enable the electric utility to recover from the consumer extra expenses it might incur for fuel costs used in generating electricity. These charges can vary each time the utility prepares the bill, without it having to apply to the regulatory agency for a rate change.

Some utilities increase their rates during the summer months when people turn on their air conditioners. This air-conditioning load taxes the utility's generating capabilities during the peak summer months. The higher rate structure also gives people an incentive to not set their thermostats too low, thus saving energy.

To encourage customers to conserve energy, a utility company might offer the following rate structure:

First 500 kWh @ 6.12 cents per kWh
Over 500 kWh @ 10.64 cents per kWh (summer rate)
Over 500 kWh @ 8.64 cents per kWh (winter rate)

The widespread introduction of smart metering worldwide has changed how utilities measure your power consumption. Smart meters communicate with the power utility and send data about your power consumption directly to the utility. It is possible for a utility to charge consumers different rates for the energy they use, based on the time of day the power was used. Some smart meters may send information about surge voltages and harmonics to the utility, enabling the utility to diagnose local power problems.

REVIEW

Note: Refer to the *CEC* or the blueprints provided with this textbook when necessary. Where applicable, responses should be written in complete sentences.

1. Where does a consumer's service start and end? ______________________________

____

____

____

____

2. What are supply service conductors? ____

____

____

____

3. Who is responsible for determining the service location? ____

____

4. a. [Circle one.] The service head must be located (above) (below) the point where the supply service conductors are spliced to the consumer's service conductors.
   b. What *CEC* rule provides the answer to (a)? ____
5. a. What size and type of conductors are installed for this service? ____

   ____

   b. What size of conduit is installed? ____

   ____

   c. What size of grounding electrode conductor (system grounding conductor) is installed? (Not neutral) ____
   d. Is the grounding electrode conductor insulated, armoured, or bare? ____

   ____

6. How and where is the grounding electrode conductor attached to the water pipe? ____

____

____

7. When a service mast is extended through a roof, must it be guyed? ____

____

8. What are the minimum distances or clearances for the following?
   a. Supply service conductors, clearance over private driveway _______________
   b. Supply service conductors, clearance over private sidewalks _______________
   c. Supply service conductors, clearance over alleys _______________
   d. Supply service conductors, clearance over a roof having a roof pitch that may be walked on. (Voltage between conductors does not exceed 300 volts.) _______________
   e. Supply service conductors, horizontal clearance from a porch _______________
   f. Supply service conductors, clearance from a window or door _______________
9. What size of ungrounded conductors is installed for each of the following residential services? (Use type RW90 XLPE copper conductors.)
   a. 60-ampere service # _____ RW90 XLPE copper
   b. 100-ampere service # _____ RW90 XLPE copper
   c. 200-ampere service # _____ RW90 XLPE copper

10. What size of grounding electrode conductors are installed for the services listed in Question 9? *(Rule 10–812)*
    a. 60-ampere service # _____ AWG grounding conductor
    b. 100-ampere service # _____ AWG grounding conductor
    c. 200-ampere service # _____ AWG grounding conductor
11. What is the recommended height of a meter socket from the ground? ______________
12. a. What is the minimum length of conductor that must be left extending out of the service head?
    ____
    ____
    b. What *CEC* rule covers this? ______________________________
13. What are the restrictions on the use of bare neutral conductors? ______________
    ____
    ____
    ____
14. How far must mechanical protection be provided when underground service conductors are carried up a pole? ______________________________
    ____
15. What is the maximum number of consumer's services allowed without special permission? Quote the *CEC* rule. ______________________________
    ____
16. Complete the following table by filling in the columns with the appropriate information.

	CIRCUIT NUMBER	AMPERE RATING	POLES	VOLTS	WIRE SIZE
A. Living room receptacle outlets					
B. Workbench receptacle outlets					
C. Water pump					
D. Attic exhaust fan					
E. Kitchen lighting					
F. Hydromassage tub					
G. Attic lighting					
H. Counter-mounted cooking unit					
I. Electric furnace					

17. a. What size of conductors supply Panel B? ______________________
    b. What size of conduit? ______________________________
    c. Is this raceway run in the form of EMT or rigid conduit? ______________
    d. What size of overcurrent device protects the feeders to Panel B? ______________
    ____

18. How many electric meters are provided for this residence? ______
19. According to the *CEC*, is it permissible to ground rural service entrance systems and equipment to driven ground rods when a metallic water system is not available? ______
20. What *CEC* table lists the sizes of grounding electrode conductors to be used for service entrances of various sizes similar to the type found in the residence discussed in this text? ______
21. Do the following conductors require mechanical protection?
    a. #8 grounding conductor ______
    b. #6 grounding conductor ______
    c. #4 grounding conductor ______
22. Why is bonding service entrance equipment necessary? ______
23. What special types of bushings are required on service entrances when metallic raceways are used? ______
24. When #8 AWG conductors or larger are installed, what additional provision is required on the conduit ends? ______
25. What minimum size copper bonding jumpers must be installed to properly bond the electric service for the residence discussed in this text? ______

26. a. What is a Type D fuse? ______________________________

b. Where could a Type D fuse be installed? ______________________________

27. a. What is the maximum voltage permitted between conductors when using plug fuses? ______________________________

b. May plug fuses be installed in a switch that disconnects a 120/240-volt clothes dryer?

____

c. Give a reason for the answer to (b). ______________________________

28. What part of a circuit breaker causes the breaker to trip

a. on an overload? ______________________________

b. on a short circuit? ______________________________

29. List the standard sizes of circuit breakers up to and including 100 amperes. __________

30. Using the method shown in this unit, what is the approximate short-circuit current available at the terminals of a 50-kVA single-phase transformer rated 120/240 volts? The transformer impedance is 1%.

a. Line-to-line? ______________________________

b. Line-to-neutral? ______________________________

31. Where is the service for this residence located? ______________________________

32. a. On what type of wall is Panel A fastened? ______________________________

b. On what type of wall is Panel B fastened? ______________________________

33. When conduits pass through the wall from outside to inside, the conduit must be ______________________________ to prevent air circulation through the conduit.

34. Briefly explain why electrical systems and equipment are grounded. ______________

____

____

____

____

35. What *CEC* rule states that all overcurrent devices must have adequate interrupting ratings for the current to be interrupted? ______________________________

36. All electrical components have some sort of "withstand rating," which indicates the ability of the component to withstand fault currents for the time required by the overcurrent device to open the circuit. What *CEC* rule refers to withstand ratings with reference to fault current protection? ______________________________

37. Arcing fault damage is closely related to the value of ______________________________.

38. In general, systems are grounded so that the maximum voltage to ground does not exceed [circle one]

   a. 120 volts

   b. 150 volts

   c. 300 volts

39. An electric clothes dryer is rated at 5700 watts. The electric rate is 10.091 cents per kWh. The dryer is used continuously for three hours. Find the cost of operation, assuming that the heating element is on continuously. ______________________

____

____

____

____

____

40. *Rule 14–012* requires that the service equipment (breakers, fuses, and the panel itself) be rated [circle one]

   a. equal to or greater than the current it must interrupt

   b. equal to or greater than the available voltage and the fault current that is available at the terminals

41. Read the meter below. Last month's reading was 22 796. How many kilowatt-hours of electricity were used for the current month? ______________________________

UNIT 5

Electrical Outlets

OBJECTIVES

After studying this unit, you should be able to

- identify and explain the electrical outlet symbols used in the plans of the single-family dwelling
- discuss the types of outlets, boxes, fixtures, and switches used in the residence
- explain the methods of mounting the various electrical devices used in the residence
- understand the meaning of the terms *receptacle outlet* and *lighting outlet*
- understand the preferred way to position receptacles in wall boxes
- position wall boxes in relation to finished wall surfaces
- make surface extensions from concealed wiring
- determine the number of wires permitted in a given size box

ELECTRICAL OUTLETS

The *Canadian Electrical Code (CEC)* defines an *outlet* as a point on a wiring system where current is taken to supply utilization equipment.

The type and location of electrical outlets used for an installation are shown on electrical floor plans (Figure 5-1).

A receptacle outlet is an outlet where one or more receptacles are installed (Figure 5-2).

A lighting outlet is intended for the direct connection of a lampholder, a lighting fixture, or a pendant cord terminating in a lampholder (Figure 5-3).

A toggle switch is *not* an outlet; however, the term *outlet* is used broadly by electricians to include non-current-consuming switches and similar control devices in a wiring system when estimating the cost of the installation. Each type of outlet is represented on the plans as a symbol.

Figure 5-1 shows the outlets by the symbols ⊖ and -◯- Standard electrical symbols are shown in Figure 2-9E in Unit 2.

The curved lines in Figure 5-1 run from the outlet to the switch or switches that control the outlet. These lines are usually curved so that they cannot be mistaken for hidden edge lines. Outlets shown on the plan without curved lines are independent and have no switch control.

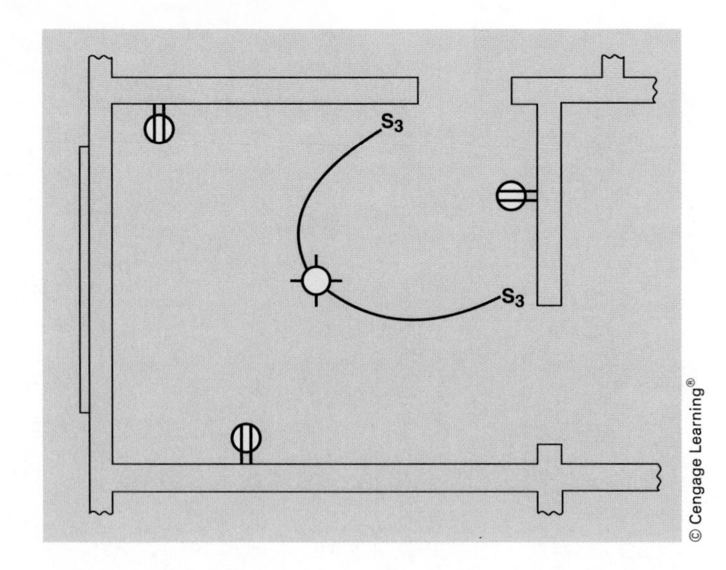

FIGURE 5-1 Use of electrical symbols and notations on a floor plan.

When preparing electrical plans, most architects, designers, and electrical engineers use symbols approved by the Standards Council of Canada (SCC) whenever possible. However, plans may contain symbols that are not found in these standards. When unlisted (non-standard) symbols are used, the electrician must refer to a legend that interprets these

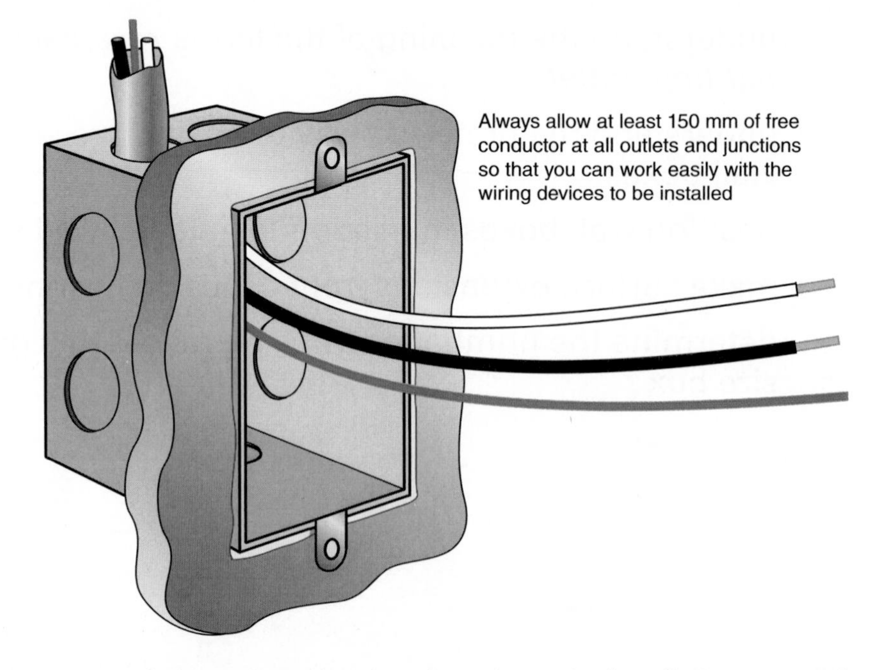

FIGURE 5-2 When a receptacle is connected to the wires, the outlet is called a *receptacle outlet.* For ease in working with wiring devices, *CEC Rule 12–3000(6)* requires that at least 150 mm of free conductor be provided.

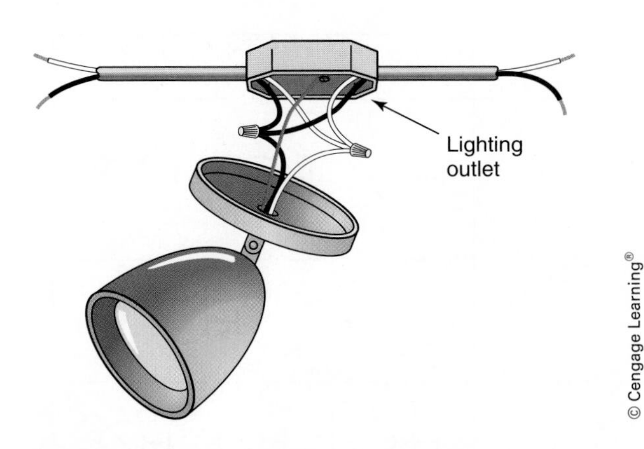

FIGURE 5-3 When a lighting fixture is connected to the wires, the outlet is called a *lighting outlet.* The *CEC* requires that at least 150 mm of free conductor be provided.

symbols. The legend may be included on the plans or in the specifications. In many instances, a notation on the plan will clarify the meaning of the symbol. See Figure 5-4.

Figures 5-5, 5-6A, 5-6B, and 5-6C show electrical symbols for outlets and their meanings. Although several symbols have the same shape, differences in the interior presentation plans indicate their different meanings. For example, different meanings are shown in Figure 5-5 for the outlet symbols. A good practice in studying symbols is to learn the basic forms first and then add the supplemental information to obtain different meanings.

SYMBOL	NOTATION
①	Plugmold entire length of workbench. Outlets 90 cm O.C. install 122 cm to centre from floor, GFCI-protected.
②	Track lighting. Provide five lampholders.
③	Two 40-watt rapid start fluorescent lamps in valance, control with dimmer switch.

FIGURE 5-4 How notations might be added to a symbol to explain its meaning. The architect or engineer can choose to fully explain the meaning directly on the plan if there is sufficient room or, if there is insufficient room, a notation can be used.

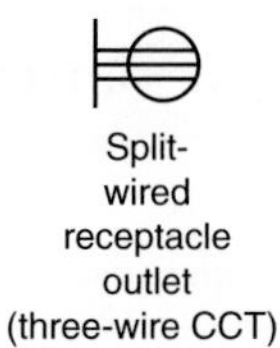

FIGURE 5-5 Variations in outlet symbols.

LIGHTING FIXTURES AND OUTLETS

Architects often include in the specifications an amount of money to purchase electrical fixtures. The electrical contractor includes this amount in the bid, and the choice of fixtures is then left to the homeowner. If the owner selects fixtures whose total cost exceeds the fixture allowance, the owner is expected to pay the difference between the actual cost and the specification allowance. If the fixtures are not selected before the roughing-in stage of wiring of the house, the electrician usually installs outlet boxes having standard fixture mounting studs.

Most modern surface-mount lighting fixtures can be fastened to an octagon box (Figures 5-7A and 5-7B) or an outlet box with a plaster ring using appropriate #8–32 metal screws and mounting strap furnished with the fixture.

A box must be installed at each outlet or switch location, *Rule 12–3000(1).* There are exceptions to this rule, but in general they relate to special manufactured wiring systems where the "box" is an integral part of the system. For standard wiring methods, such as cable or conduit, a box is usually required.

Figures 5-8A and 5-8B and Figures 5-9A and 5-9B illustrate types of non-metallic and metallic boxes.

Outlet boxes must be accessible, *Rule 12–3014(1).*

Be careful when roughing in boxes for fixtures and also when hanging ceiling fixtures to

- non-metallic device (switch) boxes
- non-metallic device (switch) plaster rings
- metallic device (switch) boxes
- metallic device (switch) plaster rings
- any non-metallic box, unless specifically marked on the box or carton for use as a fixture support, or to support other equipment, or to accommodate heat-producing equipment

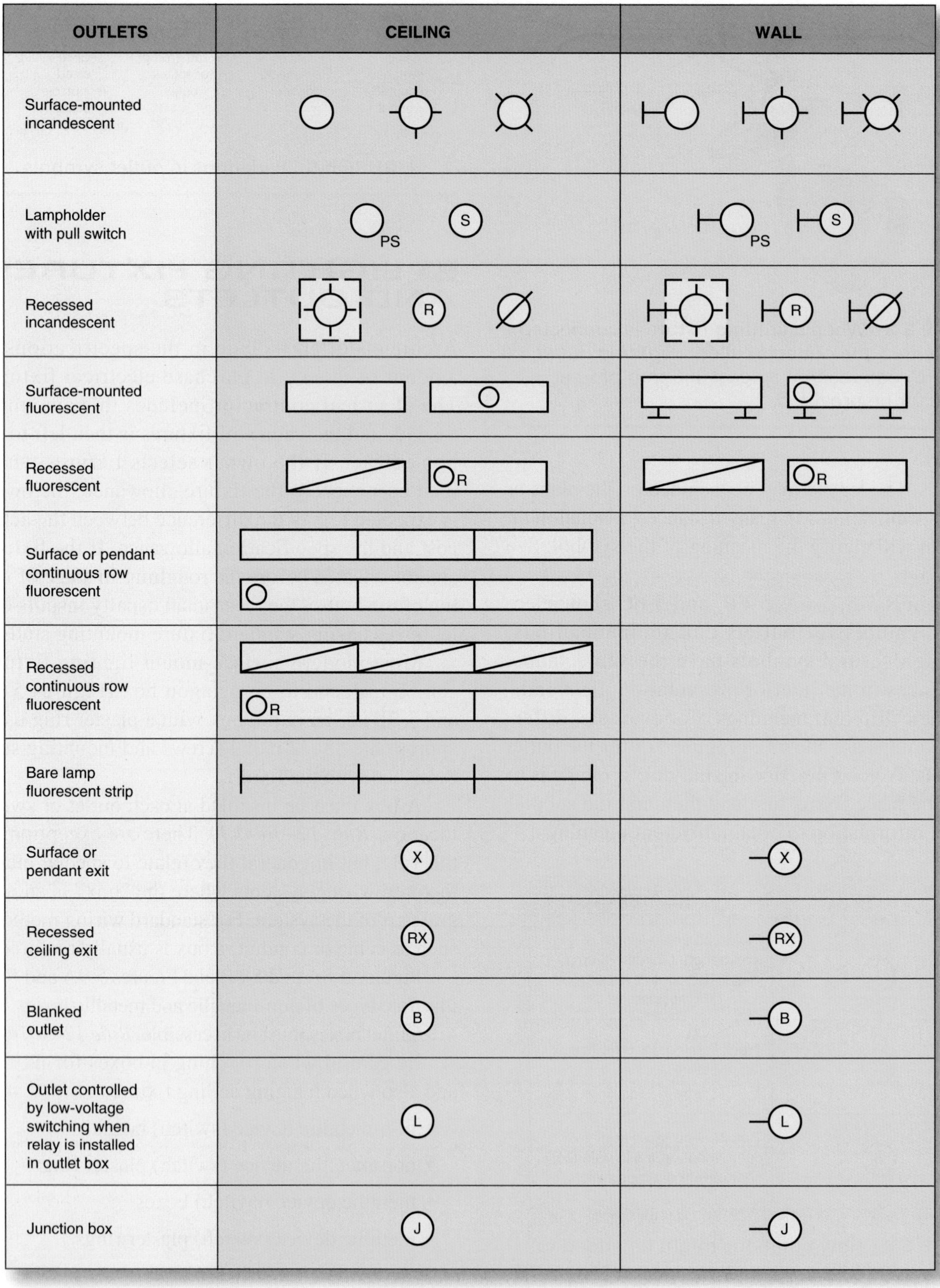

OUTLETS	CEILING	WALL
Surface-mounted incandescent		
Lampholder with pull switch	PS S	PS S
Recessed incandescent	R	R
Surface-mounted fluorescent		
Recessed fluorescent	R	R
Surface or pendant continuous row fluorescent		
Recessed continuous row fluorescent	R	
Bare lamp fluorescent strip		
Surface or pendant exit	X	X
Recessed ceiling exit	RX	RX
Blanked outlet	B	B
Outlet controlled by low-voltage switching when relay is installed in outlet box	L	L
Junction box	J	J

Courtesy of National Electrical Contractors Association

FIGURE 5-6A Symbols for different types of lighting outlets.

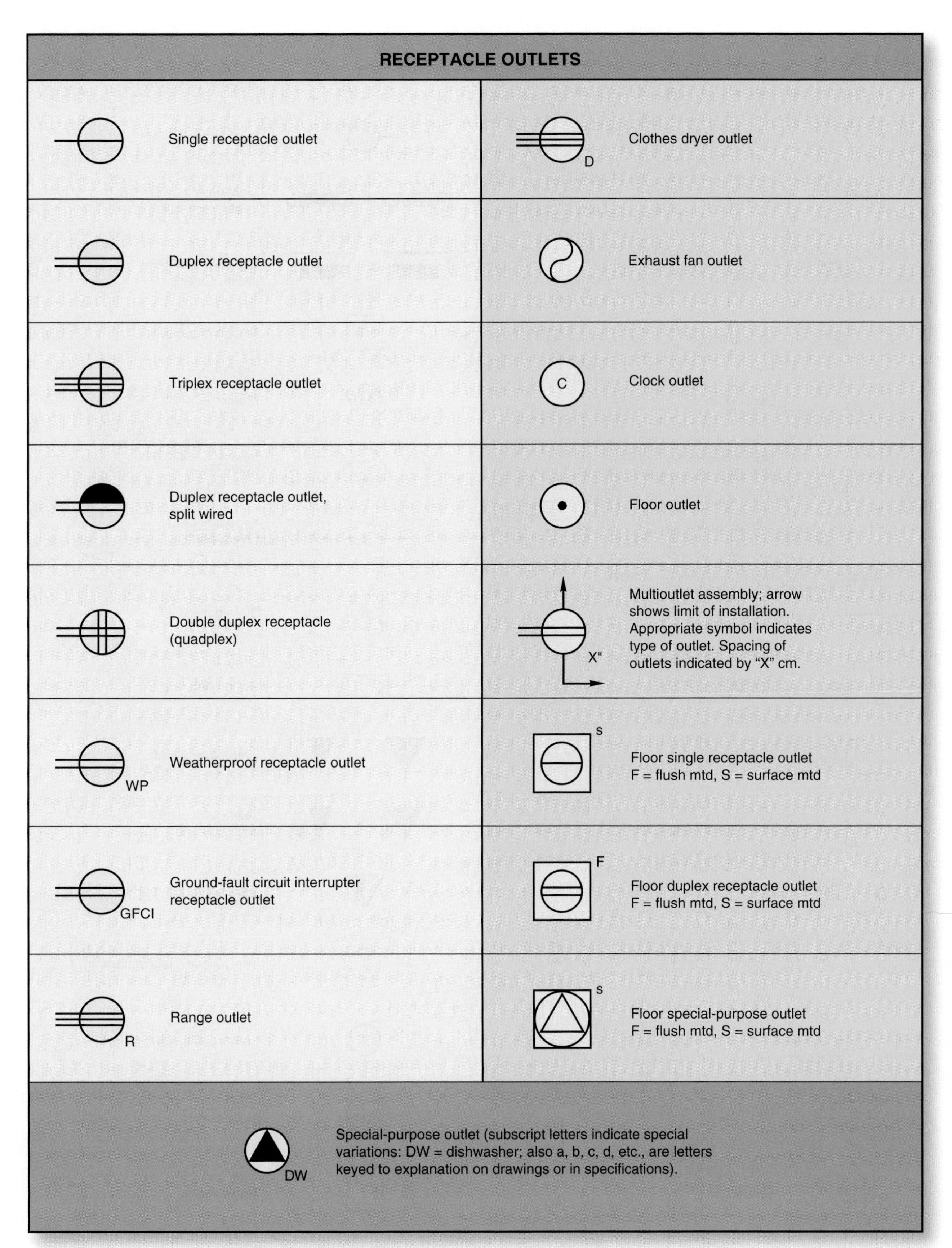

FIGURE 5-6B Symbols for receptacle outlets.

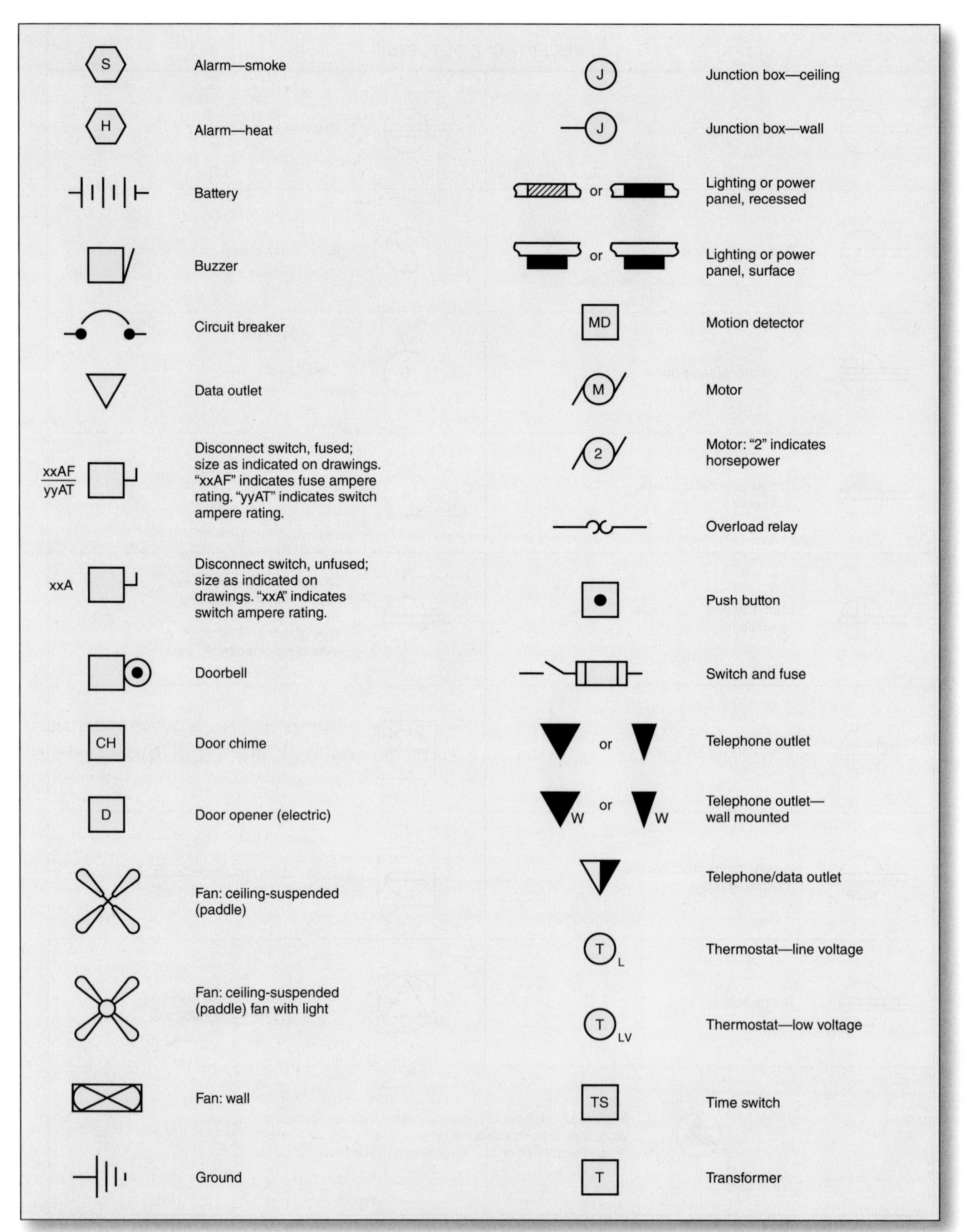

FIGURE 5-6C Miscellaneous symbols.

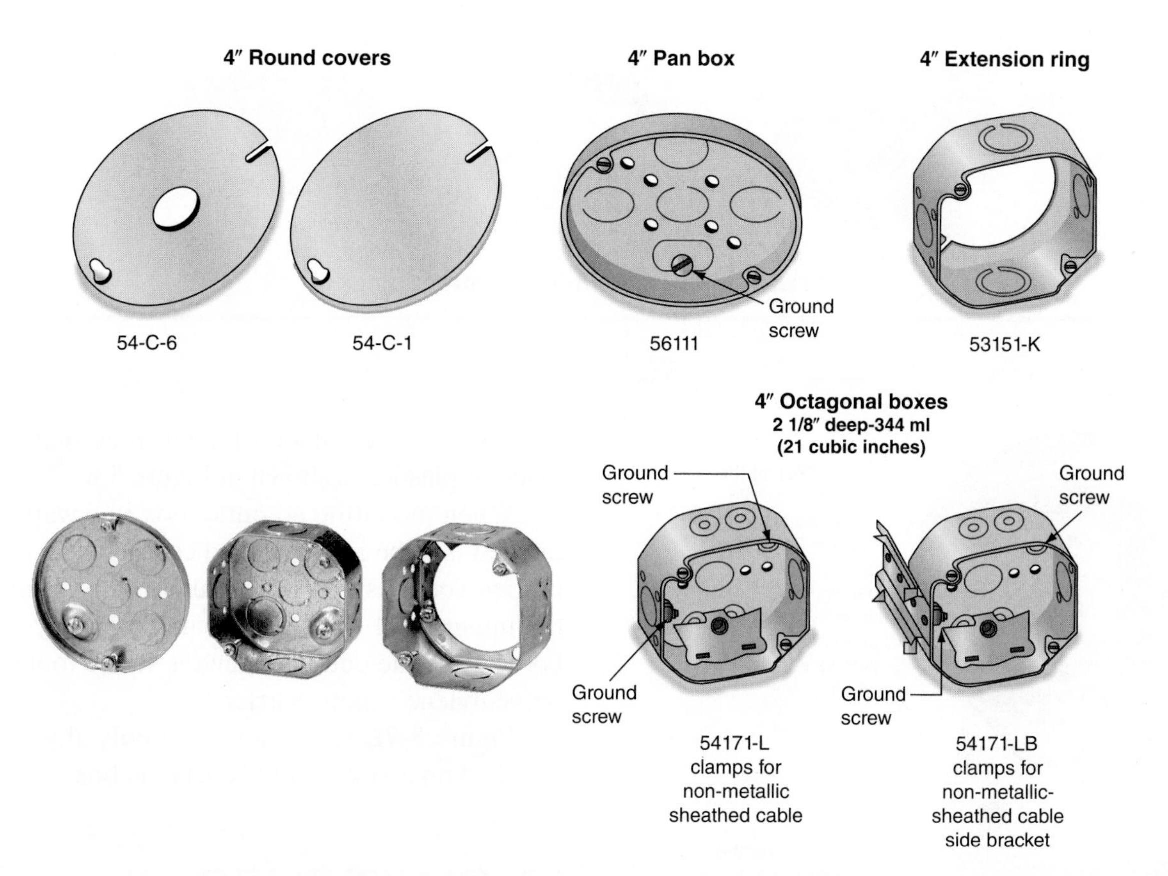

FIGURE 5-7A Octagonal outlet boxes, showing extension ring and covers.

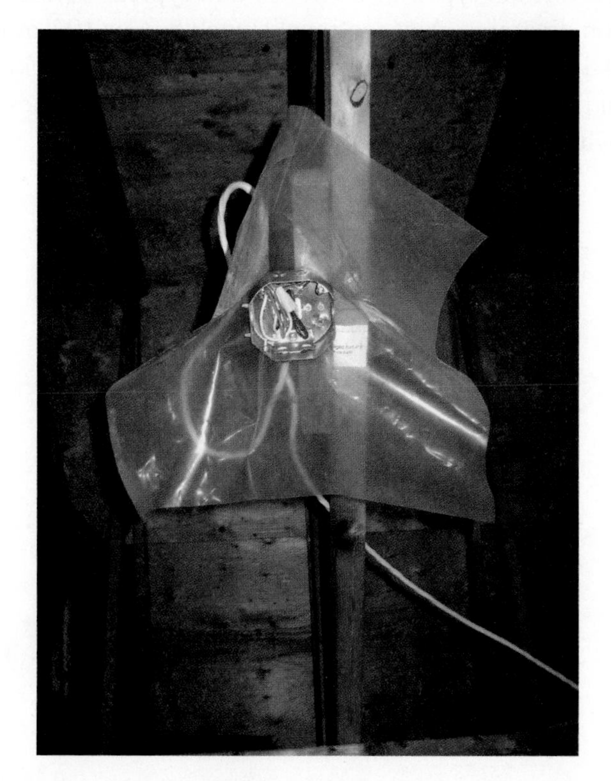

FIGURE 5-7B Outlet box complete with vapour boot.

Therefore, unless the box, device plaster ring, or carton is marked to indicate that it has been listed by the CSA for the support of fixtures, do not use where fixtures are to be hung. Refer to *Rule 12–1110.*

Be careful when installing a ceiling outlet box for the purpose of supporting a fan. See Unit 11 for a detailed discussion of paddle fans and their installation.

If the owner selects fixtures before construction, the architect can specify these fixtures in the plans and/or specifications, giving the electrician advance information on any special framing, recessing, or mounting requirements. This information must be provided in the case of recessed fixtures, which require a specific wall or ceiling opening.

Many types of lighting fixtures are available. Figure 5-10 shows several typical lighting fixtures that may be found in a dwelling unit. It also shows the electrical symbols used on plans to designate these fixtures and the type of outlet boxes or switch boxes on which the lighting fixtures can be mounted. A standard receptacle outlet is shown as well. These

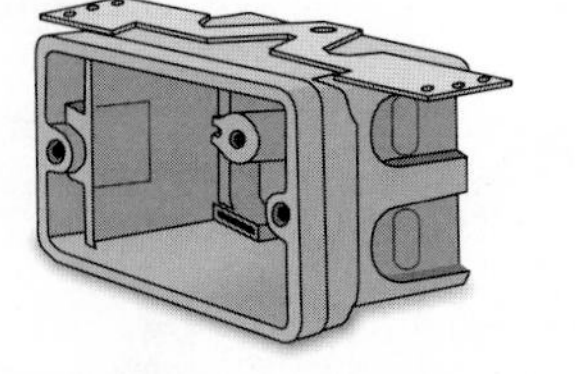
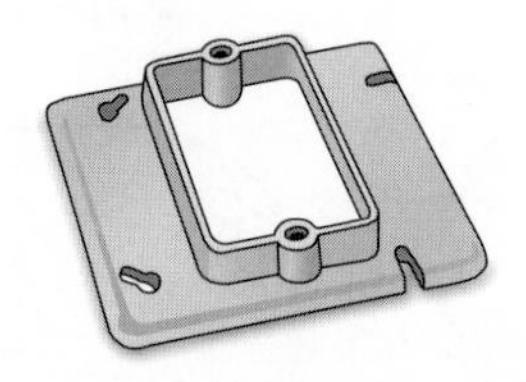
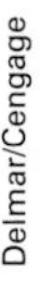

FIGURE 5-8A Non-metallic outlet box.

switch boxes are made of steel; they may also be made of plastic, as shown in Figure 5-8.

When mounting an outlet box in a wall or ceiling that is to be insulated and covered with a vapour barrier, you must make special provision to maintain the integrity of the vapour barrier. A polyethylene boot covers the outlet box and is sealed to the 6-mm polyethylene vapour barrier.

Figure 5-7B shows a typical polyethylene boot installed on a 102-mm (4″) octagon box.

SWITCHES

Figure 5-11 shows some of the standard symbols for various types of switches and typical connection diagrams. Any sectional switch box or 102-mm (4-in.) square box with a side-mounting bracket and raised switch cover can be used to install these switches (Figure 5-10).

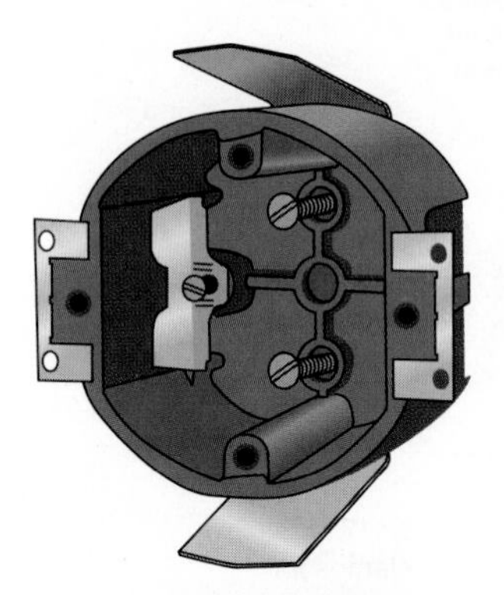

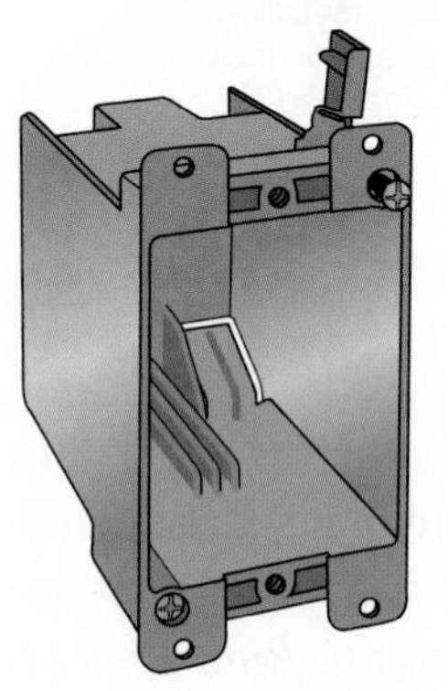

© Cengage Learning®

FIGURE 5-8B Non-metallic outlet box.

Courtesy of Thomas & Betts Corporation

FIGURE 5-9A Gang-type switch (device) box.

Courtesy of Thomas & Betts Corporation

FIGURE 5-9B Octagon box.

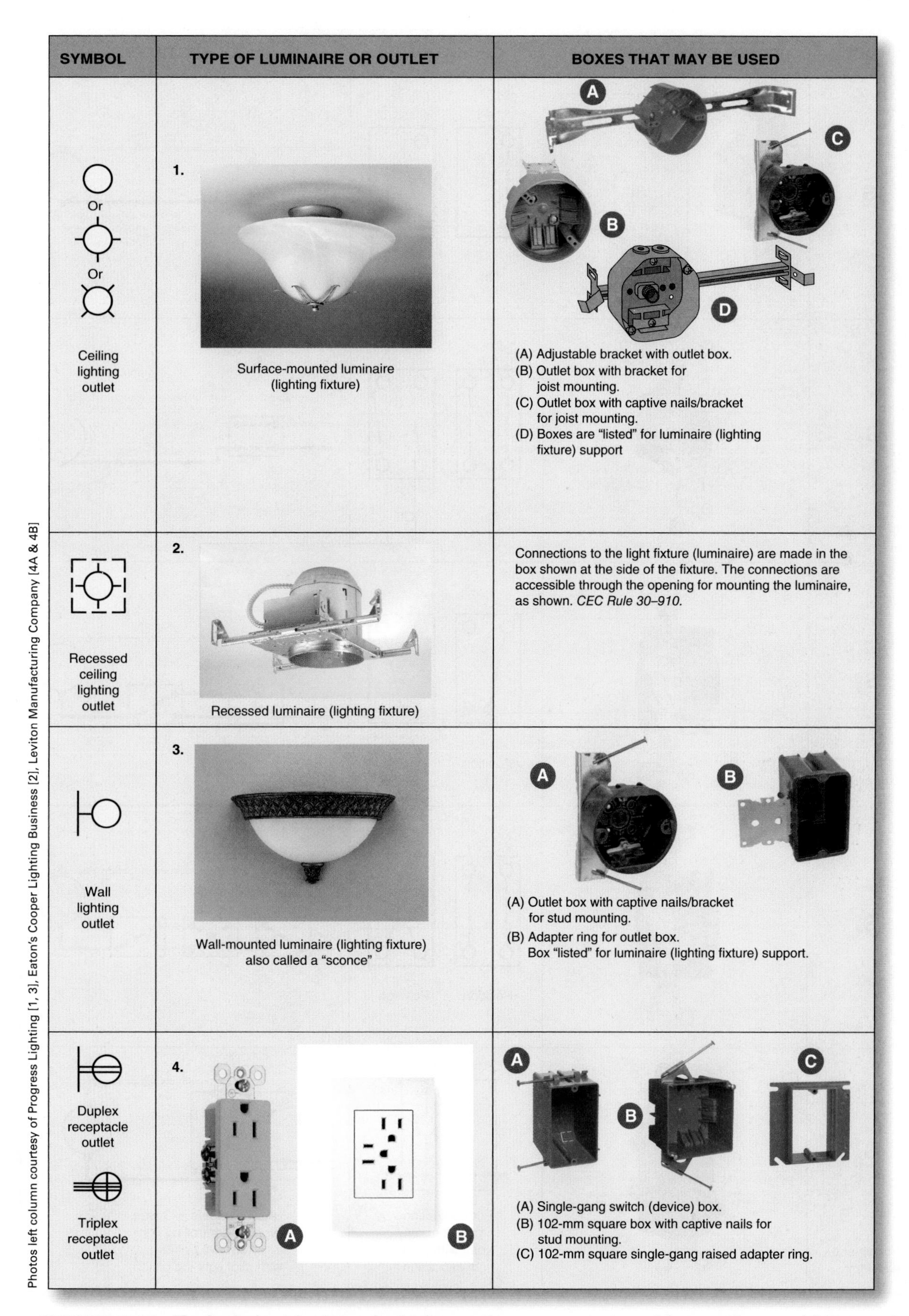

FIGURE 5-10 Typical electrical symbols, items they represent, and boxes that could be used.

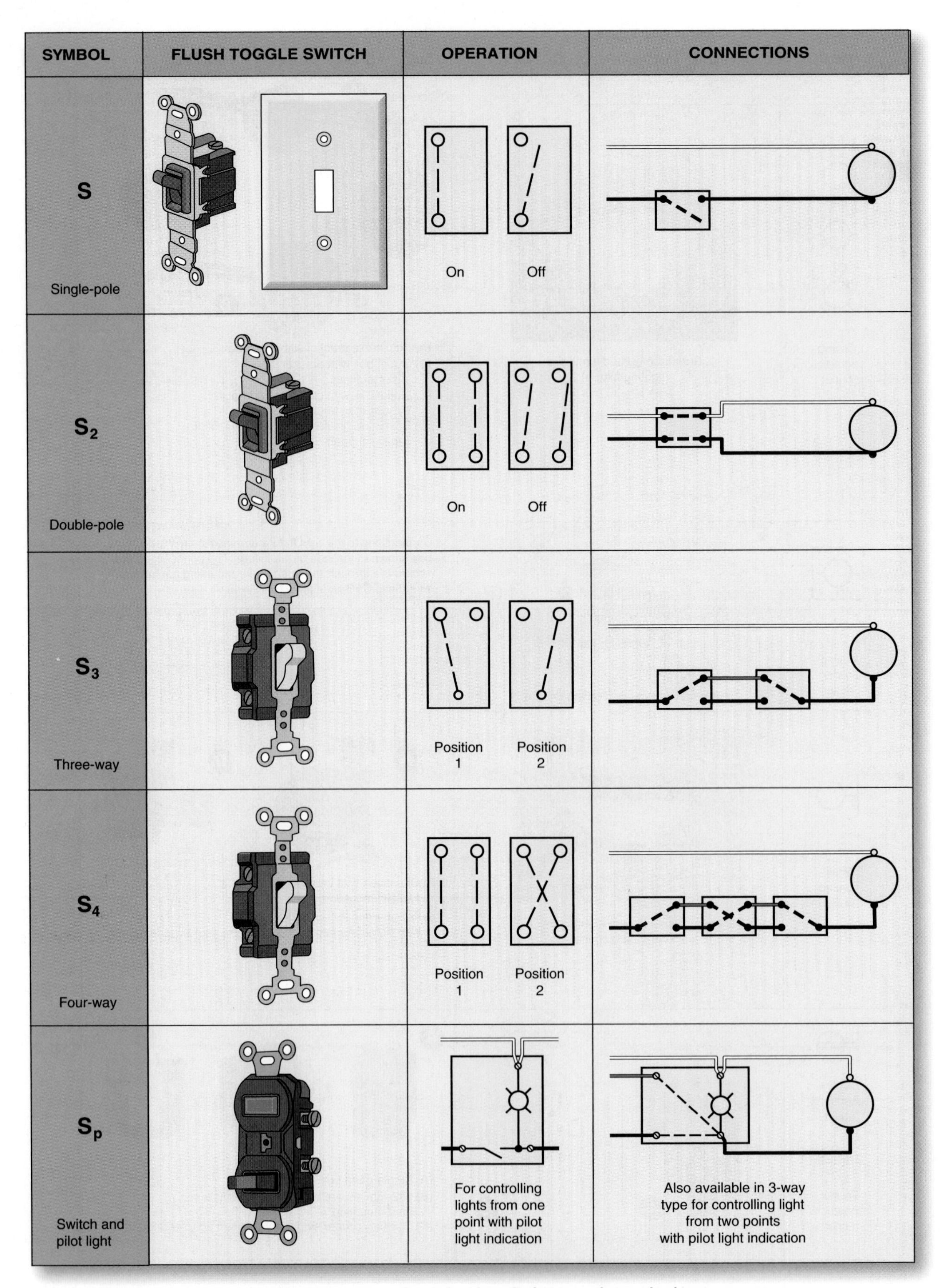

FIGURE 5-11 Standard switches and symbols.

JUNCTION BOXES AND SWITCH (DEVICE) BOXES (*RULES 12–3000* THROUGH *12–3034*)

Junction boxes are sometimes placed in a circuit for convenience in joining two or more cables or conduits. All conductors entering a junction box are joined to other conductors entering the same box to form proper connections so that the circuit will operate in its intended manner.

All electrical installations must conform to the *CEC* standards requiring that junction boxes be installed so that the wiring contained in them shall be accessible without removing any part of the building. In house wiring, this requirement limits the use of junction boxes to unimproved basements, garages, and open attic spaces because blank box covers exposed to view detract from the appearance of a room. An outlet box such as the one installed for the front hall ceiling fixture is really a junction box because it contains splices. Removing the fixture makes the box accessible, thereby meeting *CEC* requirements. Refer to Figures 5-12 and 5-13.

Rule 12–3000(1) requires that a box or fitting be installed wherever splices, switches, outlets, junction points, or pull points are required (Figure 5-12). However, sometimes a change is made from one wiring method to another, for example, armoured cable to electrical metallic tubing (EMT). The fitting where the change is made must be accessible after installation; see Figure 5-14.

Boxes shall be rigidly and securely mounted, *Rule 12–3010(1).*

NON-METALLIC OUTLET AND DEVICE BOXES

The *CEC* permits non-metallic outlet and device boxes to be installed where the wiring method is non-metallic sheathed cable, *Rule 12–524.* Non-metallic boxes may be used with armoured cable or metal raceways, but only if a proper means is provided inside the boxes to bond any and all metal raceways and/or armoured cables together, *Rule 12–3000(2).*

The house wiring system is usually formed by a number of specific circuits, each consisting of a continuous run of cable from outlet to outlet or from box to box. The residence plans show many branch circuits for general lighting, appliances, electric heating, and other requirements. The specific *CEC* rules for each of these circuits are covered in later units.

GANGED SWITCH (DEVICE) BOXES

A flush switch or receptacle outlet for residential use fits into a standard 3 × 2 × 2½-in. sectional switch box (sometimes called a *device box*). When two or more switches (or outlets) are located at the same point, single switch boxes may be ganged or fastened together to provide the required mounts (Figures 5-15A and 5-15C) or a welded (two or more) gang box may be used.

Three switch boxes can be ganged together by removing and discarding one side from both the first and third switch boxes, and both sides from the second (centre) switch box. The boxes are then joined together (Figure 5-15B). After the switches are installed, the gang is trimmed with a gang plate having the required number of switch handle or receptacle outlet openings. These plates are called *two-gang wall plates, three-gang wall plates,* and so on, depending on the number of openings.

The dimensions of a standard sectional switch box, 3 × 2 inches, are the dimensions of the opening of the box. The depth of the box may vary from 1½ inches to 3½ inches (38 to 89 mm), depending on the requirements of the building construction and the number of conductors and devices to be installed. *Rules 12–3000* through *12–3034* cover outlet, switch, and junction boxes. See Figure 5-16 for a list of common box dimensions.

Rule 12–3016(1) states that boxes must be mounted so that they will be set back not more than 6 mm from the finished surface when the boxes are mounted in noncombustible walls or ceilings made of concrete, tile, or similar materials. When the wall or ceiling construction is of combustible material (wood), the box must be set flush with the surface or project from it. See Figures 5-17A and 5-17B.

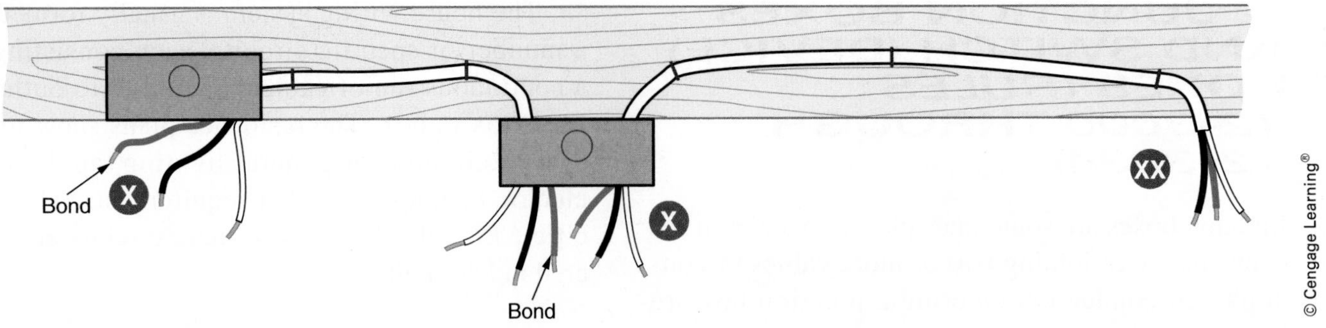

FIGURE 5-12 A box (or fitting) must be installed wherever there are splices, outlets, switches, or other junction points. Refer to the points marked X. A potential *CEC* violation is shown at point XX. However, this is a violation only if the cable is not supplying equipment with an integral connection box or if the apparatus has not been approved as a connection box, *Rules 12–3000(1)* and *12–3014(1).*

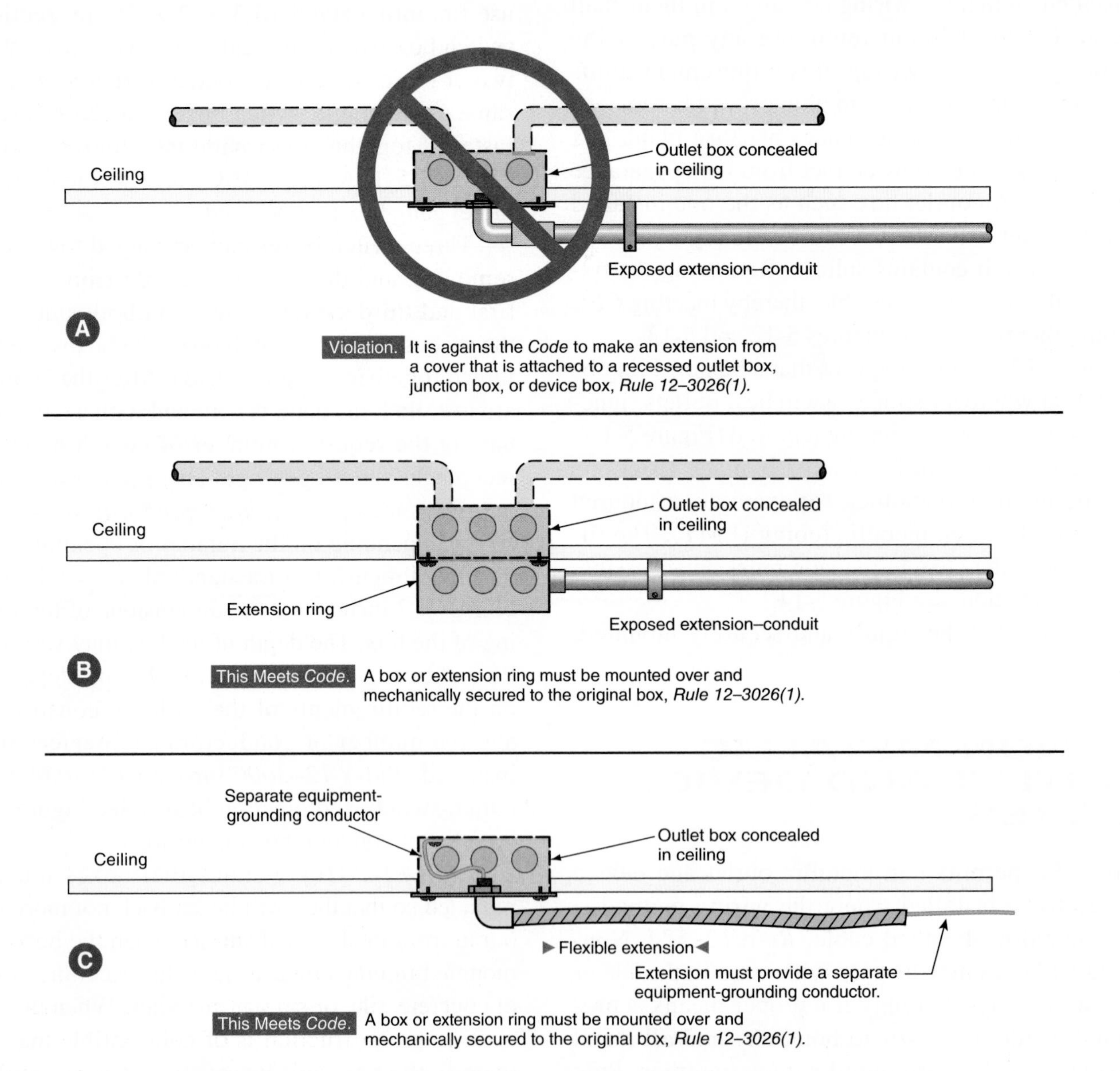

FIGURE 5-13 How to make an extension from a flush-mounted box.

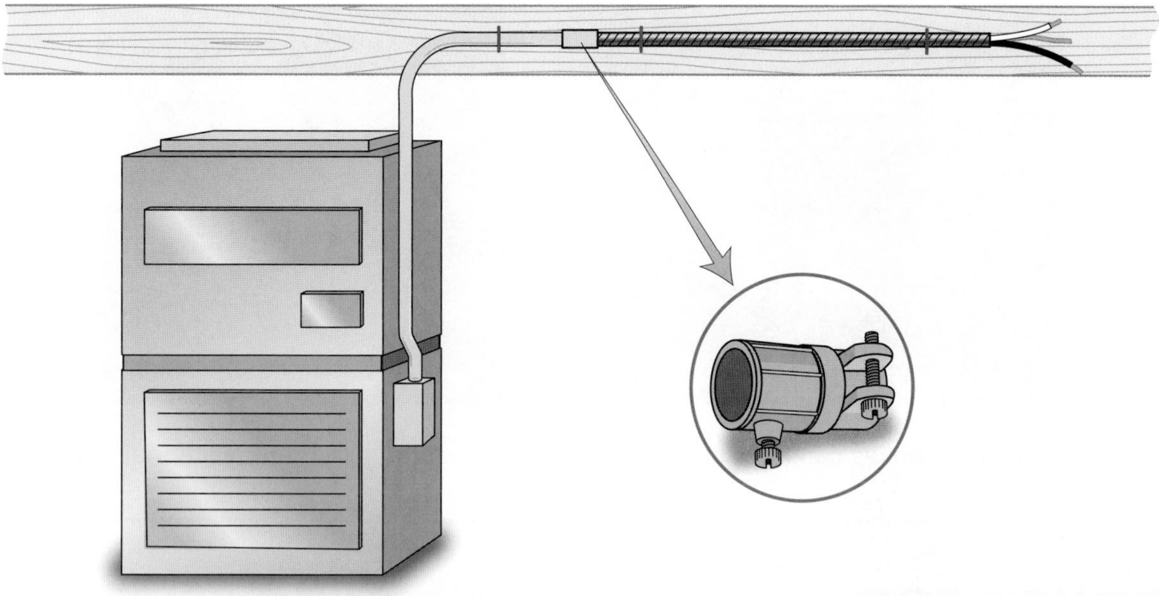

FIGURE 5-14 A transition from one wiring method to another. In this case, the armour of the AC90 cable is removed, allowing sufficient length of the conductors to be run through the conduit. A proper fitting as shown must be used at the transition point, and the fitting must be accessible after installation.

FIGURE 5-15A Device boxes. Box on left has welded sides. Box on right has removable sides, which permit several boxes to be ganged together.

FIGURE 5-15B Standard flush switches installed in ganged sectional device boxes.

These requirements are meant to prevent the spread of fire if a short circuit occurs within the box.

Single-gang switch boxes are normally mounted to a structural member of the building with nails or screws. Ganged sectional switch (device) boxes may be installed using a pair of metal mounting strips, *Rule 12–3010(2)*. These strips may also be used to install a switch box between wall studs. See Figures 5-18A and 5-18B. When an outlet box is to be mounted at a specific location between joists, as for ceiling-mounted fixtures, an offset bar hanger is used (Figure 5-10).

Older installations of switch boxes may have nails passing through the box. This is no longer permitted unless the nails are not more than 6.4 mm from the back or ends of the box. See *Rule 12–3010(5)*.

Courtesy of Sandy F. Gerolimon

FIGURE 5-15C Sectional boxes can have sides removed and secured to another sectional box, creating a two-gang box.

BOXES FOR CONDUIT WIRING

For some types of construction, conduit rather than cable wiring is used.

When conduit is installed in a residence, it is quite common to use 102-mm square boxes trimmed with suitable plaster covers (Figure 5-16). There are sufficient knockouts in the top, bottom, sides, and back of the box to permit a number of conduits to run into it. Plenty of room is available for the conductors and wiring devices. Note how easily these 102-mm square outlet boxes can be mounted back to back by installing a small fitting between them; see photos B and C in Figure 5-19.

You can trim 4-in. (102-mm) square outlet boxes with one-gang or two-gang plaster rings where wiring devices will be installed. Where lighting fixtures will be installed, use a plaster ring having a round opening.

Any unused openings in outlet and device boxes must be closed, as per *Rule 12–3024,* using knockout fillers, Figure 5-20.

The number of conductors allowed in outlet and device boxes is covered elsewhere in this textbook.

SPECIAL-PURPOSE OUTLETS

Special-purpose outlets are normally shown on floor plans. They are described by a notation and are also detailed in the specifications. The plans included with this textbook indicate special-purpose outlets by a triangle inside a circle with a subscript letter. In some cases, a subscript number is added to the letter.

When a special-purpose outlet is indicated on the plans or in the specifications, the electrician must check for special requirements, such as a separate circuit, a special 240-volt circuit, a special grounding or polarized receptacle, a device box with extra support (Figure 5-21), or other preparation. Special-purpose outlets include

- central vacuum receptacle
- weatherproof receptacle
- dedicated receptacle, perhaps for a home computer
- air-conditioning receptacle
- clock receptacle

NUMBER OF CONDUCTORS IN A BOX

Rule 12–3034 dictates that outlet boxes, switch boxes, and device boxes be large enough to provide ample room for the wires without having to jam or crowd the wires into them. *CEC Table 23* specifies the maximum number of conductors allowed in standard outlet boxes and switch boxes; see Table 5-1. A conductor running through the box is counted as one conductor. Each conductor originating outside the box and terminating inside it is counted as one conductor. Conductors that

QUICK-CHECK BOX SELECTION GUIDE
FOR BOXES GENERALLY USED FOR RESIDENTIAL WIRING

DEVICE BOXES

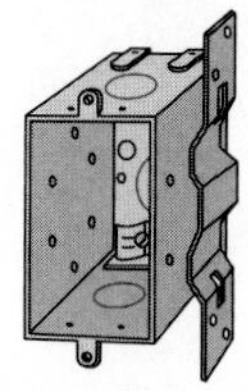

Wire size	Millilitres	3x2x1 1/2 (7.5 in.^{3})	3x2x2 (10 in.^{3})	3x2x2 1/4 (10.5 in.^{3})	3x2x2 1/2 (12.5 in.^{3})	3x2x3 (16 in.^{3})
		131	163	163	204	245
14 AWG	14	5	6	6	8	10
12 AWG	12	4	5	5	7	8

The number of conductors is "per gang"

SQUARE BOXES

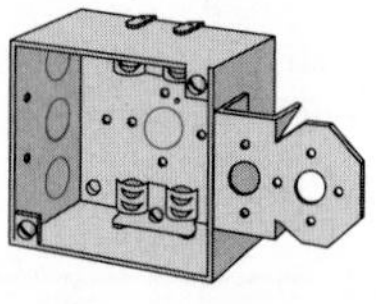

Wire size	Millilitres	4x4x1 1/2 (21 in.^{3})	4x4x2 1/8 (30.3 in.^{3})	4 11/16x4 11/16x1 1/2	4 11/16x4 11/16x2 1/8
		344	491	491	688
14 AWG	14	14	20	20	28
12 AWG	12	12	17	17	24

OCTAGON BOXES

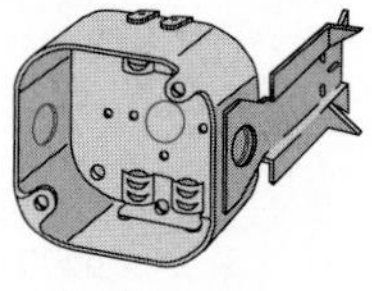

Wire size	Millilitres	4x1 1/2 (15.5 in.^{3})	4x2 1/8 (21.5 in.^{3})
		245	344
14 AWG	14	10	14
12 AWG	12	8	12

HANDY BOXES

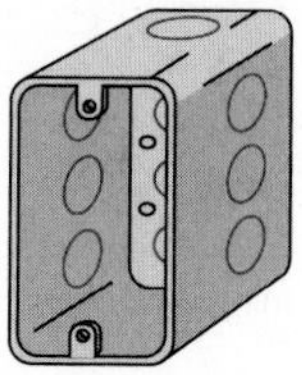

Wire size	Millilitres	4x2 1/8x1 1/2 (10.3 in.^{3})	4x2 1/8x1 7/8 (13 in.^{3})	4x2x2 1/8 (14.5 in.^{3})
		147	229	262
14 AWG	14	6	9	10
12 AWG	12	5	8	9

RAISED COVERS

Where raised covers are marked with their volume in cubic inches, that volume may be added to the box volume to determine maximum number of conductors in the combined box and raised cover. Non-metallic raised covers are available.

MASONRY BOXES

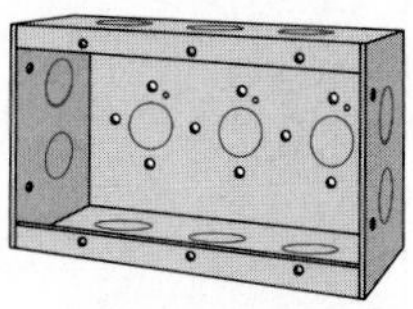

Wire size	3 3/4x2x2 1/2 (14 in.^{3})	3 3/4x2x3 1/2 (21 in.^{3})
14 AWG	7	10
12 AWG	6	9

3-gang box

Masonry boxes have lots of room. Available up to 6 gang. The conductor fill is for each gang. Refer to *Table 23*.

Note: Be sure to make deductions from the above maximum number of conductors permitted for wire connectors, wiring devices, cable clamps, fixture studs, and grounding conductors. The volume is taken directly from *Table 23* of the *CEC*.

FIGURE 5-16 Quick-check box selection guide.

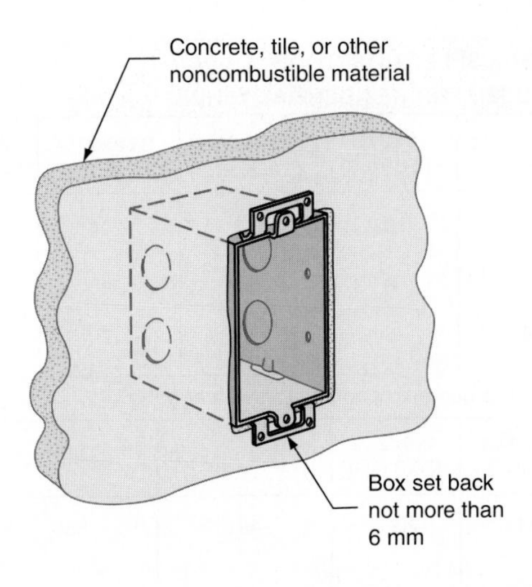

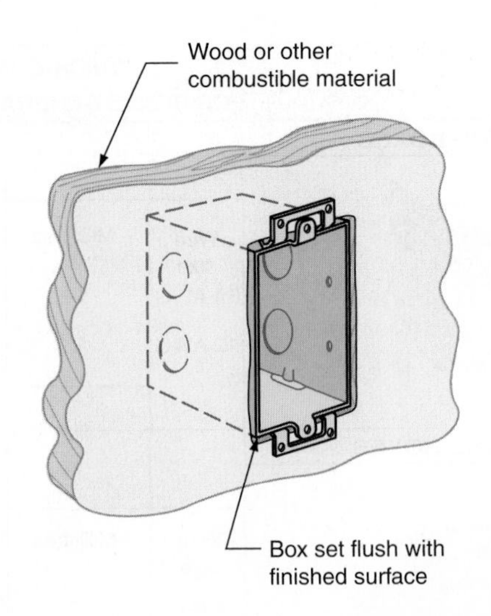

FIGURE 5-17A Box position in walls and ceilings constructed of various materials, *Rule 12–3016.*

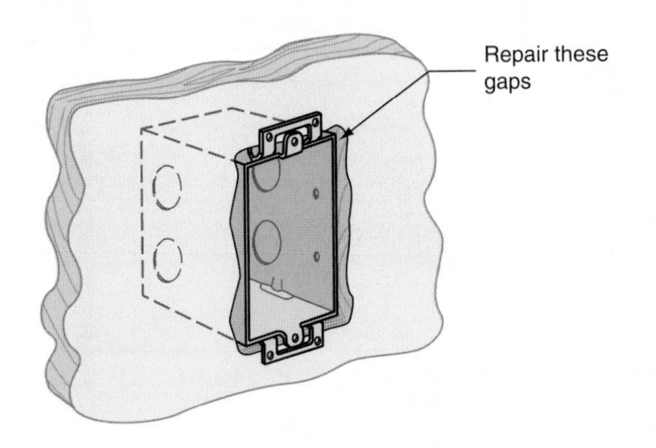

FIGURE 5-17B Repair these gaps, *Rule 26–710.*

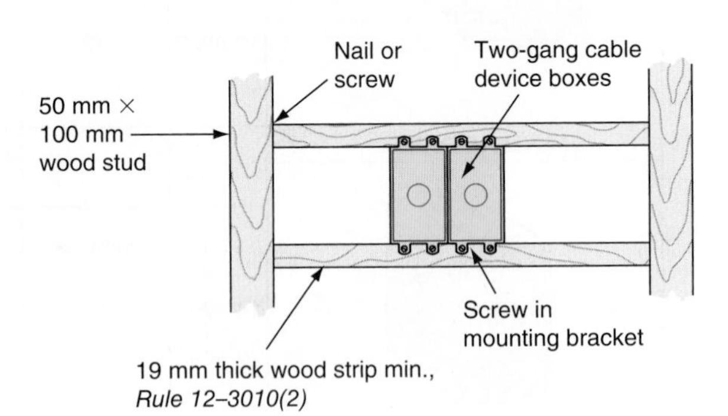

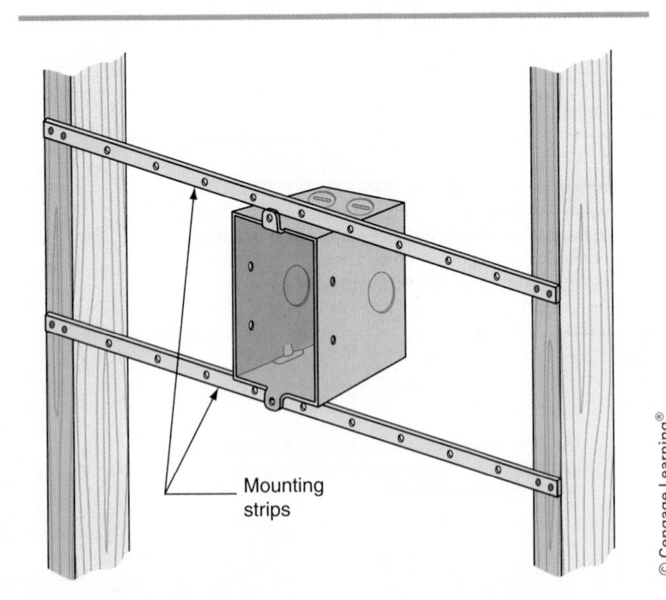

FIGURE 5-18A Switch (device) boxes installed between studs using metal mounting strips. If wooden boards are used, they must be at least 19 mm thick and securely fastened to the stud, *Rule 12–3010(2).*

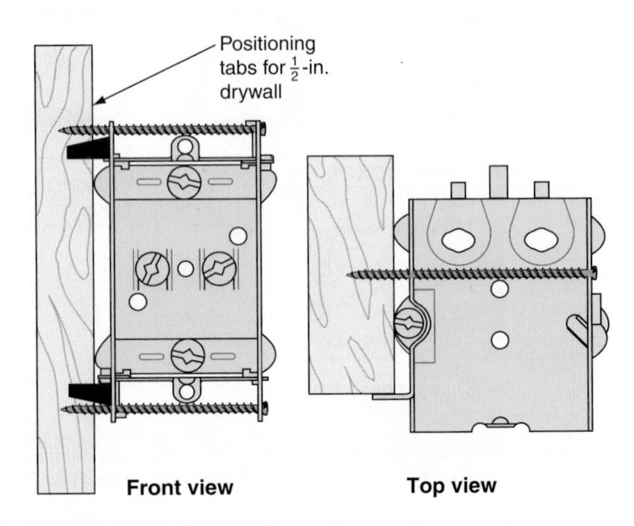

FIGURE 5-18B A method of mounting sectional boxes to wood studs.

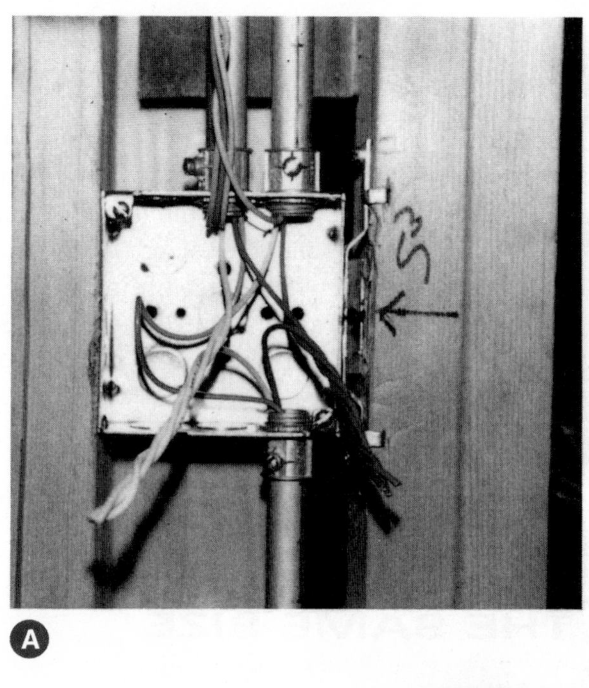

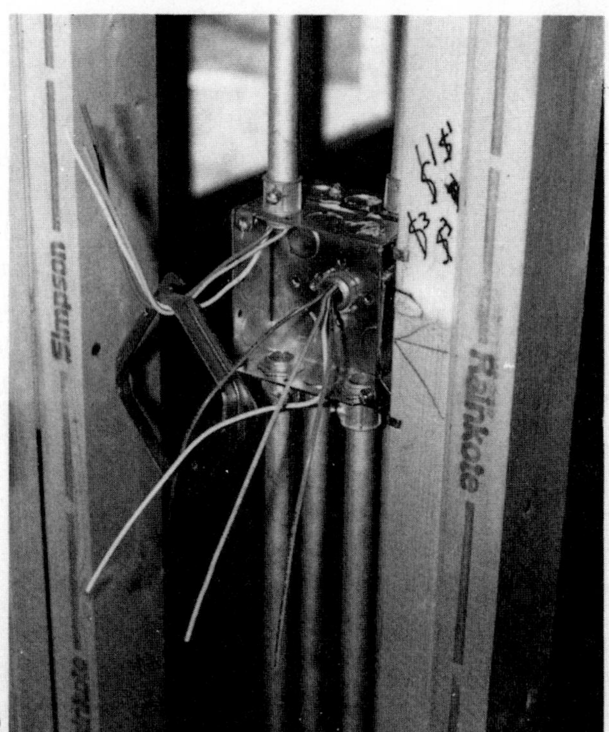

Delmar/Cengage

FIGURE 5-19 Boxes for conduit wiring.

originate and terminate within the box are not counted.

When conductors are the same size, the proper box size can be selected by referring to *CEC Table 23.* When conductors of different sizes are used, refer to *CEC Table 22;* see Table 5-2.

Tables 22 and *23* do not consider fittings or devices such as fixture studs, wire connectors, hickeys, switches, or receptacles that may be in the box. For assistance in this case, see Table 5-3). When the box contains one or more fittings (such as fixture studs or hickeys; see Figure 5-22), the number of

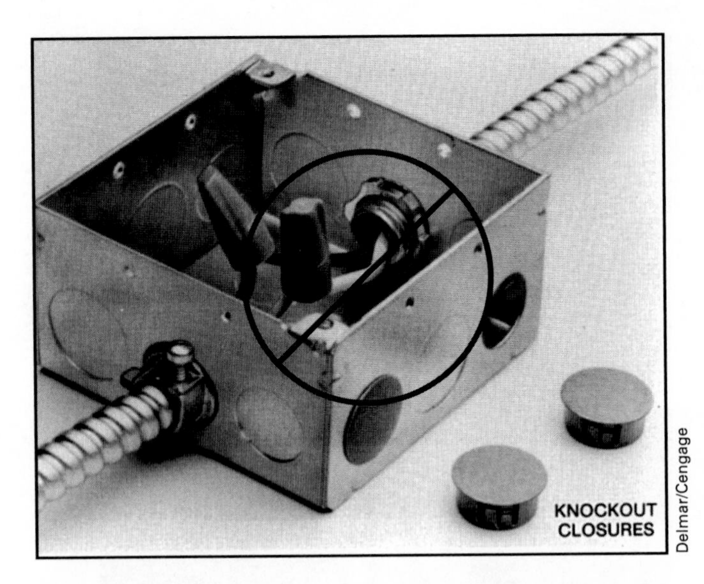

FIGURE 5-20 Unused openings in boxes must be closed according to *CEC Rule 12–3024.* This is done to contain electrical short-circuit problems inside the box or panel and to keep rodents out.

WIRES MUST BE 150 mm MINIMUM, *CEC Rule 12–3000(6)*. See Figure 5-2.

FIGURE 5-21 Flush-mounted device box for steel studs with required additional support.

conductors must be one less than shown in the table for each type of device.

Deduct two conductors for each device mounted on a single strap (switches, receptacles). The deduction of two conductors has become necessary because of severe crowding of conductors in the box when the switch or receptacle is mounted or when dimmer switches are installed. In most cases, dimmer switches are larger than conventional switches.

Deduct one conductor for every two-wire connector. For example, for two- or three-wire connectors, deduct one wire; for four or five connectors deduct two wires. There is no deduction for the first wire connector because it will probably be used for the second ground connection. Be sure to include all conductor sizes (#14, #12, #10, #8, and #6 AWG) when determining the size of the box to be installed. See Figures 5-23 and Figure 5-24 for examples of this *CEC* requirement, *Rule 12–3034(1)*.

SELECTING A BOX WHEN ALL CONDUCTORS ARE THE SAME SIZE

EXAMPLE

A box contains one fixture stud and two cable clamps. The number of conductors permitted in the box shall be one less than shown in *CEC Table 23*. (Deduct one conductor for the fixture stud.)

A further deduction of two conductors is made for each wiring device in the box, but if two more devices are mounted on the same strap, they count as a single device. For example, if a switch and a receptacle are mounted on a single strap, only two conductors are deducted.

SELECTING A BOX WHEN CONDUCTORS ARE DIFFERENT SIZES *(RULE 12–3034[4])**

When the box contains different size wires, do the following:

- Determine the size of conductors to be used and the number of each (e.g., two #10 and two #12).
- Determine if the box is to contain any wiring devices (switch or receptacle), wire connectors, or fixture studs. This volume adjustment must

* Courtesy of CSA Group.

TABLE 5-1

CEC Table 23.

CEC TABLE 23
NUMBER OF CONDUCTORS IN BOXES
(RULE 12–3034)

Box Dimensions	Trade Size	Capacity mL (in.^{3})	Maximum Number of Conductors (per AWG size)				
			14	12	10	8	6
Octagonal	4 × 1½	245 (15)	10	8	6	5	3
	4 × 2$^1/_8$	344 (21)	14	12	9	7	4
Square	4 × 1½	344 (21)	14	12	9	7	4
	4 × 2$^1/_8$	491 (30)	20	17	13	10	6
	4$^{11}/_{16}$ × 1½	491 (30)	20	17	13	10	6
	4$^{11}/_{16}$ × 2$^1/_8$	688 (42)	28	24	18	15	9
Round	4 × ½	81 (5)	3	2	2	1	1
Device	3 × 2 × 1½	131 (8)	5	4	3	2	1
	3 × 2 × 2	163 (10)	6	5	4	3	2
	3 × 2 × 2¼	163 (10)	6	5	4	3	2
	3 × 2 × 2½	204 (12.5)	8	7	5	4	2
	3 × 2 × 3	245 (15)	10	8	6	5	3
	4 × 2 × 1½	147 (9)	6	5	4	3	2
	4 × 2$^1/_8$ × 1$^7/_8$	229 (14)	9	8	6	5	3
	4 × 2$^3/_8$ × 1$^7/_8$	262 (16)	10	9	7	5	3
Masonry	3¾ × 2 × 2½	229 (14)/gang	9	8	6	5	3
	3¾ × 2 × 3½	344 (21)/gang	14	12	9	7	4
	4 × 2¼ × 2$^3/_8$	331 (20.25)/gang	13	11	9	7	4
	4 × 2¼ × 3$^3/_8$	364 (22.25)/gang	14	12	9	8	4
Through Box	3¾ × 2	3.8/mm (6/in.) depth	4	3	2	2	1
Concrete Ring	4	7.7/mm (12/in.) depth	8	6	5	4	2
FS	1 Gang	229 (14)	9	8	6	5	3
	1 Gang tandem	557 (34)	22	19	15	12	7
	2 Gang	426 (26)	17	14	11	9	5
	3 Gang	671 (41)	27	23	18	14	9
	4 Gang	917 (56)	37	32	24	20	12
FD	1 Gang	368 (22.5)	15	12	10	8	5
	2 Gang	671 (41)	27	23	18	14	9
	3 Gang	983 (60)	40	34	26	21	13
	4 Gang	1392 (85)	56	48	37	30	18

Courtesy of CSA Group

be based on the volume of the largest conductor in the box.

- Deduct one conductor for each pair of wire connectors. For example, two or three equal one wire, four or five equal two wires. *Note:* If only one wire connector is used, it is not counted.
- Deduct two conductors for each switch or receptacle that is less than 25 mm between the mounting strap and the back of the device.
- Make the necessary adjustments for devices, wire connectors, and fixture studs.
- Size the box based on the total volume required for conductors according to *Table 22.*

TABLE 5-2

CEC Table 22.

CEC TABLE 22
SPACE FOR CONDUCTORS IN BOXES
(RULE 12–3034)

Size of Conductor, AWG	Usable Space Required for Each Insulated Conductor, mL
14	24.6
12	28.7
10	36.9
8	45.1
6	73.7

Courtesy of CSA Group

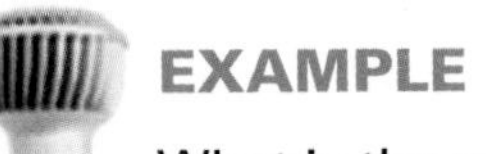

EXAMPLE

What is the minimum volume required for a box that will contain one switch, two #14 wires, two #12 wires, and three #33 wire connectors? The cable is armoured cable.

SOLUTION

Two #14 AWG wires × 24.6 cm³/wire = 49.2 cm³
Two #12 AWG wires × 28.7 cm³/wire = 57.4 cm³
Three wire connectors
Each pair counts as one wire switch = 28.7 cm³
Each switch counts as two wires = 57.4 cm³
Total volume = 192.7 cm³

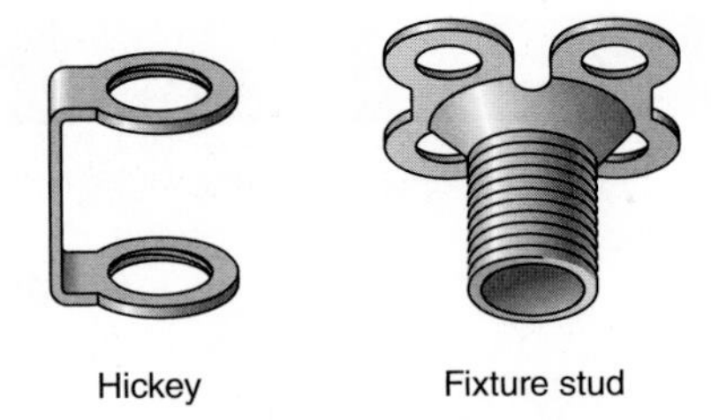

FIGURE 5-22 Hickey and fixture stud.

Therefore, select a box having a minimum volume of 192.7 mL of space. A 3 × 2 × 2½-in. device box has a volume of 204 mL from *Table 23,* and this would be sufficient. The cubic-inch volume may be marked on the box; otherwise refer to the column of *Table 23* entitled "Capacity."

When sectional boxes are ganged together, the volume to be filled is the total volume of the assembled boxes. Fittings, such as plaster rings, raised covers, and extension rings, may be used with the sectional boxes. If these fittings are marked with their volume or have dimensions comparable to the boxes in *Table 23,* their volume may be considered when determining the total volume, *Rule 12–3034(6).*

In some cases, the volume listed in *Table 23* is not equivalent to the length × width × depth of the box. When doing calculations using boxes listed in *Table 23,* use the volume found in the table. Otherwise multiply L × W × D to find the actual volume of the box. For example, a 12 × 12 × 6-in. box is not found in *Table 23.* Therefore, the box volume is 12 × 12 × 6 = 864 in.³ (14.1 L).

TABLE 5-3

Quick checklist for determining the proper size of boxes.

• If box contains NO fittings, devices, fixture studs, hickeys, switches, or receptacles . . .	• . . . refer directly to *CEC Table 23.*
• If box contains ONE or MORE fittings, fixture studs, or hickeys . . .	• . . . deduct ONE from maximum number of conductors permitted in *Table 23* for each type.
• For each device on a single strap containing ONE or MORE devices such as switches, receptacles, or pilot lights . . .	• . . . deduct ANOTHER TWO from *Table 23.*
• For ONE or MORE isolated (insulated) grounding conductors . . .	• . . . deduct ANOTHER ONE from *Table 23.*
• For each PAIR of insulated wire connectors . . .	• . . . deduct another conductor from the number permitted by *Table 23.*
• For conductors running through the box . . .	• . . . count ONE conductor for each conductor running through the box.
• For conductors that originate outside the box and terminate inside the box . . .	• . . . count ONE conductor for each conductor originating outside and terminating inside the box.
• If no part of the conductor leaves the box—for example, a "jumper" wire used to connect a receptacle . . .	• . . . don't count the conductors.

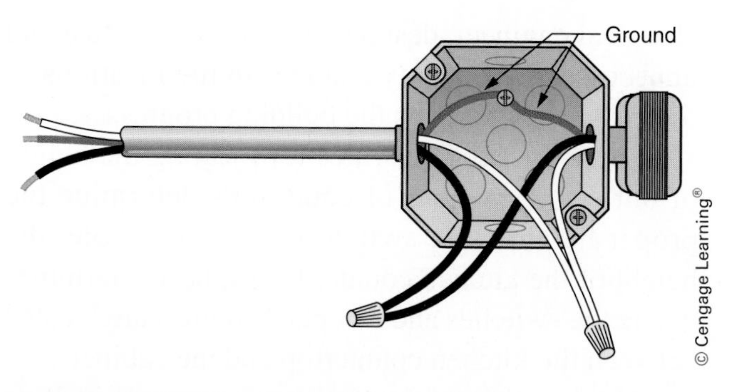

FIGURE 5-23 The transformer leads are not counted, since no part of the conductors from the transformer leaves the box.

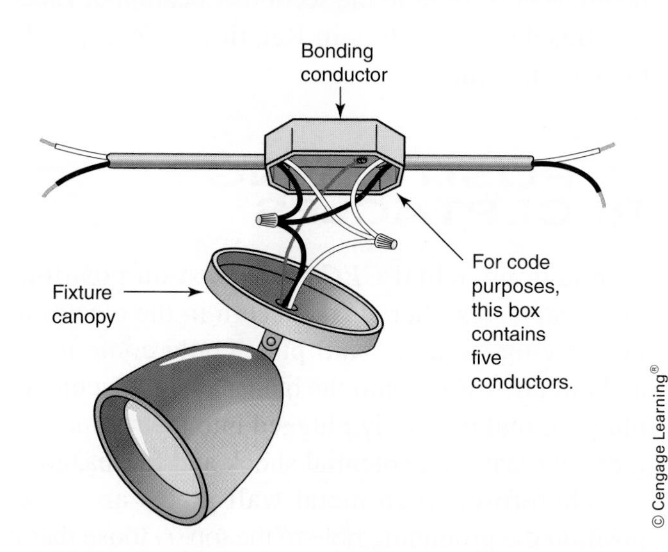

FIGURE 5-24 Fixture wires are not counted in determining correct box size. *CEC Rule 12–3034(1)(d)* specifies that #18 and #16 AWG fixture wires supplying a lighting fixture, mounted on the box containing the fixture wires, shall not be counted.

Figures 5-25 and 5-26 show how a 19-mm raised cover (plaster ring) increases the wiring space of the box to which it is attached.

Electrical inspectors are very aware that GFCI receptacles, dimmers, and certain types of timers take more space than regular receptacles. Therefore, it is a good practice to install device boxes that will provide lots of room for wires, instead of pushing and crowding the wires into the box.

EXAMPLE

How many #12 conductors are permitted in the box and raised plaster ring shown in Figure 5-26?

SOLUTION See *Rule 12–3034* and *Tables 22* and *23* for volume required per conductor.

This box and cover will take

$$\frac{418 \text{ mL}}{28.7 \text{ mL per \#12 conductor}} = 14 \text{ \#12 conductors}$$

maximum, less the deductions for devices and connectors, per *Rule 12–3034(2)*

Figure 5-23 shows how transformer leads #18 AWG or larger are treated when selecting a box.

Wire #16 or #18 AWG supplying a lighting fixture is not counted when sizing the box that the fixture is mounted on; see Figure 5-24. This is stated in *Rule 12–3034(1)(d)*.

Figure 5-16, the quick-check box selection guide, shows some of the most popular types of

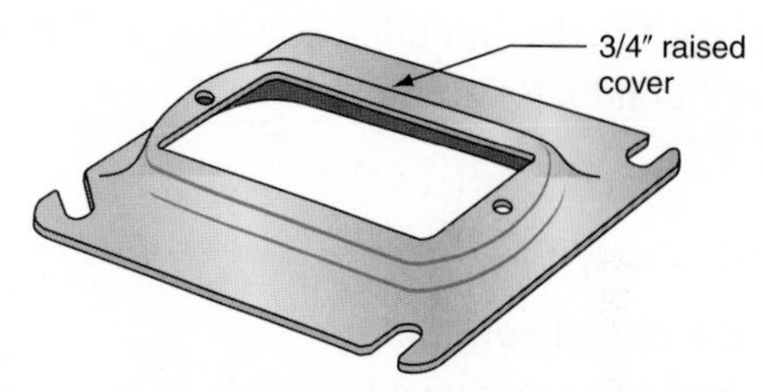

FIGURE 5-25 Raised cover. Raised covers are sometimes called *plaster rings.*

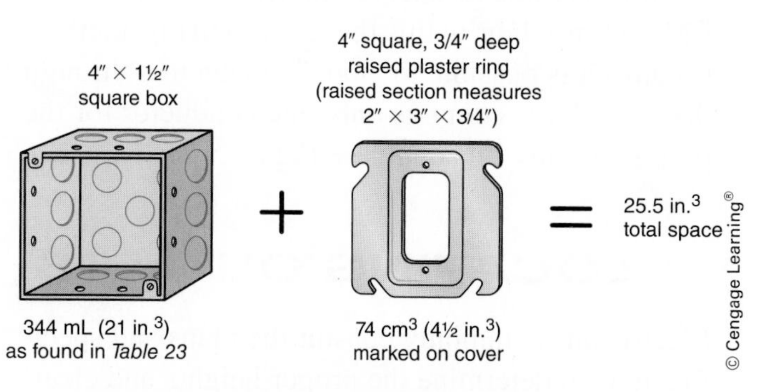

FIGURE 5-26 Area of box and plaster ring.

boxes used in residential wiring. You can refer to it as you work your way through this text.

When wiring with cable, "feeding through" a box is impossible. But, when installing the wiring using conduit, it is possible to feed straight through a box. Only one wire is deducted for a wire running straight through a box.

BOX FILL

The following method can make box fill easier to calculate when determining the proper size of junction box or wall box to install:

1. Count the number of circuit wires.
2. Add one wire for a fixture stud (if any).
3. Add one wire for each pair of wire connectors.
4. Add one wire for one or more isolated (insulated) grounding conductors (if any).
5. Add two wires for each wiring device.

Look up the total count in *Table 23* to find a box appropriate for the intended use that will hold the number of conductors required.

EXAMPLE

Six circuit conductors	6
One wiring device (switch)	+2
Three wire connectors	+1
Total	9

Therefore, select a box capable of containing nine or more conductors. See *Table 23*.

It is possible to select a smaller box when using EMT than when using the cable wiring method because it is possible to "loop" conductors through the box. These count as only one conductor for the purpose of box fill count; see Figure 5-27.

LOCATING OUTLETS

Electricians commonly consult the plans and specifications to determine the proper heights and clearances for installing electrical devices. The electrician then has these dimensions verified by the architect, electrical engineer, designer, or homeowner to avoid unnecessary and costly changes in the locations of outlets and switches as the building progresses.

There are no *Code* rules for locating the height of outlets. A number of conditions determine the proper height for a switch box. For example, the height of the kitchen counter backsplash determines where the switches and receptacle outlets are located between the kitchen countertop and the cabinets.

The location of receptacles is covered in *Rule 26–712* of the *CEC* and Unit 6 of this text.

When locating receptacles near electric baseboard heaters, refer to the section "Location of Electric Baseboard Heaters in Relation to Receptacle Outlets" in Unit 17.

POSITIONING RECEPTACLES

Although no actual *CEC* rules exist on positioning receptacles, there is a concern in the electrical industry that a *metal* wall plate could come loose and fall downward onto the blades of an attachment plug cap that is loosely plugged into the receptacle, thereby creating a potential shock and fire hazard.

Therefore, when metal wall plates are used, position the grounding hole to the top. A loose metal plate could fall onto the grounding blade of the attachment plug cap, but no hazard would result.

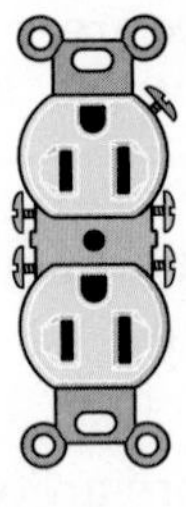

Position the grounding hole to the top.

Or when metal cover plates are used, position grounded neutral blades on top. A loose metal plate could fall onto these grounded neutral blades with no hazardous effect.

Position grounded neutral blades on top.

Metal conduit

A Conduit: Since two conductors are looped through the box, the conductor count in this illustration is four

Cable (each with two circuit conductors plus one equipment-bonding conductor)

B Cable: The conductor count in this illustration is four. The two equipment-bonding conductors are not counted.

Cable and internal clamps (each cable has two circuit conductors plus one equipment-bonding conductor)

C Cable and a box with internal cable clamps: The conductor count in this illustration is four. The two equipment-bonding conductors are not counted; The two cable clamps are not counted.

FIGURE 5-27 Conductor count for both metal conduit and cable installations.

When metal cover plates are used, do not position the "hot" terminal on top. A loose metal plate could fall onto these live blades. If this was a split-circuit receptacle fed by a three-wire, 120/240-volt circuit, the short would be across the 240-volt line.

Do not position the "hot" terminal on top.

To ensure uniform installation and safety, and in accordance with long-established custom, standard electrical outlets are located, as shown in Figure 5-28. These dimensions usually are satisfactory. However, the electrician must check the blueprints, specifications, and details for measurements that may affect the location of a particular outlet or switch.

The cabinet spacing, available space between the countertop and cabinet, and the tile height may influence the location of the outlet or switch. For example, if the top of the wall tile is exactly 1.22 m (48 in.) from the finished floor line, a wall switch should not be mounted 1.22 m (48 in.) to centre. This

SWITCHES	
	*Height above floor
Regular	1.17 m (46 in.)
Between counter and kitchen cabinets (depends on backsplash)	1.12–1.17 m (44–46 in.)
RECEPTACLE OUTLETS	
	*Height above floor
Regular (not permitted above electric baseboard heaters)	300 mm (12 in.)
Between counter and kitchen cabinets (depends on backsplash)	1.12–1.17 m (44–46 in.)
In garage	1.22 m (48 in.)
WALL BRACKETS	
	*Height above floor
Outside	1.8 m (72 in.)
Inside	1.52 m (60 in.)
Side of medicine cabinet	1.52 m (60 in.)

*Note: All dimensions given are from the finished floor to the centre of the outlet box. Verify all dimensions before roughing in.

FIGURE 5-28 Outlet locations.

is considered poor workmanship since it prevents the wall plate from sitting flat against the wall. The switch should be located entirely within the tile area or entirely out of the tile area; see Figure 5-29. This situation requires the full cooperation of all trades involved on the construction job and a careful review of the architectural drawings before roughing in the outlets.

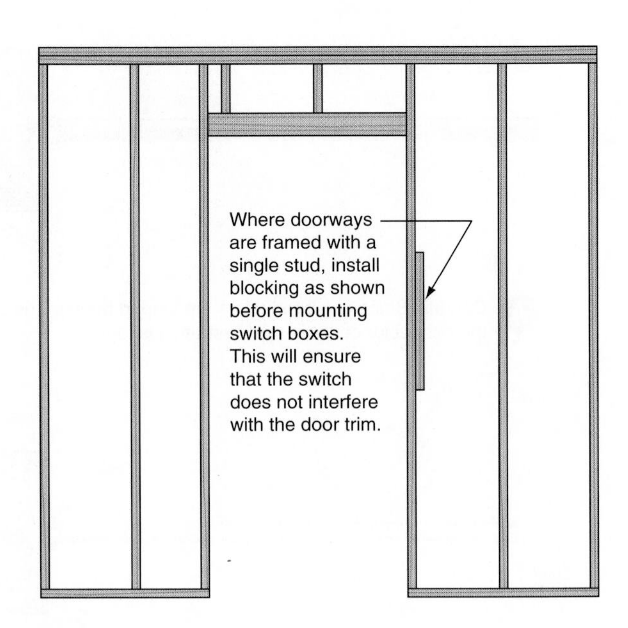

FIGURE 5-30 Locating switches next to doorways.

See Figure 5-30 for how to locate switches next to a doorway.

Faceplates

Faceplates for switches and receptacles shall be installed to completely cover the wall opening and seat against the mounting surface. The mounting surface may be the wall or the gasket of a weatherproof box.

Faceplates should be level. In addition, a minor detail that can improve the appearance of the

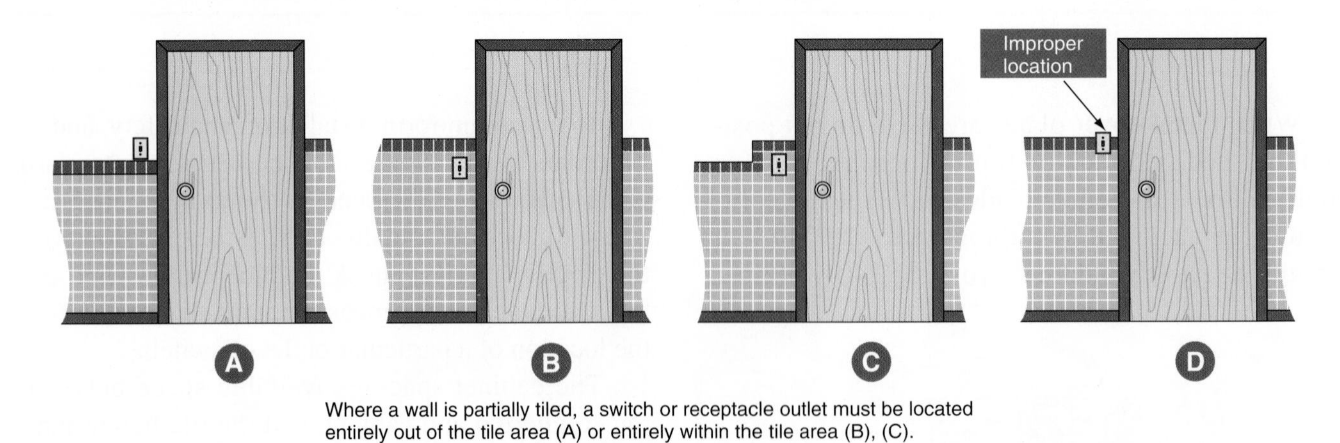

Where a wall is partially tiled, a switch or receptacle outlet must be located entirely out of the tile area (A) or entirely within the tile area (B), (C).

The faceplate in (D) does not "hug" the wall properly. This installation is considered unacceptable by most electricians (*Rule 12–3002*).

FIGURE 5-29 Locating an outlet on a tiled wall.

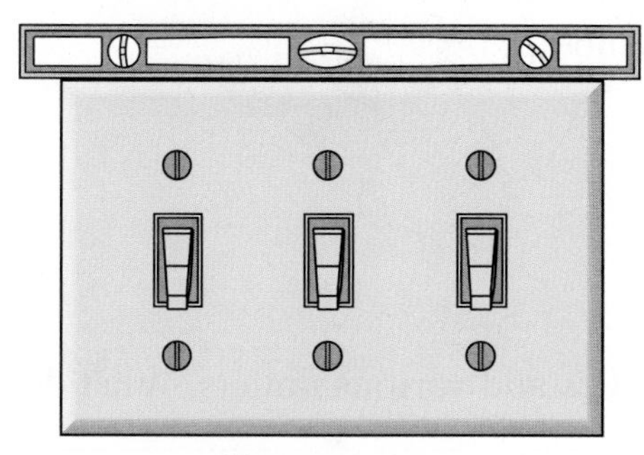

FIGURE 5-31 Aligning the slots of the faceplate mounting screws in the same direction makes the installation look neater.

installation is to align the slots in all the faceplate mounting screws in the same direction; see Figure 5-31.

The location of lighting outlets is determined by the amount and type of illumination required to provide the desired lighting effects. (It is not the intent of this text to describe how proper and adequate lighting is determined; rather, it covers the proper methods of installing the circuits for such lighting.)

BONDING

Bonding is the joining together of all the non-current-carrying metal parts of an electrical system with a conductor that has a low impedance (opposition to the flow of current). Bonding provides protection from shock hazards by maintaining all of the non-current metal parts of the electrical system at the same potential (voltage). It prevents a voltage from developing between (the enclosures of) two pieces of electrical equipment due to static buildup or the failure of insulation on conductors.

Bonding is the most important part of the installation of electrical outlets. It will permit the system to "fail safe." Your electrical inspector will look closely at the workmanship involved in the installation of the bonding conductors.

It is important that the continuity of the bonding conductor be maintained. Therefore, the bonding conductor is brought into the box and attached to the bonding screw in the box before it is attached to the green bonding screw on a receptacle. With such connections, if someone cut the connections to the receptacle to remove it, it would not interfere with the bonding connection to the box.

When connecting the bonding conductor to the bonding screws in outlet boxes, ensure that the conductor does not get pinched between the head of the screw and the shoulder of the box (which prevents the conductor from slipping out from under the head of the screw as the screw is tightened down). This can damage the bonding conductor, which reduces its size (cross-sectional area) and its effectiveness.

REVIEW

Note: Refer to the *CEC* or the blueprints provided with this textbook when necessary. Where applicable, responses should be written in complete sentences.

1. What does a plan show about electrical outlets? ______________________________

____

____

2. What is an outlet? ____

____

____

____

3. Match the following switch types with the proper symbol.
   a. single-pole S_p
   b. three-way S_4
   c. four-way S
   d. single-pole with pilot light S_3
4. The plans show curved lines running between switches and various outlets. What do these curved lines indicate? ______
5. Why are the lines mentioned in Question 4 usually curved? ______
6. a. What are junction boxes used for? ______
   b. Are junction boxes normally used in wiring the first floor? ______
   c. Are junction boxes normally used to wire exposed portions of the basement? ______
7. How are standard sectional switch (device) boxes mounted? ______
8. a. What is an offset bar hanger? ______
   b. What types of boxes may be used with offset bar hangers? ______
9. What methods may be used to mount lighting fixtures to an outlet box fastened to an offset bar hanger? ______
10. What is the size of the opening of a switch (device) box for a single device? ______

11. The space between a door casing and a window casing is 89 mm. Two switches are to be installed at this location. What problems could you encounter when placing the switches in this location? What would you recommend as a possible solution?

____

____

____

____

____

____

12. Three switches are mounted in a three-gang switch (device) box. The wall plate for this assembly is called a ______________________ plate.

13. [Circle the correct answer.] For each fixture stud inside a box, (increase) (decrease) the number of conductors allowed by one.

14. a. How high above the finished floor are the switches located in the garage of this dwelling? ______________________

    b. In the living room of this dwelling? ______________________

15. How high above the finished floor are the receptacle outlets in the garage located?

    ______________________ In the living room? ______________________

16. Outdoor receptacle outlets in this dwelling are located ______________ mm above grade.

17. In the spaces provided, draw the correct symbol for each of these items:

a. ____ Lighting panel	j. ____ Special-purpose outlet
b. ____ Clock outlet	k. ____ Fan outlet
c. ____ Duplex outlet	l. ____ Range outlet
d. ____ Outside-line telephone	m. ____ Power panel
e. ____ Single-pole switch	n. ____ Three-way switch
f. ____ Four-way switch	o. ____ Pushbutton
g. ____ Duplex outlet, split-circuit	p. ____ Thermostat
h. ____ Lampholder with pull switch	q. ____ Electric door opener
i. ____ Weatherproof outlet	r. ____ Multioutlet assembly

18. The front edge of a box installed in a combustible wall must be ______________ with the finished surface. Rule number? ______________________

19. List the maximum number of #12 AWG conductors permitted in a

    a. 4 × 1½-in. (102 × 38-mm) octagon box. ______________________

    b. $4^{11}/_{16}$ × 1½-in. (119 × 38-mm) square box. ______________________

    c. 3 × 2 × 2½-in. (76 × 51 × 64-mm) device box. ______________________

20. Hanging a ceiling fixture directly from a plastic outlet box is permitted only if ______
______
______
______
______

21. It is necessary to count fixture wires when counting the permitted number of conductors in a box according to *Rule 12–3034(1)(d).*
[Underline or circle the correct answer.] (True) (False)

22. *CEC Table 23* allows a maximum of ten wires in a certain box. However, the box will have two wire connectors and one fixture stud in it. What is the maximum number of wires allowed in this box? ______
______

23. When laying out a job, the electrician will usually make a layout of the circuit, taking into consideration the best way to run the cables and/or conduits and how to make up the electrical connections. Doing this ahead of time, the electrician determines exactly how many conductors will be fed into each box. With experience, the electrician will probably select two or three sizes and types of boxes that will provide adequate space to meet *Code* requirements. *CEC Table 23* shows the maximum number of conductors permitted in a given size box. However, the number of conductors shown in the table must be reduced
by ______ conductor(s) for two wire connectors
by ______ conductor(s) for the fixture stud
by ______ conductor(s) for each wiring device mounted on a single strap
by ______ conductor(s) for one or more bare copper bonding conductors

24. Define the term *bonding*. Why is the installation of bonding conductors important?
______
______
______
______
______

UNIT 6

Determining the Number and Location of Lighting and Receptacle Branch Circuits

OBJECTIVES

After studying this unit, you should be able to

- determine the number and location of lighting outlets in a dwelling unit
- determine the minimum number of and location of receptacles required in a dwelling unit
- identify *CEC* requirements for lighting and receptacle branch circuits

In the design and planning of dwelling units, it is standard practice to permit the electrician to plan the circuits. Thus, the residence plans do not include layouts for the various branch circuits. The electrician may follow the general guidelines established by the architect; however, all wiring systems must conform to the standards of the *Canadian Electrical Code (CEC)* and the *National Building Code of Canada (NBC)*, as well as local and provincial requirements.

This unit focuses on lighting branch circuits and small-appliance circuits. The circuits supplying the electric range, oven, clothes dryer, and other specific circuitry not considered to be a lighting branch circuit or a small-appliance circuit are covered later in this text.

LIGHTING

In this unit, we look at where lighting fixtures and receptacles must be located to meet the minimum requirements of the *CEC*. Lighting fixtures (luminaires), lamps, and ballasts will be covered in Unit 10.

The provisions for supplying a light fixture in a room or area and whether it is to be controlled by a wall switch are found in *Part 9* of the *National Building Code of Canada*. The *CEC* has adopted these requirements, which are found in *Rules 30–500* through *30–510*. Table 6-1 outlines these requirements for lighting in dwelling units. Figure 6-1 shows them graphically.

TABLE 6-1

CEC requirements for lighting in dwelling units.

ROOM OR AREA	PROVIDE	*CEC* REFERENCE RULE
Entrance	A luminaire controlled by a wall switch inside the building	*30–500*
Living room, bedroom	At least one luminaire controlled by a wall switch or a receptacle controlled by a wall switch	*30–502*
Kitchen, dining room, hall, bathroom, vestibule, utility room	At least one luminaire controlled by a wall switch	*30–502*
Stairway	All stairways must be lighted A luminaire controlled by a switch at the top and bottom of the stairway when the stairway has four or more risers If the stairway leads to an unfinished basement that does not have an outside exit, a single switch at the top of the stairs may be used	*30–504*
Basement	One luminaire for each 30 m^2 or portion thereof of unfinished basement The luminaire closest to the stairs must be controlled by a switch at the top of the stairs	*30–506*
Storage room	At least one luminaire in the storage room	*30–508*
Garage and carport	At least one luminaire controlled by a wall switch near the doorway When a wall-mounted or ceiling luminaire that is not located above a space that would be occupied by a car is used, a built-in switch is permitted on the luminaire A carport that is lighted by a luminaire at the entrance to the building does not require additional lighting	*30–510*

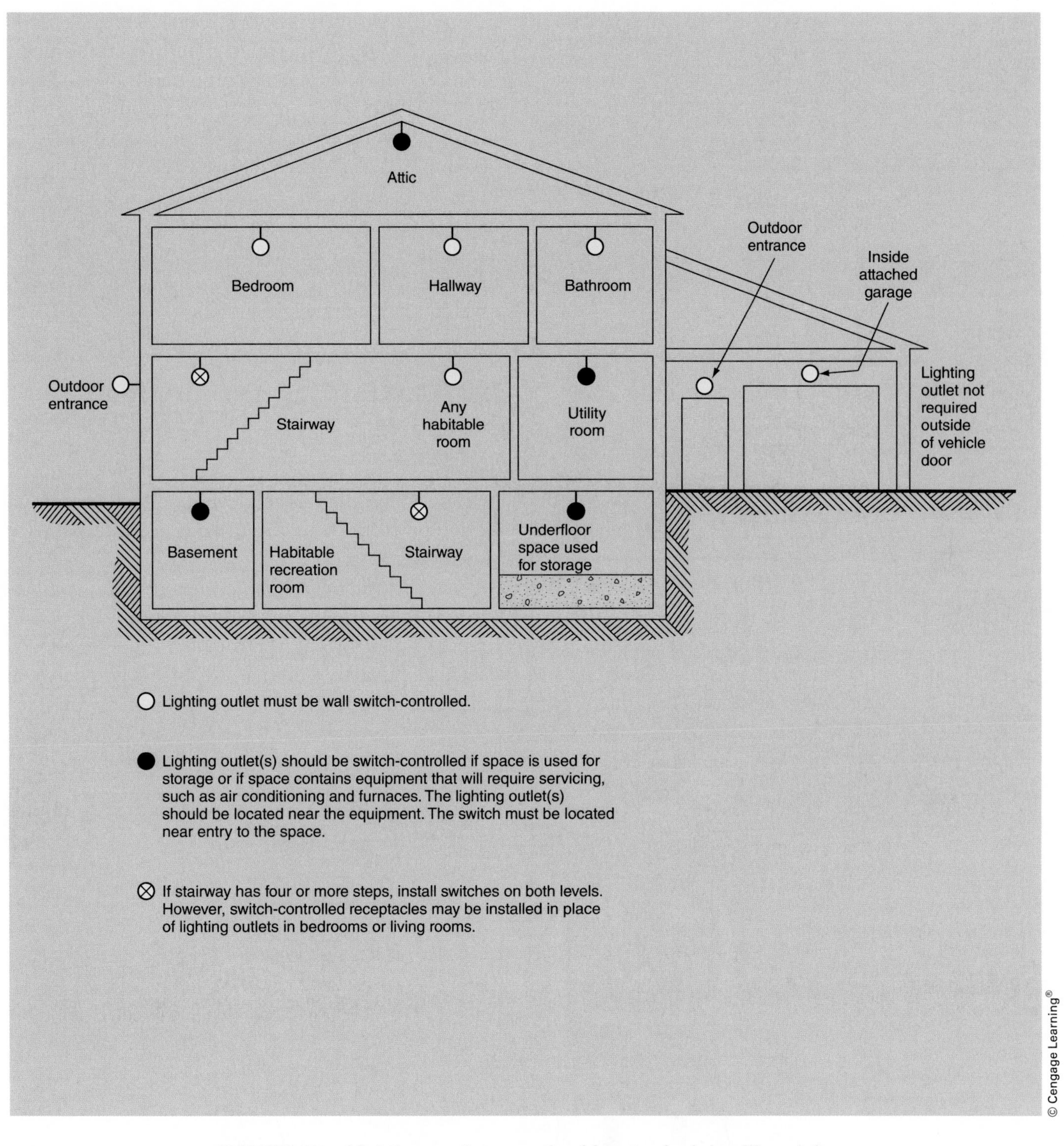

FIGURE 6-1 Lighting outlets required in a typical dwelling unit.

RECEPTACLES

Receptacles are required in dwelling units for the connection of large and small appliances. Receptacles installed for small appliances are 15 ampere, 125 volt (two pole, three wire) or 20 ampere, T slot, 125 volt (two pole, three wire). See Figure 6-2.

Some residences are lighted by table and floor lamps plugged into switch-controlled wall receptacle outlets. In bedrooms and living rooms,

This duplex receptacle meets the requirements of CSA configuration 5-20RA Diagram 1.

This duplex receptacle meets the requirements of CSA configuration 5-15R Diagram 1.

FIGURE 6-2 These are 15- and 20-ampere, 125-volt (two-pole, three-wire) receptacle configurations.

switch-controlled wall receptacles are permitted in lieu of ceiling and/or wall lighting outlets. Only half of the receptacle may be switched if the receptacle is needed to satisfy the minimum number of receptacles required by *Rule 26–712(a)*. This is achieved by removing the tab on the "hot" side of the receptacle (Figures 6-3 through 6-5).

This unit will identify the locations where the *CEC* requires 15-ampere, 125-volt (5-15R) and 20-ampere, 125-volt (5-20RA) receptacles. For information on special applications of receptacles or their branch circuits, see Units 11 through 20.

Table 6-2 and Figures 6-6 through 6-8 outline the minimum number of receptacles required in a dwelling unit and their locations.

BASICS OF WIRE SIZING AND LOADING

The *CEC* defines a *branch circuit* as "that portion of the wiring installation between the final overcurrent device protecting the circuit and the outlet(s)"; see Figure 6-9.

The *CEC* defines a *feeder* as the circuit conductors between the service equipment and the final branch-circuit overcurrent device. In the residence used as an example in this text, the wiring between Panel A and Panel B is a feeder.

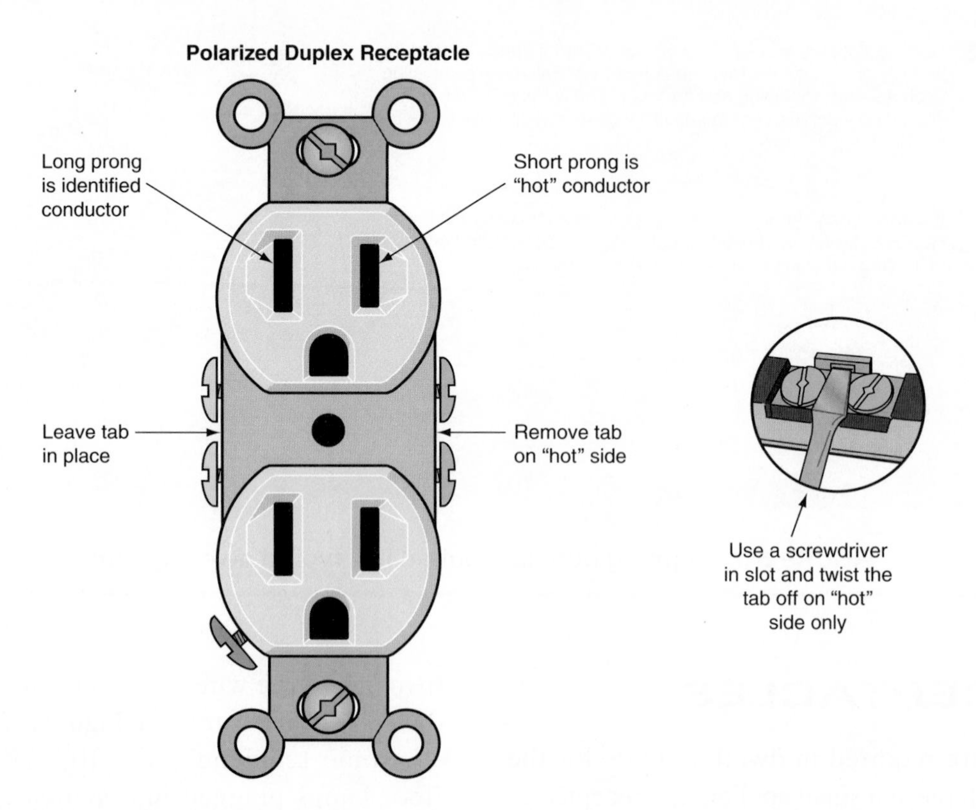

FIGURE 6-3 Duplex receptacle showing method for removing tab on the hot side.

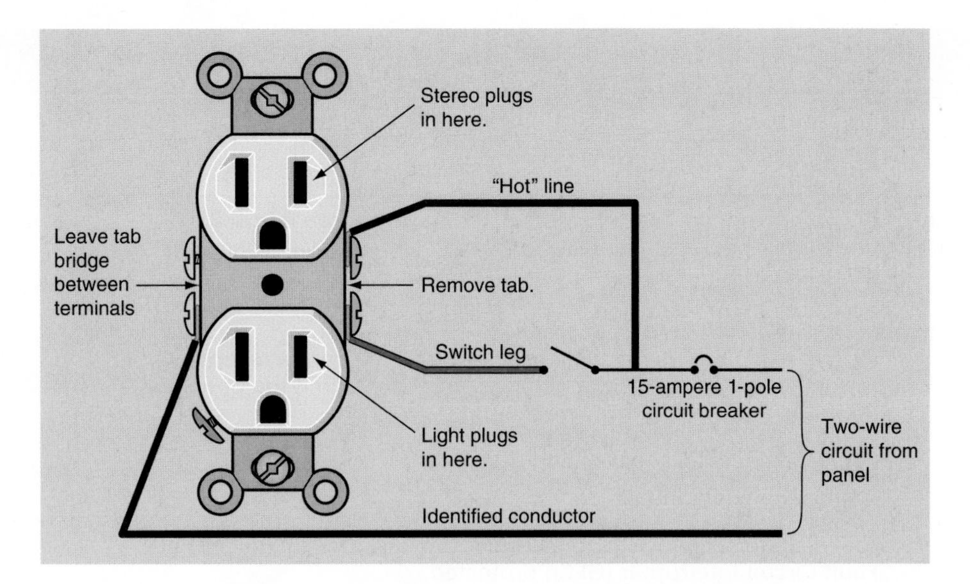

FIGURE 6-4 Split-switched receptacle wiring diagram.

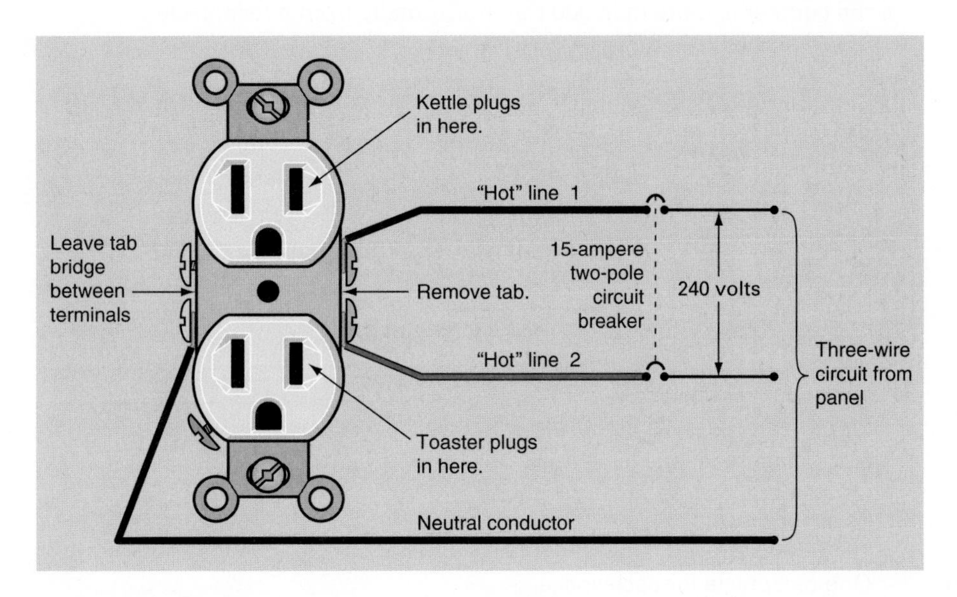

FIGURE 6-5 Split-wired receptacle (split receptacle) on a two-pole 15-ampere breaker connected to a three-wire circuit for use in a kitchen.

The ampacity (current-carrying capacity) of a conductor must not be less than the rating of the overcurrent device protecting that conductor, *Rule 14–104*. An exception is a motor branch circuit, where it is common to have overcurrent devices (fuses or breakers) sized larger than the ampacity of the conductor. The *CEC* covers motors and motor circuits specifically in *Section 28*. The rating of the branch circuit is determined by the lesser of the rating of either the overcurrent device or the ampacity of the wire. For example, if a 20-ampere conductor is protected by a 15-ampere fuse, then the circuit is a 15-ampere branch circuit, *Rule 8–104(1)*.

The ampacity of branch-circuit conductors must not be less than the maximum load to be served. Where the circuit supplies receptacle outlets, as is typical in houses, the conductor's ampacity must be rated to carry the load, *Rule 8–104(2)*. The load current

TABLE 6-2

CEC requirements for receptacles in dwelling units.

ROOM OR AREA	PROVIDE	*CEC* REFERENCE RULE
Living room, dining room, family room, bedroom	No point along the floor line of a wall may be more than 1.8 m horizontally from a receptacle, including isolated wall spaces 900 mm or longer. See the rule for exclusions.	*26–712(a)(c)*
Hallway	No point in a hallway may be more than 4.5 m from a receptacle. Receptacles in rooms off the hallway may be used if the opening to the room from the hallway is not fitted with a door.	*26–712(f)*
Laundry room	One receptacle for the washing machine. At least one additional receptacle.	*26–710(e)*
Bathroom	At least one receptacle, within 1 m of the washbasin and Ground-Fault Circuit Interrupter (GFCI) protected.	*26–710(f)*
Kitchen	One receptacle for a refrigerator. One receptacle for a gas range if gas piping has been provided. Counter receptacles such that no point measured along the back of the counter is more than 900 mm horizontally from a receptacle. Counter receptacles may be 15-A, three-conductor split circuit, or 20-AT slot. See Unit 14 for more detail. Isolated work surfaces 300 mm or more require a receptacle. At least one receptacle for a permanently installed island. At least one receptacle for counter peninsula. At least one receptacle in a dining area forming part of a kitchen. One receptacle for a microwave oven.	*26–712(d)*
Utility room	At least one receptacle.	*26–712(a)*
Unfinished basement areas	At least one receptacle.	*26–710(e)*
Balcony, porch, veranda	At least one receptacle.	*26–712(e)*
Outdoors	At least one receptacle for the use of outdoor appliances.	*26–714(a)*
Garage, carport	One receptacle for each space.	*26–714(b)*
Miscellaneous	One receptacle for a cord-connected central vacuum if ducting system is installed.	*26–710(m)*

Courtesy of CSA Group

must not exceed 80% of the rating of the overcurrent device for that circuit, *Rules 8–304* and *8–104(6)(a).*

Where the ampacity of the conductor does not match up with a standard rating of a fuse or breaker, the next higher standard size of overcurrent device may be used, provided the overcurrent device does not exceed 600 amperes, *Rule 14–104(a)*. This exception is not applicable to residential branch circuits because the standard overcurrent device sizes listed in *CEC Table 13* (Table 6-3 in this textbook) correspond directly with the ampacities listed in *Table 2* for #14 and #12 gauge wire. *Table 13* shows that the maximum overcurrent protection is 15 amperes when #14 AWG copper wire is used.

Residential branch circuits are fed from panelboards having overcurrent devices that are rated to be run continuously at a maximum of 80% of their ampere rating. The conductors attached to these overcurrent devices may not be loaded to more than 80% of the ampere rating listed in *Table 2*. For example,

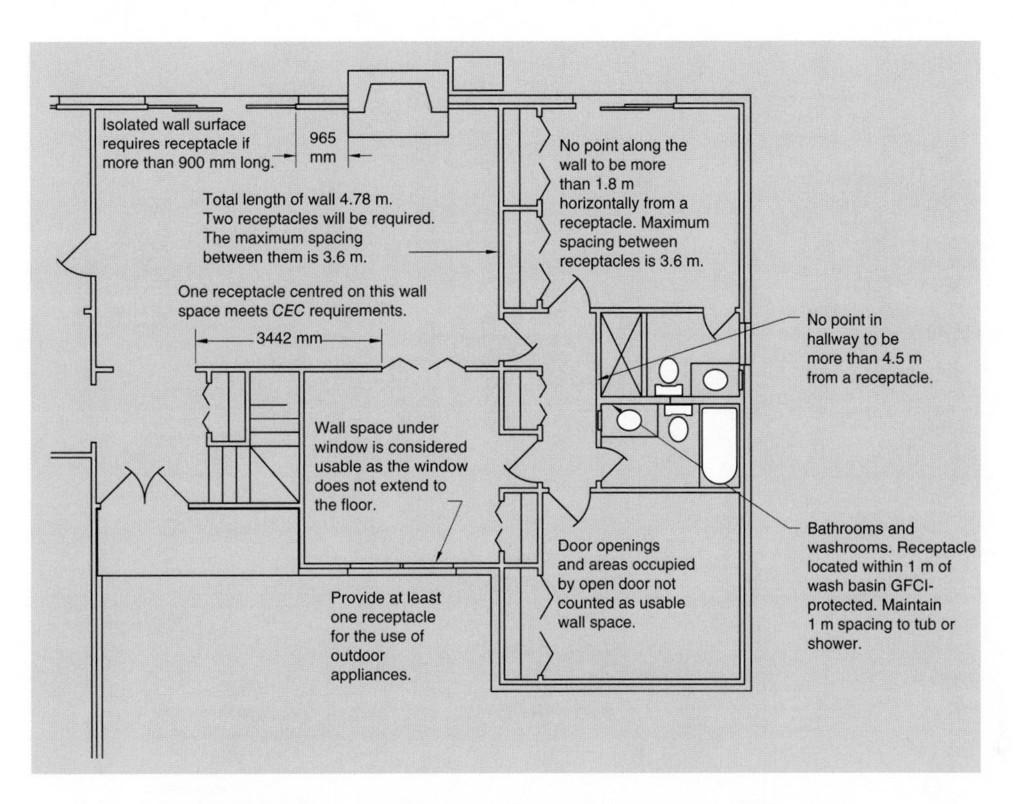

FIGURE 6-6 Locating receptacles on the first floor.

a 15-ampere circuit breaker has a maximum load of 12 amperes (15 amperes × 0.8). Therefore, the wire must be rated for 15 amperes in *Table 2;* a #14 R90 XLPE is suitable.

The ampacities of conductors commonly used in residential occupancies are listed in *Table 2.* These are subject to *correction factors,* which must be applied where there are high ambient temperatures, for example, in attics. See *Note 1* to *Table 5A.*

Conductor ampacities are also *de-rated* when four or more conductors are installed in a single raceway or cable. See *CEC Table 5C* (Table 6-4 in this textbook).

A continuous load should not exceed 80% of the ampere rating of the branch circuit. The *CEC* defines a *continuous load* as one that persists under normal operation for "more than one hour in any two-hour period if the load does not exceed 225 amperes," *Rule 8–104 (3)(a).* The *CEC* permits 100% loading on an overcurrent device *only* if that overcurrent device is *listed* for 100% loading. The CSA has *no* listing of circuit breakers of the moulded-case type used in residential installations that are suitable for 100% loading.

This calls for the judgment of the electrician and/or the electrical inspector. Will a branch circuit supplying a heating cable in a driveway likely be on for one hour or more? The answer is "probably." Good design practice and experience tell us to never load a circuit conductor or overcurrent protective device to more than 80% of its rating.

The branch-circuit rating shall *not* be less than any noncontinuous load plus 125% of the continuous load, *Rule 8–104.* For residential applications, all loads are considered to be continuous except for the load on the service or feeder conductors, *Rule 8–200(3).*

The *CEC* considers most receptacle outlets in a residence part of the basic load, and additional load calculations are *not* necessary. Appliance outlets with loads greater than 12 amperes are not considered part of the basic load; these larger loads are calculated individually. The power requirements for these receptacle outlets will be discussed in detail later in this textbook.

To determine the number of lighting circuits required, divide the total watts by the voltage rating of the circuit(s) to be used.

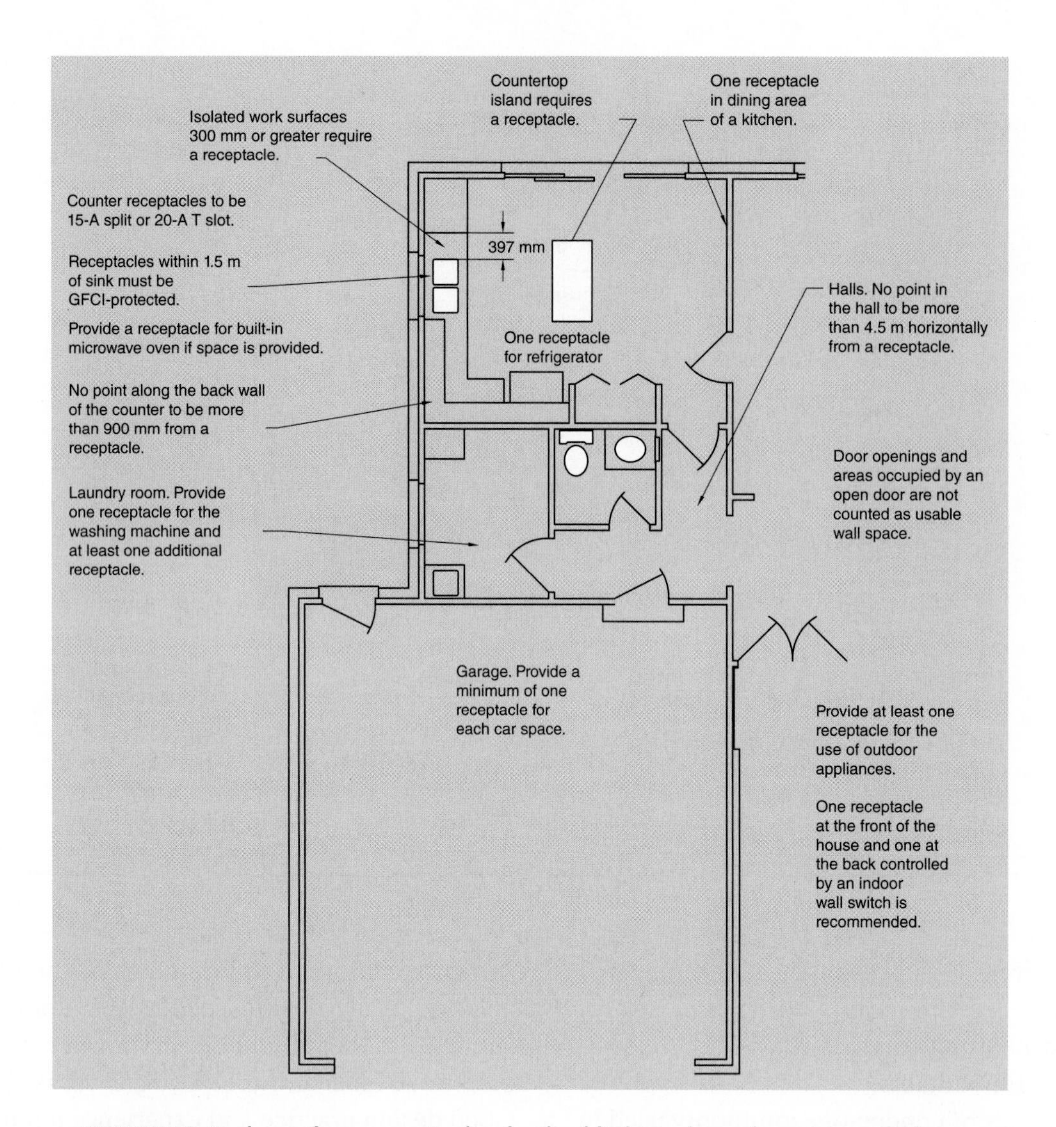

FIGURE 6-7 Locating receptacles in the kitchen, laundry room, and garage.

In this residence, all lighting circuits are rated at 15 amperes. The maximum loading of these circuits is 12 amperes (15 amperes × 0.8). Therefore, the minimum number of circuits for general lighting and power is

$$\frac{\text{Watts}}{120 \text{ volts}} = \text{Amperes}$$

$$\frac{\text{Amperes}}{12} = \text{Minimum number of circuits required}$$

Note that the *CEC* provides the minimum number of branch circuits required. For various purposes, you may install more.

Load calculations for electric water heaters, ranges, motors, electric heat, and other specific loads are in addition to the general lighting load, and are discussed throughout this text as they appear.

VOLTAGE

All calculations throughout this text use voltages of 120 volts and 240 volts. *CEC Rule 8–100,* entitled "Current Calculations," states that the voltage divisors to use shall be 120, 208, and 240, as applicable. This voltage is used for calculation even though the actual voltage may be 220, 245, or 247. This provides us with a uniformity, to avoid misleading and confusing results in our computations.

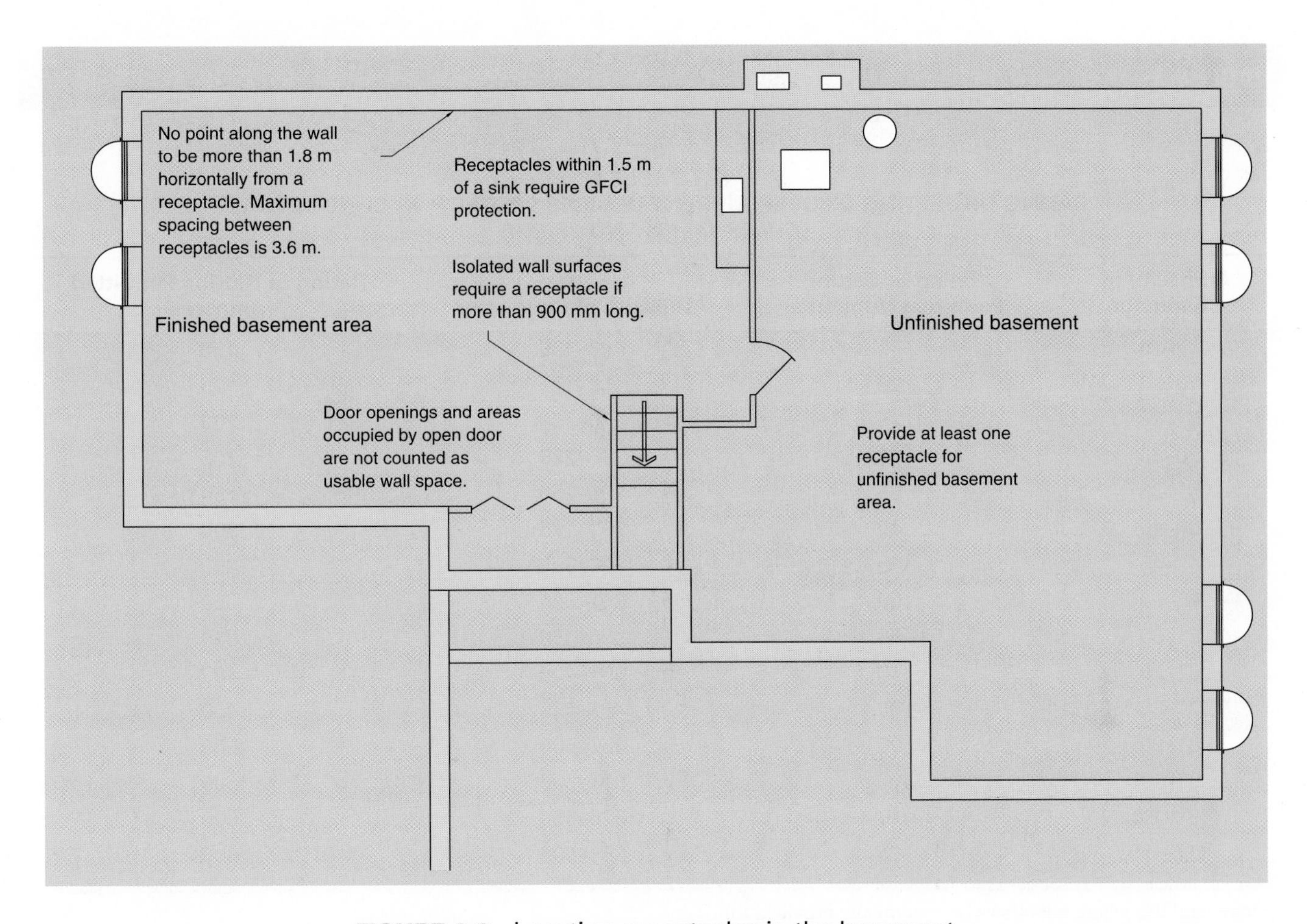

FIGURE 6-8 Locating receptacles in the basement.

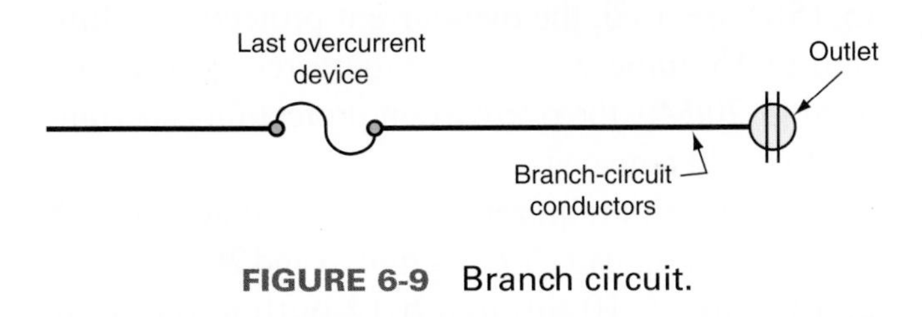

FIGURE 6-9 Branch circuit.

LIGHTING BRANCH CIRCUITS

The *CEC* limits overcurrent protection for lighting branch circuits to a maximum of 15 amperes in dwelling units, *Rule 30–104*. When a branch is loaded it will normally have a maximum of 12 outlets, *Rule 8–304*. If the **actual** load on the circuit is known, the number of lighting outlets may be increased as long as the load on the circuit does not exceed 80% of the rating of the circuit breaker or fuse protecting the branch circuit. It is normally impossible to tell what the **actual** load of each lighting fixture on a branch circuit is when doing the rough-in, unless a large number of recessed luminaires are installed; therefore, the maximum number of outlets on a lighting branch circuit is 12.

When you install track lighting with movable fixtures, the authors recommend that you count every 600 mm as one outlet.

RECEPTACLE BRANCH CIRCUITS

Branch circuits in dwelling units can consist of lighting only, receptacles only, or a combination of lighting and receptacles. When a circuit consists of a combination of lighting and receptacles, the protection for the circuit is limited to 15 amperes, the same as it would be if the circuit had lighting only.

When a circuit contains receptacles only, the protection for the circuit cannot exceed the rating of

TABLE 6-3

CEC Table 13.

CEC TABLE 13
RATING OR SETTING OF OVERCURRENT DEVICES PROTECTING CONDUCTORS*
(*RULES 14–104* AND *28–204*)

Ampacity of Conductor	Rating or Setting Permitted (Amperes)	Ampacity of Conductor	Rating or Setting Permitted (Amperes)
0–15	15	126–150	150
16–20	20	151–175	175
21–25	25	176–200	200
26–30	30	201–225	225
31–35	35	226–250	250
36–40	40	251–275	300
41–45	45	276–300	300
46–50	50	301–325	350
51–60	60	326–350	350
61–70	70	351–400	400
71–80	80	401–450	450
81–90	90	451–500	500
91–100	100	501–525	600
101–110	110	526–550	600
111–125	125	551–600	600

* For general use where not otherwise specifically provided for.

Courtesy of CSA Group

TABLE 6-4

CEC Table 5C.

CEC TABLE 5C
AMPACITY CORRECTION FACTORS FOR *TABLES 2 AND 4*
(*RULES 4–004* AND *12–2210* AND *TABLES 2* AND *4*)

Number of Conductors	Ampacity Correction Factor
1–3	1.00
4–6	0.80
7–24	0.70
25–42	0.60
43 and up	0.50

Courtesy of CSA Group

the receptacle, *Rule 14–600*. Receptacles for general applications in dwelling units (convenience receptacles) may be CSA configuration 5-15R or 5-20RA (T slot) and are found in *CEC Diagram 1* (see Figure 6-2 in this textbook). If 15-ampere receptacles (5-15R) are used, the overcurrent protection is limited to 15 amperes. If 20-ampere receptacles are used (5-20RA), the overcurrent protection is permitted to be 20 amperes.

The general requirements for receptacle branch circuits are found in *CEC Sections 8* and 26. Table 6-5 and Figures 6-10 through 6-12 outline the basic *CEC* requirements for branch circuits supplying receptacles in dwelling units. Units 11 through 20 give more information about receptacles and their branch circuits.

PREPARING A LIGHTING AND RECEPTACLE LAYOUT

Two blank full-size drawings of the floor plan are provided with this textbook to allow you to design and prepare your own electrical layout.

TABLE 6-5

CEC requirements for branch circuits supplying receptacles in dwelling units.

ROOM OR AREA	PROVIDE	*CEC* REFERENCE RULE
Living room, dining room, family room, hallway	Maximum of 12 outlets per circuit.	*8–304*
Bedroom	Maximum of 12 outlets per circuit. Receptacles in bedrooms must be Arc Fault Circuit Interrupter (AFCI) protected. See Unit 9.	*26–722*
Laundry room, utility room	At least one circuit solely for receptacles in laundry rooms. At least one circuit solely for receptacles in utility room or areas.	*26–722*
Bathroom	No special requirements in CEC. Consult with your local electrical inspector for recommendations.	
Kitchen	One circuit for the receptacle for a refrigerator. A receptacle provided for a clock outlet is permitted on this circuit.	*26–722*
	A minimum of two circuits to be provided for counter receptacles, when more than one counter receptacle is provided. No more than two counter receptacles are permitted on a circuit except that receptacles provided for people with disabilities may be added.	*26–724*
	One circuit for the dining area of a kitchen. A receptacle for a gas range may be on the same circuit.	*26–724*
	One circuit when a receptacle has been supplied for a microwave oven.	*26–722*
Balcony, porch, veranda	One circuit if for a single dwelling unit and accessible from grade level.	*26–726*
Outdoors	One circuit for receptacles located outdoors.	*26–726*
Garage, carport	One circuit for receptacles in garages or carports. The garage lighting and door opener may be on the same circuit.	*26–726*
Miscellaneous	One receptacle for a cord-connected central vacuum if ducting system is installed.	*26–722*

Courtesy of CSA Group

Begin by using Table 6-1 to identify where luminaires are required. Locate the luminaires on the floor plan. Add any extra luminaires (such as in closets) that you feel enhance the design. (Just remember that the lowest bid will normally get the job). Determine where you would like to locate the switches controlling the luminaires. Write “S” on the floor plan if you want to control the luminaires in a room from one location. Write “S_3” on the floor plan if you want to control the lights from two locations. If you want to control the luminaires from more than two locations, see Unit 8.

Use Table 6-2 to determine the location of receptacles required by the *CEC* and locate them on the floor plans. Use Figures 6-6 through 6-8 to assist you. Add any extra receptacles that you feel enhance the design.

When you have completed the layout of lighting and receptacle outlets, group the outlets into circuits. In general, 12 outlets are permitted on a circuit. There can be a mixture of lights and receptacles in most cases. Use Table 6-3 and Figures 6-10 through 6-12 to help you.

Further information about additional loads that will be added to the layout are found in later units.

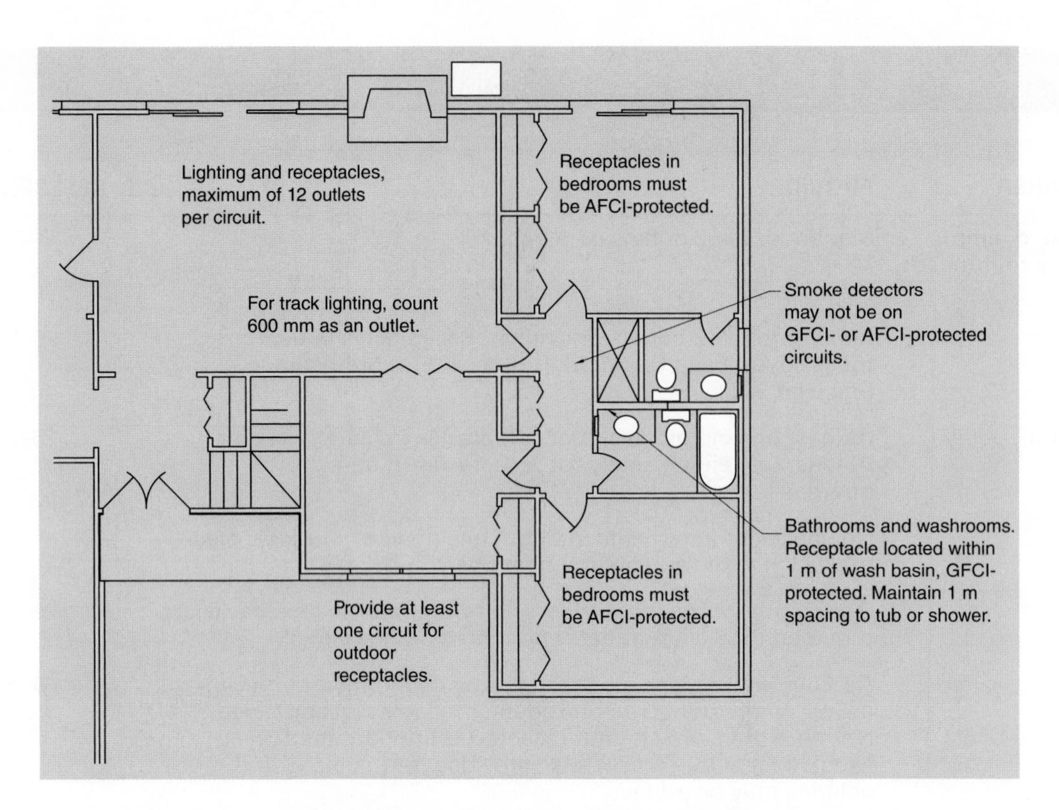

FIGURE 6-10 Circuiting for receptacles.

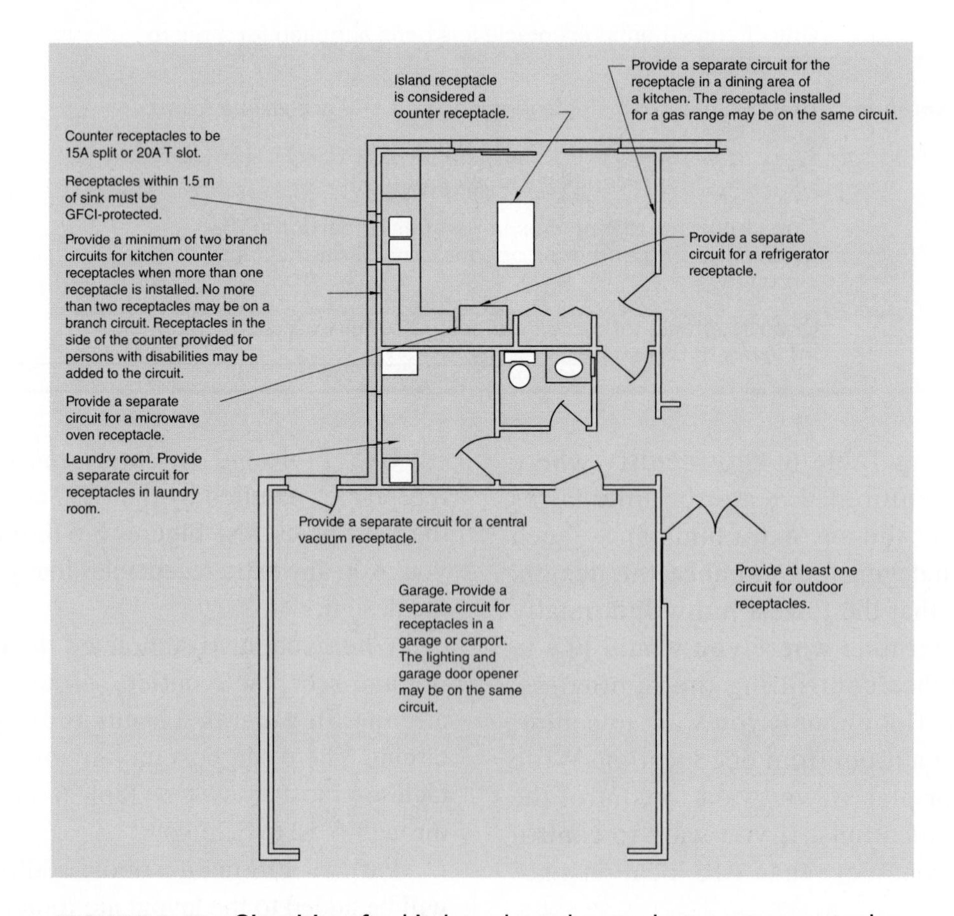

FIGURE 6-11 Circuiting for kitchen, laundry, and garage receptacles.

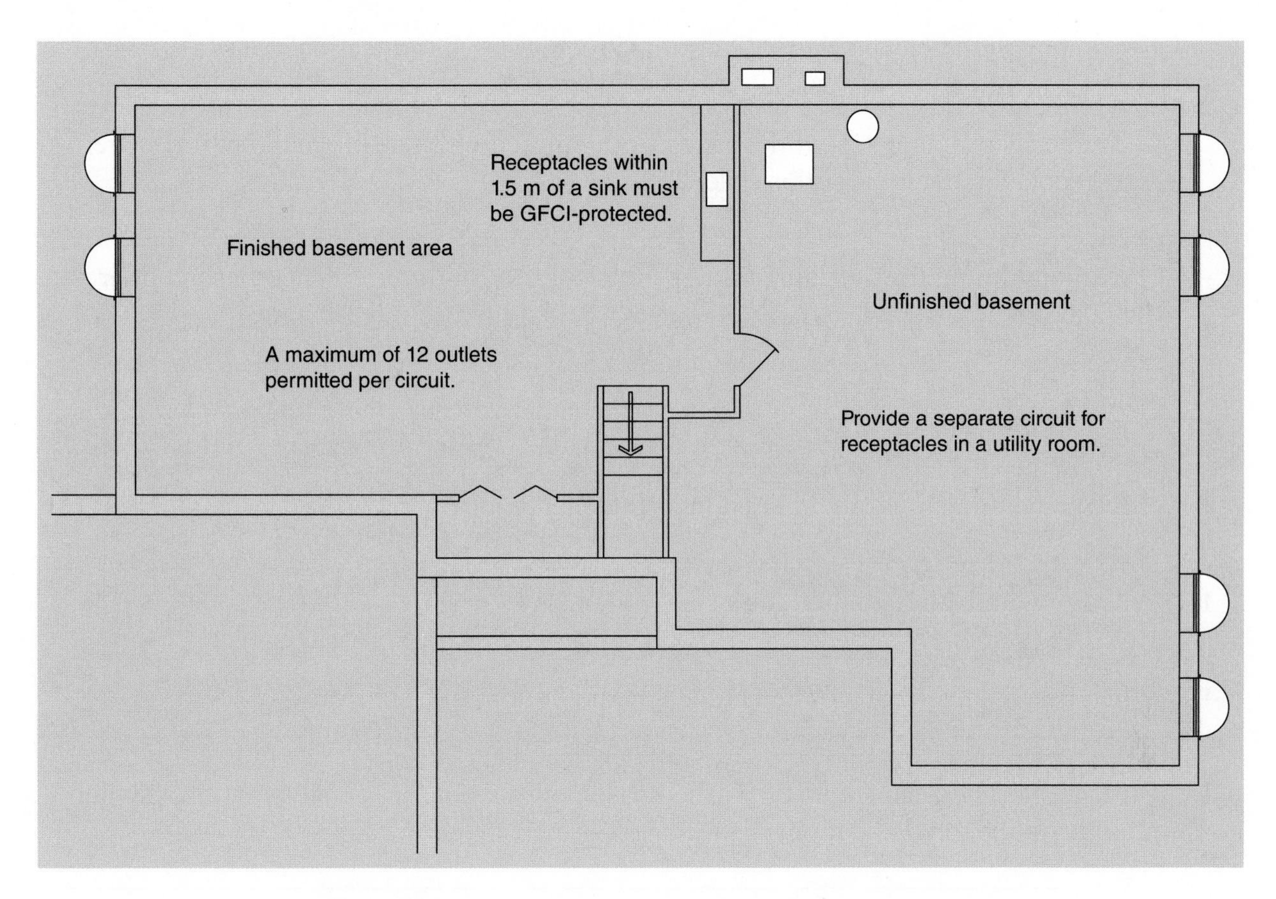

FIGURE 6-12 Circuiting for basement receptacles.

REVIEW

Where applicable, responses should be written in complete sentences.

1. A living room wall is 6 m long. How many receptacles are required for the room?

2. A hallway is 3 m long and has a doorway fitted with a door at each end. Does this hallway require a receptacle?

3. A hallway is 6 m long and has an open doorway at each end. The hall opens onto two rooms that each has one receptacle within 1 m of the doorway. Does this hallway require a receptacle?

4. How many receptacles are required in an unfinished basement with an area of 63 m^2?

5. A kitchen counter is 3 m long. A sink that is 600 mm is located in the exact centre of the counter. How many receptacles are required for the counter?

6. How many of the receptacles in Question 5 require GFCI protection?

7. A counter has an isolated work space of 600 mm between a sink and a built-in oven. Is a receptacle required for this space?

8. How many luminaires or lampholders are required for an unfinished basement area that is 63 m^2?

9. How many receptacles are required in an attached two-car garage?

10. How many luminaires are required in an attached two-car garage?

11. When is a luminaire required for a stairway?

12. When is switch control required at the top and bottom of a stairway?

13. When a stairway leads to an unfinished basement that has an outside exit, is a switch required at the top and bottom of the stairs?

14. When should a receptacle be provided for a central vacuum unit?

15. A receptacle in a utility room is supplied from its own circuit. Can the wiring from the receptacle be extended to supply eight outlets in a finished family room?

16. Why shouldn't a freezer be placed on a circuit protected by a GFCI?

17. What outlets in a dwelling unit must have AFCI protection?

18. A counter has three 5-20RA receptacles. What is the minimum number of branch circuits that may be used for these receptacles?

19. What is the minimum number of circuits that may be run to a laundry room?

20. Does the receptacle located next to a washbasin in a bathroom have to be on its own circuit?

UNIT 7

Conductor Sizes and Types, Wiring Methods, Wire Connections, Voltage Drop, and Neutral Sizing for Services

OBJECTIVES

After studying this unit, you should be able to

- define the terms used to size and rate conductors
- discuss aluminum conductors
- describe the types of cables used in most dwelling unit installations
- list the installation requirements for each type of cable
- describe the uses and installation requirements for electrical conduit systems
- describe the requirements for service grounding conductors
- describe the use and installation requirements for flexible metal conduit
- make voltage drop calculations
- make calculations using *CEC, Table D3*
- understand the fundamentals of the markings found on the terminals of wiring devices and wire connectors
- understand why the ampacity of high-temperature insulated conductors may not be suitable at that same rating on a particular terminal on a wire device, breaker, or switch
- determine when a reduced neutral can be used on residential services

CONDUCTORS

Throughout this text, all references to conductors are for copper conductors, unless otherwise stated.

Wire Size

The copper wire used in electrical installations is graded for size according to the American Wire Gauge (AWG) standard. The wire diameter in the AWG standard is expressed as a whole number. AWG sizes vary from fine, hairlike wire used in coils and small transformers to very large-diameter wire required in industrial wiring to handle heavy loads.

The wire may be a single strand (solid conductor) or consist of many strands. Each strand of wire acts as a separate conducting unit. Conductors that are #8 AWG and larger generally are stranded. The wire size used for a circuit depends on the maximum current to be carried. The *Canadian Electrical Code (CEC)* states that the minimum conductor size permitted in house wiring is #14 AWG. (Exceptions to this rule are covered in *CEC Section 16* and *Table 12* for the wires used in lighting fixtures, bell wiring, and remote-control low-energy circuits.) See the conductor application chart in Table 7-1.

Ampacity

Ampacity is defined as the current in amperes a conductor can carry continuously under the condition of use without exceeding its temperature rating. This value depends on the cross-sectional area of the wire. Since the area of a circle is proportional to the square of the diameter, the ampacity of a round wire varies with the diameter squared (d^2). The diameter of a wire is measured in mils; a *mil* is defined as one-thousandth of an inch (0.001 in.). A *circular mil* is the area of a circle that is one mil in diameter. The circular mil area (CMA) of a wire determines the current-carrying capacity of the wire. This can be expressed as a formula

$$\text{CMA} = d^2$$

Therefore, the larger the CMA of a wire, the greater its current-carrying capacity.

AWG sizes are also expressed in mils and range from #40 (10 circular mils) to #4/0 (211 600 circular mils). Wire sizes larger than 4/0 are expressed in circular mils only. Since the letter "k" designates 1000, large conductors, such as 500 000 circular mils, can also be expressed as 500 kcmils. However, many texts and electricians continue to use an older term, MCM, in which the first "M" is a roman numeral representing 1000. Thus, 500 MCM also means 500 000 circular mils.

Conductors must have an ampacity not less than the maximum load they are supplying; see Figure 7-1. All conductors of a specific branch circuit must have an ampacity of the branch circuit's rating; see Figure 7-2. There are some exceptions to this rule, such as taps for electric ranges (Unit 15).

CEC Tables 1 through *4* list the allowable ampacities of conductors in free air or when not more than three are installed in a raceway or cable. *Table 12* lists the allowable ampacities of equipment wire and flexible cords. These ampacities are reduced if the conductors are installed in areas of high ambient (surrounding) temperature *(Table 5A)*. Ampacities

TABLE 7-1

Conductor applications chart.

CONDUCTOR SIZE	APPLICATIONS
#16, #18 AWG	Cords, low-voltage control circuits, bell and chime wiring
#14, #12 AWG	Normal lighting circuits, circuits supplying receptacle outlets
#10, #8, #6, #4 AWG	Clothes dryers, ovens, ranges, cooktops, water heaters, heat pumps, central air conditioners, furnaces, feeders to subpanels
#3, #2, #1 (and larger) AWG	Main service entrances, feeders to subpanels

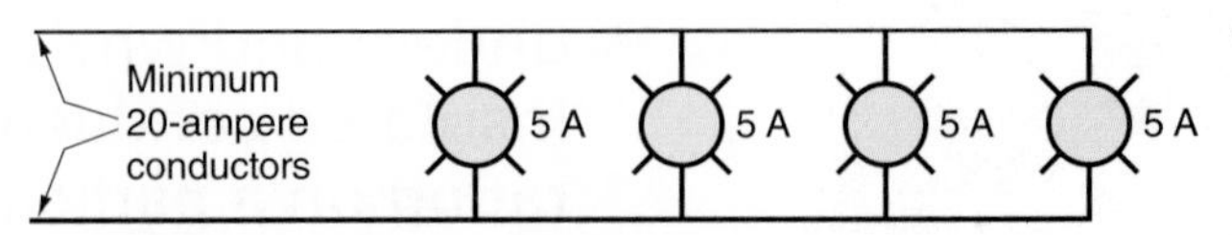

FIGURE 7-1 Branch-circuit conductors are required to have an ampacity not less than the maximum load to be served.

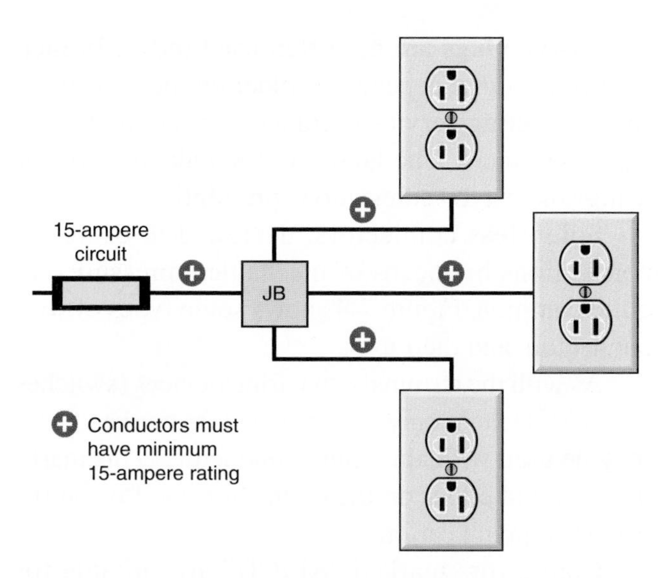

FIGURE 7-2 All conductors in this circuit supplying readily accessible receptacles shall have an ampacity of not less than the rating of the branch circuit. In this example, 15-ampere conductors must be used, *Rule 8–104(1).*

are also reduced when there are more than three conductors in a raceway or cable *(Table 5C).*

Circuit Rating

The rating of the overcurrent device (OCD) and the ampacity of the wire must be compared. The branch-circuit ampacity rating is the lesser of these two values (Figure 7-3).

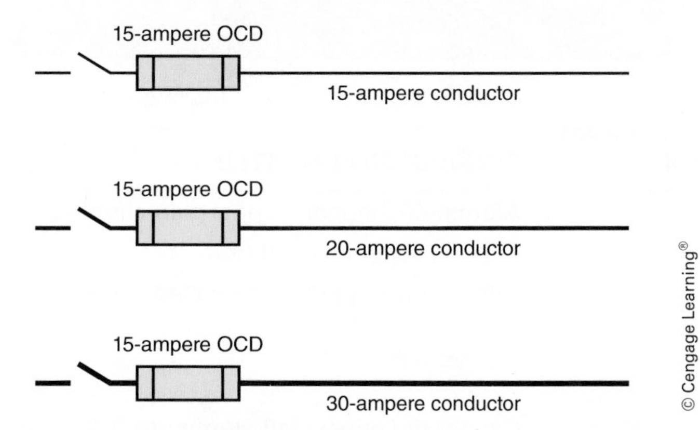

FIGURE 7-3 These are all classified as 15-ampere circuits even though larger conductors were used for some other reason, such as solving a voltage drop problem. The lesser of the amperage rating of the overcurrent device or the wire determines the rating of a circuit, *Rule 8–104(1).*

Aluminum Conductors

The conductivity of aluminum wire is not as great as that of copper wire for a given size. For example, a #8 90°C copper wire has an ampacity of 55 amperes according to *Table 2.* Aluminum wire has a significant weight advantage, but a #8 90°C aluminum wire has an ampacity of only 45 amperes, *Table 4.*

It is important to consider resistance when selecting a conductor for a long run. An aluminum conductor has a higher resistance than a copper conductor for a given wire size, which therefore causes a greater voltage drop.

$$\text{Voltage drop } (E_d) = \text{Amperes (I)} \times \text{Resistance (R)}$$

Although aluminum conductors are approved for branch-circuit wiring, they are seldom used for this application. They are, however, widely used in such applications as apartment feeders and utility distribution systems. They are also found on such equipment as ballasts and luminaires.

Common Connection Problems

Problems often arise with aluminum conductors *that are not properly connected*, as follows:

- A corrosive action is set up when dissimilar wires come into contact with each other, especially if moisture is present.
- The surface of aluminum oxidizes as soon as it is exposed to air. If this oxidized surface is not broken through, a poor connection results. When installing aluminum conductors, particularly in large sizes, you must brush an inhibitor onto the aluminum conductor and scrape the conductor with a stiff brush where the connection is to be made. Scraping the conductor breaks through the oxidation, and the inhibitor keeps the air from coming into contact with the conductor, preventing further oxidation. Aluminum connectors of the compression type usually have an inhibitor paste factory-installed inside the connector.
- Aluminum wire expands and contracts more than copper wire for an equal load, another possible cause of a poor connection. Crimp connectors for aluminum conductors are

usually longer than those for comparable copper conductors, resulting in greater contact surface of the conductor in the connector.

Proper Installation Procedures

Proper, trouble-free connections for aluminum conductors require terminals, lugs, and/or connectors that are suitable for the type of conductor being installed.

Terminals on receptacles and switches must be suitable for the conductors being attached. Table 7-2 shows how the electrician can identify these terminals. Listed connectors provide proper connection when properly installed. See *Rules 12–116* and *12–118.*

Wire Connections

When splicing wires or connecting to a switch, fixture, circuit breaker, panelboard, meter socket, or other electrical equipment, you must twist the wires together and then use a wire connector, *Rule 12–112*.

Wire connectors are known in the trade by such names as screw terminal, pressure terminal connector, wire connector, Wing-Nut™, Wire-Nut™, Scotchlok™, split-bolt connector, pressure cable connector, solderless lug, and solder lug. Solder-type lugs are not often used today. In fact, connections that depend on solder are not permitted for connecting service entrance conductors to service equipment. The labour costs make the cost of using solder-type connections prohibitive.

Solderless connectors, designed to establish connections by means of mechanical pressure, are quite common. Figure 7-4 shows some types of wire connectors and their uses.

As with the terminals on wiring devices (switches and receptacles), only wire connectors marked "AL" may be used with aluminum conductors. This marking is found either on the connector itself or on (or in) the shipping carton.

Connectors marked "AL/CU" are suitable for use with aluminum, copper, or copper-clad aluminum conductors. This marking also appears on the connector itself or on (or in) the shipping carton.

Connectors not marked "AL" or "AL/CU" are for use with copper conductors only.

Conductors made of copper, aluminum, or copper-clad aluminum may not be used in combination in the same connector unless the conductor is specially identified for the purpose either on the connector itself or on (or in) the shipping carton. The conditions of use will also be indicated. Such connectors are usually limited to dry locations.

Rule 12–118 covers termination and splicing of aluminum conductors.

TABLE 7-2

Approved equipment and conductor combinations.

TYPE OF DEVICE	MARKING ON TERMINAL OR CONNECTOR	CONDUCTOR PERMITTED
15- or 20-ampere receptacle and switch	CO/ALR	Aluminum, copper, copper-clad aluminum
15- and 20-ampere receptacle and switch	None	Copper, copper-clad aluminum
30-ampere and greater receptacle and switch	AL/CU	Aluminum, copper, copper-clad aluminum
30-ampere and greater receptacle and switch	None	Copper only
Screwless pressure terminal connector of the push-in type	None	Copper or copper-clad aluminum
Wire connector	AL/CU	Aluminum, copper, copper-clad aluminum
Wire connector	None	Copper only
Wire connector	AL	Aluminum only
Any of the above devices	Copper or CU only	Copper only

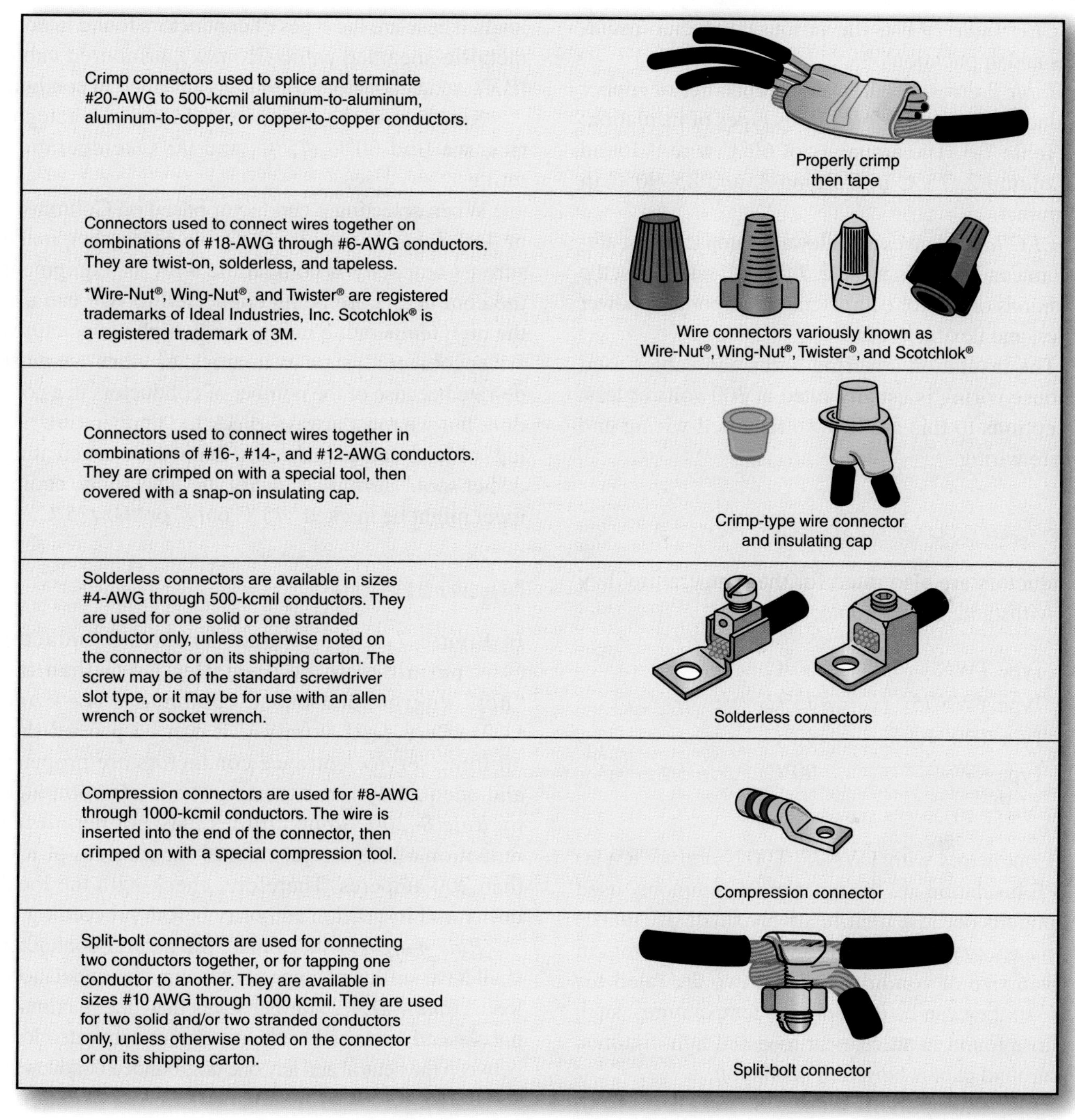

FIGURE 7-4 Types of wire connectors.

Insulation of Wires

Rule 12–100 requires that insulation on conductors be suitable for the conditions in which they are installed. The insulation must completely surround the metal conductor, have a uniform thickness, and run the entire length of the wire or cable.

One type of insulation used on wires is a thermoplastic material (T90 Nylon or TEW), although thermoset (rubber) compounds are also widely used.

A thermoplastic will deform and melt when excess heat is applied to it. Any insulation with a designation that starts with a "T," such as T90 Nylon or TEW, is a thermoplastic.

A thermoset material will not soften and deform when heat is applied to it. When heated above its rated temperature, it will char and crack. Any insulation that starts with "R," such as R90 XLPE, is a thermoset material.

CEC Table 19 lists the various conductor insulations and applications.

Table 2 gives the allowable ampacities of copper conductors in a cable for various types of insulation; see Table 7-3. The ampacity of 60°C wire is found in Column 2, 75°C in Column 3, and 85–90°C in Column 4.

CEC Table 4 gives the allowable ampacities of aluminum conductors in a cable. *Tables 11* and *19* list the conditions of use for equipment wires, portable power cables, and flexible cords.

The insulation covering wires and cables used in house wiring is usually rated at 300 volts or less. Exceptions to this are low-voltage bell wiring and fixture wiring.

Temperature Considerations

Conductors are also rated for the temperature they can withstand. For example:

Type TWN	60°C
Type TWN75	75°C
Type T90 Nylon	90°C
Type RW90 XLPE	90°C

Conductors with TWN75, T90 Nylon, or RW90 XLPE insulation are the types most commonly used in conduits because their relatively small size makes them easy to handle and allows more conductors in a given size of conduit. The last two are rated for 90°C so they can be used in high temperatures, such as those found in attics, near recessed light fixtures, and around cables buried in insulation.

The most commonly used cable is NMD 90, a non-metallic-sheathed type. Easy to handle, it is rated for 90°C.

The permitted ampacity of all 90°C-rated conductors used in ceiling light outlet boxes is reduced to that of 60°C-rated conductor, *Rule 30–408.*

Conductor Temperature Ratings

CEC Table 2 shows the types of insulations available on conductors for building wire used in most electrical installations. This is the table that electricians refer to most often when selecting wire sizes for specific loads. These are the types of conductors found in non-metallic-sheathed cable (Romex), armoured cable (BX), and conductors commonly installed in conduit.

Note that in the common building wire categories, we find 60°C, 75°C, and 90°C temperature ratings.

When selecting a conductor based on Columns 3 or 4 of *Table 2* (see also the table footnotes), make sure its ampacity is compatible with the equipment the conductors are being connected to. We can use the high temperature ratings when high temperatures are encountered, such as in attics, or when we must de-rate because of the number of conductors in a conduit, but we must always check the temperature ratings of terminals to make sure that we are not creating a "hot spot." Terminals and/or the label in the equipment might be marked "75°C only" or "60°/75°C."

Neutral Size

In Figure 7-5, the grounded neutral conductor Ⓝ is permitted to be a smaller AWG than the "hot" ungrounded phase conductors (L–1 and L–2) *(Rule 4–024)* only if it can be proved that all three service entrance conductors are properly and adequately sized to carry the loads computed by *Rule 8–200.* Some utilities may not permit the reduction of the service neutral for services of less than 200 amperes. Therefore, check with the local utility and inspection authority before proceeding.

Rule 4–024(1) states that "the neutral conductor shall have sufficient ampacity to carry the unbalanced load." *Rule 4–024(2)* further states that "the maximum unbalanced load shall be the maximum connected load between the neutral and any one ungrounded conductor."

In this residence, a number of loads carry little or no neutral currents, including the electric water heater, electric clothes dryer, electric oven and range, electric furnace, and electric air conditioner (Figure 7-5). These appliances often draw their biggest current through the "hot" conductors because they are fed with 240-volt Ⓒ or 120/240-volt circuits Ⓔ. Loads Ⓑ are connected line-to-neutral only.

Thus, we find logic in the *CEC,* which permits reducing the neutral size on services, feeders, and branch circuits where the computations prove that the neutral will be carrying less current than the "hot" phase conductors.

TABLE 7-3

CEC Table 2 extract.

CEC TABLE 2

ALLOWABLE AMPACITIES FOR NOT MORE THAN THREE COPPER CONDUCTORS, RATED NOT MORE THAN 5000 V AND UNSHIELDED, IN RACEWAY OR CABLE (BASED ON AN AMBIENT TEMPERATURE OF 30°C)*
(*RULES 4–004, 8–104, 12–2210, 12–2260, 12–3034, 26–142, 42–008,* AND *42–016* AND *TABLES SA, SC, 19, 39,* AND *D3*)

	Allowable Ampacity† ††					
Size AWG or kcmil	**60°C‡**	**75°C‡**	**90°C‡ ****	**110°C‡ See Note**	**125°C‡ See Note**	**200°C‡ See Note**
14§	15	20	25	25	30	35
12§	20	25	30	30	35	40
10§	30	35	40	45	45	60
8	40	50	55	65	65	80
6	55	65	75	80	90	110
4	70	85	95	105	115	140
3	85	100	115	125	135	165
2	95	115	130	145	155	190
1	110	130	145	165	175	215
0	125	150	170	190	200	245
00	145	175	195	220	235	290
000	165	200	225	255	270	330
0000	195	230	260	290	310	380
250	215	255	290	320	345	—
300	240	285	320	360	385	—
350	260	310	350	390	420	—
400	280	335	380	425	450	—
500	320	380	430	480	510	—
600	350	420	475	530	565	—
700	385	460	520	580	620	—
750	400	475	535	600	640	—
800	410	490	555	620	660	—
900	435	520	585	655	700	—
1000	455	545	615	690	735	—
1250	495	590	665	745	—	—
1500	525	625	705	790	—	—
1750	545	650	735	820	—	—
2000	555	665	750	840	—	—
Col. 1	Col. 2	Col. 3	Col. 4	Col. 5	Col. 6	Col. 7

* See *Table 5A* for the correction factors to be applied to the values in Columns 2 through 7 for ambient temperatures over 30°C.
† The ampacity of aluminum-sheathed cable is based on the type of insulation used on the copper conductors.
‡ These are maximum allowable conductor temperatures for one, two, or three conductors run in a raceway, or two or three conductors run in a cable, and may be used in determining the ampacity of other conductor types listed in *Table 19,* which are so run, as follows: From *Table 19* determine the maximum allowable conductor temperature for that particular type, then from this table determine the ampacity under the column of corresponding temperature rating.
§ See *Rule 14–104(2).*
** For mineral-insulated cables, these ratings are based on the use of 90°C insulation on the emerging conductors and for sealing. Where a deviation has been allowed in accordance with *Rule 2–030*, mineral-insulated cable may be used at a higher temperature without a decrease in allowable ampacity, provided that insulation and sealing material approved for the higher temperature are used.
†† See *Table 5C* for the correction factors to be applied to the values in Columns 2 through 7 where there are more than three conductors in a run of raceway or cable.

Note: These ampacities apply to bare wire or under special circumstances where the use of insulated conductors having this temperature rating is acceptable.

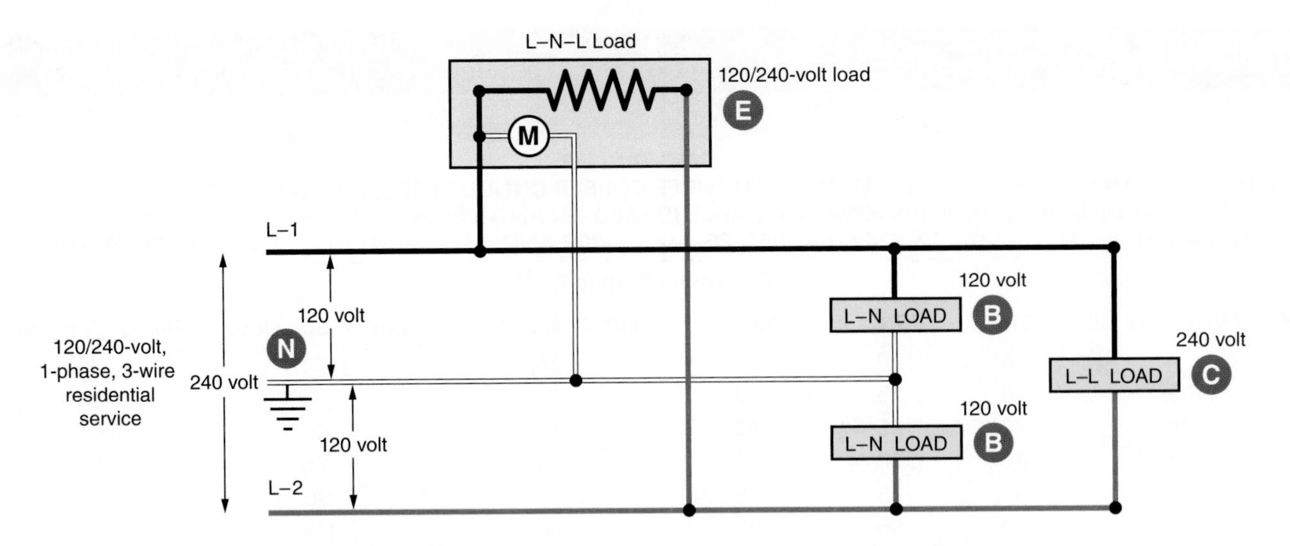

FIGURE 7-5 Line-to-line and line-to-neutral loads.

VOLTAGE DROP

Low voltage can cause lights to dim, television pictures to shrink, motors to run hot, electric heaters to produce less than their rated heat output, and appliances to operate improperly.

Low voltage in a home can be caused by

- wire that is too small for the load being served
- a circuit that is too long
- poor connections at the terminals
- conductors operating at high temperatures having higher resistance than when operating at lower temperatures

A simple formula for calculating voltage drop on single-phase systems considers only the direct-current (DC) resistance of the conductors and the temperature of the conductor. The more accurate formulas consider alternating-current (AC) resistance, reactance, three-phase installations, temperature, and spacing in metal conduit and in non-metallic conduit. Voltage drop is covered in great detail this book's companion textbook, *Electrical Wiring: Commercial,* also published by Nelson Education.

The simple voltage drop formula is more accurate with smaller conductors, but loses accuracy as conductor size increases. It is sufficiently accurate for the voltage drop calculations necessary for residential wiring.

To find voltage drop

$$E_d = \frac{K \times I \times L \times 2}{CMA}$$

where

E_d = Allowable voltage drop in volts

K = Resistance of conductor at 75°C:

- For copper conductors:
  - About 12 ohms per circular mil foot
  - About 39.4 ohms per circular mil metre
- For aluminum conductors:
  - About 19 ohms per circular mil foot
  - About 63.3 ohms per circular mil metre*

I = Current in amperes flowing through the conductors

L = Length of the conductor from the beginning of the circuit to the load:

- In feet if the circular mil foot is used for K
- In metres if the circular mil metre is used for K

*Derived from the circular mil foot.

Then multiply the length of the run × 2 = Total length of wire

CMA = Cross-sectional area of the conductor in circular mils (Table 7-4).

To find conductor size:

$$CMA = \frac{K \times I \times L \times 2}{E_d}$$

Code Reference to Voltage Drop

Rule 8–102 states the maximum allowed voltage drop percentages. The voltage drop shall not exceed 3% in any feeder or branch circuit (Figure 7-6) and 5% overall.

Figure 7-7 explains the requirements for a voltage drop that involves both the feeder and a branch circuit.

Caution Using High-Temperature Wire

When trying to use smaller conduits, be careful about selecting conductors on the ability of their insulation to withstand high temperatures.

TABLE 7-4

Cross-sectional area for building wire and cable.

NOMINAL	
AWG	**CM Area**
14*	4 110
12*	6 530
10*	10 380
8	16 510
6	26 240
4	41 740
3	52 630
2	66 360
1	83 690
0	105 600
00	133 100
000	167 800
0000	211 600

* These sizes are customarily supplied with solid conductors.

Reference to *Table 2* will confirm that the ampacity of high-temperature conductors is greater than the ampacity of lower-temperature conductors. For instance:

- #8 90°C copper has an ampacity of 55 amperes
- #8 60°C copper has an ampacity of 40 amperes

Always calculate a voltage drop when your decision to use the higher-temperature-rated conductors was based on the fact that the use of smaller-size conduit would be possible. Select the proper size of conductor to use for the job according to the load requirements and then complete a voltage drop calculation to see that the maximum permitted voltage drop is not exceeded.

EXAMPLE

What is the approximate voltage drop on a 120-volt, single-phase circuit consisting of #14 AWG copper conductors where the load is 11 amperes and the distance of the circuit from the panel to actual load is 25.9 metres?

SOLUTION

$$E_d = \frac{K \times I \times L \times 2}{CMA}$$

$$= \frac{39.4 \times 11 \times 25.9 \times 2}{4110}$$

$$= 5.46 \text{ volts drop}$$

Note: Refer to Table 7-4 for the CMA value.

The 5.46 voltage drop in Example 1 exceeds the voltage drop permitted by the CEC, which is

$$3\% \text{ of } 120 \text{ volts} = 3.6 \text{ volts}$$

Using #12 AWG:

$$E_d = \frac{39.4 \times 11 \times 25.9 \times 2}{6530}$$

$$= 3.44 \text{ volts drop}$$

This meets *Code.*

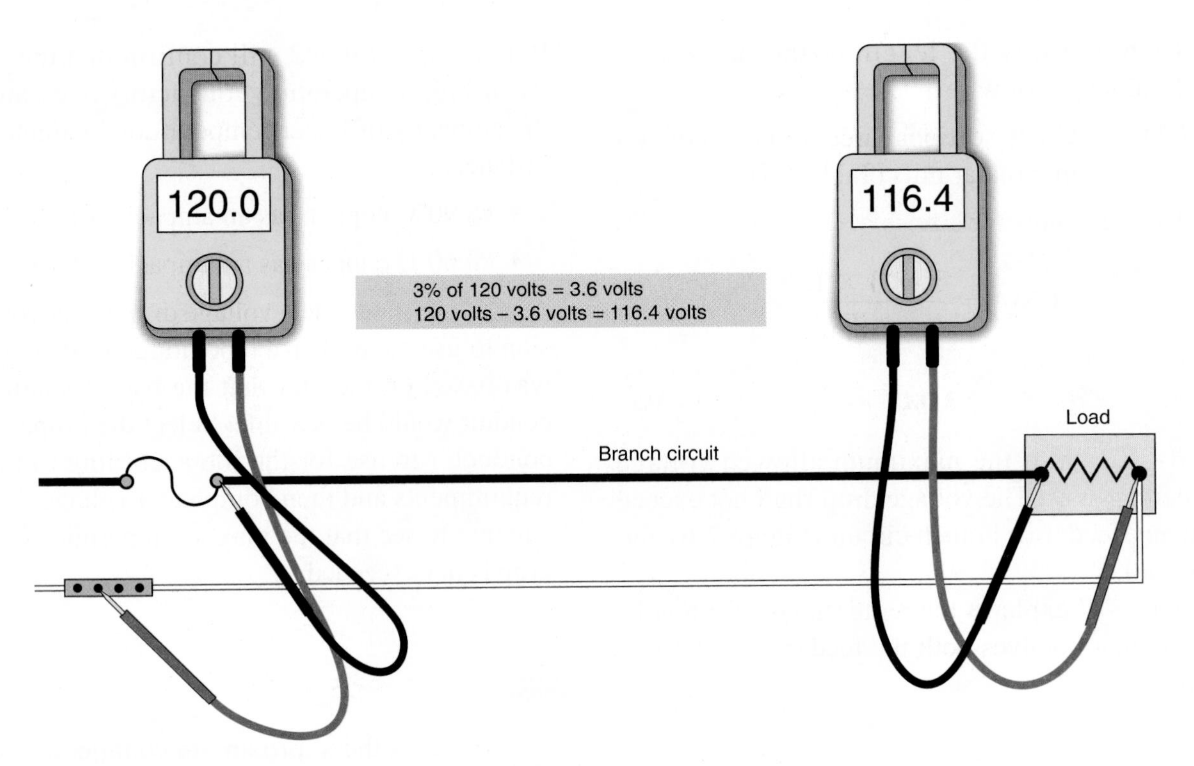

FIGURE 7-6 Maximum allowable voltage drop on a branch is 3%, *Rule 8–102(1)(c).*

EXAMPLE

Find the wire size needed to keep the voltage drop to no more than 3% on a single-phase 240-volt circuit that feeds a 44-ampere air conditioner. The circuit originates at the main panel, which is located about 19.8 m from the air conditioner. No neutral is required for this equipment.

SOLUTION First, size the conductors at 125% of the air-conditioner load:

$44 \times 1.25 = 55$-ampere conductors are required.

Checking *Table 2,* the copper conductors would be #6 rated at 75°C.

The permitted voltage drop is

$$E_d = 240 \times 0.03$$

$$= 7.2 \text{ volts}$$

$$\text{CMA} = \frac{K \times I \times L \times 2}{E_d}$$

$$= \frac{39.4 \times 44 \times 19.8 \times 2}{7.2}$$

$$= 9535 \text{ circular mils minimum}$$

According to Table 7-4, this would be a #10 AWG conductor, but this size of conductor would be too small for the load. So, you are back to a #6 conductor.

If you compare the calculation of voltage drop to the size of the wire or current, note that the larger wire gauge was from the current calculation. If the voltage drop calculation required a larger wire than the current table, the larger wire is always installed.

APPROXIMATE CONDUCTOR SIZE

Rule One

For wire sizes up through 0000, every third size doubles or halves in circular mil area. A #1 AWG conductor is two times larger than a #4 AWG conductor (83 690 versus 41 470), and a "0" wire is one-half the size of "0000" wire (105 600 versus 211 600).

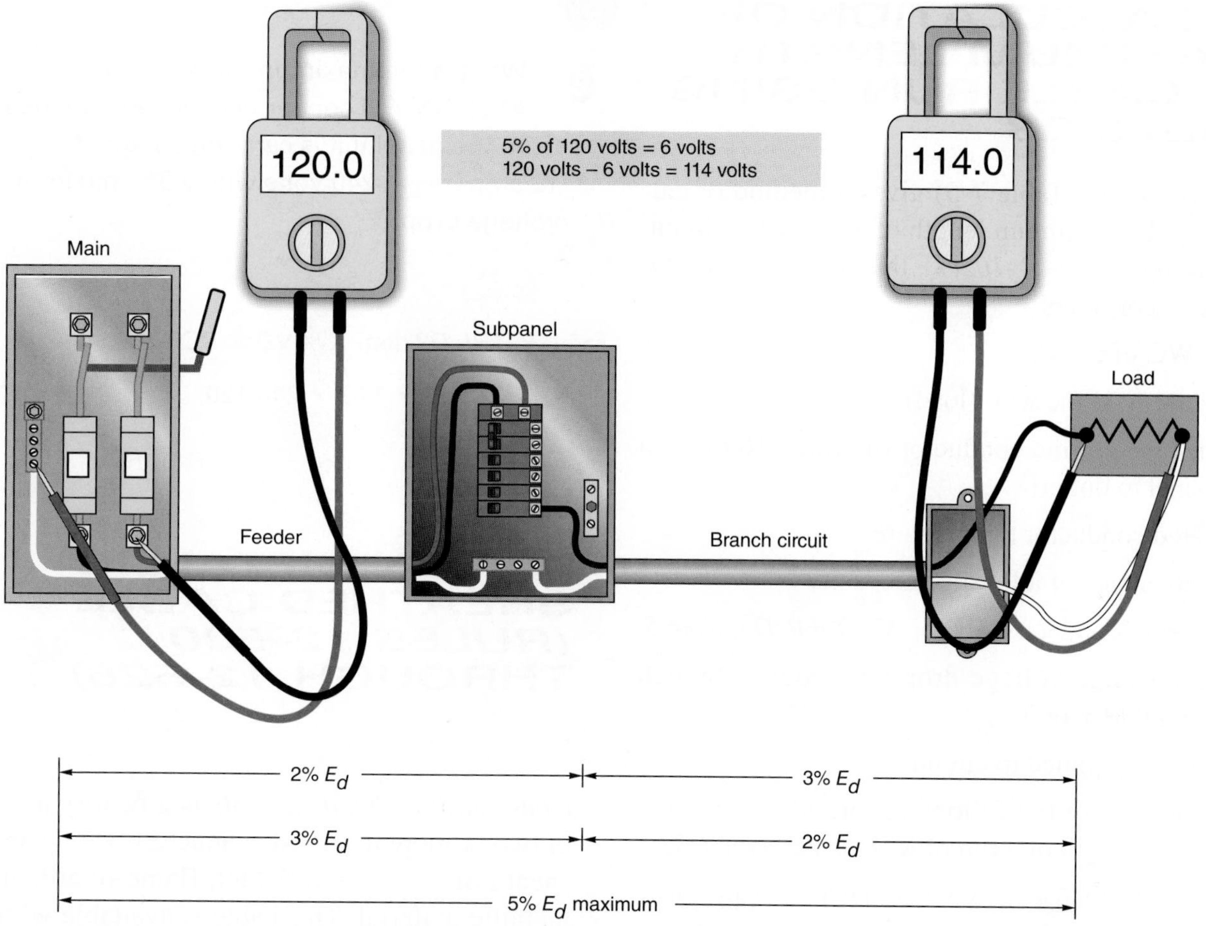

120.0
5% of 120 volts = 6 volts
120 volts – 6 volts = 114 volts
114.0
Main
Subpanel
Load
Feeder
Branch circuit
2% E_d
3% E_d
3% E_d
2% E_d
5% E_d maximum

FIGURE 7-7 *Rule 8–102(1)(b)* states that the total voltage drop from the beginning of a feeder to the farthest outlet on a branch circuit should not exceed 5%. In this figure, if the voltage drop in the feeder is 3%, do not exceed 2% voltage drop in the branch circuit. If the voltage drop in the feeder is 2%, do not exceed 3% voltage drop in the branch circuit.

Rule Two

For wire sizes up through 0000, every consecutive wire size is about 1.26 times larger or smaller than the preceding wire size. A #3 AWG conductor is about 1.26 times larger than a #4 AWG conductor (41 740 × 1.26 = 52 592). Thus, a #2 AWG wire is about 1.26 times smaller than a #1 AWG wire (66 360 ÷ 1.26 = 52 667).

A #10 AWG conductor has a cross-sectional area of 10 380 CM and a resistance of 1.2 ohms per 1000 ft. The resistance of aluminum wire is about 2 ohms per 1000 ft. (300 m). By remembering these numbers, you will be able to perform voltage drop calculations without having the wire tables readily available.

EXAMPLE

What are the approximate CMA and resistance of a #6 AWG copper conductor?

SOLUTION

WIRE SIZE	CMA	OHMS PER 1000 FEET (300 m)
10	10 380	1.2
9		
8		
7	20 760	0.6
6	26 158	0.0476

When the CMA of a wire is doubled, its resistance is cut in half. And inversely, when a given wire size is reduced to one-half, its resistance doubles.

CALCULATION OF MAXIMUM LENGTH OF CABLE RUN USING *TABLE D3*

CEC Table D3 (Table 7-5) gives a method of calculating the maximum length of a two-wire circuit. According to *Rule 8–102(1),* this may be either 2% or 3%. It considers

- AWG of the wire
- current on the wire (load)
- distance of the conductor run (e.g., 10 m from panel to outlet)
- rated conductor temperature
- percentage of allowable ampacity
- type of conductor, CU or AL, *Table D3, Note 5*
- percentage voltage drop on feeder or branch circuit (2% or 3%)
- voltage applied to circuit

You can use the following formula to find the maximum length of a run of wire using *Table D3*:

L = *TBL D3* dist. × % VD × DCF × Volts/120

where

L = Maximum length of two-wire CU conductor run in metres

TBL D3 dist. = Distance shown in *Table D3* for the size of wire (AWG heading at top of table) and the actual load in amperes ("Current, A" column)

% VD = Maximum percentage voltage drop allowed on the circuit (value should be shown as a whole number, not as a percentage)

DCF = Distance correction factor, calculated by the distance correction factor table found in *Table D3, Note 3,* using "Percentage of allowable ampacity"

$$\text{\% allowable ampacity} = \frac{\text{Actual load amperes}}{\text{Allowable conductor ampacity}} \times 100$$

and rated conductor temperature

Volts = Voltage that the circuit operates at (e.g., 120, 208, 240)

EXAMPLE

What is the maximum distance that a #10/2 NMWU copper wire circuit may be run from the panel if it is carrying a load of 18 amperes at 240 volts with a 3% maximum voltage drop?

SOLUTION

L = *TBL D3* dist. × % VD × DCF × Volts/120

= 7.8 × 3 × 1.07 × 240/120

= 50.08 m

NON-METALLIC-SHEATHED CABLE *(RULES 12–500 THROUGH 12–526)*

Description

Non-metallic-sheathed cable is a factory assembly of two or more insulated conductors having an outer sheath of moisture-resistant, flame-retardant non-metallic material. This cable is available with two, three, or four current-carrying conductors, ranging in size from #14 through #2 for copper or aluminum. Two-wire cable contains one black conductor, one white conductor, and one bare grounding conductor or green insulated conductor. Three-wire cable contains one black, one white, one red, and one bare grounding conductor or green insulated conductor.

The Canadian Standards Association (CSA) lists three classifications of non-metallic-sheathed cable:

- Type NMD 90 has a flame-retardant, moisture-resistant non-metallic covering over the conductors and is suitable for both dry and damp locations.
- Types NMW and NMWU have a flame-retardant, moisture-resistant, corrosion-resistant, fungus-resistant non-metallic covering over the conductors. NMWU is designed especially for use in wet and direct burial locations.

Electricians once referred to non-metallic-sheathed cable as *Romex* or *Loomex. Table 19* indicates that the conductors in Type NMD 90 cable are rated for 90°C (194°F), making the cable suitable for the

TABLE 7-5

CEC Table D3.

CEC TABLE D3

DISTANCE TO CENTRE OF DISTRIBUTION FOR A 1% DROP IN VOLTAGE ON NOMINAL 120-V, 2-CONDUCTOR COPPER CIRCUITS

(SEE APPENDIX B NOTE TO RULE 4–004.)

Current, A	Copper Conductor Size, AWG														
	18	16	14	12	10	8	6	4	3	2	1	1/0	2/0	3/0	4/0
	Distance to centre of distribution measured along the conductor run, m (calculated for conductor temperature of 60°C)														
1.00	24.2	38.5	61.4												
1.25	19.4	30.8	49.1												
1.6	15.1	24.1	38.4	61.0											
2.0	12.1	19.3	30.7	48.8											
2.5	9.7	15.4	24.6	39.0	62.0										
3.2	7.6	12.0	19.2	30.5	48.5										
4.0	6.1	9.6	15.3	24.4	38.8	61.7									
5.0	4.8	7.7	12.3	19.5	31.0	49.3									
6.3	3.8	6.1	9.7	15.5	24.6	39.1	62.2								
8.0	3.0	4.8	7.7	12.2	19.4	30.8	49.0								
10.0	2.4	3.9	6.1	9.8	15.5	24.7	39.2	62.4							
12.5		3.1	4.9	7.8	12.4	19.7	31.4	49.9	62.9						
16		2.4	3.8	6.1	9.7	15.4	24.5	39.0	49.1	62.0					
20			3.1	4.9	7.8	12.3	19.6	31.2	39.3	49.6	62.5				
25				3.9	6.2	9.9	15.7	24.9	31.4	39.7	50.0	63.1			
32					4.8	7.7	12.2	19.6	24.6	31.0	39.1	49.3	62.1		
40					3.9	6.2	9.8	15.6	19.7	24.8	31.3	39.4	49.7	62.7	
50						4.9	7.8	12.5	15.7	19.8	25.0	31.5	39.8	50.1	63.2
63						3.9	6.2	9.9	12.5	15.7	19.8	25.0	31.6	39.8	50.2
80						3.1	4.9	7.8	9.8	12.4	15.6	19.7	24.8	31.3	39.5
100							3.9	6.2	7.9	9.9	12.5	15.8	19.9	25.1	31.6
125								5.0	6.3	7.9	10.0	12.6	15.9	20.1	25.3
160									4.9	6.2	7.8	9.9	12.4	15.7	19.8
200										5.0	6.3	7.9	9.9	12.5	15.8
250												6.3	8.0	10.0	12.6
320													6.2	7.8	9.9

Courtesy of CSA Group

Notes: 1. *Table D3* is calculated for copper wire sizes No. 18 AWG to No. 4/0 AWG and, for each size specified, gives the approximate distance in metres to the centre of distribution measured along the conductor run for a 1% drop in voltage at a given current, with the conductor at a temperature of 60°C. Inductive reactance has not been included because it is a function of conductor size and spacing.
2. The distances for a 3% or 5% voltage drop are 3 or 5 times those for a 1% voltage drop.
3. Because the distances in *Table D3* are based on conductor resistances at 60°C, these distances must be multiplied by the correction factors in the following table according to the temperature rating of the conductor used and the percentage load with respect to the allowable ampacity. Where the calculation and the allowable ampacity fall between two columns, the factor in the higher percentage column shall be used.

Distance Correction Factor

Rated Conductor Temperature	Percentage of Allowable Ampacity						
	100	90	80	70	60	50	40
60°C	1.00	1.02	1.04	1.06	1.07	1.09	1.10
75°C	0.96	1.00	1.00	1.03	1.06	1.07	1.09
85–90°C	0.91	0.95	1.00	1.00	1.04	1.06	1.08
110°C	0.85	0.90	0.95	1.00	1.02	1.05	1.07
125°C	0.82	0.87	0.92	0.97	1.00	1.04	1.07
200°C	0.68	0.76	0.83	0.90	0.96	1.00	1.04

4. For other nominal voltages, multiply the distances in metres by the other nominal voltage (in volts) and divide by 120.
5. Aluminum conductors have equivalent resistance per unit length to copper conductors that are smaller in area by two AWG sizes. *Table D3* may be used for aluminum conductors because of this relationship (e.g., for No. 6 AWG aluminum, use the distances listed for No. 8 AWG copper in *Table D3*). Similarly, for No. 2/0 AWG aluminum use the distances for No. 1 AWG copper.
6. The distances and currents listed in *Table D3* follow a pattern. When the current, for any conductor size, is increased by a factor of 10, the corresponding distance decreases by a factor of 10. This relationship can be used when no value is shown in the table. In that case, look at a current 10 times larger. The distance to the centre of distribution is then 10 times larger than the listed value.
7. For multi-conductor cables, ensure that the wire size obtained from this table is suitable for ampacity from *Table 2* or *4*, and *Rule 4–004*.
8. For currents intermediate to listed values, use the next higher current value.
9. Example of the use of this table:

Consider a 2-conductor circuit of No. 12 AWG copper NMD90 carrying 16 A at nominal 240 V under maximum ambient of 30°C.

The maximum run distance from the centre of distribution to the load without exceeding a 3% voltage drop is:

Maximum run length for No. 12 AWG, 16 A, 1% voltage drop at nominal 120 V from table is 6.1 m

Distance correction factor to be used is:

From *Table 2,* allowable ampacity for 2-conductor No. 12 AWG NMD90 (90°C rating per *Table19*) is 30 A. The given current is 16 A or 53% (16/30) of the allowable ampacity.

The distance correction factor to be used, from Note (3), 85–90°C row, 60% column, is 1.04.

The maximum run length is:

$$6.1\text{m} \times 3(\%) \times 1.04 \times \frac{240\text{V}}{120\text{V}} = 38\text{m}$$

If the distance is between 38 and 60.5 m, a larger size of conductor is required, e.g., No. 10 AWG (40 A allowable ampacity) 9.7 m × 3(%) × 1.08 × 240 V/120 V = 62.9 m.

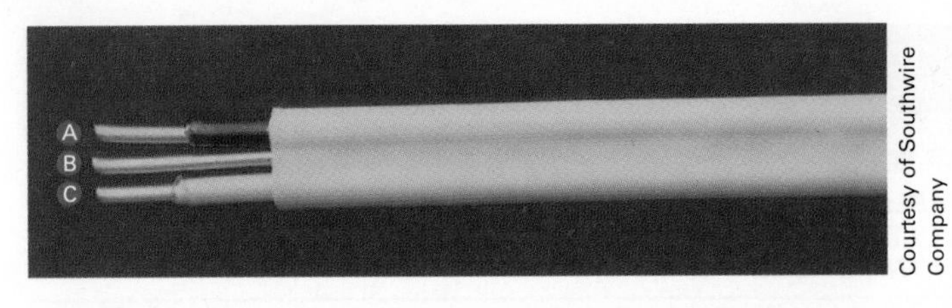

FIGURE 7-8 Non-metallic-sheathed Type NMD 90 cable showing (top to bottom) a black "ungrounded" (hot) conductor, a bare equipment-bonding conductor, and a white "grounded" conductor.

high temperatures found in attics, near recessed lighting fixtures, and around cables buried in insulation.

Non-metallic-sheathed cable has an uninsulated copper conductor that is used for bonding purposes only (Figure 7-8). This bonding conductor is not intended for use as a current-carrying circuit wire.

Equipment-bonding requirements are specified in *Rules 10–400* through *408, 10–808,* and *10–814.* These rules require that all boxes and fixtures in the residence be bonded to ground.

Table 16A (Table 7-6 below) lists the sizes of the bonding conductors used in cable assemblies. Note that the copper bonding conductor shown is a smaller size than the circuit conductors for 15-, 20-, and 30-ampere ratings.

Table 7-7 shows the uses permitted for Type NMD 90 and Type NMW and NMWU cable. Refer also to conditions of use in *CEC Table 19.*

Installation

Non-metallic-sheathed cable is the least expensive of the wiring methods. It is relatively light in weight and easy to install. It is widely used for dwelling unit installations on circuits of 300 volts or less. The installation of all types of non-metallic-sheathed cable must conform to the requirements of *Rules 12–500* through *12–526* (Figure 7-9):

- The cable must be strapped or stapled not more than 300 mm from a box or fitting. See Figure 7-10 for permitted fasteners.
- The intervals between straps or staples must not exceed 1.5 m.
- The cable must be protected against physical damage where necessary.
- The cable must not be bent or stapled so that the outer covering or the wires are damaged.
- The cable must not be used in circuits of more than 300 volts between conductors.
- Wherever non-metallic-sheathed cable is less than 1.5 metres above a floor, *Rule 12–518* requires that the cable be protected by rigid conduit, EMT, or other means (Figure 7-11). A fitting (bushing or connector) must be used at both ends of the conduit to protect the cable from abrasion, *Rule 12–906.* A metal conduit must be bonded to ground.
- When cables are "bundled" together for distances longer than 600 mm (Figure 7-12), the heat generated by the conductors cannot easily dissipate. These conductors *must be* de-rated according to *CEC Table 5C.*

When non-metallic-sheathed cable is run close to a heating source, an air space must be maintained to minimize the transfer of heat to the cable. The spacing requirements are

- 25 mm to heating ducts
- 50 mm to masonry chimneys
- 150 mm to chimney and flue cleanouts, *Rule 12–506(4)(c)*

The spacing may be reduced if an approved thermal barrier is placed between the conductor and the heat source that will maintain an ambient conductor temperature of 30°C.

Figure 7-13 shows the *CEC* requirements for securing non-metallic-sheathed cable.

Rule 10–904(3) requires that all metal boxes for use with non-metallic cable must have provisions for attaching the bonding conductor (Figure 7-14). Figure 7-15 shows a gangable switch (device) box that is tapped for a screw by which the bonding conductor may be connected underneath.

ARMOURED CABLE *(RULES 12–600* THROUGH *12–618)*

Description

The *CEC* describes the construction of armoured cable in *Table D1.* It lists three types: AC90 (commonly called "BX"), ACWU 90, and TECK 90. AC90 is approved for dry locations only. At one

TABLE 7-6

CEC Table 16A.

CEC TABLE 16A
MINIMUM SIZE CONDUCTORS FOR BONDING CONDUCTORS
(*RULES 10–204, 10–626, 10–814, 10–816, 12–1814, 24–104, 24–202, 30–1030, 68–058,* AND *68–406*)

Size of largest ungrounded conductor	Size of bonding conductor	
Copper, AWG or kcmil	**Copper, AWG or kcmil**	**Aluminum, AWG or kcmil**
14 and 12	14	12
10	12	10
8	10	8
6–4	8	6
3–2/0	6	4
3/0–300	4	2
350–500	3	1
600–750	2	1/0
800–1000	1	2/0
1250–2000	1/0	3/0
Aluminum, AWG or kcmil	**Aluminum, AWG or kcmil**	**Copper, AWG or kcmil**
12	12	14
10 and 8	10	12
6	8	10
4–2	6	8
1–4/0	4	6
250–400	2	4
500–700	1	3
750–1000	1/0	2
1250–1500	2/0	1
1750–2000	3/0	1/0

Note: Where multiple ungrounded conductors are used in parallel runs, parallel bonding conductors should be used in accordance with *Rule 10–814* and installed in close proximity to the corresponding ungrounded conductors to minimize increased impedance in the bonding conductor(s).

time BX was a trademark owned by the General Electric Company; it is now a generic term.

Armoured cable is an assembly of insulated conductors in a flexible metallic enclosure; see Figure 7-16. Armoured cable may be used in a variety of locations depending on the type of jacket on top of the armoured cable and the location for the cable.

All three types may be used for a service entrance above grade in dry locations, but ACWU 90 and TECK 90 are approved for service entrance below grade. In all instances, verify with the local supply authority which cables you can use for service entrances.

Armoured cable is generally available with two, three, or four conductors in sizes from #14 AWG to 2000 kcmil, inclusive:

- Two wire = One black, one white + one bare
- Three wire = One black, one red, one white + one bare
- Four wire = One black, one red, one blue, one white + one bare

Armoured cable must have an internal bonding wire of copper or aluminum to meet the bonding requirements of the *CEC*.

TABLE 7-7

Uses permitted in typical residential wiring for type NMD 90, NMW, and NMWU non-metallic-sheathed cable.

FOR TYPICAL RESIDENTIAL WIRING TYPE NMD 90, NMW, AND NMWU CABLE	TYPE NMD 90	TYPE NMW	TYPE NMWU
• May be used on circuits of 300 volts or less	Yes	Yes	Yes
• Has flame-retardant and moisture-resistant outer covering	Yes	Yes	Yes
• Has fungus-resistant and corrosion-resistant outer covering	No	Yes	Yes
• May be used to wire one- and two-family dwellings or multifamily dwellings	Yes	Yes	Yes
• May be installed exposed or concealed in dry or damp locations	Yes	Yes	Yes
• May be embedded in masonry, concrete, plaster, fill	No	No	No
• May be used for concealed wiring in Category 1 and Category 2 locations, *Rules 22–200, 22–202*	No	Yes	Yes
• May be installed in wet locations	No	No	Yes
• May be used for direct burial locations	No	No	Yes
• May be used as service entrance cable	No	No	No
• Must be protected against damage	Yes	Yes	Yes

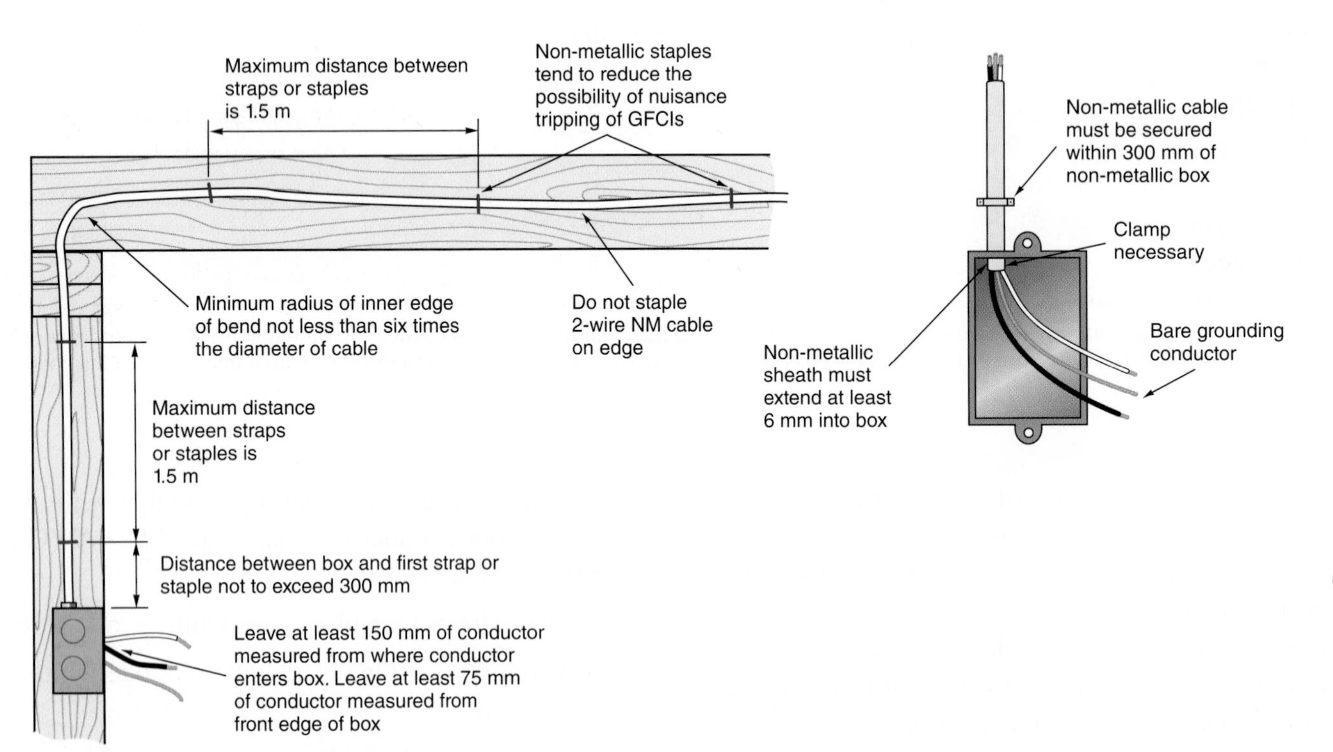

FIGURE 7-9 Installation of non-metallic-sheathed cable.

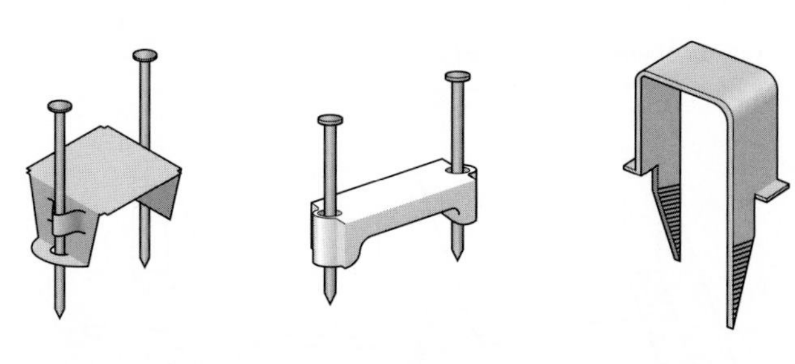

FIGURE 7-10 Devices used for attaching NMSC cable to wood surfaces, *Rule 12–510(1)*.

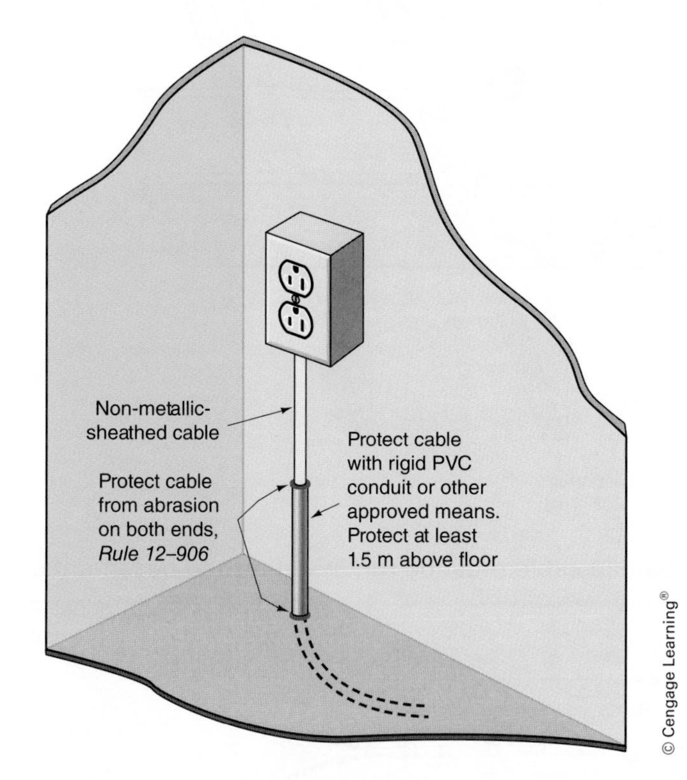

FIGURE 7-11 Installation of exposed non-metallic-sheathed cable where passing through a floor.

Use and Installation of Armoured Cable*

Armoured cable can be used in more applications than non-metallic-sheathed cable. *Rules 12–600* through *12–618* govern the use of armoured cable in dwelling unit installations. AC90 and TECK 90 armoured cable

- may be used on circuits and feeders for applications of 600 volts or less

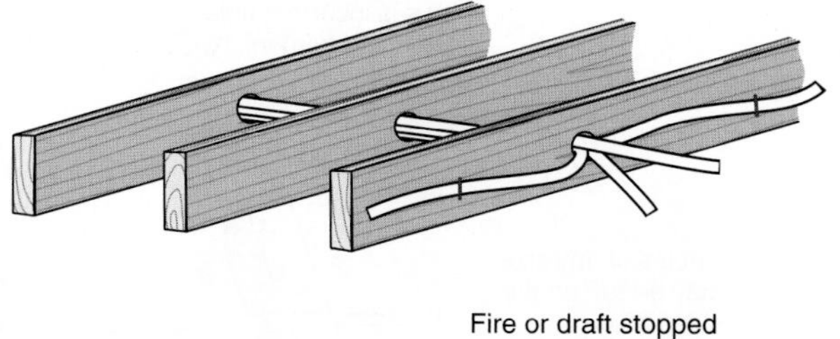

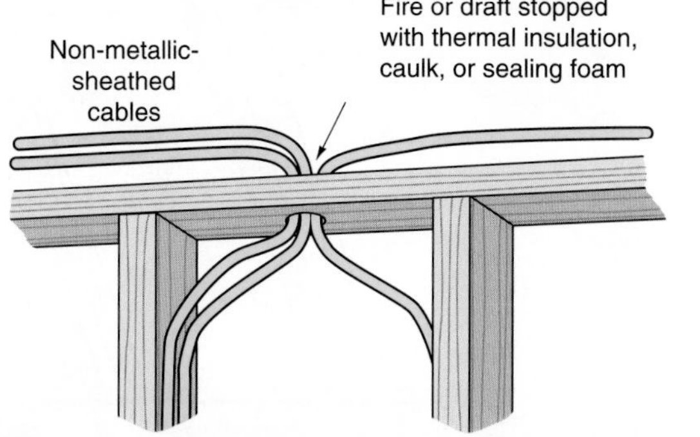

FIGURE 7-12 Bundled cables. When cables are bundled for distances over 600 mm, the correction factor in *CEC Table 5C* applies.

- may be used for open and concealed work in dry locations
- may be run through walls and partitions
- may be laid on the face of masonry walls and buried in plaster if the conductors are not larger than #10 AWG and used for the extension of existing outlets
- must be secured within 300 mm of every outlet box or fitting and at intervals not exceeding 1.5 m
- must not be bent so that the radius of the curve of the inner edge is less than six times the external diameter of the cable
- must have an approved fibre or plastic insulating bushing (anti-short) at the cable ends to protect the conductor insulation (Figure 7-17)

Armoured cable (Types ACWU 90 and TECK 90) is approved for use in these situations:

- underground installations
- burying in masonry or concrete during construction with minimum 50 mm of covering

*Courtesy of CSA Group

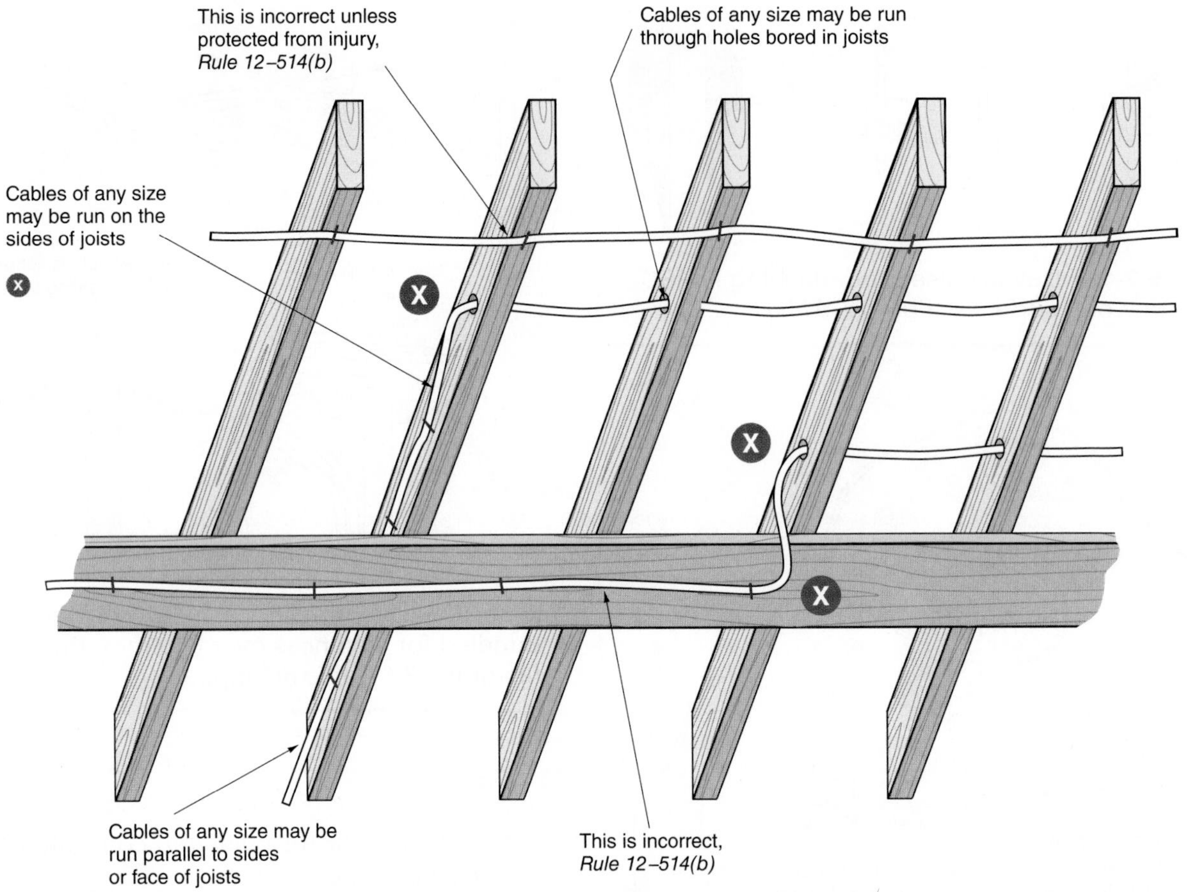

This is incorrect unless protected from injury, *Rule 12–514(b)*
Cables of any size may be run through holes bored in joists
Cables of any size may be run on the sides of joists
Cables of any size may be run parallel to sides or face of joists
This is incorrect, *Rule 12–514(b)*
© Cengage Learning®

FIGURE 7–13 *Rule 12–514* states how non-metallic-sheathed cable must be run in unfinished basements.

Ceiling joist
Non-metallic-sheathed cable
Ceiling boxes must have provisions for securing the non-metallic-sheathed cable

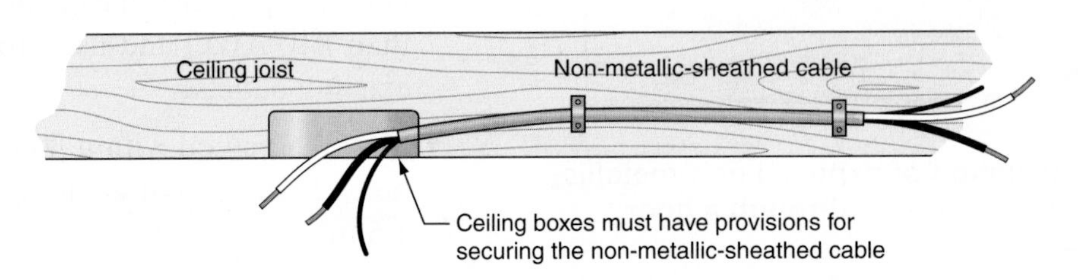

FIGURE 7-14 *Rule 12–524* requires boxes to be of a type for use with NMSC.

- installation in any location exposed to weather
- installation in any location exposed to oil, gasoline, or other materials that have a destructive effect on the insulation

To remove the outer metal cable armour, you should use a tool like the one shown in Figure 7-18A or a hacksaw (Figure 7-18B) to cut through one of the raised convolutions of the cable armour. Do not cut too deep or you will cut into the conductor insulation. Then bend the cable armour at the cut. It will snap off easily, exposing the desired amount of conductor.

To prevent cutting of the conductor insulation by the sharp metal armour, insert an anti-short bushing at the cable ends (Figure 7-17).

Both non-metallic-sheathed cable and armoured cable have advantages that make them suitable for particular types of installations. However, the type of cable to be used in a specific situation depends largely on the wiring method permitted or required by the local building code.

Ground screw tapped hole

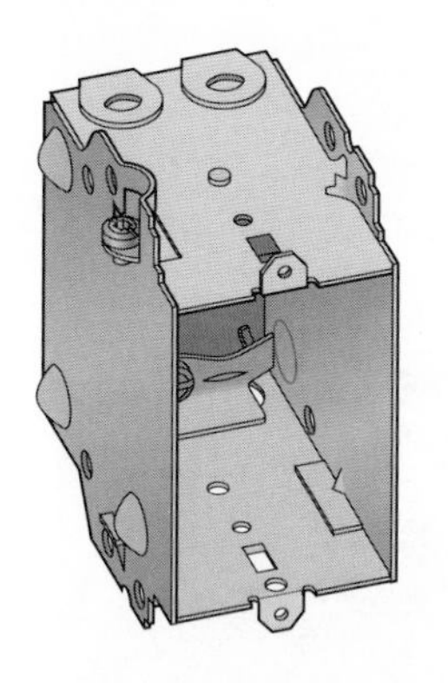

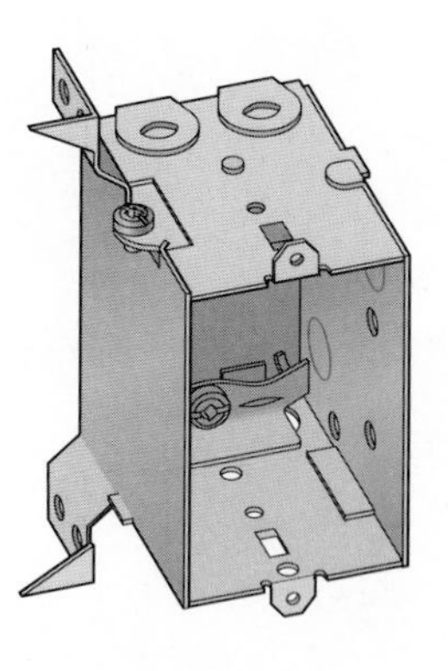

Courtesy of Thomas & Betts Corporation

FIGURE 7-15 Gangable device boxes with clamps for NMSC.

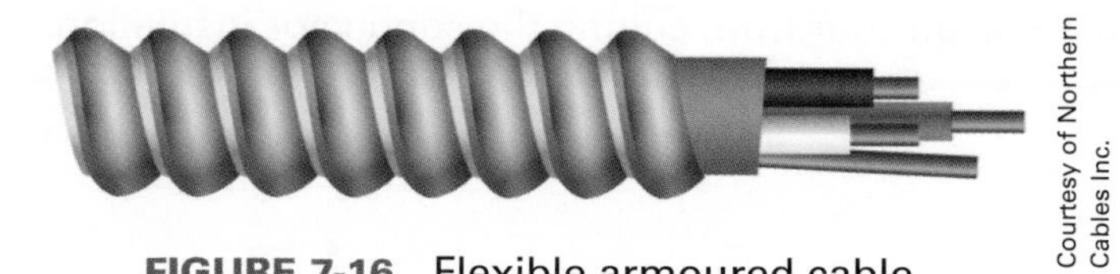

Courtesy of Northern Cables Inc.

FIGURE 7-16 Flexible armoured cable.

INSTALLING CABLES THROUGH WOOD AND METAL FRAMING MEMBERS (*RULE 12–516*)

To complete the wiring of a residence, you must run cables through studs, joists, and rafters. One method is to run the cables through holes drilled at the approximate centres of wooden building members, or at least 32 mm from the nearest edge. Holes bored through the centre of standard 2 × 4s meet the requirements of *Rule 12–516* (Figure 7-19).

Where cables and certain types of raceways are run concealed or exposed parallel to a stud, joist, rafter, or other building framing member, the cables and raceways must be supported and installed so that there is at least 32 mm from the edge of the stud, joist, or other framing member to the cable or raceway. If the 32-mm clearance cannot be maintained, a steel plate or equivalent at least 1.59 mm thick must be used to protect the cable or raceway from damage that can occur when nails or screws are driven into the wall or ceiling, *Rule 12–516;* see Figures 7-19 and 7-20.

This additional protection is not required when the raceway is rigid metal conduit, rigid PVC conduit, or EMT.

You cannot install additional protection inside the walls or ceilings of an existing building where the walls and ceilings are already closed in. *Rule 12–520* permits armoured cables to be "fished" between boxes or other access points without additional protection. The logic is similar to the exception that permits non-metallic-sheathed cable to be fished between outlets without requiring supports as stated in *Rule 12–520.*

The recreation room of this residence requires adequate protection against physical damage for the wiring. The walls in the recreation room are to be panelled. Therefore, if the carpenter uses 1 × 2 or 2 × 2 furring strips (Figure 7-21A), non-metallic-sheathed cable would require the additional 1.59 mm steel plate protection up the walls from the receptacle outlet and switch. This could be quite costly and time-consuming, and a good case could be made for doing the installation in EMT.

Watch out for any wall partitions where 2 × 4 studs are installed "flat"; see Figure 7-21B. In this case, you can either provide mechanical protection as required by *Rule 12–516(2)* or install the wiring in EMT.

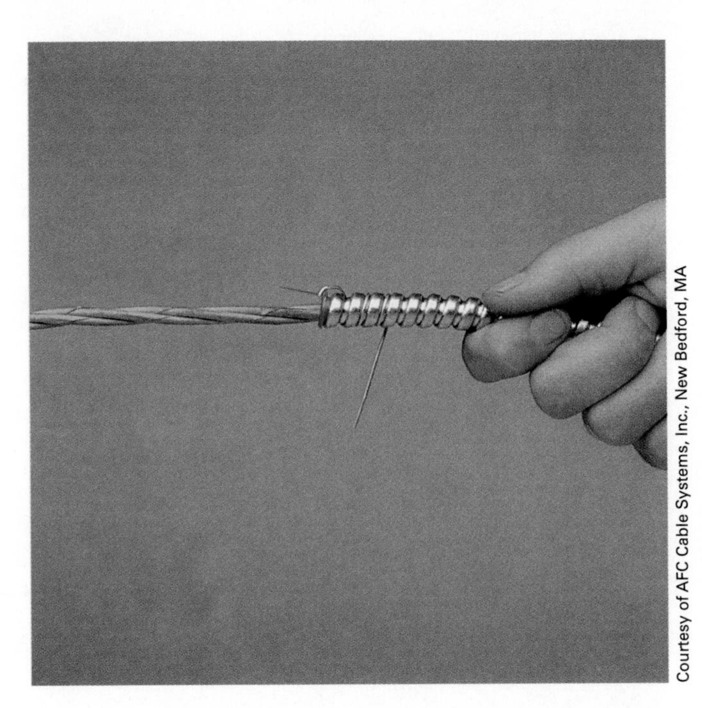

Courtesy of AFC Cable Systems, Inc., New Bedford, MA

FIGURE 7-17 Anti-short bushing prevents the sharp metal armour from cutting the conductor insulation.

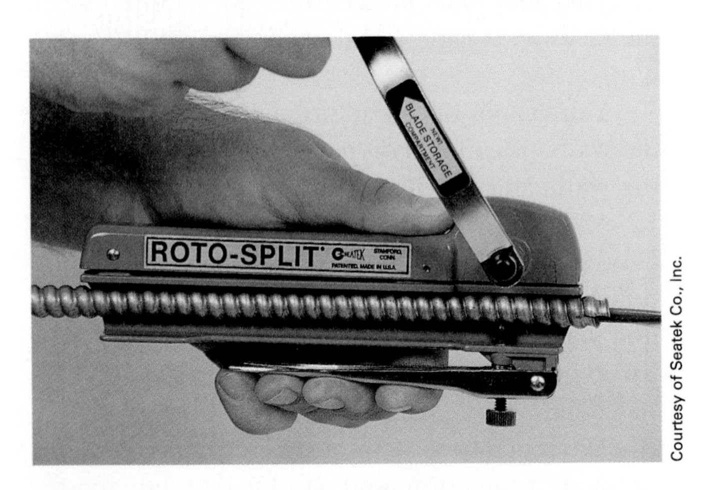

Courtesy of Seatek Co., Inc.

FIGURE 7-18A This tool precisely cuts the outer armour of armoured cable, making it easy to remove with a few turns of the handle.

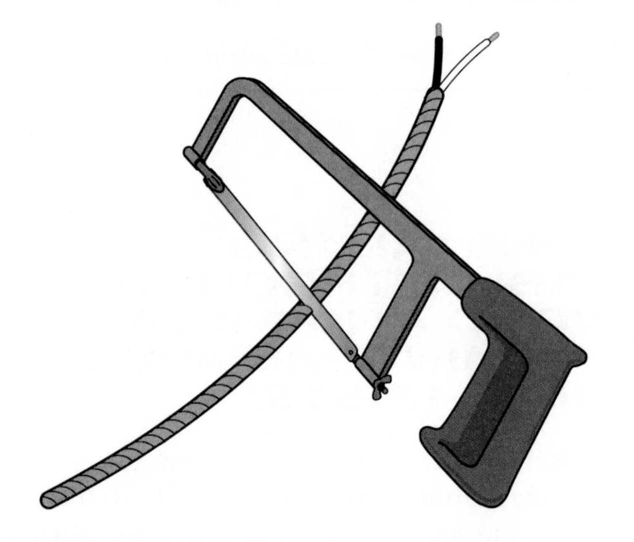

© Cengage Learning®

FIGURE 7-18B A hacksaw can be used to cut through a raised convolution of the cable armour.

Steel framing is used in a wide variety of residential applications, including single and multi-dwelling buildings. The framing is galvanized sheet steel that is pre-punched to accommodate the installation of electrical and mechanical piping and cabling. Steel framing is used for non-structural (light-gauge) and structural (heavy-gauge) applications.

When installing metal boxes on steel framing, you can use self-piercing screws on light-gauge framing, but self-drilling screws will be required on heavy-gauge framing. A minimum of three threads should penetrate the framing. Steel studs do not provide as much support as wooden studs; therefore, extra support must be provided to prevent movement of the box after the drywall has been installed, *Rule 12–3010.*

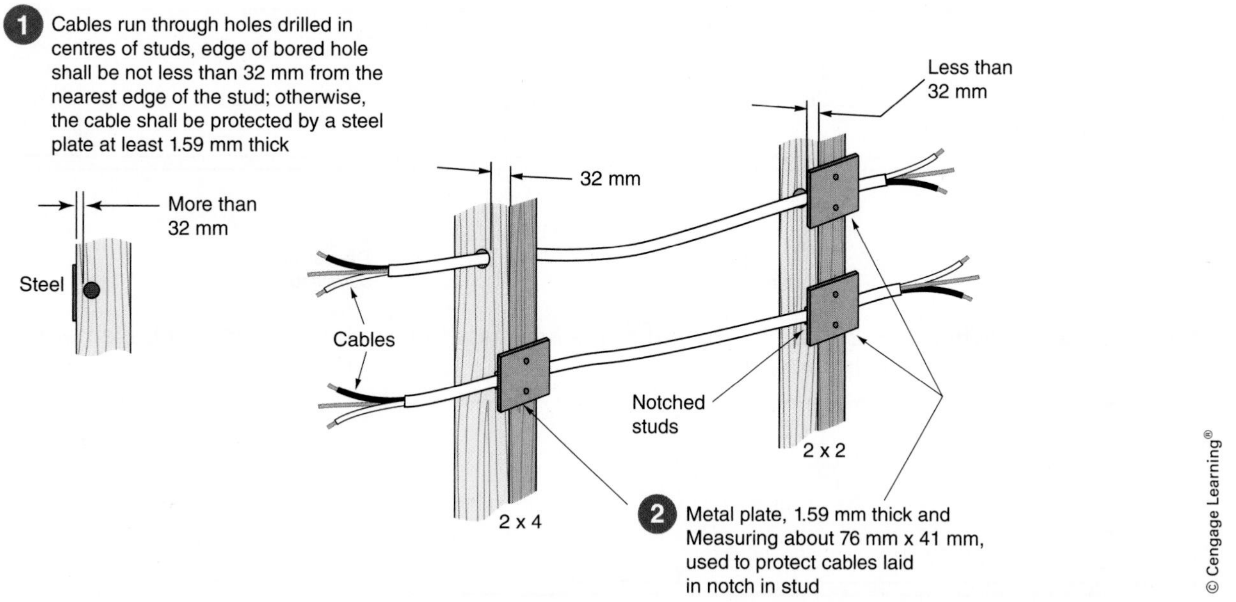

FIGURE 7-19 Methods of protecting cables, *Rule 12–516.* This rule does not apply to rigid metal conduit and EMT.

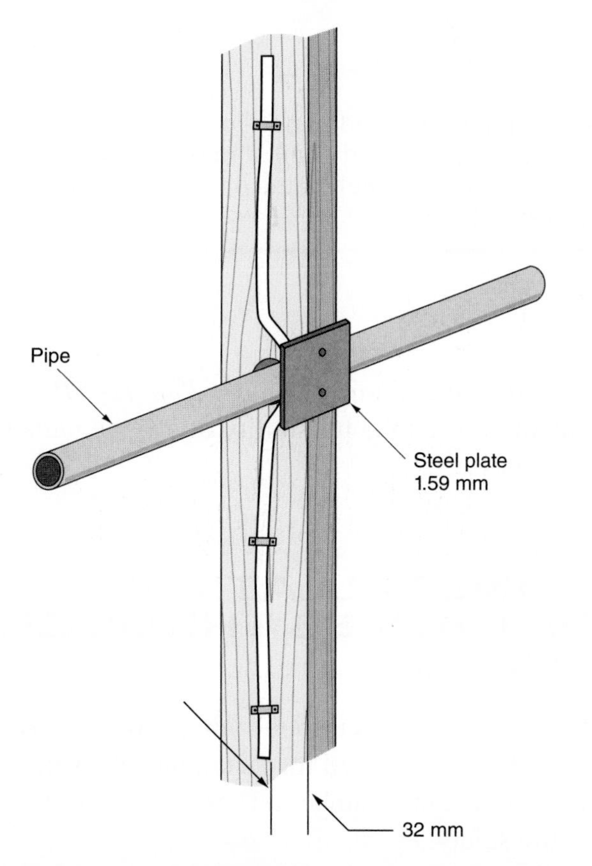

FIGURE 7-20 Where cables or certain types of raceways are run parallel to a joist or stud (any framing member), keep the cable 32 mm from the edge of the framing member. Where this clearance is impossible to maintain, install a steel plate at least 1.59 mm thick, *Rule 12–516.*

When running NMSC through metal framing, you must protect the cable from mechanical injury by an approved insert where it passes through a framing member, *Rule 12–516* (Figure 7-22).

INSTALLATION OF CABLE IN ATTICS

Wiring in the attic must be done in cable and meet the requirements of *Rules 12–500* through *12–526* for *non-metallic-sheathed* cable.

In accessible attics (see Part A of Figure 7-23), cables must not be installed

- across the top of floor or ceiling joists ①,
- run across the face of studs ② or rafters ③ that have a vertical distance between the floor or floor joists exceeding one metre.

Guard strips are not required if the cable is run along the sides of rafters, studs, or floor joists ④, provided *Rule 12–516(1, 2)* is met.

Part C of Figure 7-23 illustrates a cable installation that most electrical inspectors consider to be safe. Because the cables are installed close to the point where the ceiling joists and the roof rafters meet, they are protected from physical

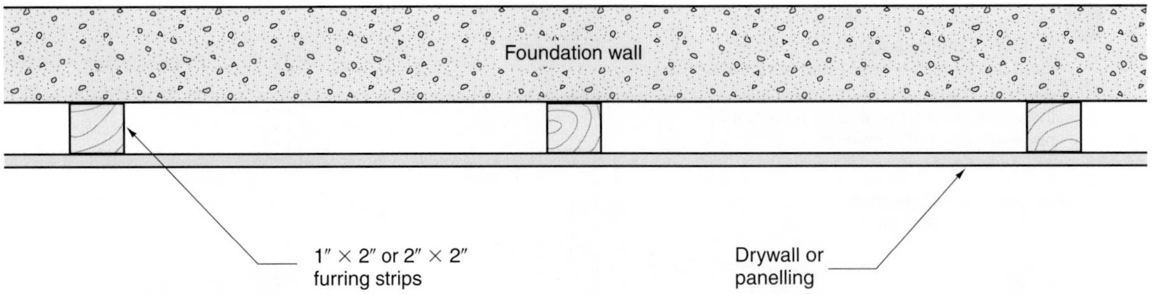

Foundation wall
1″ × 2″ or 2″ × 2″ furring strips
Drywall or panelling

FIGURE 7-21A When 1 × 2 or 2 × 2 furring strips are installed on the surface of a basement wall, additional protection must be provided.

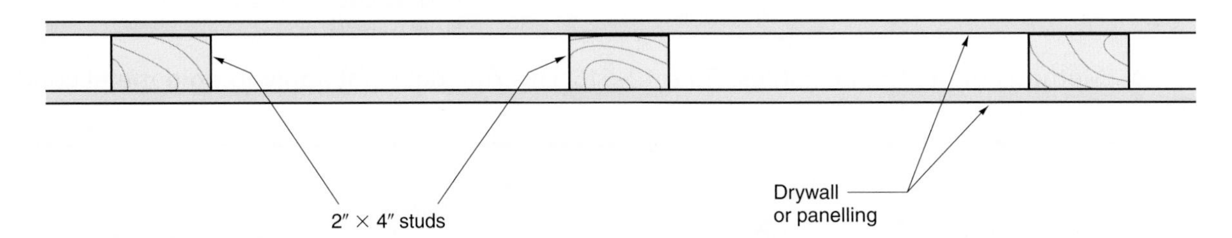

FIGURE 7-21B When 2 × 4 studs are installed "flat," such as might be found in non-load-bearing partitions, a cable running parallel to or through the studs must be protected against nails or screws being driven through the cable. This means protecting the cable with 1.59-mm steel plates or equivalent for its entire length, or installing the wiring in a metallic raceway system.

damage. It would be very difficult for a person to crawl into this space, or to store cartons in an area with a clearance of less than one metre (Part B of Figure 7-23). Although the plans for this residence show a 600-mm-wide catwalk in the attic, the owner may decide to install additional flooring in the attic to obtain more storage space. Because of the large number of cables required to complete the circuits, it would interfere with flooring to install guard strips wherever the cables run across the tops of the joists; see Figure 7-24. However, the cables can be run through holes bored in the joists and along the sides of the joists and rafters. In this way, the cables do not interfere with the flooring. Note that most building codes will not allow manufactured roof systems to be drilled. This includes roof trusses.

When running cables parallel to framing members, maintain at least 32 mm between the cable and the edge of the framing member, *Rule 12–516 (1, 2)*. This minimizes the possibility of driving nails into the cable.

INSTALLATION OF CABLES THROUGH DUCTS

Rule 2–128 is strict about what types of wiring methods are permitted for installing cables through ducts or plenum chambers. These stringent rules are for fire safety.

The *CEC* permits Type NMD 90 cable to be installed in joist and stud spaces (e.g., cold air returns) in dwellings. It recommends that Type NMD 90 pass through such spaces only if the cable is run perpendicular to its long dimensions (Figure 7-25).

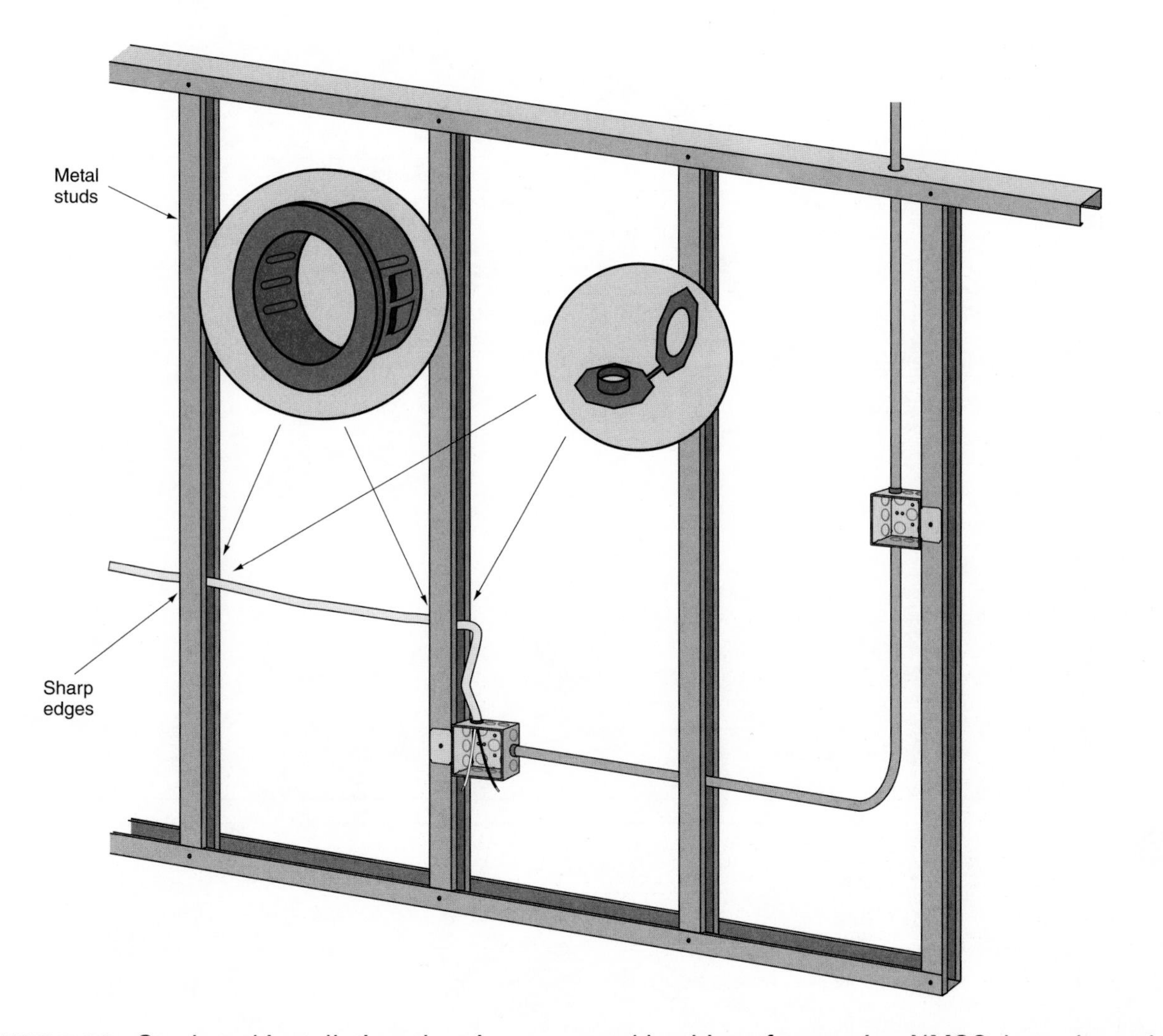

FIGURE 7-22 Steel stud installation showing approved bushings for running NMSC through steel studs.

CONNECTORS FOR INSTALLING NON-METALLIC-SHEATHED AND ARMOURED CABLE

The connectors shown in Figures 7-26A and 7-26B are used to fasten non-metallic-sheathed cable and armoured cable to the boxes and panels in which they terminate. These connectors clamp the cable securely to each outlet box. Many boxes have built-in clamps (Figure 7-26C) and do not require the separate connectors. Knockouts and pryouts on the sides of boxes allow installation of cables and connectors; see Figure 7-26D.

Unless the CSA listing of a specific cable connector indicates that the connector has been tested for use with more than one cable, the rule is *one cable, one connector.*

ELECTRICAL METALLIC TUBING *(RULE 12–1400)*, RIGID METAL CONDUIT *(RULE 12–1000)*, AND RIGID PVC CONDUIT *(RULES 12–1100 THROUGH 12–1124)*

In a building constructed of cement block, cinder block, or poured concrete, it is necessary to make the electrical installation in conduit.

According to the *CEC* installation requirements, EMT and rigid metal conduit:

- may be used for open or concealed work
- may be buried in concrete or masonry (with certain provisions)

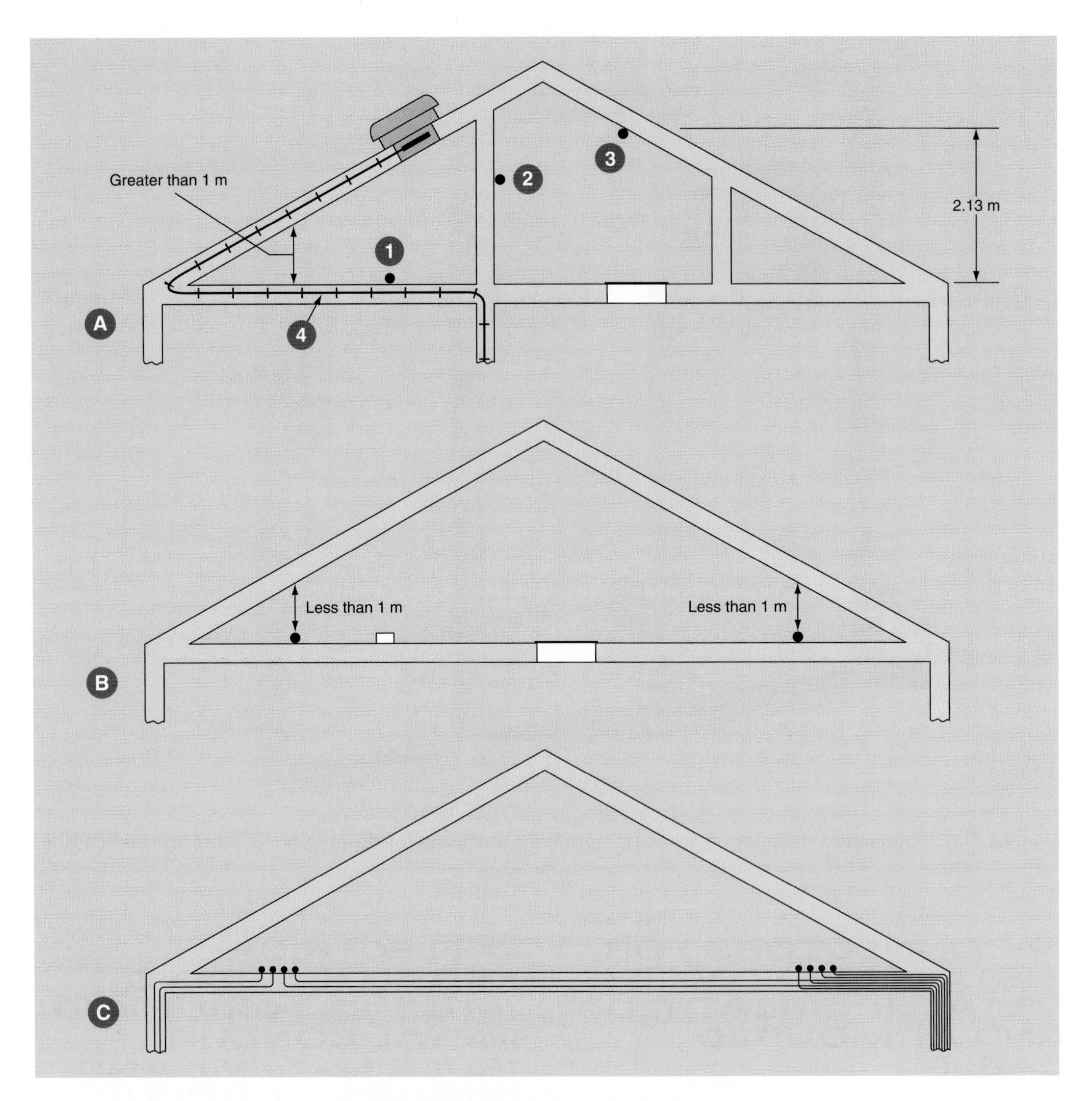

FIGURE 7-23 Protection of cable in attic.

- may be used in combustible or noncombustible construction

In general, EMT must be supported within 1 m of each box or fitting and at 1.5 m-intervals along runs for 16-mm and 21-mm conduit; see Figure 7-27 and *Rule 12–1010.*

The number of conductors permitted in EMT and rigid conduit is given in *Table 6.* Heavy-wall rigid conduit provides greater protection against mechanical injury to conductors than does EMT, which has a thinner wall but is more expensive.

The residence specifications in this text indicate that a meter base is to be supplied and installed by the electrical contractor. Main Panel A, a combination panel, is located in the workshop. The meter base is located on the outside of the house as

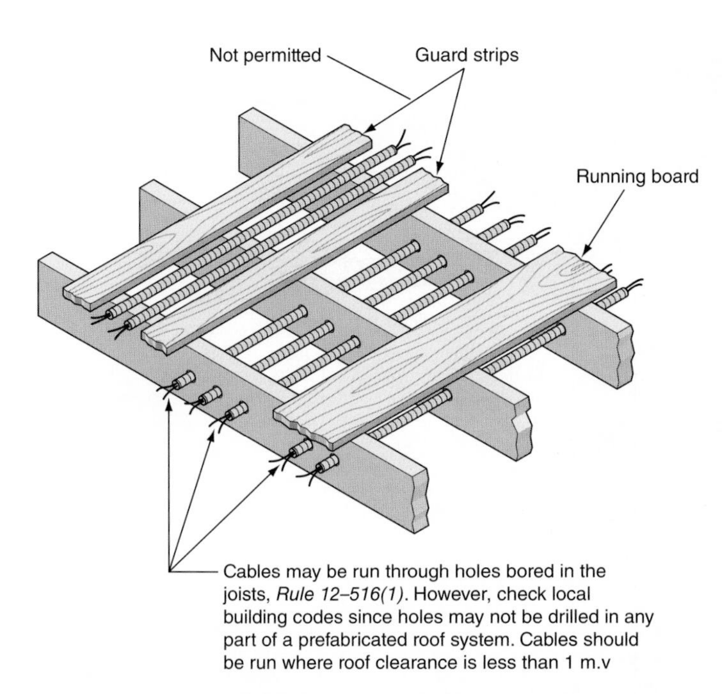

FIGURE 7-24 Methods of protecting cable installations in attics.

indicated on the electrical plan. The electrical contractor must furnish and install a 53-mm rigid PVC conduit between the meter base and Main Panel A.

Rule 12–906 requires that insulating bushings or equivalent be used where No. 8 or larger conductors enter a raceway. Also, if rigid metallic conduit is used between the meter base and the combination panel, bonding-type bushings must be used on the service entrance conduit. (See *Rule 10–606* for methods of bonding at service equipment.)

The plans indicate that EMT is to be run from the workshop to the attic.

Properly installed EMT provides good continuity for equipment bonding, a means of withdrawing or pulling in additional conductors, and excellent protection against physical damage to the conductors.

RIGID PVC CONDUIT (*RULES 12–1100* THROUGH *12–1124)*

PVC conduit is easily handled, stored, cut, and joined. It is commonly used for service entrance conduit in residential services in both overhead and underground applications. Easy to work with, it provides a rigid, watertight conduit for the consumer's service conductors.

When joining PVC to a fitting, you must cut, ream, clean, and glue the conduit. To cut PVC, use a hacksaw; for larger sizes of conduit, use a mitre box so that the cut is square. *Measure twice, cut once.* After cutting PVC conduit, ream out the burrs with a knife and clean the end of the pipe and inside of the fitting with an approved PVC pipe cleaner. Then apply a liberal coat of solvent cement to both surfaces, slide together, and give a quarter turn to make sure the solvent is spread evenly on the material. Hold together for a few seconds, and the joint is made. See Figure 7-28, Parts A and B.

Do not use an open flame to bend PVC, *Rule 12–1108* and *Appendix B.* Instead, use an approved heat gun, heat blanket, or other flameless heat source to heat the pipe to about 125°C; see Parts C and D of Figure 7-28. Always heat an area of at least 10 times the diameter of the conduit before attempting to bend it. If not heated enough, the pipe will kink or collapse. When you make a 90° bend, the centre radius of the bend must meet the minimum radius requirements of *Table 7, Rule 12–922.*

Thus, for 16-mm conduit, the minimum radius is 102 mm. If the radius is ten times the internal diameter, it is a lot easier to fish and pull in wires after the conduit is installed. Draw a template of the bend first, and then heat the conduit and shape it to the template; see Part E of Figure 7-28.

To provide adequate support for the conduit system, you must use proper straps. The two-hole strap in Figure 7-29, Part O, allows the conduit to move as it expands or contracts. Never clamp the conduit tightly, *Rule 12–1114(3).* The spacing for supports is about half the distance for other types of conduit; therefore, twice the number of straps are needed. See *Rule 12–1114* for maximum support spacing.

Rule 12–1118 and *Appendix B* give a formula for calculating the linear expansion of various materials. Since PVC has one of the highest coefficients of linear expansion, check that the maximum expansion of the conduit due to temperature change will not exceed 45 mm, *Rule 12–1118.* The formula is

$$\Delta l = L \times \Delta T \times C$$

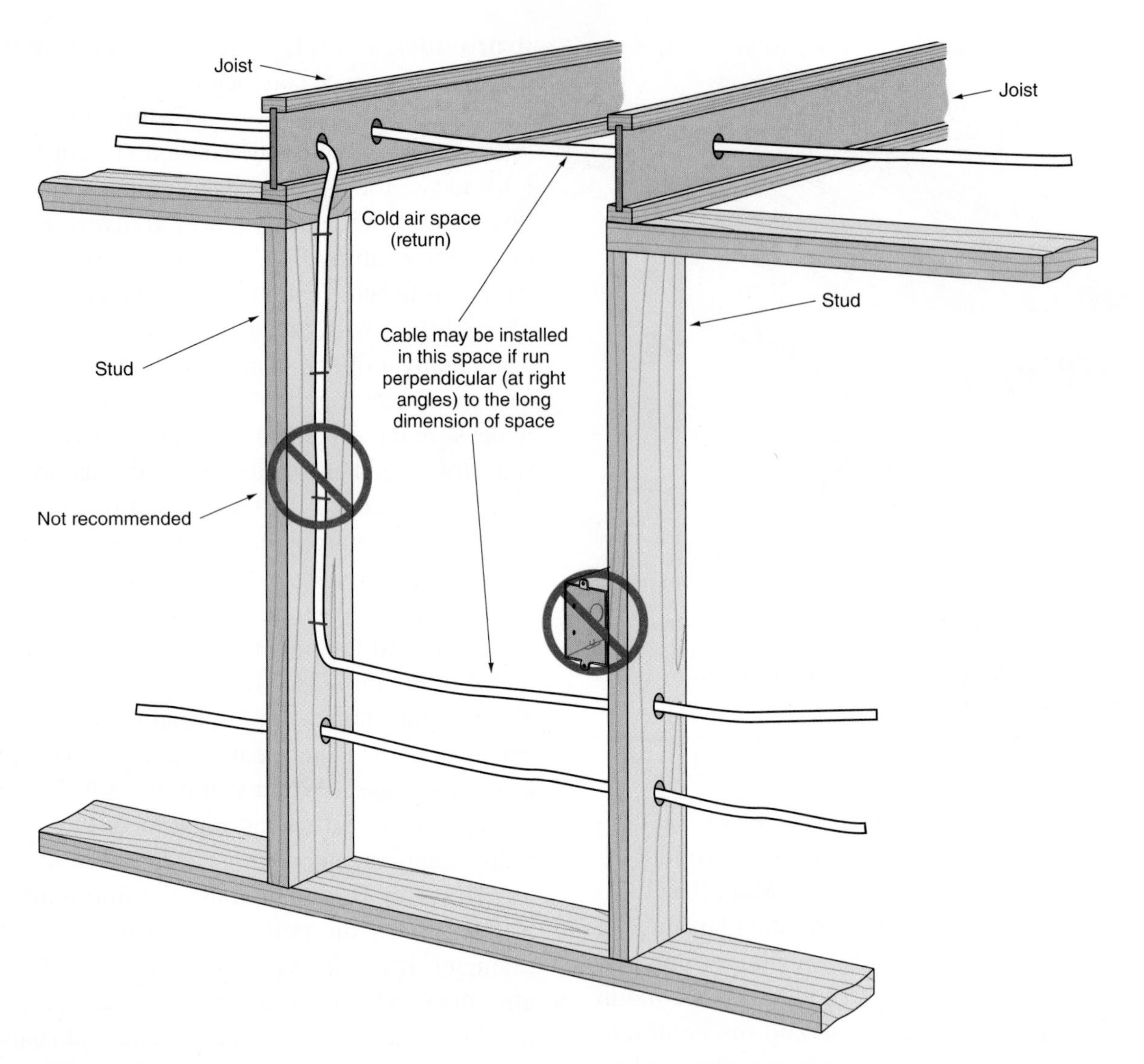

FIGURE 7-25 Non-metallic-sheathed cable may pass through cold air return, joist, or stud spaces in a dwelling.

FIGURE 7-26A Various cable connectors for non-metallic-sheathed cable.

FIGURE 7-26B Various cable connectors for armoured cable and flexible conduit.

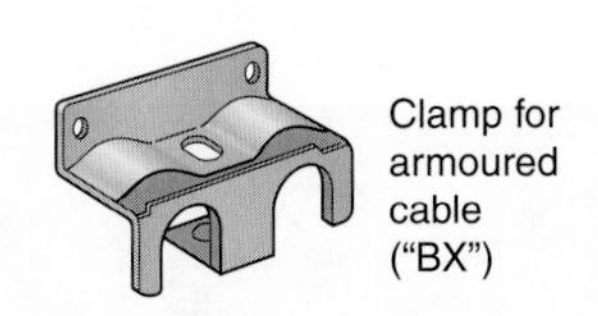

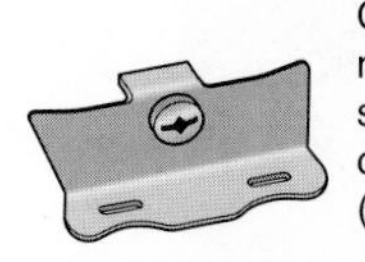

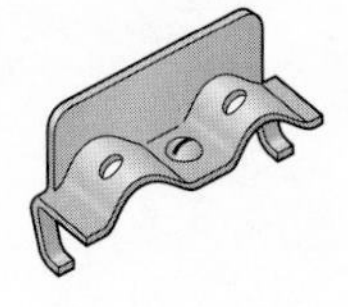

FIGURE 7-26C Typical cable clamps for boxes.

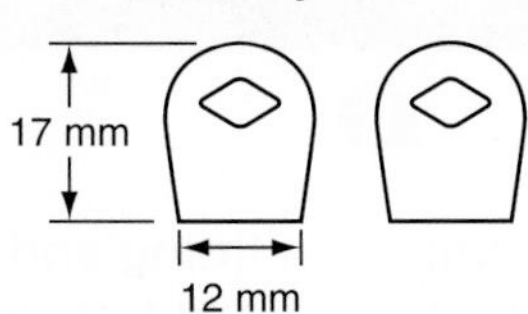

FIGURE 7-26D Knockouts and pryouts. Note the difference between the nominal trade sizes of the knockouts and the actual diameters.

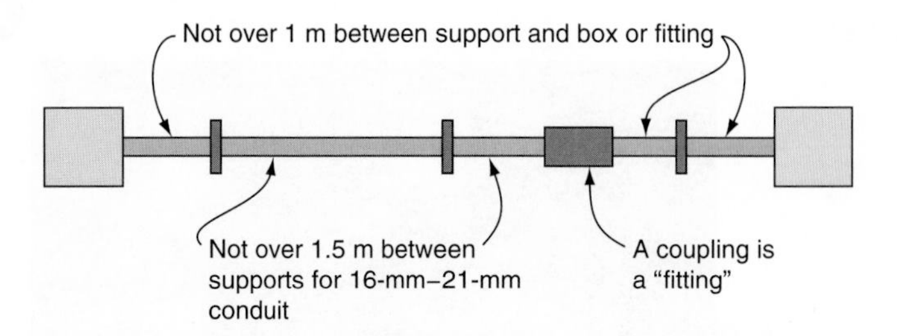

FIGURE 7-27 *CEC* requirements for support of rigid metal conduit *(Rule 12–1010)* and EMT *(Rule 12–1406).* For supporting rigid PVC conduit, refer to *Rule 12–1114.*

where

Δl = Change in length, in millimetres, of a run of rigid PVC conduit, due to the maximum expected variation in temperature

L = Length of the run of conduit in metres

ΔT = Maximum expected temperature change, in degrees Celsius

C = Coefficient of linear expansion

EXAMPLE

A 25-m length ($L = 25$) of rigid PVC conduit is run outside where the minimum expected temperature is −40°C and the maximum expected temperature is −40°C ($\Delta T = 80$). The coefficient of linear expansion for PVC conduit is 0.0520 mm per m per °C ($C = 0.052$), *Rule 12–1118* and *Appendix B.*

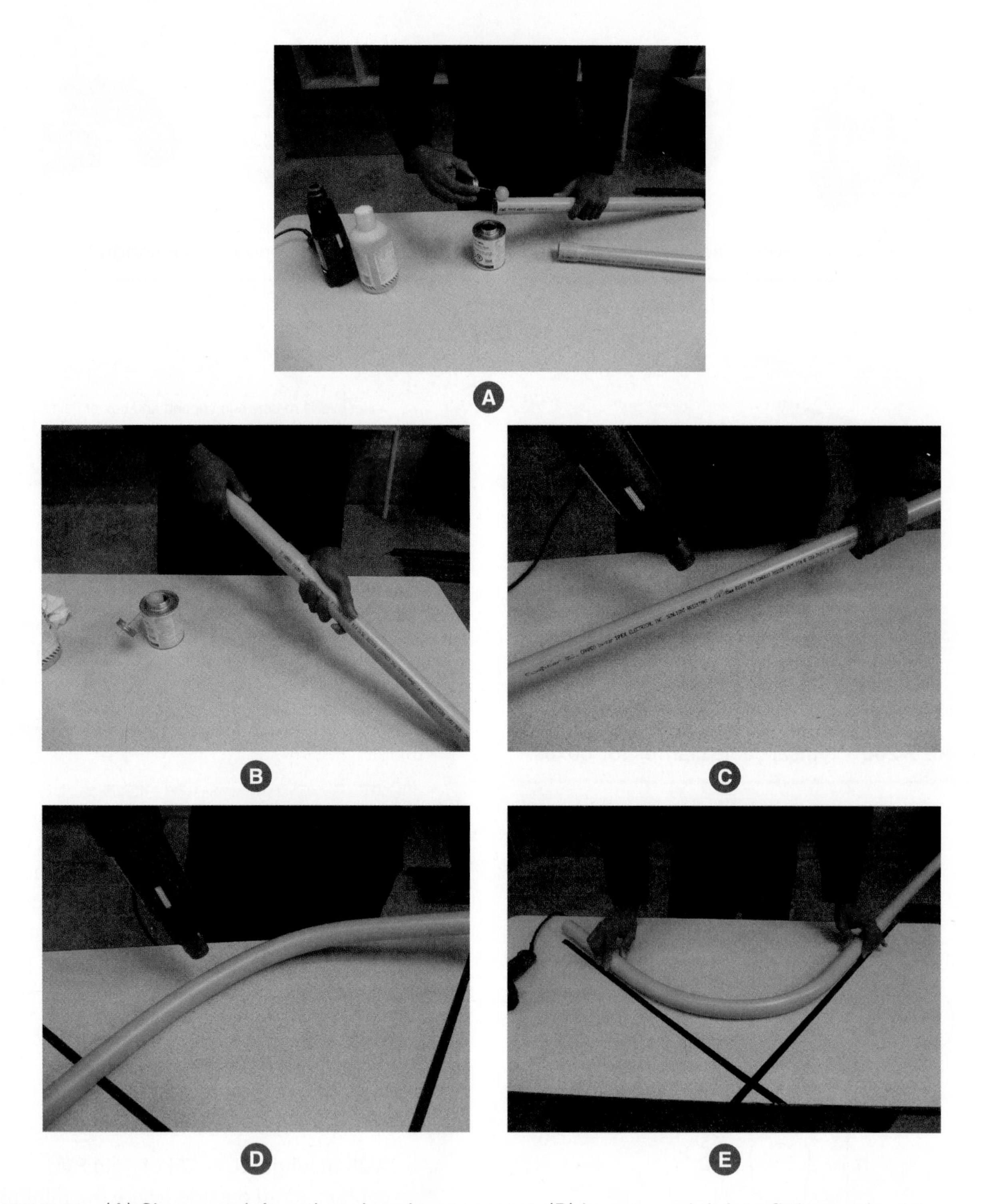

Photos A–E: Courtesy of Sandy F. Gerolimon

FIGURE 7-28 (A) Clean conduit and apply solvent cement. (B) Insert conduit into fitting and rotate a quarter turn. (C) Heat conduit until soft using approved heater or heat gun (D) for bending conduit. (E) Use template for 90° bends, *Rule 12–1108* and *Appendix B.*

SOLUTION

$$\Delta l = L \times \Delta T \times C$$

$$= 25 \times 80 \times 0.052$$

$$= 104 \text{ mm}$$

The expansion and contraction of the conduit exceeds the 45-mm maximum permitted in *Rule 12–1118;* therefore, an "O" ring expansion joint with 104 mm of expansion will be required; see Figure 7-30. Generally, where the expected temperature change is more than 14°C, you should use expansion joints.

FLEXIBLE CONNECTIONS *(RULES 12–1000 AND 12–1300)*

Some equipment installations require flexible connections, both to simplify the installation and to stop the transfer of vibrations. In residential wiring, flexible connections are used to wire attic fans, food waste disposers, dishwashers, air conditioners, heat pumps, recessed fixtures, and similar equipment.

The three types of flexible conduit used for these connections are

- flexible metal conduit; Figure 7-31, Part A
- liquidtight flexible metal conduit; Figure 7-31, Part B
- liquidtight flexible non-metallic conduit; Figure 7-31, Part C

Figure 7-32 shows types of connectors used with flexible conduit.

FLEXIBLE METAL CONDUIT *(RULES 12–1002 THROUGH 12–1014)*

Rule 12–1002 covers the use and installation of flexible metal conduit. This wiring method is similar to armoured cable, except that the conductors are installed by the electrician. For armoured cable, the cable armour is wrapped around the conductors at the factory to form a complete cable assembly.

Figure 7-33 shows some installations that commonly use flexible metal conduit. The flexibility required to make the installations is provided by the flexible metal conduit, and a separate bonding conductor is required.

Summarizing *Rules 12–1002* through *12–1014* concerning flexible metal conduit:

- Do not install in wet locations unless the conductors are suitable for that use (RW90, TWN75). Ensure that water will not enter the enclosure or other raceways to which the flex is connected.
- Do not bury in concrete.
- Do not bury underground.
- Do not use in locations subject to corrosive conditions.
- Do not use the metal armour of flexible conduit as a bonding means; a separate bonding conductor must be installed inside flexible conduit, *Rule 12–1306.*
- Runs of 12-mm trade size flexible metal conduit are not permitted to be longer than 1.5 m, *Rule 12–1004.* Fixture "drops" are a good example. (See Unit 19 for a discussion of fixture drops.)
- Support conduit every 1.5 m.
- Support conduit within 300 mm of box, cabinet, or fitting.
- If flexibility is needed, support conduit within 900 mm of termination.
- Do not conceal angle-type fittings because access will be needed when pulling in wires.
- You may use conduit for exposed or concealed installations.
- Same conductor fill rules as regular conduit, *Table 6A–K.* The cross-sectional area of 12-mm trade size conduit is 118 mm^2, *Rule 12–1304(2).*
- Use fittings approved for use with flexible metal conduit.

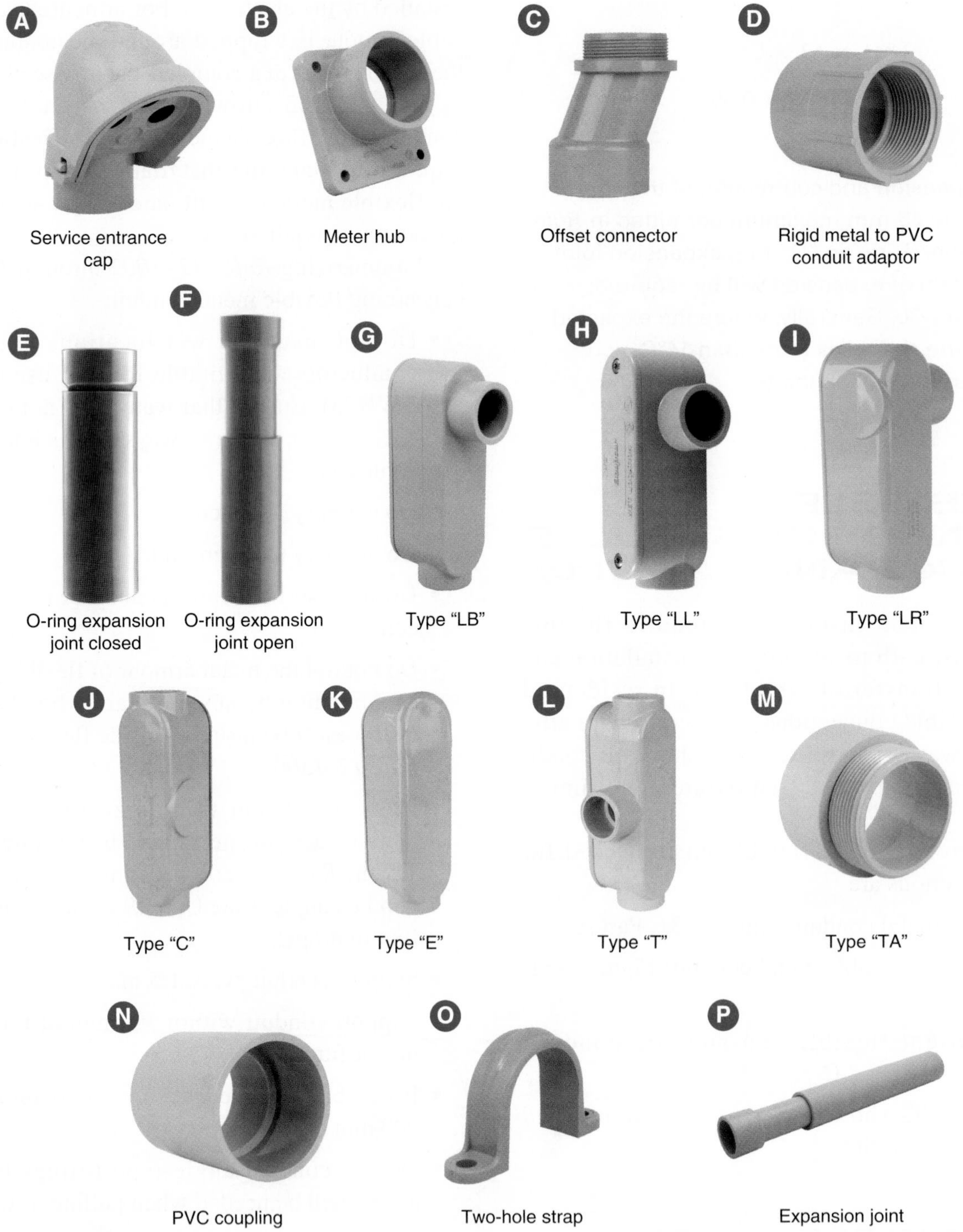

A Service entrance cap
B Meter hub
C Offset connector
D Rigid metal to PVC conduit adaptor
E O-ring expansion joint closed
F O-ring expansion joint open
G Type "LB"
H Type "LL"
I Type "LR"
J Type "C"
K Type "E"
L Type "T"
M Type "TA"
N PVC coupling
O Two-hole strap
P Expansion joint

Photos A–P: Courtesy IPEX Inc.

FIGURE 7-29 Various PVC fittings: Access fittings, a TA fitting for connection to a panel or a box, a coupling, and a two-hole PVC strap for support.

FIGURE 7-30 PVC "O" ring expansion joint.

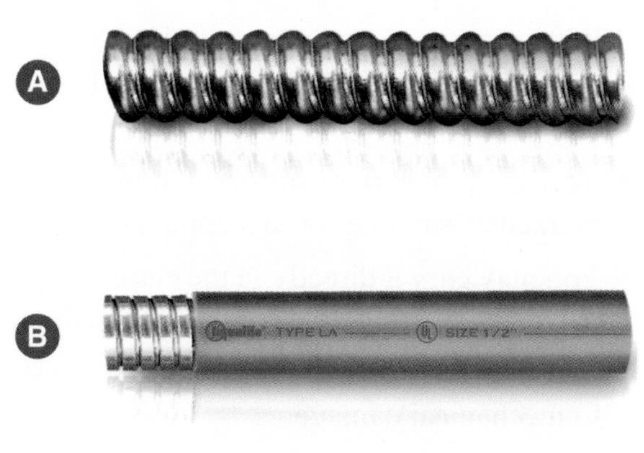

FIGURE 7-31 (A) Flexible metal conduit.
(B) Liquidtight flexible metal conduit.
(C) Liquidtight flexible non-metallic conduit.

Photos A–C: Courtesy of Electri-Flex Co.

LIQUIDTIGHT FLEXIBLE METAL CONDUIT *(RULES 12–1300* THROUGH *12–1306)**

The use and installation of liquidtight flexible metal conduit are described in *Rules 12–1300* through *12–1306*. Liquidtight flexible metal conduit has a "tighter" fit to its spiral turns than standard flexible metal conduit, as well as a thermoplastic outer jacket that is liquidtight. It is commonly used as the flexible connection to a central air-conditioning unit located outdoors.

*Courtesy of CSA Group

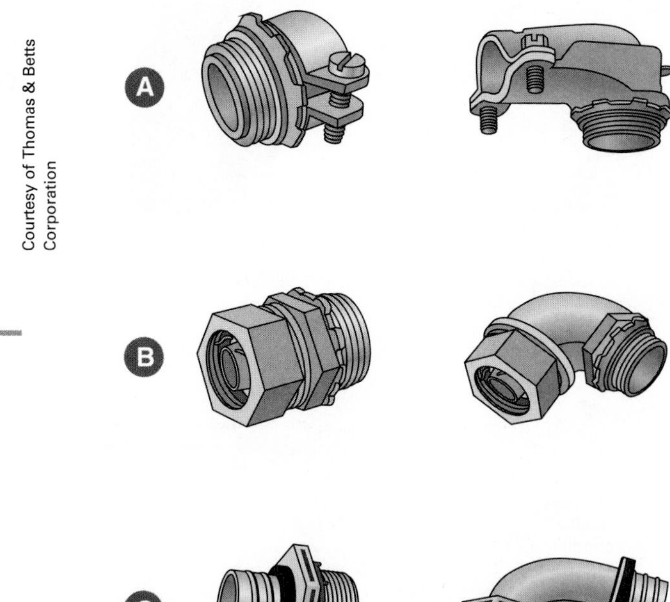

Courtesy of Thomas & Betts Corporation

FIGURE 7-32 Fittings for (A) flexible metal conduit, (B) liquidtight flexible metal conduit, and (C) liquidtight flexible non-metallic conduit.

Figure 7-34 shows the *CEC* rules for the use of liquidtight flexible metal conduit.

Liquidtight flexible metal conduit:

- May be used for exposed and concealed installations.
- May be buried directly in the ground if so listed and marked.
- May not be used where subject to physical abuse.
- May not be used if ambient temperature and heat from conductors will exceed 60°C unless marked for a higher temperature.
- If 12-mm trade size, is not permitted longer than 1.5 m.
- Must have the same conductor fill as regular conduit, *Rule 12–1014*.
- Must have conductor fill for 12-mm size based on cross-sectional area of 118 mm^2, *Rule 12–1304(2)*.
- Must be used only with approved fittings.

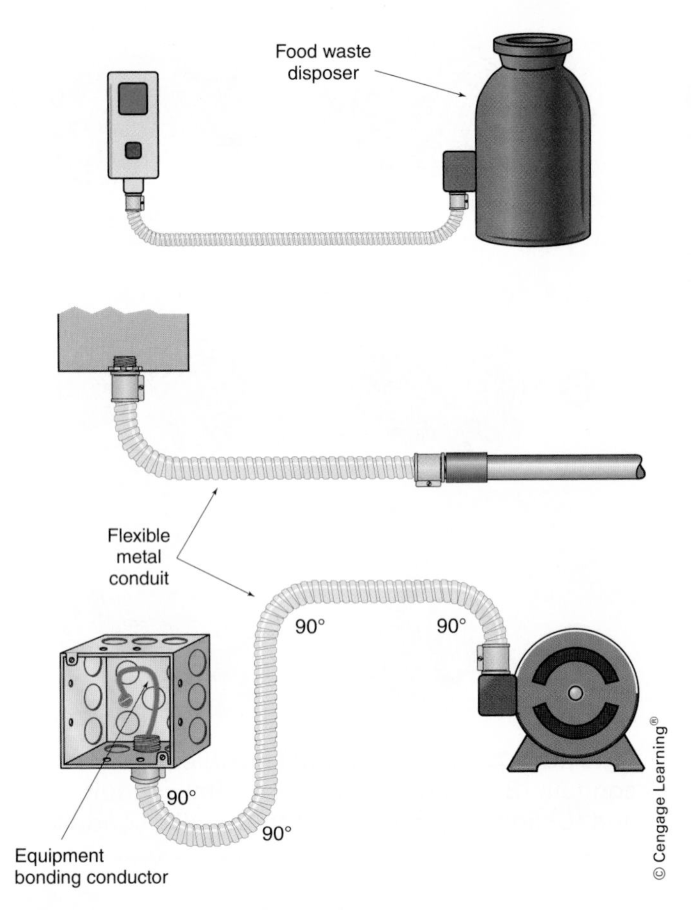

FIGURE 7-33 Some common places where flexible metal conduit may be used.

- Must not be used when flexing at low temperatures will cause injury.
- Must not be used as general-purpose raceway.
- Is not acceptable as a grounding means, *Rule 10–618(3).*
- Must not have concealed angle-type fittings.

LIQUIDTIGHT FLEXIBLE NON-METALLIC CONDUIT *(RULES 12–1300 THROUGH 12–1306)**

Summarizing *CEC* rules pertaining to liquidtight flexible non-metallic conduit:

- Do not use in direct sunlight unless specifically marked for use in direct sunlight.
- You may use it in exposed or concealed installations.
- It can become brittle in extreme cold applications.
- You may use it outdoors when listed and marked as suitable for this application.
- You may bury it directly in the earth when listed and marked as suitable for this application.
- You may not be use it where it might be subject to mechanical damage.
- You may not use it if ambient temperature and heat from the conductors will exceed the temperature limitation of the non-metallic material.
- 12-mm trade size is allowed for enclosing the leads to a motor, *Rule 12–1302(2).*
- Conductor fill is the same as for regular conduit, *Table 6, Rule 12–1014.*
- It must be used with approved fittings.
- A separate bonding conductor is required and should be sized with reference to *Table 16A* and installed according to *Rule 10–808.*

*Courtesy of CSA Group

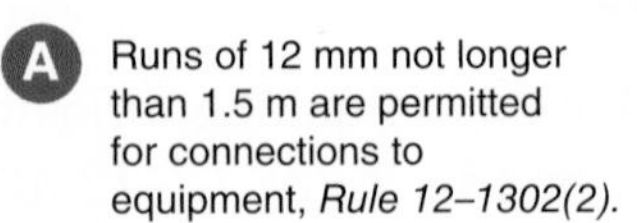

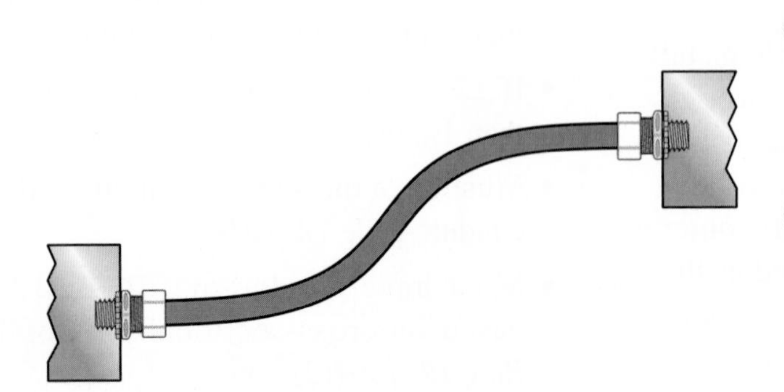

FIGURE 7-34 Liquidtight flexible metal conduit rules, *Rules 12–1300* through *12–1306.*

REVIEW

Note: Refer to the *CEC* or the blueprints provided with this textbook when necessary. Where applicable, responses should be written in complete sentences.

1. What is the largest size of solid wire that is commonly used for branch circuits and feeders?

 ____

2. What is the minimum size of branch-circuit conductor that may be installed in a dwelling unit?

 ____

3. What exceptions, if any, are there to the answer for Question 2? ______________________________

 ____

4. Define the term *ampacity.* ______________________________

 ____

5. What is the maximum voltage rating of all NMSC? ______________________________

6. Indicate the ampacity of these Type T90 Nylon (copper) conductors. Refer to *Table 2.*

 a. 14 AWG ______ amperes

 b. 12 AWG ______ amperes

 c. 10 AWG ______ amperes

 d. 8 AWG ______ amperes

 e. 6 AWG ______ amperes

 f. 4 AWG ______ amperes

7. What is the maximum operating temperature of these conductors? Give the answer in Celsius. (Use *Tables 11* and *19.*)

 a. Type TEW ______________________________

 b. Type R90 ______________________________

 c. Type LVT ______________________________

 d. Type DRT ______________________________

8. What are the colours of the conductors in non-metallic-sheathed cable for

 a. two-wire cable? ______________________________

 b. three-wire cable? ______________________________

9. For non-metallic-sheathed (Type NMD 90) cable, can the uninsulated conductor be used for purposes other than bonding? ______________________________

10. Under what condition may non-metallic-sheathed cable (Type NMD 90) be fished in the hollow voids of stud walls? ______________________________

11. a. What is the maximum distance permitted between straps on a cable installation?

 ____

b. What is the maximum distance permitted between a box and the first strap in a cable installation? ______________________

12. What is the difference between Type AC90 and Type TECK 90 cable?

____

13. [Fill in the blank and then circle the correct answer.] Type AC90 cable may be bent to a radius of not less than ______ times the diameter of the cable, measured to the (inside, outside) edge of the cable.

14. When armoured cable is used, what protection is provided at the cable ends? ______

____

____

____

15. What protection must be provided when installing a cable in a notched stud or joist?

____

____

____

____

____

____

16. Cables passing through a stud where the edge of the bored hole is less than ___ mm from the edge of the stud require additional protection.

17. a. Is non-metallic-sheathed cable permitted in your area for residential occupancies?

____

b. From what source is this information obtained? ______________________

18. [Circle the correct answer.] Is it permitted to use flexible metal conduit over 2 m long as a bonding means? (Yes) (No)

19. [Circle the correct answer.] Liquidtight flexible metal conduit (may) (may not) serve as a bonding means.

20. The allowable current-carrying capacity (ampacity) of aluminum wire is less than that of an equivalent copper wire. Use *Rule 4–002* and *Tables 2, 4,* and *13* to complete the following table. Enter both ampacity and maximum overcurrent protection values.

	COPPER		ALUMINUM	
WIRE	**Ampacity**	**Overcurrent Protection**	**Ampacity**	**Overcurrent Protection**
#12 R90				
#10 R90				
#3 TW75				
0000 TW				
500 kcmil RW75				

21. [Circle the correct answer.] All solderless wire connectors are approved to connect aluminum and copper conductors together in the same connector. (True) (False)

22. Terminals of switches and receptacles marked CO/ALR are suitable for use with _______ and ________ conductors.

23. When non-metallic-sheathed cables are bunched or bundled together for distances longer than 600 mm, what happens to their current-carrying ability?

____

____

24. A 120-volt branch circuit supplies a resistive heating load of 10 amperes. The distance from the panel to the heater is about 43 metres. Calculate the voltage drop using (a) #14, (b) #12, (c) #10, and (d) #8 AWG copper conductors.

25. In Question 24, it is desired to keep the voltage drop to 3% maximum. What is the minimum size of wire that would be installed to accomplish this 3% maximum voltage drop?

____

____

____

UNIT 8

Switch Control of Lighting Circuits, Receptacle Bonding, and Induction Heating Resulting from Unusual Switch Connections

OBJECTIVES

After studying this unit, you should be able to

- identify the grounded and ungrounded conductors in cable or conduit (colour coding)
- identify the types of toggle switches for lighting circuit control
- select a switch with the proper rating for the specific installation conditions
- describe the operation that each type of toggle switch performs in typical lighting circuit installations
- demonstrate the correct wiring connections the *CEC* requires for each type of switch
- understand the various ways to bond wiring devices to the outlet box
- understand how to design circuits to avoid heating by induction

The electrician installs and connects various types of lighting switches. To do this, the electrician must know both the operation and method of connection of each type of switch and understand the meanings of the current and voltage ratings marked on lighting switches and the *Canadian Electrical Code (CEC)* requirements for installing them.

CONDUCTOR IDENTIFICATION (*RULE 4–036*)

Before making any wiring connections to devices, the electrician must be familiar with the ways in which conductors are identified. For alternating-current (AC) circuits, the *CEC* requires that the grounded (identified) circuit conductor have an outer covering that is either white or grey. The *CEC* also recognizes that installations may require that identified conductors from different systems are to be installed in the same raceway, box, or enclosure. In this case, the identified conductors of the other system shall be permitted to be white with a coloured stripe, other than green, along the insulation as indicated in *Rule 4–030(2)*. In multiwire branch circuits, the grounded circuit conductor is also called a *neutral* conductor, *Rule 4–026* and *Section 0*.

The ungrounded (unidentified) conductor of a circuit must be marked in a colour other than green, white, or grey. This conductor generally is called the *hot* conductor. You feel a shock if you hold this conductor and the grounded conductor at the same time, or touch this conductor and a grounded surface such as a water pipe at the same time.

Neutral Conductor

In all residential wiring, the grounded neutral conductor's insulation is white or natural grey. It may also be white with a coloured stripe as permitted by *Rule 4–030(2)*. The grounded circuit conductor is also called a *neutral conductor* when it is part of a multiwire branch circuit.

When a two-wire circuit is used, the white grounded circuit conductor is not truly a neutral conductor unless

- It is the conductor that carries only the unbalanced current from the other conductors, as in the case of multiwire circuits of three or more conductors, *Rule 4–004(3)*.
- It is the conductor where the voltage from every other conductor to it is equal under normal operating conditions.

Therefore, the white grounded circuit conductor of a two-wire circuit is not really a "neutral," even though many electricians refer to it as such. See Figures 8-1A and 8-1B.

Colour Coding (Cable Wiring)

Non-metallic-sheathed cable (Romex) and armoured cable (BX) are colour-coded as follows:

- Two-wire: One black ("shot" phase conductor)
  One white (grounded "identified" conductor)
  One bare (equipment-bonding conductor)
  or
  One black ("hot" phase conductor)
  One red ("hot" phase conductor)
  One bare (equipment-bonding conductor)

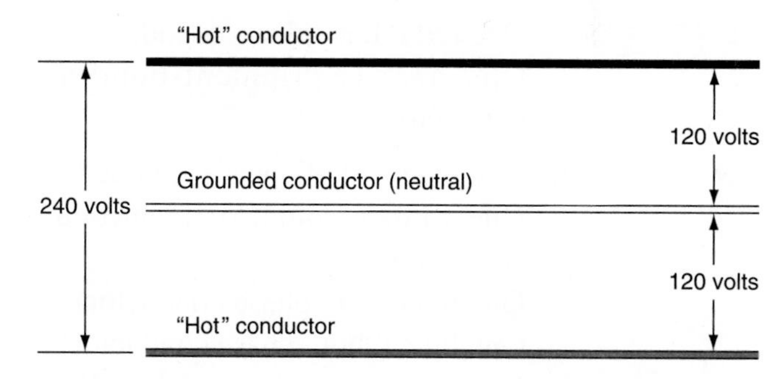

This is a three-wire circuit. The grounded conductor can be termed a neutral conductor.

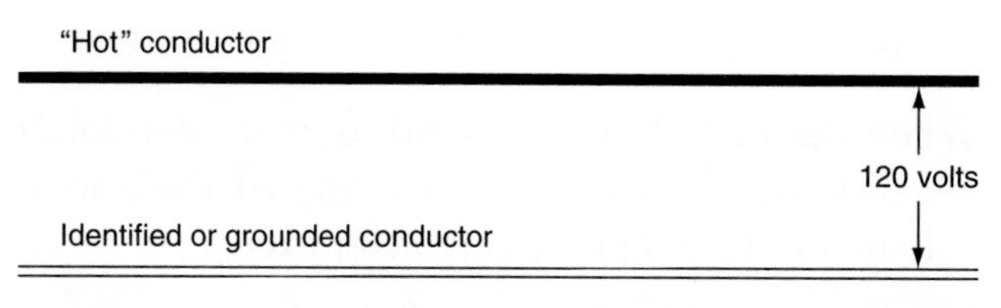

This is a two-wire circuit. The grounded conductor is not truly a neutral conductor.

FIGURE 8-1A Definition of a true neutral conductor.

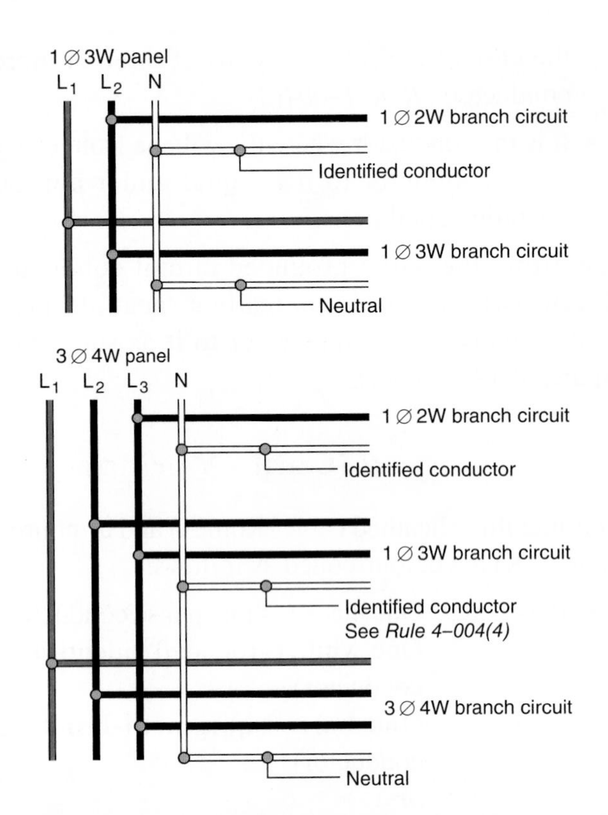

FIGURE 8-1B The difference between an identified conductor and a neutral conductor in single-phase three-wire and three-phase four-wire panels.

- Three-wire: One black ("hot" phase conductor)
  One white (grounded "identified" conductor)
  One red ("hot" phase conductor)
  One bare (equipment-bonding conductor)
- Four-wire: One black ("hot" phase conductor)
  One white (grounded "identified" conductor)
  One red ("hot" phase conductor)
  One blue ("hot" phase conductor)
  One bare (equipment-bonding conductor)

Colour Coding (Conduit Wiring)

When the installation is conduit, *Rule 4–038(3)* permits the electrician to use any of the above colours for the hot phase conductor except:

- Green: Reserved for use as a bonding conductor only
- White or grey: Reserved for use as the grounded identified circuit conductor

Changing Colours

Should it become necessary to change the actual colour of the conductor to meet *CEC* requirements, the electrician may change the colours as indicated in Table 8-1.

For cable installations only, *Rule 4–036(2)* permits you to use the white conductor as a switch "loop" for switch drops, but states that the white conductor must feed the switch. Therefore, you must make up the connection so that the return conductor from the switch is a coloured or unidentified conductor. When the connections are done in this way, the *CEC* does not require re-identification, but instead accepts the splices and attachment of the conductors to the terminals of the switch as adequate identification. Therefore, if a white wire is used in way that makes it a "hot" conductor, such as in a switch circuit, it should be wired so that it is *always* "hot." It should be the feed to the switch, never the return.

Look ahead to Figure 8-6, which illustrates re-identification where the white wire running from the octagon box to the switch box is a "hot" phase conductor.

Although *Rule 4–036(2)* states that re-identification is *not* required, some electrical inspectors may require that these white wires be rendered permanently unidentified to avoid any confusion; see Figure 8-2. *Acceptable* is defined as "acceptable to the authority enforcing this code."

TABLE 8-1

Changing colour of wire.

FROM	TO	DO
Red, black, blue	Bonding conductor	For conductors larger than #2 AWG, strip off insulation to make it bare—or paint it green where exposed in the box—or mark exposed insulation with green tape, *Rule 4–038(1)(b)*.
Red, black, blue	Grounded identified conductor	Re-mark with coloured tape or paint white or grey. For conductors larger than #2 AWG, *Rule 4–032.*
White, grey, or green	Red, black, blue	Re-identify with coloured tape or paint, Figure 8-2, *Rule 4–038(3)(c)*.

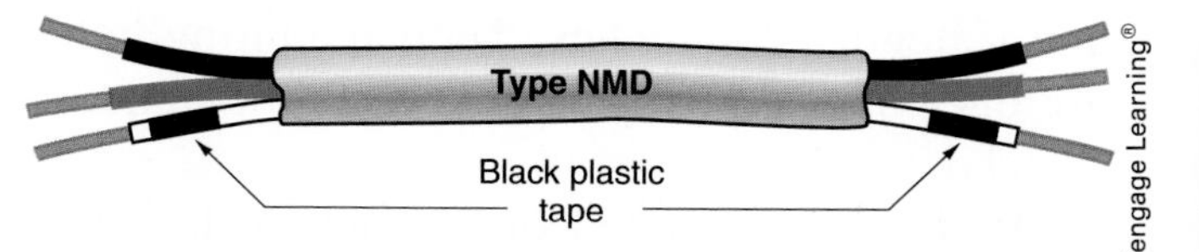

FIGURE 8-2 You may wrap the white wire with a piece of black plastic tape so that everyone will know that this wire is *not* a grounded conductor. Use red tape if you want to make the conductor red, *Rule 4–038(3).*

When using two-conductor non-metallic-sheathed cable for applications that do not require an identified conductor other than switch lines, the authority enforcing the code may require the use of a cable that has a red conductor in place of the white. In residential applications, this would be for loads that operate at 240 volts, such as electric heaters and hot water tanks. This cable usually has a red jacket so that the inspection authority can easily identify it.

Many electricians choose to use non-metallic-sheathed cable with different coloured jackets. For example, the 12/2 cable used for the 20-ampere T slot receptacles may have a yellow jacket, or the 14/2 cable used for the Arc Fault Circuit Interrupter (AFCI)-protected circuit may have a blue jacket. Different coloured cables for specific applications aid the electrician when the conductors are terminated in the panel, clearly identifying the ampacity and type of circuit breaker required.

TOGGLE SWITCHES (*RULES 14–500* THROUGH *14–514*)

The most frequently used switch in lighting circuits is the toggle flush switch; see Figure 8-3. When mounted in a flush switch box, the switch is concealed in the wall with only the insulated handle or toggle protruding.

Figure 8-4 shows how switches and receptacles are weatherproofed.

Toggle Switch Ratings

The Canadian Standards Association (CSA) classifies toggle switches for lighting circuits as *general-use switches.* These switches are divided into three categories.

FIGURE 8-3 Toggle flush switches.

FIGURE 8-4 Receptacle outlet and toggle switch are protected by a weatherproof cover.

Category 1 contains general-use switches used to control

- AC or direct-current (DC) circuits
- resistive loads not to exceed the ampere rating of the switch at rated voltage
- inductive loads not to exceed one-half the ampere rating of the switch at rated voltage unless the switch is marked with the letter “F”
- tungsten filament lamp loads not to exceed the ampere rating of the switch at 125 volts when the switch is marked with the letter “T”

A tungsten filament lamp draws a very high momentary inrush current at the instant the circuit is energized because the *cold resistance* of tungsten is very low. For instance, the cold resistance of a typical 100-watt lamp is about 9.5 ohms. This same lamp has a *hot resistance* of 144 ohms when operating at 100% of its rated voltage.

Normal operating current would be

$$I = \frac{E}{R} = \frac{120}{144} = 0.83 \text{ amperes}$$

But, *maximum* instantaneous inrush current could be as high as

$$I = \frac{E}{R} = \frac{170(\text{peak voltage})}{9.5} = 17.9 \text{ amperes}$$

This inrush current drops off to normal operating current in about six cycles (0.10 seconds). The contacts of T-rated switches are designed to handle these momentary high inrush currents.

The AC/DC general-use switch normally is not marked AC/DC. However, it is always marked with the current and voltage rating, such as 10 A–125 V or 5 A–250 V–T, *Rule 14–508(b)(iii).*

Category 2 contains those AC general-use switches used to control

- alternating currents only
- resistive, inductive, and tungsten filament lamp loads not to exceed the ampere rating of the switch at 120 volts
- motor loads not to exceed 80% of the ampere rating of the switch at rated voltage, but not exceeding two horsepower

AC general-use snap switches are marked "AC only" in addition to identifying their current and voltage ratings. A typical switch marking is 15 amperes, 120–277 volts AC. The 277-volt rating is required on 277/480-volt systems.

Category 3 contains those 347-volt AC switches used to control

- AC loads only
- resistive and inductive loads not to exceed the ampere rating of the switch at 347 volts

In addition, the current rating shall not be less than 15 amperes, 347 volts. The switches are designed so they cannot be mounted in boxes for Category 1 and Category 2 switches: Their mounting holes are 6.4 mm farther apart than those in standard switches.

If they are grouped or ganged in the same box, AC general-use switches operating in circuits exceeding 300 volts to ground must have permanent barriers installed between them. A typical switch is 15 amperes, 347 volts. The 347-volt rating is required on 347/600-volts systems. Switches and devices rated at 347 volts will not be installed in dwelling units because the maximum allowable voltage is 150 volts to ground, *Rule 2–104.*

Terminals of switches rated at 15 or 20 amperes, when marked CO/ALR, are suitable for use with aluminum, copper, and copper-clad aluminum conductors. Switches not marked CO/ALR are suitable for use with copper and copper-clad conductors only.

Screwless pressure terminals of the conductor push-in type may be used with copper and copper-clad aluminum conductors only. These push-in type terminals are not permitted for use with ordinary aluminum conductors, *Rule 12–118(6).*

Further information on switch ratings is given in *Rule 14–508* and in CSA Standard C22.2 No. 111, "General-Use Switches."

Table 7-2 in Unit 7 lists the markings on wiring devices indicating the types of conductors for which the devices are used.

Toggle Switch Types

Toggle switches are available in four types: single-pole, three-way, four-way, and double-pole.

Single-Pole Switch. A single-pole switch is used when a light or group of lights or other load is to be controlled from one switching point. The switch is identified by its two terminals and the toggle marked ON/OFF. The single-pole switch is connected in series with the ungrounded or hot wire feeding the load.

Figure 8-5 shows a single-pole switch controlling a light from one switching point. The 120-volt source feeds directly through the switch location. The identified white wire continues directly to the load, and the unidentified black wire is broken at the single-pole switch.

In Figure 8-6, the 120-volt source feeds the light outlet directly, and a two-wire cable with black and

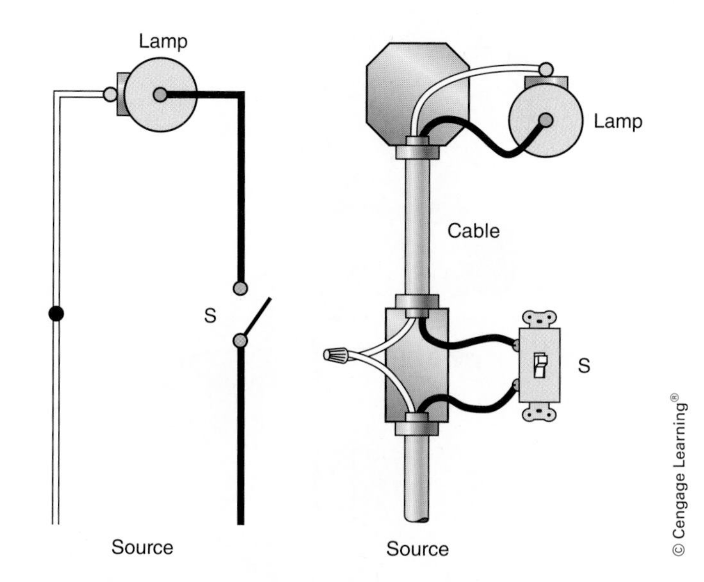

FIGURE 8-5 Single-pole switch in circuit with feed at switch.

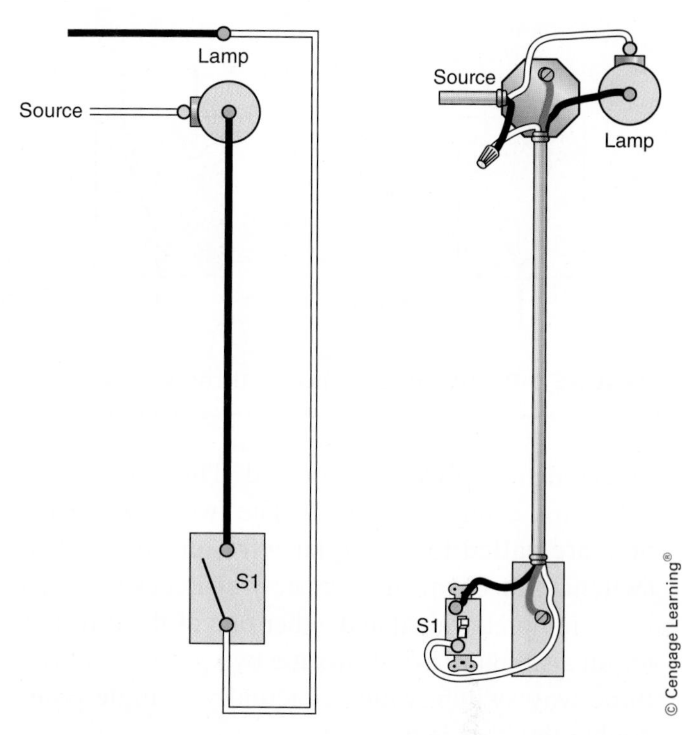

FIGURE 8-6 Single-pole switch in circuit with feed at light.

white wires is used as a switch loop between the light outlet and the single-pole switch. *Rule 4–036(2)* permits the use of a white wire in a single-pole switch loop. The unidentified or black conductor must connect between the switch and the load.

Figure 8-7 shows another application of a single-pole switch. The feed is at the switch that controls the light outlet. The receptacle outlet is independent of the switch.

Three-Way Switch. The term three-way is somewhat misleading, since a three-way switch is used to control a light from two locations. A three-way switch has one terminal to which the internal switch

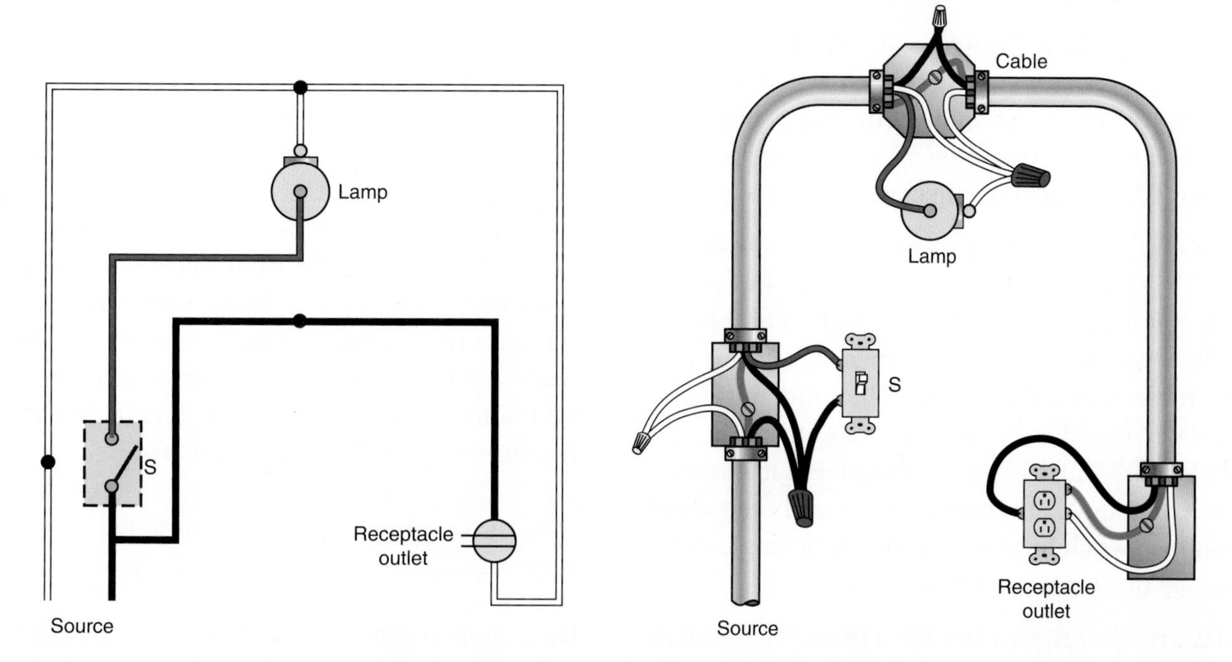

FIGURE 8-7 Ceiling outlet controlled by single-pole switch with live receptacle outlet and feed at switch.

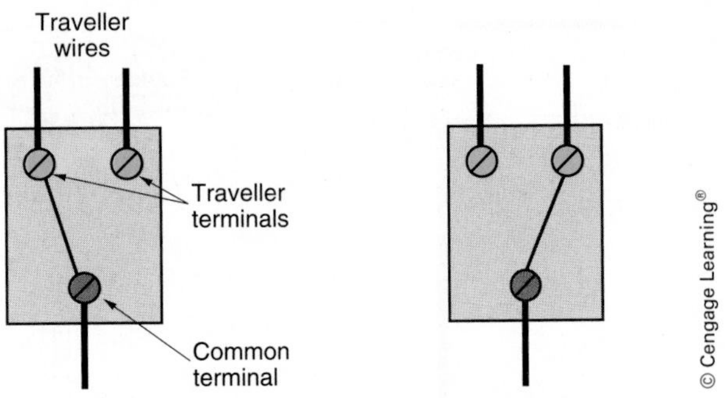

Traveller wires
Traveller terminals
Common terminal

© Cengage Learning®

FIGURE 8-8 Two positions of a three-way switch.

mechanism is always connected. This terminal is called the common terminal. The two other terminals are called the traveller wire terminals. The switching mechanism alternately switches between the common terminal and either one of the traveller terminals. Figure 8-8 shows the two positions of the three-way switch, which is actually a single-pole, double-throw switch.

A three-way switch differs from a single-pole switch in that the three-way switch does not have an ON or OFF position. Thus, the switch handle does not have ON/OFF markings (Figure 8-9) and cannot be used as a disconnecting means, *Rule 2–304(2)*. The three-way switch can be identified further by its three terminals. The common terminal is darker in colour than the two traveller wire terminals, which are usually natural brass in colour.

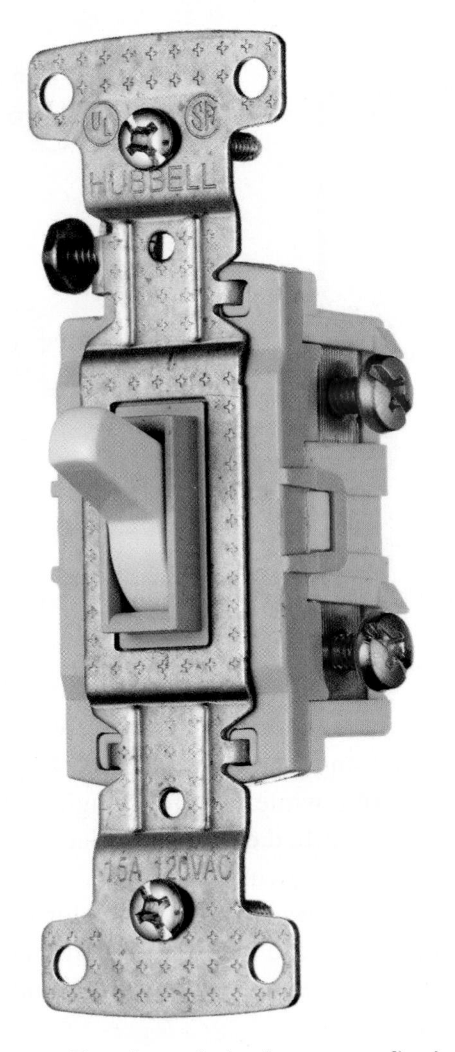

Courtesy of Hubbell Canada LP

FIGURE 8-9 Toggle switch: three-way flush switch.

Three-Way Switch Control with Feed at Switch. Three-way switches are used when a load is to be controlled from two different switching points. Figure 8-10 shows how two three-way switches are used. Note that the feed is at the first switch control point.

Three-Way Switch Control with Feed at Light. The circuit in Figure 8-11 uses three-way switch control with the feed at the light. For this circuit, the white wire in the cable must be used as part of the three-way switch loop. The unidentified or black wire is used as the return wire to the light outlet, Rule 4–036(2). This exception makes it unnecessary to paint, tape, or otherwise re-identify the white wire at the switch location when wiring single-pole, three-way or four-way switch loops.

Alternative Configuration for Three-Way Switch Control with Feed at the Light. Figure 8-12 shows another arrangement of components in a three-way switch circuit. The feed is at the light with cable runs from the ceiling outlet to each of the three-way switch control points located on either side of the light outlet.

Three-way switches are needed in rooms that have entry from more than one location. In this residence, the garage lighting, entry hall lighting, living room receptacle outlets, rear outdoor lighting located between the living room sliding doors and the kitchen sliding doors, and receptacle outlets in the study require three-way switches.

Conductor Colour Coding for Switch Connections

The colour coding for the travellers (messenger) for three-way and four-way switch connections when using cable can vary. *Rule 30–600* requires that the

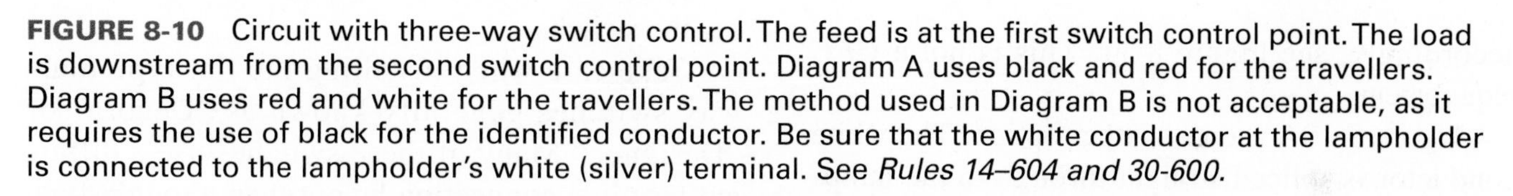

FIGURE 8-10 Circuit with three-way switch control. The feed is at the first switch control point. The load is downstream from the second switch control point. Diagram A uses black and red for the travellers. Diagram B uses red and white for the travellers. The method used in Diagram B is not acceptable, as it requires the use of black for the identified conductor. Be sure that the white conductor at the lampholder is connected to the lampholder's white (silver) terminal. See *Rules 14–604 and 30-600.*

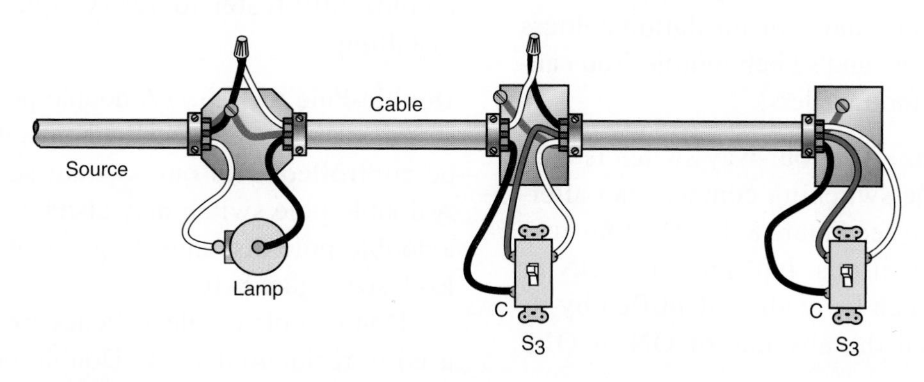

FIGURE 8-11 Circuit with three-way switch control and feed at the light.

white (identified) conductor always be connected to the white (silver) terminal of a lampholder or receptacle. Because of the many ways that three-way or four-way switches can be connected, the choice of colours to use for the travellers is left to the electrician. Most electrical contractors and electricians establish some sort of colour coding that works well for them.

One option is to establish white and red for the travellers and make up the electrical splices

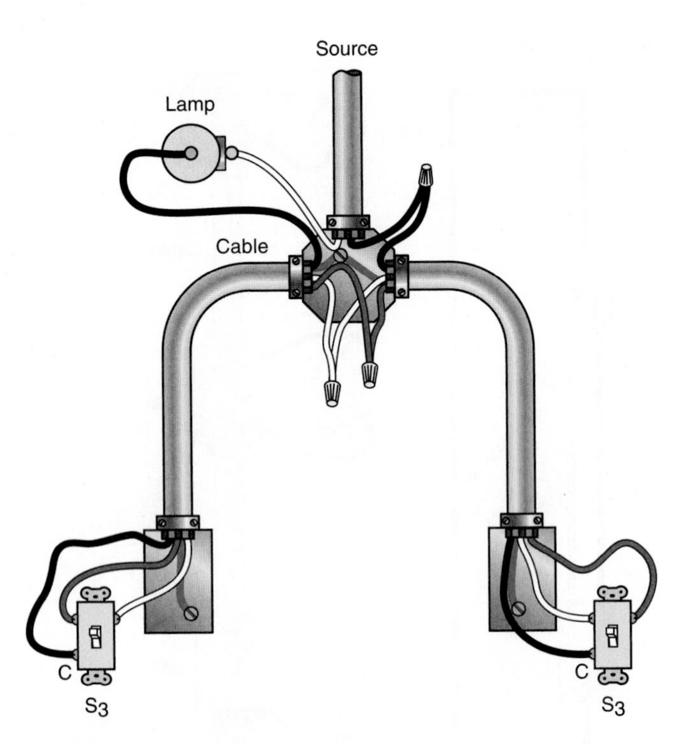

FIGURE 8-12 Alternative circuit with three-way switch control and feed at the light.

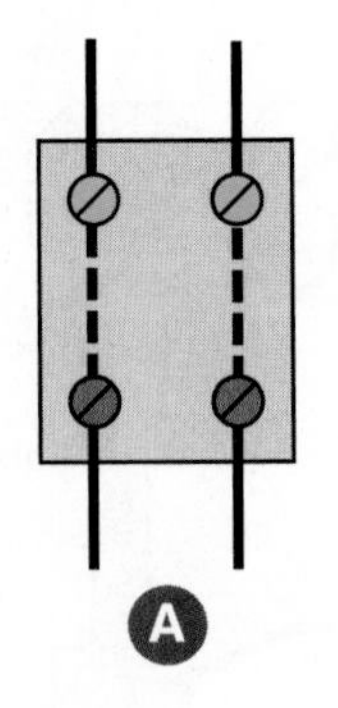

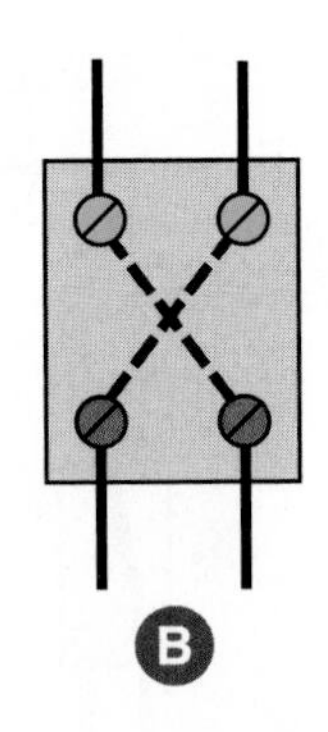

FIGURE 8-13 Two positions of a typical four-way switch.

accordingly; see Figure 8-11. This is not a *CEC* requirement.

Figure 8-10A shows another option: The white conductor is spliced straight through to the lampholder; the black and red are the travellers.

This issue does not arise when using conduit, because a variety of conductor insulation colours exist for the travellers and switch returns. You cannot use white here for travellers!

Four-Way Switch. The four-way switch is constructed so that the switching contacts can alternate their positions; see Figure 8-13. The four-way switch has two positions, but neither is ON or OFF. This switch can be readily identified by its four terminals and the absence of ON or OFF markings.

Four-way switches are used when a load must be controlled from more than two switching points. To do this, three-way switches are connected to the source and to the load. The switches at all other control points, however, must be four-way switches.

Figure 8-14 shows how a lamp can be controlled from any one of three switching points. Take care in connecting the traveller wires to the proper terminals of the four-way switch. The two traveller wires from one three-way switch are connected to the two terminals on the top of the four-way switch, and the other two traveller wires from the second three-way switch are connected to the two terminals on the bottom of the four-way switch.

Some four-way switches have the connections from side to side on the switch. The electrician must verify this using a continuity tester before connecting the switch.

Take care when installing the flat-type four-way switches generally known as "Decora" or "Designer." Many of these switches use a different terminal connection layout than a toggle-type four-way switch. Check the packaging or use a continuity tester to verify connections before installing.

Double-Pole Switch. A double-pole, or two-pole, switch may be used when two separate circuits must be controlled with one switch; see Figure 8-15. A double-pole switch may also be used to provide a double-pole disconnecting means for a 240-volt load; see Figure 8-16.

Double-pole toggle switches are not commonly used in residential work. Double-pole disconnect switches, however, are used quite often in residences for the furnace, water pump motors, and other 240-volt feeders.

Switches with Pilot Lights. Sometimes a pilot light is desired at the switch location, as is the case for the attic lighting in this residence. Pilot light switches are available in several styles; see Figures 8-17 through 8-19.

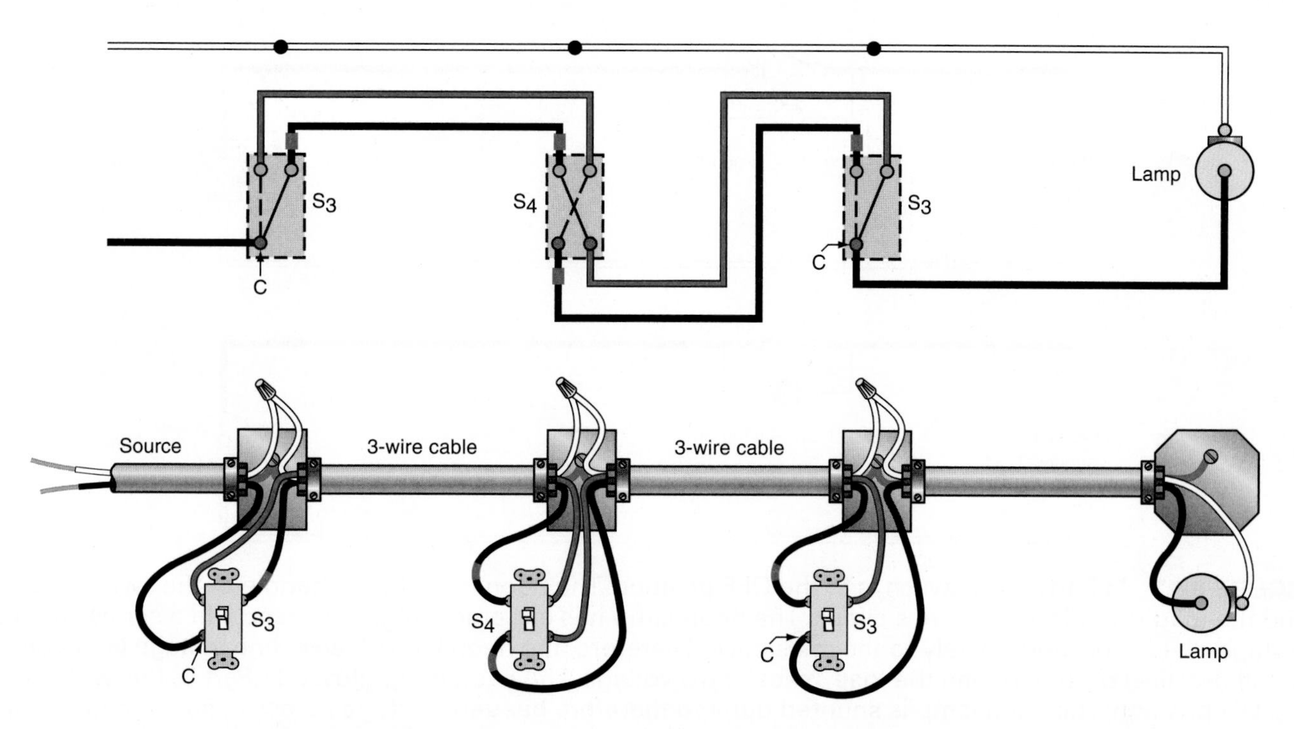

FIGURE 8-14 Circuit with switch control at three locations.

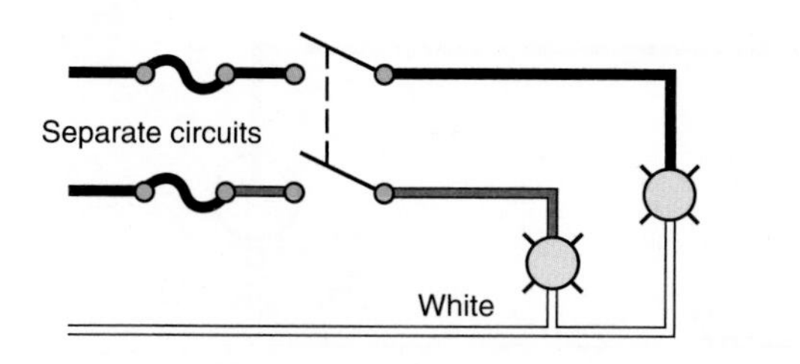

FIGURE 8-15 Application of a double-pole switch.

BONDING AT RECEPTACLES

Ground-fault protection is covered in detail in Unit 9. However, providing bonding of the equipment, bonding of the conductor to a metal box and to the receptacle's bonding terminal, is important.

To wire bonding-type receptacles with armoured cable, the cable must contain a separate bonding connector to bond the metal outlet box and the bonding terminal of the receptacle. If non-metallic-sheathed cable is used, it too must contain a separate bonding conductor, *Rules 10–606* and *10–610*. Figure 8-20 shows the bonding conductor of non-metallic-sheathed cable attached to the receptacle, *Rule 10–808(2)*.

Almost all bonding terminals of the type shown in Figure 8-20 are bonded to the metal yoke of the receptacle. However, this connection might not provide reliable continuity between a bonded outlet box and the bonding circuit of the receptacle.

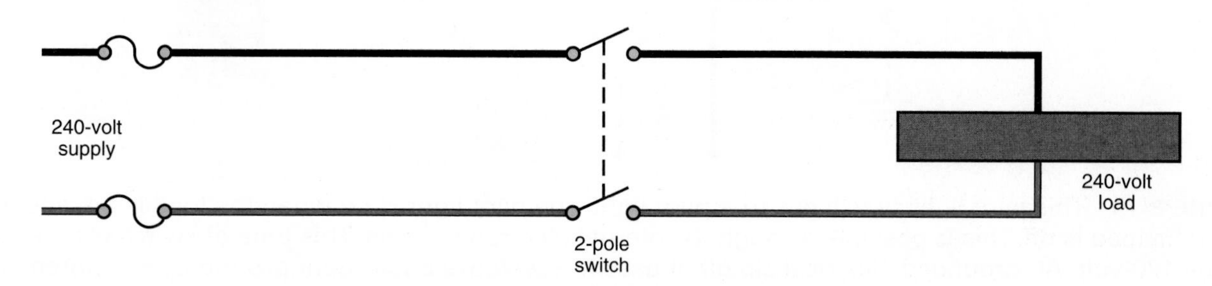

FIGURE 8-16 Double-pole (two-pole) disconnect switch.

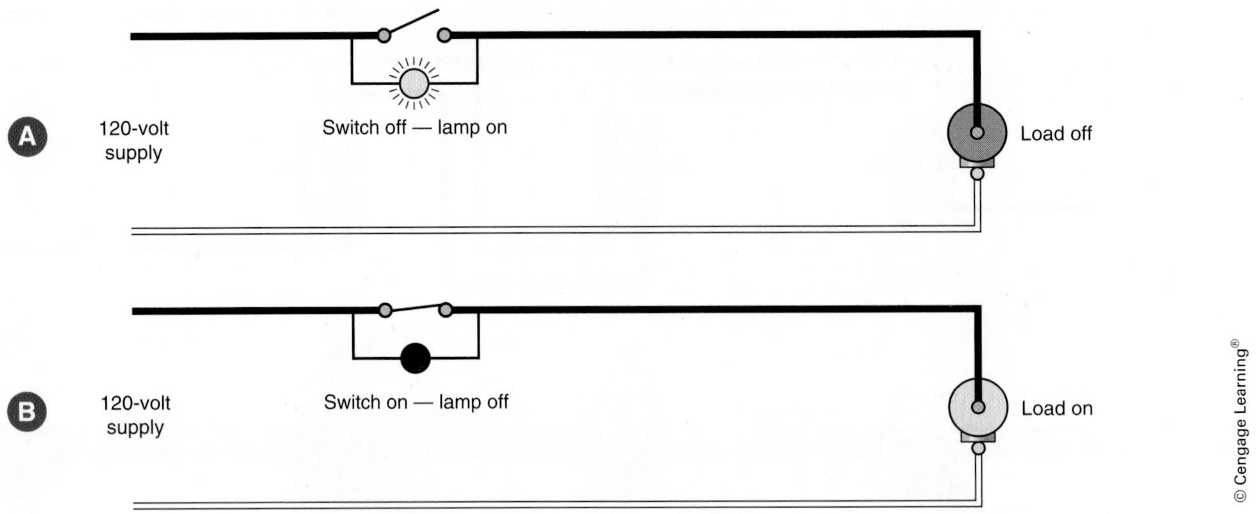

A 120-volt supply — Switch off — lamp on — Load off

B 120-volt supply — Switch on — lamp off — Load on

© Cengage Learning®

FIGURE 8-17 In Part A, the switch is in the OFF position: The neon lamp in the handle of the switch glows, and the load is off. This is a series circuit. The neon lamp has extremely high resistance. In a series circuit, voltage divides proportionately to the resistance. Therefore, the neon lamp "sees" line voltage (120 volts) for all practical purposes, and the load "sees" zero voltage. The neon lamp glows. In Part B, the switch is in the ON position: The neon lamp is shunted out and therefore has zero voltage across it, so the neon lamp does not glow. Full voltage is supplied to the load. This type of switch might be referred to as a *locator* because it glows when the load is *off* and does not glow when the load is *on*.

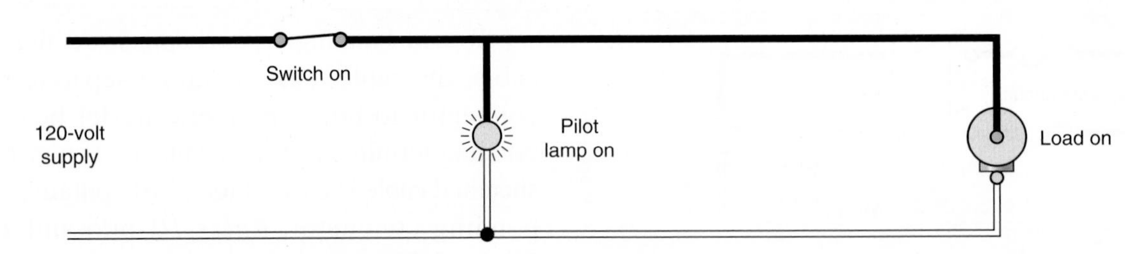

© Cengage Learning®

FIGURE 8-18 A true pilot light. The pilot lamp is an integral part of the switch. When the load is turned on, the pilot light is also on. When the load is turned off, the pilot light is also off. This switch has three terminals because it requires a grounded neutral circuit conductor.

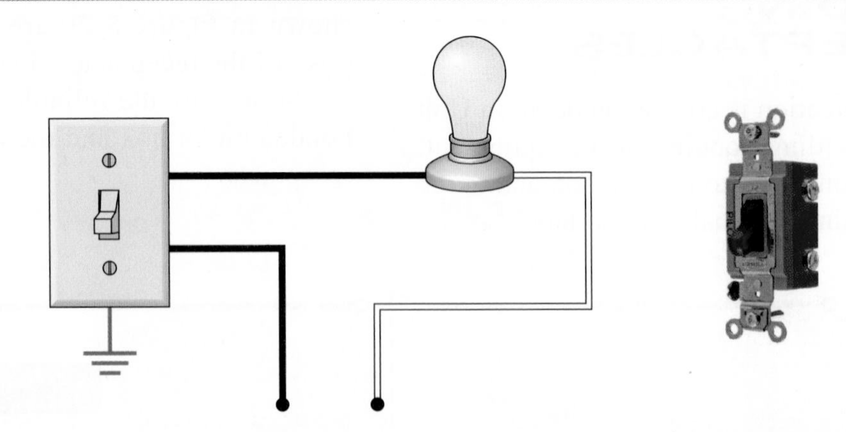

Courtesy of Hubbell Canada LP

FIGURE 8-19 This pilot light switch is a true pilot light. The pilot light is on when the load is on, and is off when the load is off. This is possible through its internal electronic circuit. This type of switch can be used on any 120-volt, AC grounded electrical circuit. It uses the system's equipment ground as its reference to permit the pilot light to glow. It does *not* require a grounded neutral conductor. The pilot light "pulsates" about 60 times per minute.

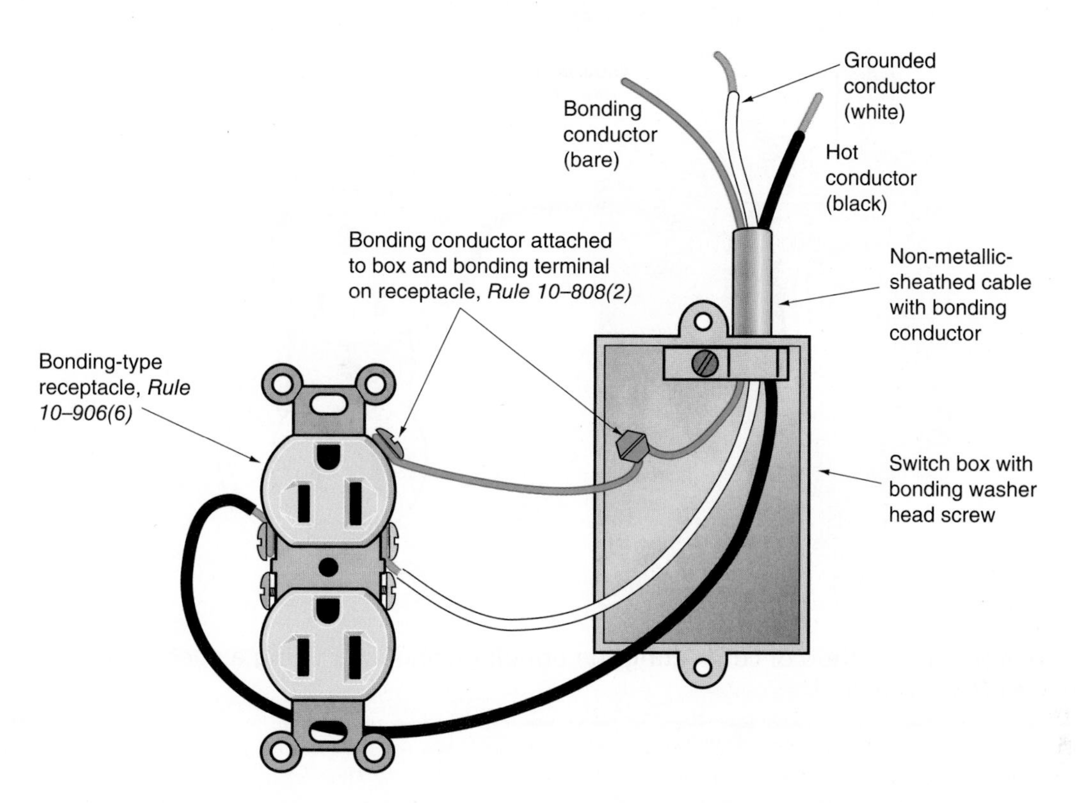

FIGURE 8-20 Connections for bonding-type receptacles. See *Rules 10–808(2)* and *10–904(6)*.

An outlet box is said to be bonded when it is connected to a bonding conductor that is part of a cable or when it is connected to a bonded metal raceway, *Rule 10–808*. When the receptacle is fastened to a bonded device box, the bonding terminal may well be bonded through the threads of the #6–32 mounting screws. For a 102-mm square outlet box, bonding to the raised cover could take place through the threads of the #8–32 cover mounting screws. However, it is doubtful that the continuity of the bonding circuit is reliable under these conditions. Therefore, *Rule 10–904(1)* requires the installation of a separate bonding jumper between all receptacles and the device box bonding connection, whatever wiring method is used—metal conduit, armoured cable, or non-metallic-sheathed cable with a bonding conductor. The metal outlet box must be bonded in addition to using the bonding terminal of the receptacle; see Figure 8-20.

To ensure the continuity of the equipment-bonding conductor path, *Rule 10–808(2)* requires that when more than one bonding conductor enters a box, they shall be secured under bonding screws or connected using solderless connectors. Splices shall not depend on solder, *Rule 10–904(2)*. The splicing must be done so that if a receptacle or other wiring device is removed, the continuity of the equipment-bonding path shall not be interrupted, as clearly shown in Figures 8-21 and 8-22. When wiring with non-metallic-sheathed cable, the equipment-bonding conductor is a non-insulated, bare conductor. When wiring with conduit, an equipment-bonding conductor, when required, must be a green insulated conductor, sized per *Table 16A*.

Splices must be made in accordance with *Rules 10–808(1)* and *(2)*.

A bonding jumper (Figure 8-20) is required between all receptacles and the device box bonding connection, *Rule 10–904(6)*. The #6-32 mounting screws of most receptacles and switches are held captive in the yoke by a small fibre or cardboard washer. Removing this washer gives additional metal-to-metal contact; however, this is not adequate as a bonding means.

Some outlet and switch-box manufacturers use washer head screws so that a bonding conductor can be placed under them. This method is useful for installing armoured cable or non-metallic-sheathed

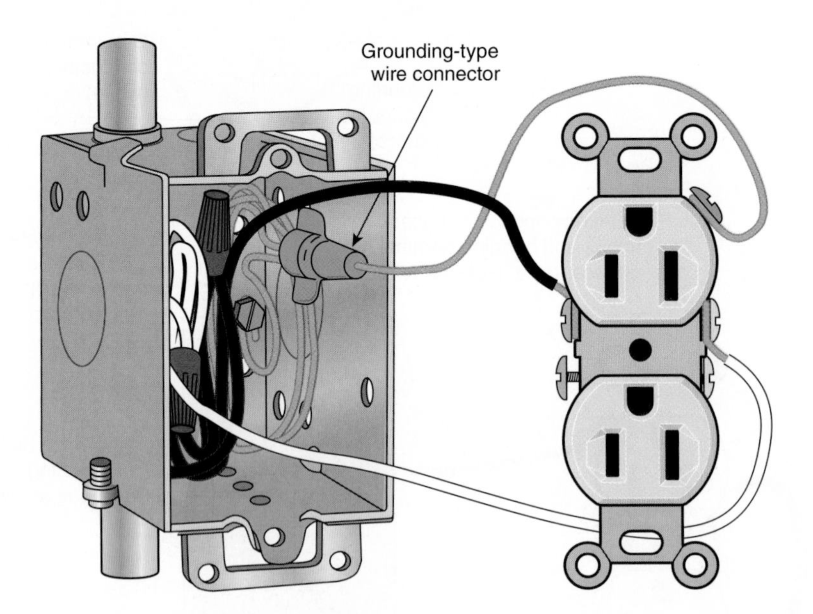

FIGURE 8-21 An approved method of connecting the bonding conductor using a special bonding-type wire connector. See *Rule 10–808(2)*.

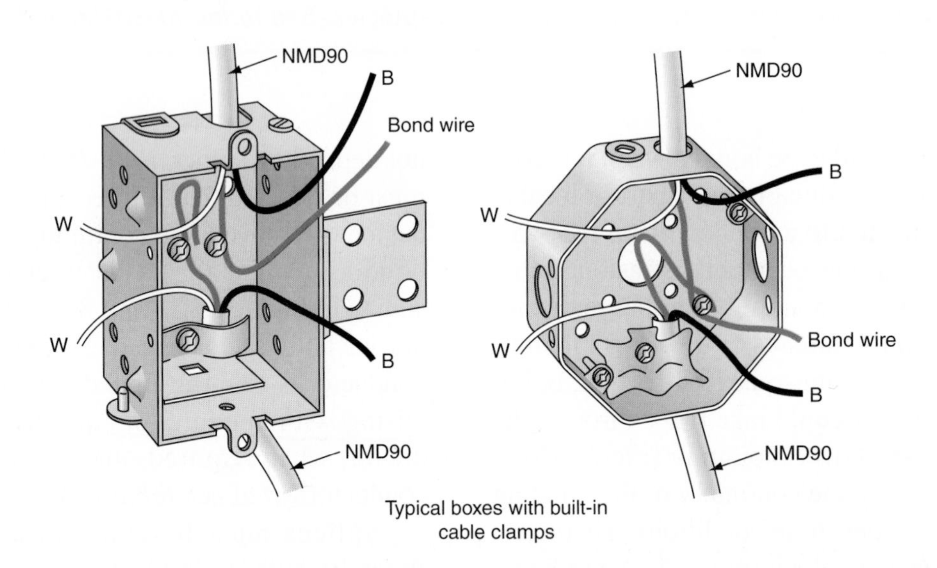

FIGURE 8-22 Method of attaching bonding wires to device box or octagon box. See *Rule 10–808(2)*.

cable, as it provides a means to terminate the bonding conductor in the metal outlet or switch box.

Rule 10–904(3) states that the bonding jumper must be secured to every metal box even if the receptacle has a special self-bonding strap; see Figures 8-23 and 8-24. *Rule 10–904(6)* specifies that when you install such a receptacle, you must wire a bonding jumper from the receptacle to the bonded outlet box in such a manner that disconnection or removal of the receptacle will not interrupt the grounding continuity.

NONBONDING-TYPE RECEPTACLES

Modern appliances and portable hand-held tools have a three-wire cord and a three-wire bonding-type attachment plug cap. Frequently, these tools and

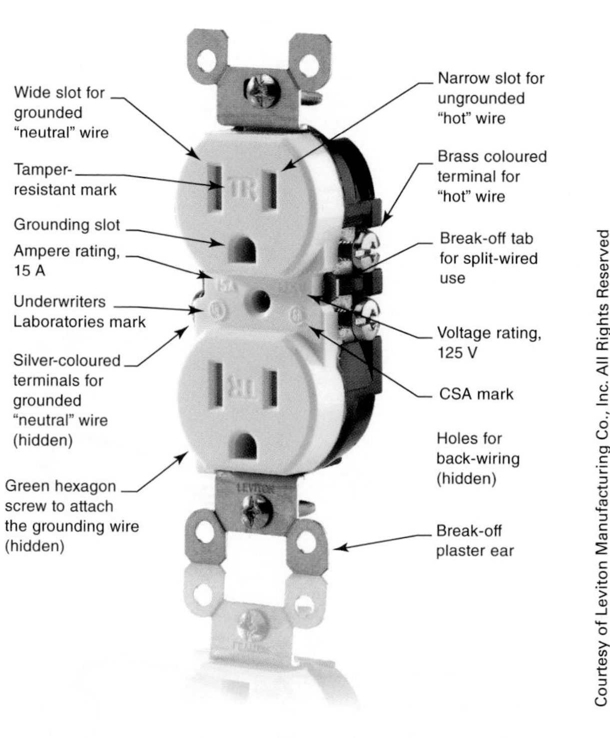

FIGURE 8-23 Bonding-type receptacle detailing parts of the receptacle.

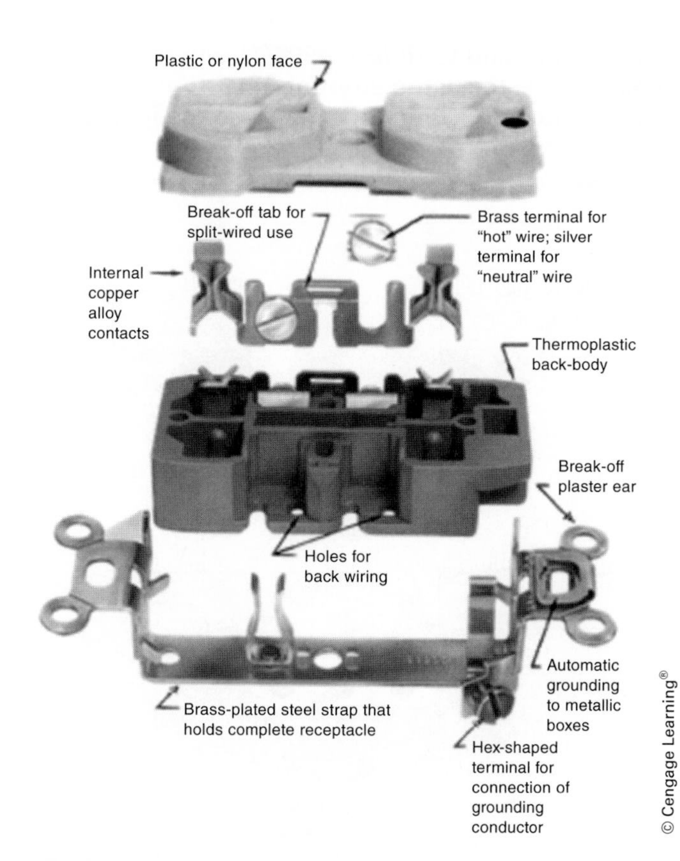

FIGURE 8-24 Exploded view of a bonding-type receptacle, showing all internal parts.

appliances are plugged into nonbonding-type receptacles with only two prongs. In these circumstances, the appliance is not bonded and the user has probably cut off or bent the grounding prong to use the appliance. **This is an incorrect and unsafe practice.** As a result, it will later be impossible to bond the appliance even when it is plugged into a properly connected three-wire bonding-type receptacle.

In these circumstances *Rules 10–408(3)* and *(4)* permit two methods of connection:

- the use of double-insulated tools and appliances, *Rule 10–408(3)* and *Appendix B*.
- the use of a portable double-insulated Ground-Fault Circuit Interrupter (GFCI) receptacle to supply the equipment.

In neither case is the equipment bonded to ground.

In many cases, it may be necessary to replace existing nonbonding-type receptacles due to damage or because a three-wire bonding-type attachment plug is to be used. A bonding-type receptacle may be used if the receptacle is bonded to ground by one of these methods:

- If the device box is bonded, attach a bonding conductor from the device to the device box.
- Bond the receptacle to the system grounding conductor with a separate bonding conductor, *Rule 26–700(7)*.
- Connect to a metal cable sheath or raceway that is bonded to ground, *Rule 26–700(7)(a)*.
- Bond the device to a metal cold-water line, *Rule 26–700(7)(c)*.

If it is not possible to use one of these methods, the new receptacle can be protected by a Class A-type GFCI breaker, Class A-type GFCI receptacle, or connection to the load side of a feed through a Class A-type GFCI receptacle, *Rule 26–700(8)(c)*.

INDUCTION HEATING

When AC circuits are run in metal raceways or trenches and through openings in metal boxes, the circuiting must be arranged to prevent *induction heating* of the metal, *Rule 12–904(1)*. This means that the conductors of a given circuit must

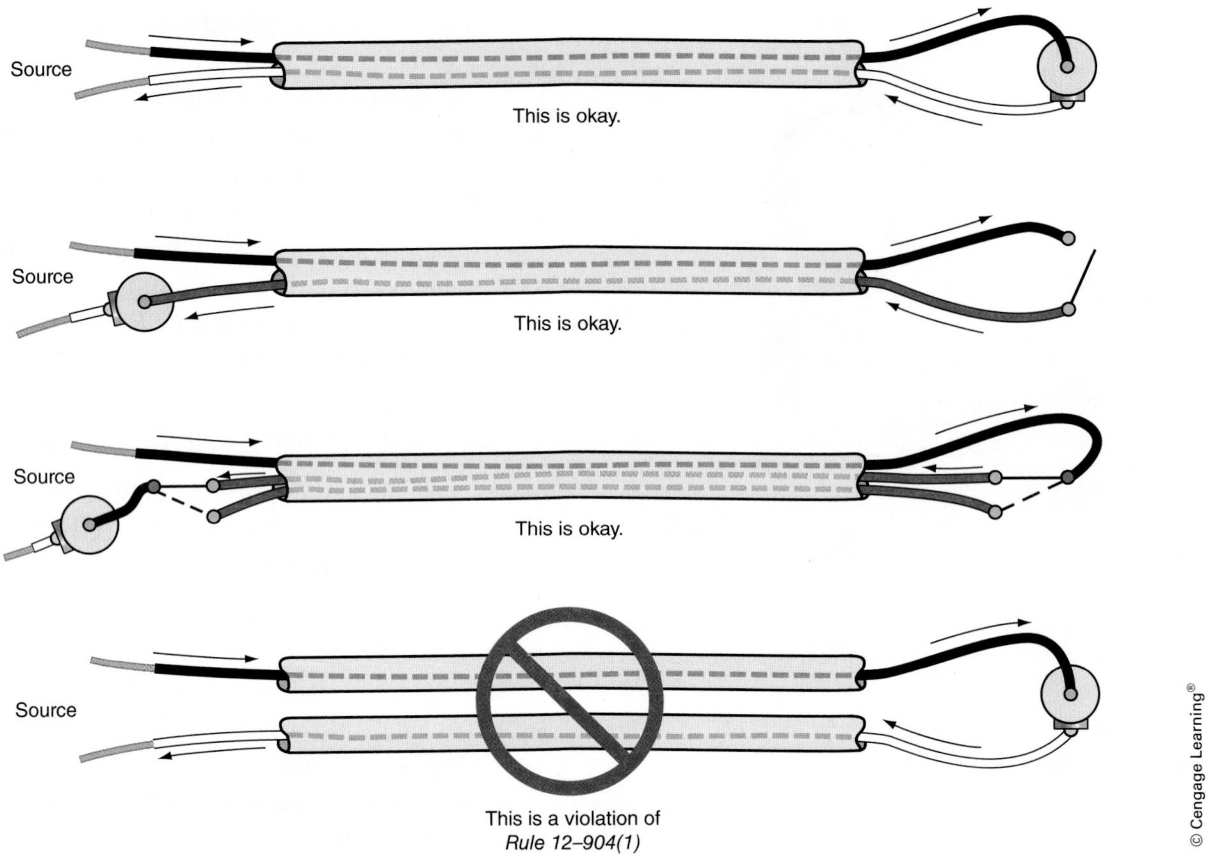

FIGURE 8-25 Arranging circuitry to avoid induction heating, according to *Rule 12–904(1)*, when metal raceway or armoured cables (BX) are used. Always ask yourself, "Is the same amount of current flowing in both directions in the metal raceway?" If the answer is "No," induction heating can damage the insulation on the conductors.

be arranged so that the magnetic flux surrounding each conductor will cancel that of the other(s); see Figure 8-25. This also means that when individual conductors of the same circuit are run in trenches, they should be kept together, not spaced far apart (Figures 8-26 through 8-28, and *Rule 12–012(4)*).

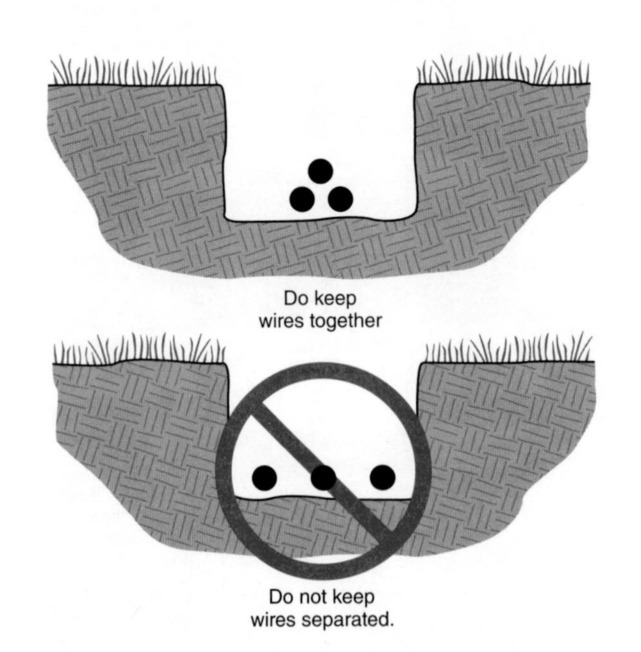

FIGURE 8-26 Arrangement of conductors in trenches.

TAMPER-RESISTANT RECEPTACLES

Tamper-resistant receptacles are designed to minimize the occurrence of electrical shock and arc-flash incidents that occur when conductive objects are inserted into the receptacle slots. Most of these incidents take place in homes. So, tamper-resistant receptacles were designed to prevent contact with live electrical components when an object is inserted into one of the

FIGURE 8-27 Induction heating can occur when conductors of the same circuit are run through different openings of a metal box.

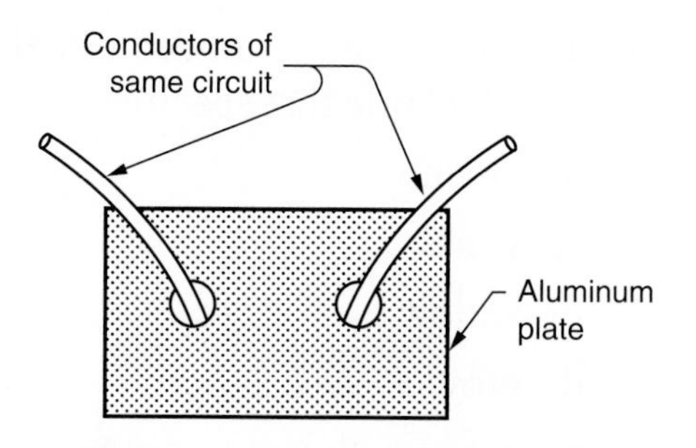

FIGURE 8-28 To reduce the inductive effects of Figure 8–27, a plate of aluminum or other nonmagnetic metal should be used, *Rule 4–008* and *Appendix B*.

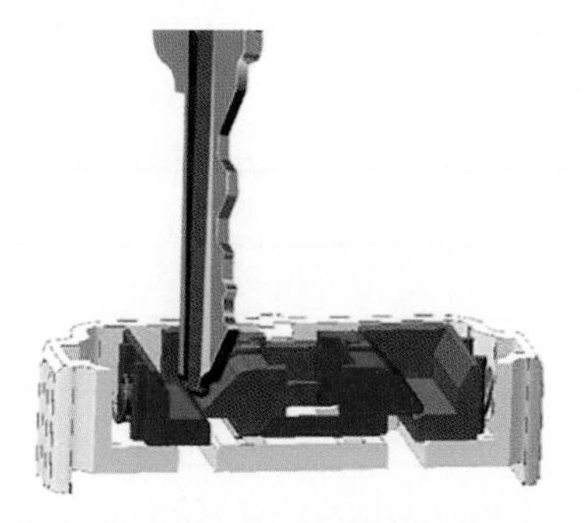

Spring-loaded shutter mechanism restricts access to an object in any one side of the receptacle.

Insertion of a two or three bladed plug will open the shutters, allowing electrical contact.

Courtesy of Progress Lighting

FIGURE 8-29 Tamper-resistant principle overview.

receptacle slots. In one manufactured design, the slots for the ungrounded (live) conductor and the grounded (identified) conductor are closed by shutters that must open simultaneously. By inserting simultaneous pressure from the two blades of the male plug cap equally against each shutter, the shutters will open. See Figure 8-29. *CEC Rule 26–712*, items (*g*) and (*h*), specify that, in a dwelling, unless a receptacle is rendered inaccessible or is located more than 2 m above the floor or finished grade, all receptacles of CSA configuration 5-15R and 5-20R shall be tamper-resistant receptacles and shall be so marked.

REVIEW

Note: Refer to the *CEC* or the blueprints provided with this textbook when necessary. Where applicable, responses should be written in complete sentences.

1. The identified grounded circuit conductor may be which colour? [Circle one.] State the *CEC* rule that specifies this.

   a. green

   b. white

   c. yellow

   d. grey

   e. green with yellow stripes

   Rule ____________________

2. A T-rated switch may be used to its __________ current capacity when controlling an incandescent lighting load.

3. What switch type and rating are required to control five 300-watt tungsten filament lamps on a 120-volt circuit? Show calculations.

4. To control a lighting load from one control point, what type of switch would you use?

5. Single-pole switches are always connected to the ____________________ wire.

6. Complete the connections in the following arrangement so that both ceiling light outlets are controlled from the single-pole switch. Assume the installation is in cable.

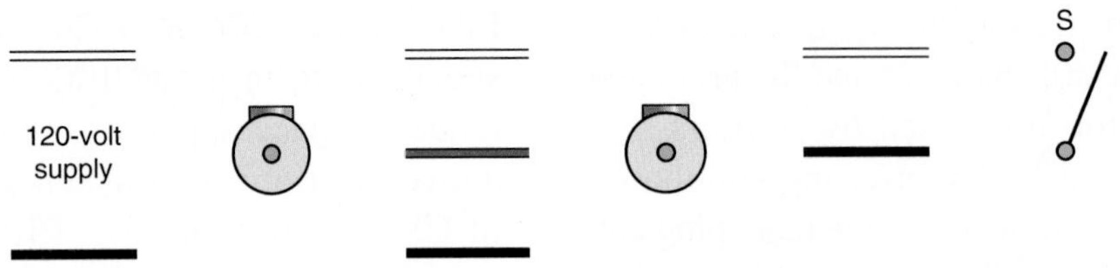

120-volt
supply

S

7. a. Complete the connections for the diagram. Installation is cable.

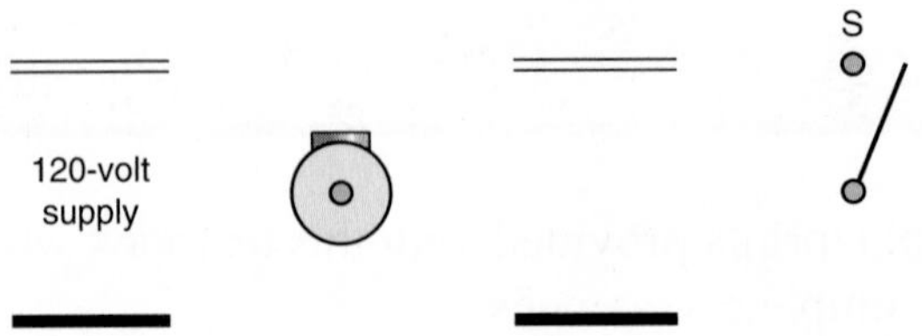

120-volt
supply

S

   b. Which conductor of the cable feeds the switch? ____________________

   c. Which conductor is used as the return wire? ____________________

d. From which wire does the switch feed tap? ______________________________

e. What are the colours of the conductors connected to the fixture? ______________

____

8. What type of switch is installed to control a lighting fixture from two control points? How many switches are needed? ______________________________

____

____

9. Complete the connections in the following diagram so that the lamp may be controlled from either three-way switch.

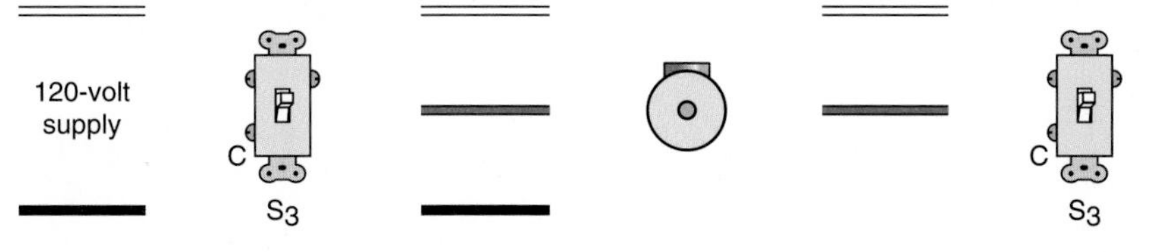

10. Show the connections for a ceiling outlet to be controlled from any one of three switch locations. The 120-volt feed is at the light, *Rule 4–036(2)*. Label the conductor colours. Assume the installation is in cable.

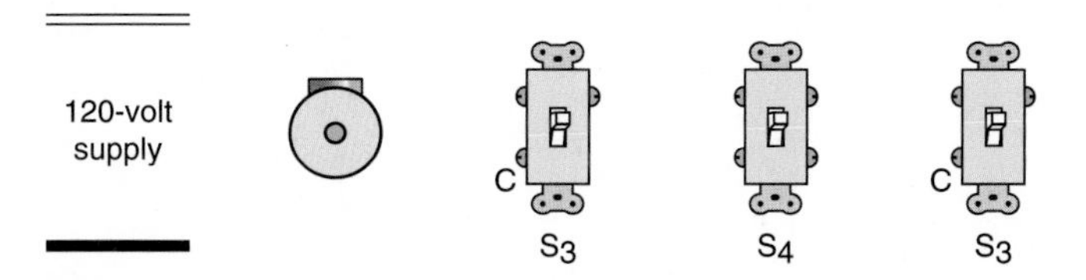

11. Match the following switch types with the correct number of terminals for each.

Three-way switch	Two terminals
Single-pole switch	Four terminals
Four-way switch	Three terminals

12. When connecting single-pole, three-way, and four-way switches, they must be wired so that all switching is done in the ____________________ circuit conductor.

13. If you had to install an underground three-wire feeder to a remote building using three individual conductors, which of the following installations "meets code"? [Circle correct installation.]

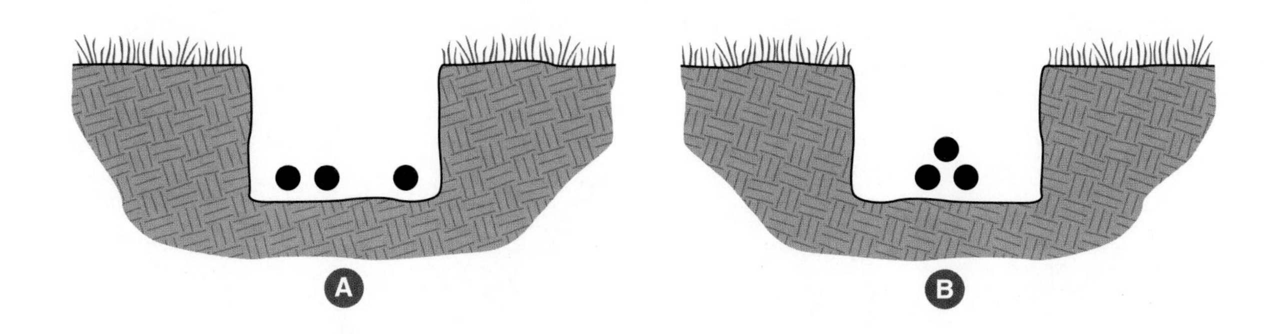

14. Is it always necessary to attach the bare equipment-bonding conductor of a non-metallic-sheathed cable to the green hexagonal bonding screw on a receptacle? Explain.
____
____
____
____
____

15. When two non-metallic-sheathed cables enter a box, is it permitted to bring both bare bonding conductors directly to the bonding terminal of a receptacle, using the terminal as a splice point? _______________

UNIT 9

Ground-Fault Circuit Interrupters, Arc-Fault Circuit Interrupters, Transient Voltage Surge Suppressors, and Isolated Ground Receptacles

OBJECTIVES

After studying this unit, you should be able to

- understand the theory of ground-fault circuit interrupters (GFCI) devices
- explain the operation and connection of GFCIs
- explain why GFCIs are required
- discuss locations where GFCIs must be installed in homes
- discuss the *Canadian Electrical Code (CEC)* rules relating to the replacement of existing receptacles with GFCI receptacles
- understand the *CEC* requirements for GFCI protection on construction sites
- understand the logic of the exemptions to GFCI mandatory requirements for receptacles in certain locations in kitchens, garages, and basements
- discuss the *CEC* requirements for arc-fault circuit interrupters (AFCIs)
- understand *CEC* requirements for the installation of GFCIs and AFCIs
- understand and discuss the basics of transient voltage surge suppressors and isolated ground receptacles

ELECTRICAL HAZARDS

The most important consideration in the installation of an electrical system is that it does not kill anyone. Many lives have been lost because of electric shock from an appliance or a piece of equipment that is "hot." This means that the "hot" circuit conductor in the appliance is contacting the metal frame of the appliance. This condition may be due to the breakdown of the insulation because of wear and tear, defective construction, or accidental misuse of the equipment.

The shock hazard exists whenever the user can touch both the defective equipment and grounded surfaces, such as water pipes, metal sink rims, grounded metal lighting fixtures, earth, concrete in contact with the earth, water, or any other grounded surface.

Figure 9-1 shows a time–current curve that indicates the amount of current that a normal healthy adult can stand for a given time. Just as in overcurrent protection for branch circuits, motors, appliances, and so on, it is always a matter of *how much for how long.*

CODE REQUIREMENTS FOR GROUND-FAULT CIRCUIT INTERRUPTERS (GFCIs)

To provide supplementary protection against shock hazard, the *CEC* requires that GFCIs be provided for the receptacle outlets of dwellings in these situations:

- For *all* 125-volt, single-phase, 15- or 20-ampere receptacles in bathrooms and washrooms, a Class A-type GFCI-protected receptacle is required. This receptacle, where practical, shall be located at least 1 m, but in no case less than 500 mm, from the bathtub or shower stall. A bathroom/washroom is defined in *CEC Section 0, Rule 26–700(11).*
- For *all* 125-volt, 15- or 20-ampere receptacles installed within 1.5 m of a sink, GFCI protection is required. For exceptions, see *Rule 26–700* and *Appendices B* and *I.*
- For *all* 125-volt, single-phase, 15-ampere receptacles located outdoors and within 2.5 m of ground

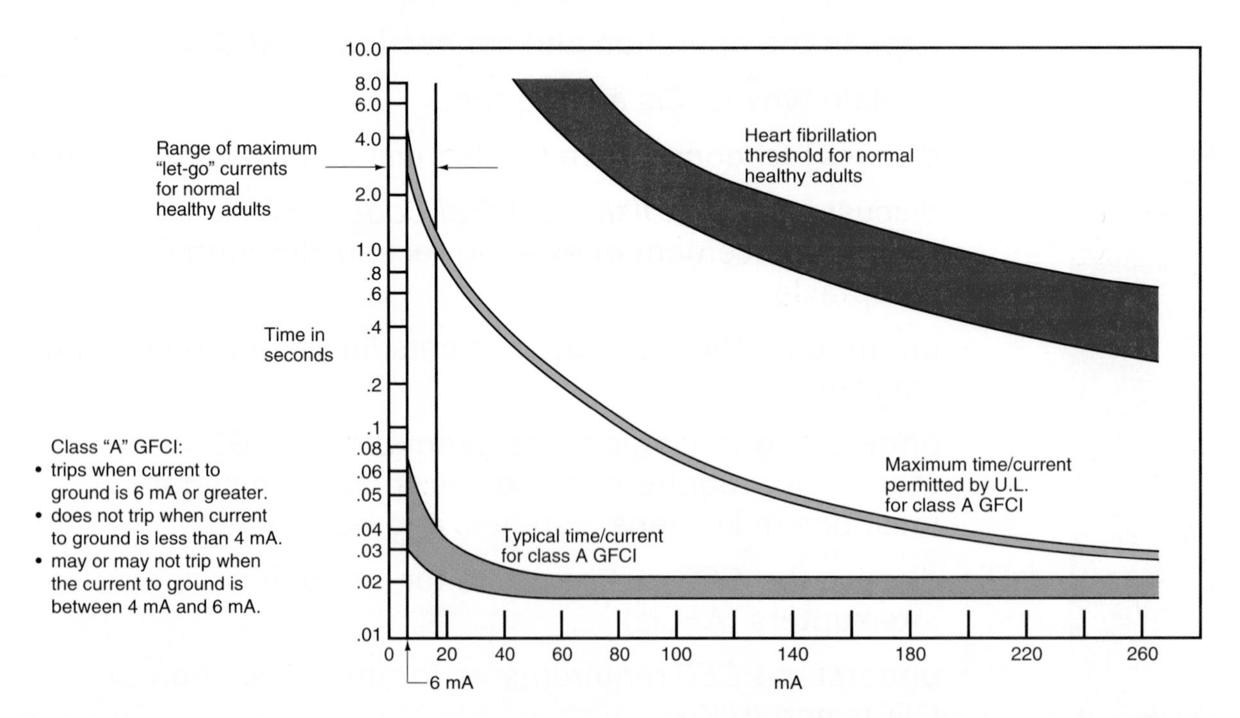

FIGURE 9-1 The time–current curve shows the tripping characteristics of a typical Class A GFCI. If you follow the 6-mA line vertically to the cross-hatched typical time–current curve, you find that the GFCI will open in from about 0.035 to 0.08 seconds. One electrical cycle is 1/60 of a second (0.0167 seconds).

level, a Class A-type GFCI-protected receptacle is required. Accordingly, a GFCI-protected receptacle is required for a receptacle that may be used for an electric lawn mower or trimmer, *Rule 26–710(n),* and one receptacle shall be provided for each car space in the garage or carport, *Rule 26–714(a)(b).*

- In boathouses, all 125-volt, single-phase, 15-ampere receptacles *must* have GFCI protection.

The *CEC* does *not* require GFCI receptacles for

- the receptacle for laundry equipment
- a single-outlet receptacle that supplies a permanently installed sump pump
- a single-outlet receptacle that is supplied by a dedicated branch circuit located and identified for specific use by a cord-and-plug-connected appliance, such as a refrigerator or freezer
- habitable, finished rooms in basements
- the receptacle installed in the ceiling of a garage and designed for the overhead garage door opener
- the receptacle installed for a cord-connected appliance occupying a dedicated space, such as a freezer
- a receptacle installed below the sink for plug-in connection of a food waste disposal
- a receptacle installed solely for a clock
- on a second-floor balcony, such as in a condominium or apartment

The *CEC* does not permit GFCIs to be used to protect circuits containing smoke detectors.

Section 68 gives the electrical requirements for swimming pools and the use of GFCI devices. (See Unit 22.)

The *CEC* requirements for ground-fault circuit protection can be met in several ways. The two most common types of GFCI protection are in the form of receptacles and circuit breakers (Figure 9-2). The circuit breakers are available in either single-pole or double-pole configurations for 120-volt, 240-volt, or 120/240-volt applications.

Figure 9-3 illustrates a Class A GFCI circuit breaker installed on a circuit. A fault or current of six milliamperes (6 mA) or more will shut off the entire circuit. For example, a ground fault at any point on

Ground-fault circuit interrupter as an integral part of a duplex grounding-type convenience receptacle.

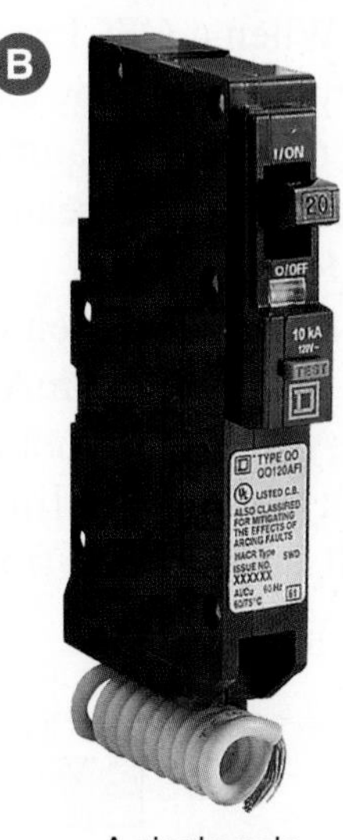

A single-pole ground-fault circuit-interrupter circuit-breaker. Double-pole GFCI circuit breakers are also available.

FIGURE 9-2 (A) A ground-fault circuit-interrupter receptacle, and (B) a ground-fault circuit-interrupter circuit breaker. The switching mechanism of a GFCI receptacle opens both the ungrounded "hot" conductor and the grounded conductor. The switching mechanism of a GFCI circuit breaker opens the ungrounded "hot" conductor only.

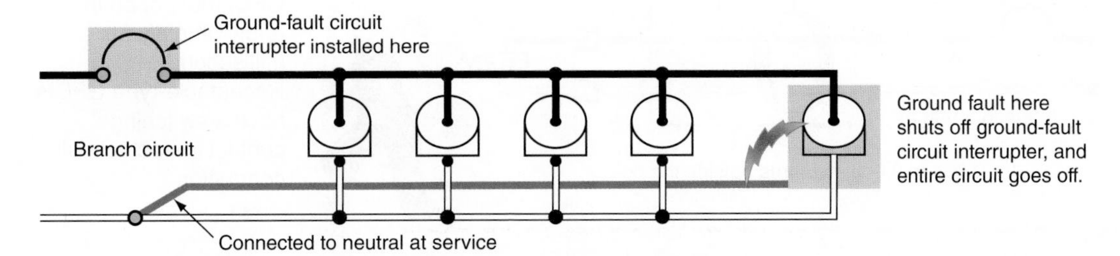

FIGURE 9-3 A GFCI as a part of the branch-circuit overcurrent device.

Branch circuit

Connected to neutral at service

Ground-fault circuit interrupter installed here

Ground fault here shuts off this outlet only. The rest of the circuit is not affected.

FIGURE 9-4 A GFCI as an integral part of a receptacle outlet.

circuit B20 will shut off the washroom and bathroom receptacles together with any other devices that are connected to the outlets on this circuit.

When a GFCI receptacle is installed, only that receptacle and downstream receptacles are shut off when a ground fault of 6 mA or more occurs; see Figure 9-4. Upstream receptacles are not affected.

Regardless of the location of the GFCI in the circuit, it must open the circuit when the current to ground exceeds 6 mA (0.006 ampere).

A GFCI does not limit the magnitude of ground-fault current. It limits the time that a ground-fault current will flow. A GFCI does not provide protection against shock hazard should a person make contact with two of the normal circuit conductors on the load side of the GFCI (e.g., line and neutral).

Figure 9-5 shows how a ground-fault circuit interrupter operates. Figure 9-6 shows the internal wiring of a GFCI.

The GFCI monitors the current balance between the hot conductor and the identified conductor. As soon as the current in the identified conductor is less than the current in the hot conductor, the GFCI senses this imbalance and opens the circuit. The imbalance indicates that part of the current in the

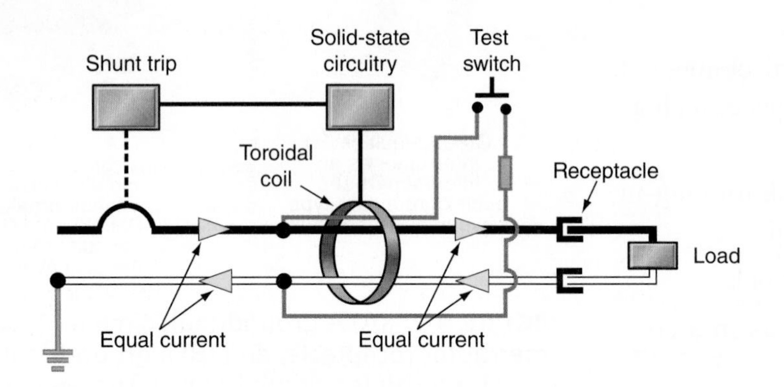

No current is induced in the toroidal coil because both circuit wires are carrying equal current. The contacts remain closed.

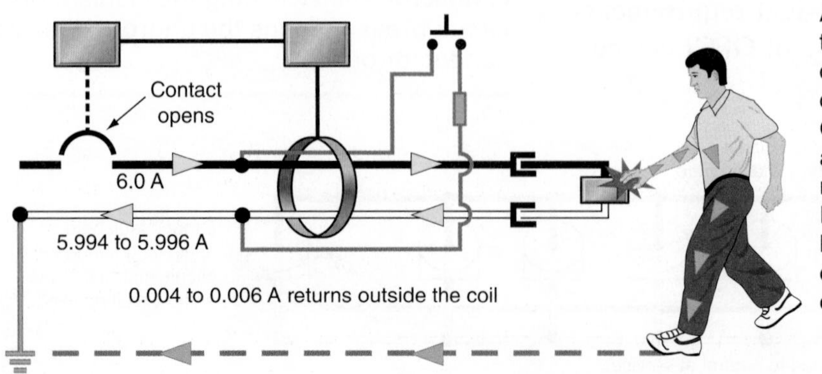

An imbalance of from 4 to 6 milliamperes in the coil will cause the contacts to open. The GFCI must open in approximately 25 milliseconds. Receptacle-type GFCIs have a switching contact in each circuit conductor.

FIGURE 9-5 Basic principle of how a GFCI operates.

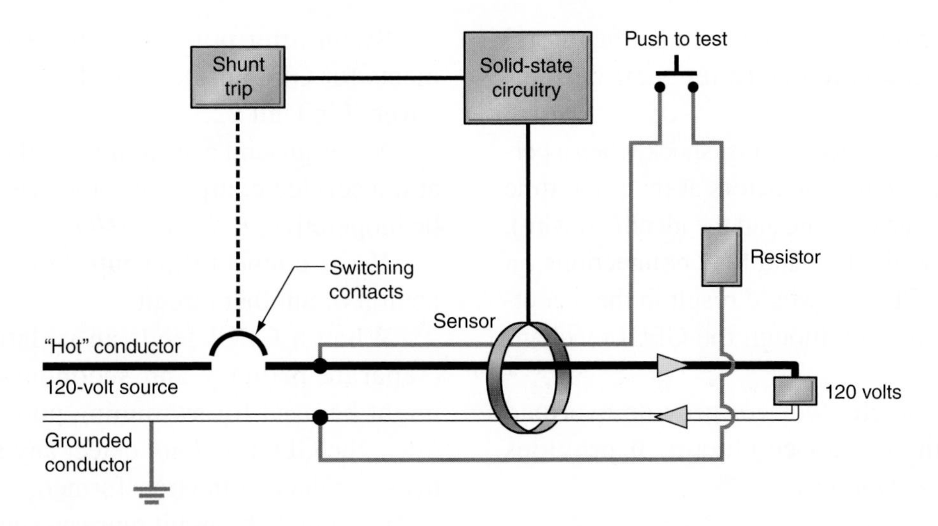

FIGURE 9-6 GFCI internal components and connections. Receptacle-type GFCIs switch both the phase ("hot") and grounded conductors. When the test button is pushed, the test current passes through the sensor and the test button, then back around (bypasses, outside of) the sensor to the opposite circuit conductor. This is how the imbalance is created and then monitored by the solid-state circuitry to signal the GFCI's contacts to open. Since both "load" currents pass through the sensor, no imbalance exists under normal receptacle use.

circuit is being diverted to some path other than the normal return path along the identified conductor. Thus, if the GFCI trips off, it is an indication of a possible shock hazard from a line-to-ground fault.

Nuisance tripping of GFCIs is known to occur. Sometimes this can be attributed to extremely long runs of cable for the protected circuit. Consult the manufacturer's instructions to determine if maximum lengths for a protected circuit are recommended. Some electricians advise using non-metallic staples or non-metallic straps instead of metallic to prevent nuisance tripping.

PRECAUTIONS FOR GROUND-FAULT CIRCUIT INTERRUPTERS

Unit 4 covers *interrupting ratings* for overcurrent devices. For life safety reasons, electricians and electrical inspectors should always concern themselves with the interrupting ratings of fuses and circuit breakers, particularly at the main service entrance equipment of a residence, condo, or apartment building. A GFCI receptacle also has the ability to interrupt current when a fault occurs. According to the Canadian Standards Association (CSA) standard for GFCI receptacles, the current-withstand rating is 5000 amperes, RMS symmetrical.

So, unless the overcurrent protective device protecting the circuit that feeds the GFCI receptacle can limit the line-to-ground fault current to a maximum of 5000 amperes RMS symmetrical, installation of any GFCI receptacles close to the main panel should be done only after a short-circuit study has been made to ensure that the available line-to-ground fault current is 5000 amperes or less.

Without this precaution, the GFCI sensing and tripping mechanisms could be rendered inoperable should a line-to-ground fault occur. When the GFCI is called on to prevent injury or electrocution of a person, it might not operate safely. This is a very important reason for periodically testing all GFCI devices. Instructions furnished with GFCIs emphasize "operate upon installation and at least monthly and record the date and results of the test on the form provided by the manufacturer." ***Never*** test a GFCI receptacle by shorting out line-to-neutral. The mechanism will be damaged and become inoperable.

On single-phase systems, the line-to-ground fault current can exceed the line-to-line fault current. See Unit 4.

Check fuse and circuit-breaker manufacturers' current-limiting charts to determine their current-limiting ability.

A GFCI does *not* protect against shock when a person touches both circuit conductors at the same time (two hot wires, or one hot wire and the identified wire).

Do not reverse the line and load connections on a feedthrough GFCI. This would result in the receptacle still being live even though the GFCI mechanism has tripped.

A GFCI receptacle *does not* provide overload protection for the circuit conductor. It provides *ground-fault protection only.*

GROUND-FAULT CIRCUIT INTERRUPTER IN RESIDENCE CIRCUITS

In this residence, the receptacle outlets installed outdoors, in the bathrooms, on the kitchen counter near the sink, and specific ones in the basement are protected by GFCIs (Figure 9-7). These receptacle outlets are connected to the circuits listed in the panel schedule in Unit 4 and identified on the floor plans.

Swimming pools also have special requirements for GFCI protection. These requirements are covered in Unit 22.

Never ground a system neutral conductor except at the service equipment; otherwise, the GFCI will be inoperative, *Rule 10–204(1).*

Never connect the neutral of one circuit to the neutral of another circuit.

When a GFCI feeds an isolation transformer (separate primary and secondary windings), as might be used for swimming pool underwater fixtures, the GFCI *will not* detect any ground faults on the secondary of the transformer.

Long branch-circuit runs can cause nuisance tripping of the GFCI due to leakage currents in the circuit wiring. GFCI receptacles at the point of use tend to minimize this problem. A circuit supplied by a GFCI circuit breaker in the main panel could cause nuisance tripping if the branch circuit is long: Some electricians say 15 m or more can cause nuisance tripping of the GFCI breaker.

In older houses that were wired with knob-and-tube wiring or early forms of non-metallic-sheathed cable (which did not include a bonding conductor), nonbonding receptacles were used.

Courtesy of Legrand/Pass & Seymour, Inc.

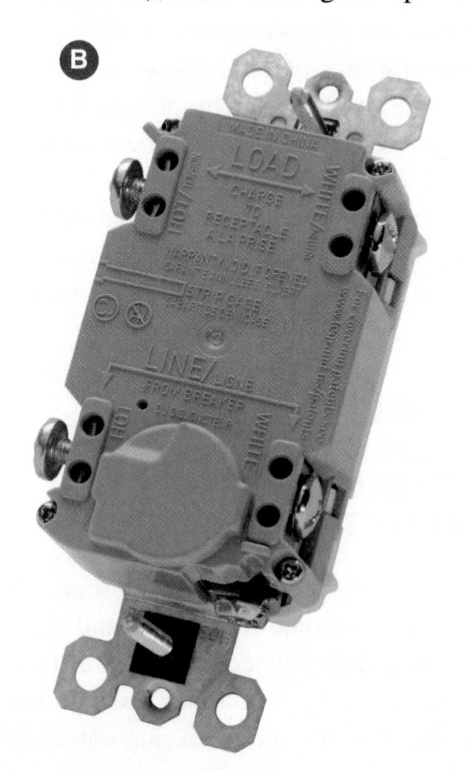

Courtesy of Legrand/Pass & Seymour, Inc.

FIGURE 9-7 (A) The front side of a ground fault circuit Interrupter. (B) The reverse side of a ground fault circuit interrupter. Note the line side connection terminal and load side connection terminals.

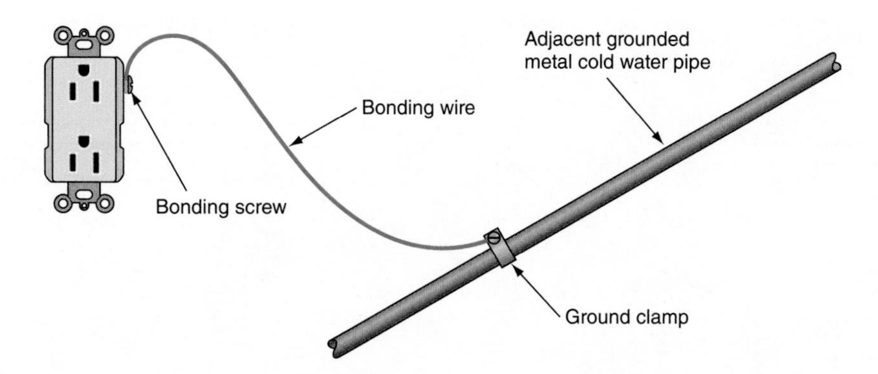

FIGURE 9-8 In existing installations that do not have bonding wire as part of the branch-circuit wiring, the *CEC* permits replacing an old-style nonbonding-type receptacle with a bonding-type receptacle, *Rule 26–700(7)*. The bonding terminal on the receptacle must be properly bonded. One acceptable way to bond the bonding terminal is to run a conductor to an effectively grounded water pipe. See *Rule 26–700(7)* for other acceptable means of properly bonding the receptacle's bonding terminal.

Rule 26–700(7) allows a bonding receptacle to be used to replace older nonbonding receptacles, provided the receptacle bonding terminal is connected to ground by one of these means:

- bonding to the nearest metallic cold water pipe (Figure 9-8)
- running a separate bonding conductor to the system ground
- connecting to a metallic raceway

Rule 26–700(8) also permits a bonding receptacle to replace an existing nonbonding receptacle if each replacement receptacle is protected by a Class A-type GFCI and no bonding conductors are extended from the receptacle to any other outlet, *Rule 26–700(9)*.

FEEDTHROUGH GROUND-FAULT CIRCUIT INTERRUPTER

The decision to use more GFCIs rather than trying to protect many receptacles through one GFCI becomes one of economy and practicality. GFCI receptacles are more expensive than regular receptacles. You must make the decision for each installation, keeping in mind that GFCI protection is a safety issue recognized and clearly stated in the *CEC*. However, the actual circuit layout is left up to the electrician.

Figure 9-9 illustrates how a feedthrough GFCI receptacle supplies many other receptacles. Should a ground fault occur anywhere on this circuit, all 11 receptacles lose power—not a good circuit layout. Attempting to locate the ground-fault problem—unless it is obvious—can be very time-consuming. Using more GFCI receptacles is generally the more practical approach.

Figure 9-10 shows how the feedthrough GFCI receptacle is connected into the circuit. When connected in this manner, the feedthrough GFCI receptacle and all downstream outlets on the same circuit will have ground-fault protection; see Figure 9-11.

The most important factors to consider are the continuity of electrical power and the economy of the installation.

CAUTION: *Do not* connect the freezer receptacle in the workshop or the refrigerator receptacle in the kitchen to a GFCI-protected circuit. These devices are not required to be connected to a GFCI-protected circuit, since any nuisance-tripping of the circuit could result in spoiled food. The leakage current tolerances allowed by the CSA approach the minimum trip settings of Class A GFCIs (4 to 6 mA). Nuisance-tripping can shut off the power to the refrigerator or freezer. Coming home to spoiled food is not a pleasant experience.

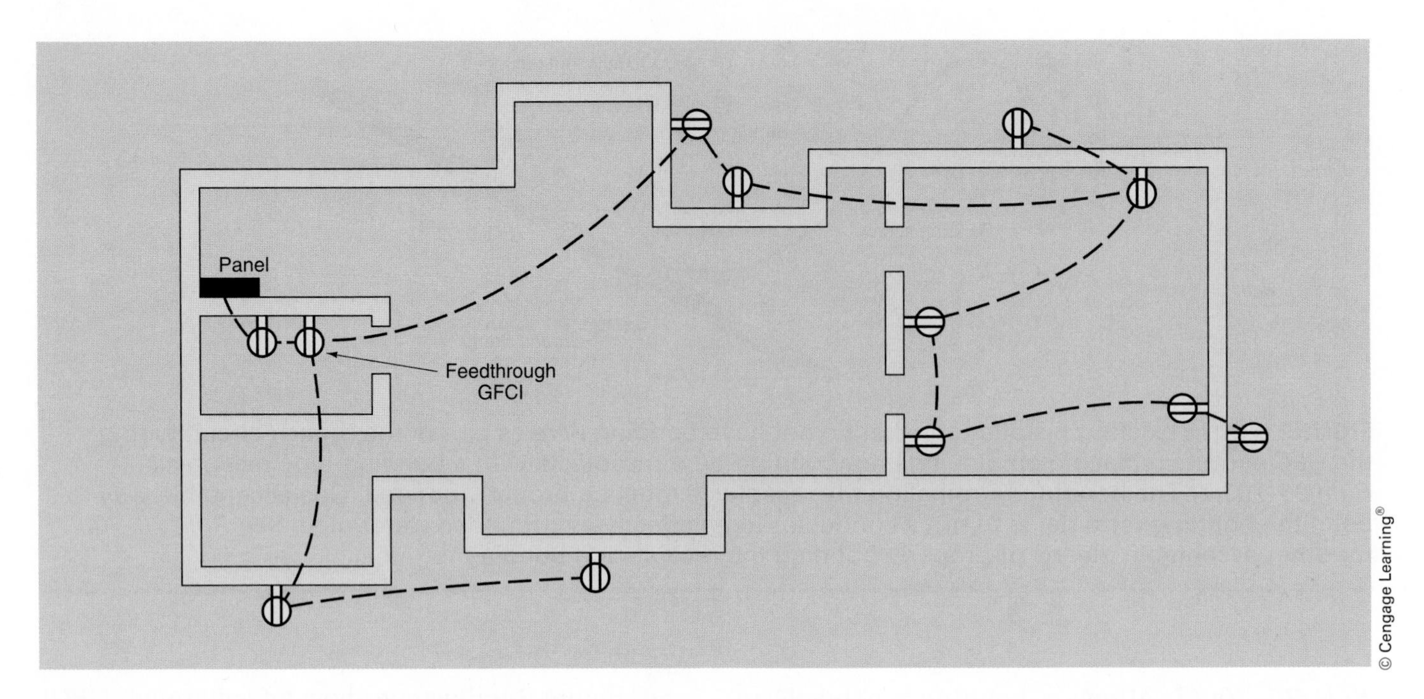

FIGURE 9-9 One GFCI feedthrough receptacle protecting 10 other receptacles. Although this might be an economical way to protect many receptacle outlets with only one GFCI feedthrough receptacle, the GFCI will probably nuisance-trip because of leakage currents. Use common sense and the manufacturer's recommendations as to how many devices should be protected by one GFCI.

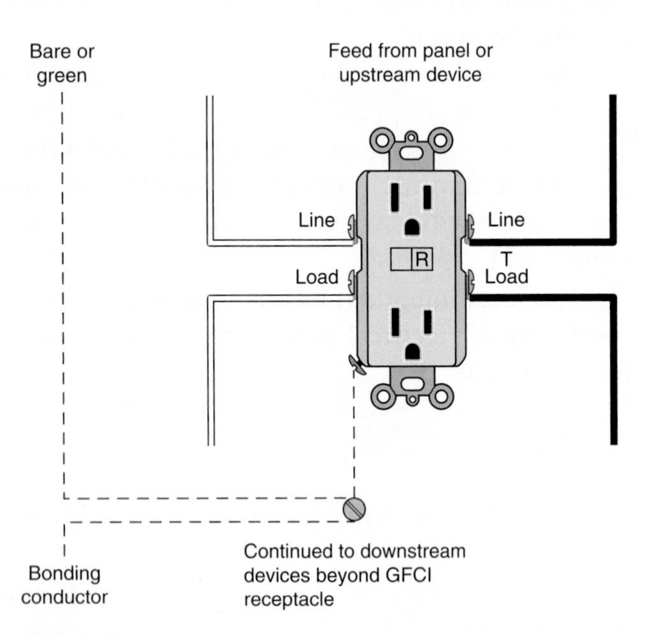

FIGURE 9-10 Connecting a feedthrough GFCI into a circuit.

A receptacle that is installed specifically for the washing machine in a combined laundry/bathroom is not required to be GFCI-protected if this receptacle is located behind the machine and is not more than 600 mm above the finished floor, *Rule 26–700, Appendix B*.

IDENTIFICATION, TESTING, AND RECORDING OF GFCI RECEPTACLES

Rule 68–068(5) requires warning signs to be placed beside switches that control circuits protected by GFCIs. The signs will alert the user to the fact the circuits controlled by the switches are GFCI-protected and will specify how often the GFCIs should be tested; see Figure 9-12.

Figure 9-2 shows the GFCI receptacle Test and Reset buttons (identified as "T" and "R" in Figure 9-11). Pushing the test button places a small ground fault on the circuit. If operating properly, the GFCI receptacle should trip to the OFF position. The operation is the same for the GFCI circuit breakers. Pushing the reset button will restore power.

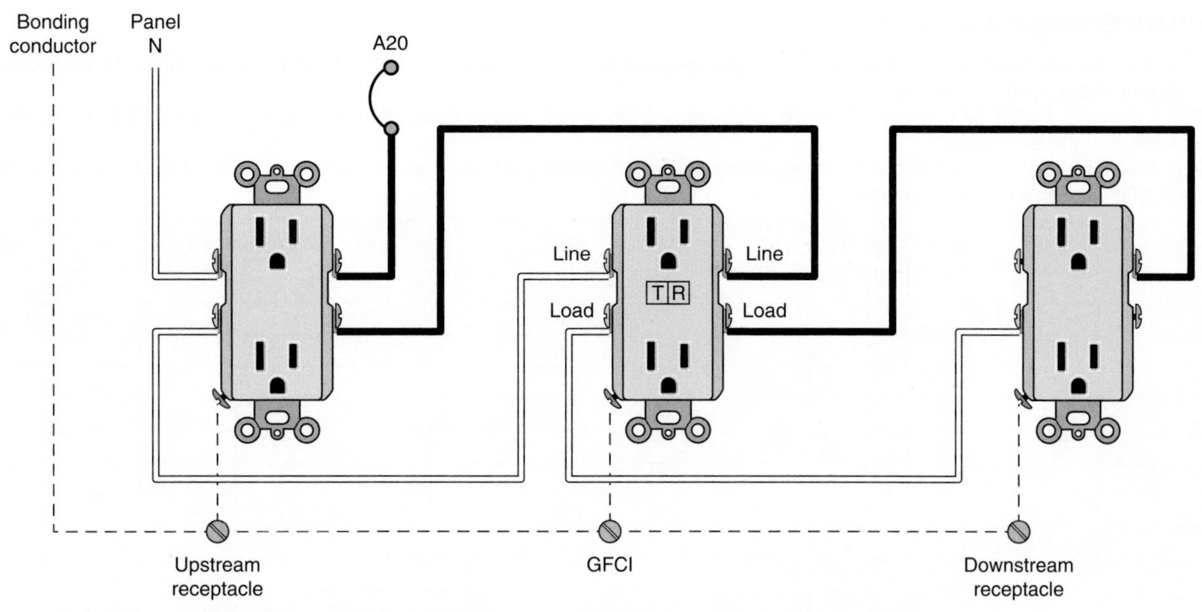

FIGURE 9-11 GFCI receptacle protecting downstream receptacles.

THIS RECEPTACLE PROVIDES PROTECTION AGAINST ELECTRIC SHOCK

THIS RECEPTACLE IS GROUND-FAULT PROTECTED

GROUND-FAULT PROTECTED Instructions at power panel

FIGURE 9-12 GFCI warning signs.

The detailed information and testing instructions are furnished with GFCI receptacles and GFCI circuit breakers. Monthly testing is recommended to ensure that the GFCI mechanism will operate properly if a human being is subjected to an electric shock.

- Will the homeowner remember to do the monthly testing?
- Will the homeowner recognize and understand the purpose and function of the GFCI receptacle?
- Should all of the bathroom GFCI receptacles be identified?
- Will the homeowner recognize a receptacle if protected by a GFCI circuit breaker located in an electrical panel far from the actual receptacle?

These are important questions and a possible reason why the *CEC* requires identification, and why the CSA requires detailed installation and testing instructions to be included in the packaging for the GFCI receptacle or circuit breaker. Instructions are not only for the electrician but also for the homeowner, and must be left in a conspicuous place so that homeowners can familiarize themselves with the receptacle, its operation, and its need for testing. Figure 9-13 shows a chart the homeowner can use to record monthly GFCI testing.

REPLACING EXISTING RECEPTACLES

The *CEC* is very specific on the type of receptacle permitted as a replacement for an existing receptacle, *Rule 26–700*.

OCCUPANT'S TEST RECORD

TO TEST, depress the "TEST" button; the "RESET" button should extend. Should the "RESET" button not extend, the GFCI will not protect against electric shock. Call qualified electrician.
TO RESET, depress the "RESET" button firmly into GFCI unit until an audible click is heard. If reset properly, the RESET button will be flush with the surface of the test button.
This label should be retained and placed in a conspicuous location to remind the occupants that for maximum protection against electric shock, each GFCI should be tested monthly.

YEAR	JAN	FEB	MAR	APR	MAY	JUN	JUL	AUG	SEP	OCT	NOV	DEC

FIGURE 9-13 Homeowner's testing chart for recording GFCI testing dates, as recommended in *Appendix B Rule 26–400.*

Replacing Existing Receptacles Where Bonding Means Exist (Refer to Figure 9-14)

When replacing an existing receptacle (Figure 9-14, Part A) where the wall box is properly bonded (E) or where the branch-circuit wiring contains an equipment-bonding conductor (D), the replacement receptacle must be of the bonding type (B) unless the replacement receptacle is of the GFCI-type (C).

In either case, be sure that only the bonding conductor of the circuit is connected to the receptacle's green hexagon bonding terminal.

Do not connect the white grounded circuit conductor to the green hexagon bonding terminal of the receptacle.

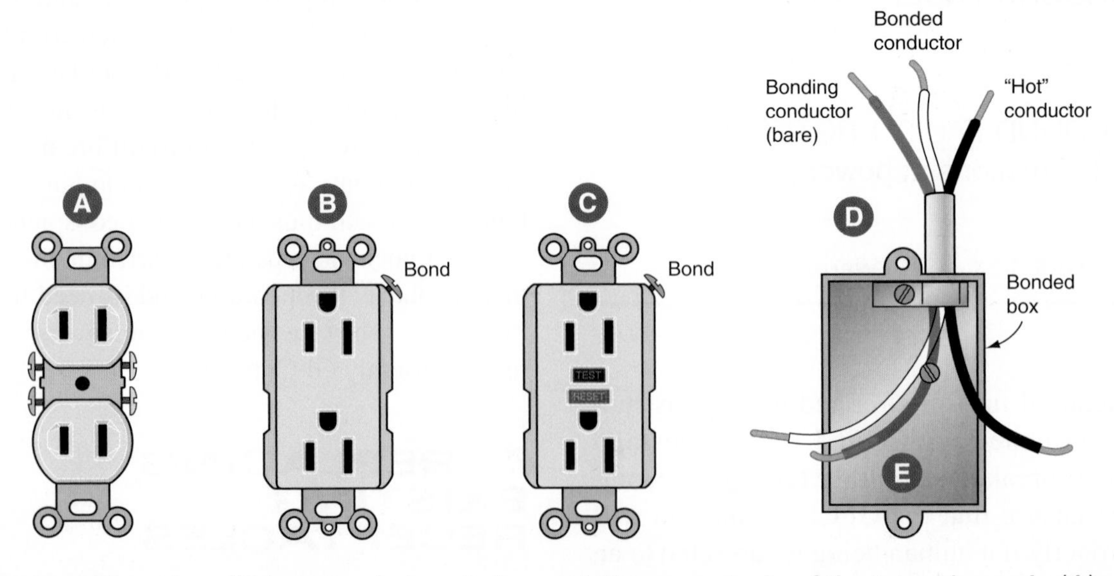

FIGURE 9-14 Where box (E) is properly bonded, an existing receptacle of the type shown in (A) *must* be replaced with a bonding-type receptacle (B). It is also acceptable to replace (A) with a GFCI receptacle (C). See the text for further discussion.

Do not connect the white grounded conductor to the wall box.

The white conductor is the identified conductor and may be connected only to the identified terminal on the receptacle.

Replacing Existing Receptacles Where Bonding Means Do Not Exist (Refer to Figures 9-15 and 9-16)

When replacing an existing nonbonding-type receptacle (Figure 9-15, Part A) where the box is not bonded (E) and an equipment-bonding conductor has not been run with the circuit conductors (D), four choices are possible for selecting the replacement receptacle:

1. The replacement receptacle may be a nonbonding-type (A).
2. The replacement receptacle may be a GFCI-type (C).
   - The green hexagon terminal of the GFCI replacement receptacle does not have to be bonded to ground. It must be left "unconnected." The GFCI's trip mechanism will operate properly when ground faults occur anywhere on the load side of the GFCI replacement receptacle. Ground-fault protection is still there; see Figure 9-6.
   - *Do not* connect a bonding conductor from the green hexagon bonding terminal of a replacement GFCI receptacle (C) to any other downstream receptacles that are fed through the replacement GFCI receptacle, *Rule 26–700(9)* (see Figure 9-16).

     The reason this is not permitted is that if someone later saw the conductor connected to the green hexagon bonding terminal of the downstream receptacle, they would immediately assume that the other end of that conductor had been properly connected to an acceptable grounding point of the electrical system. The fact is that the so-called bonding conductor had been connected to the replacement GFCI receptacle's green hexagon bonding terminal that was not bonded in the first place. This creates a false sense of security—a real shock hazard.
3. The replacement receptacle may be a bonding type (B) if it is protected by a Class A-type GFCI (C), *Rule 26–700(8)*.

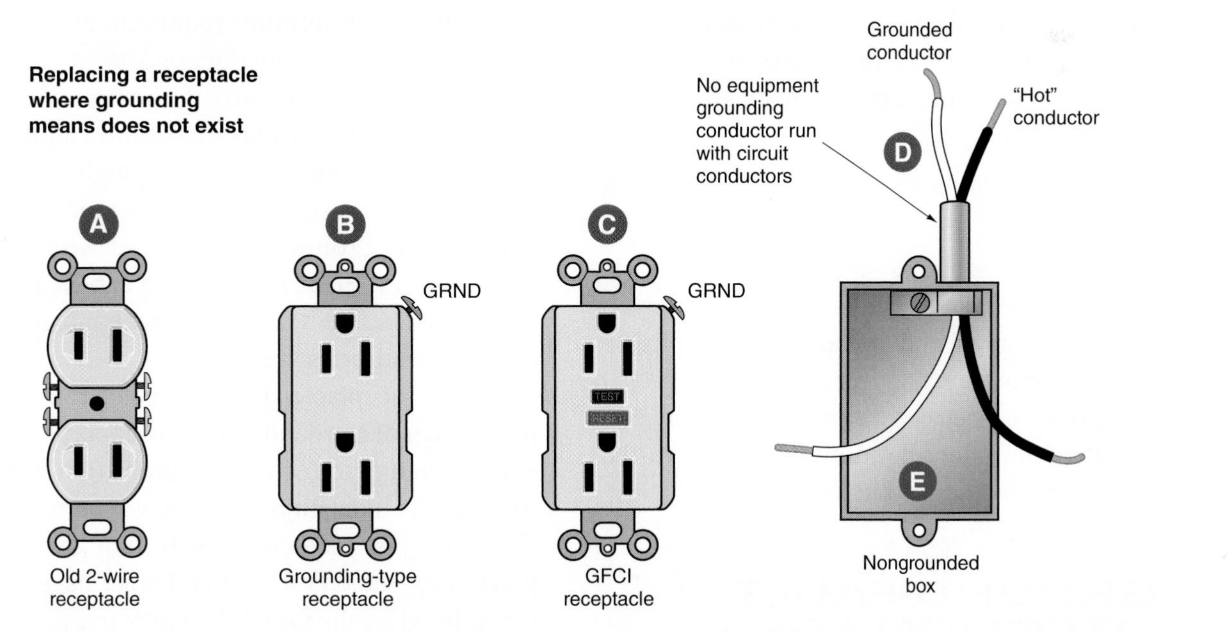

FIGURE 9-15 Where box (E) is *not* bonded and a bonding conductor has *not* been run with the circuit conductors (D), an existing nonbonding-type receptacle (A) may be replaced with a nonbonding-type receptacle (A), a GFCI-type receptacle (C), a bonding-type receptacle (B) if supplied through a GFCI receptacle (C), or a grounding-type receptacle (B) if a separate bonding conductor is run from the receptacle to a cold water pipe or other effective grounding means, *Rule 26–700(7, 8)*.

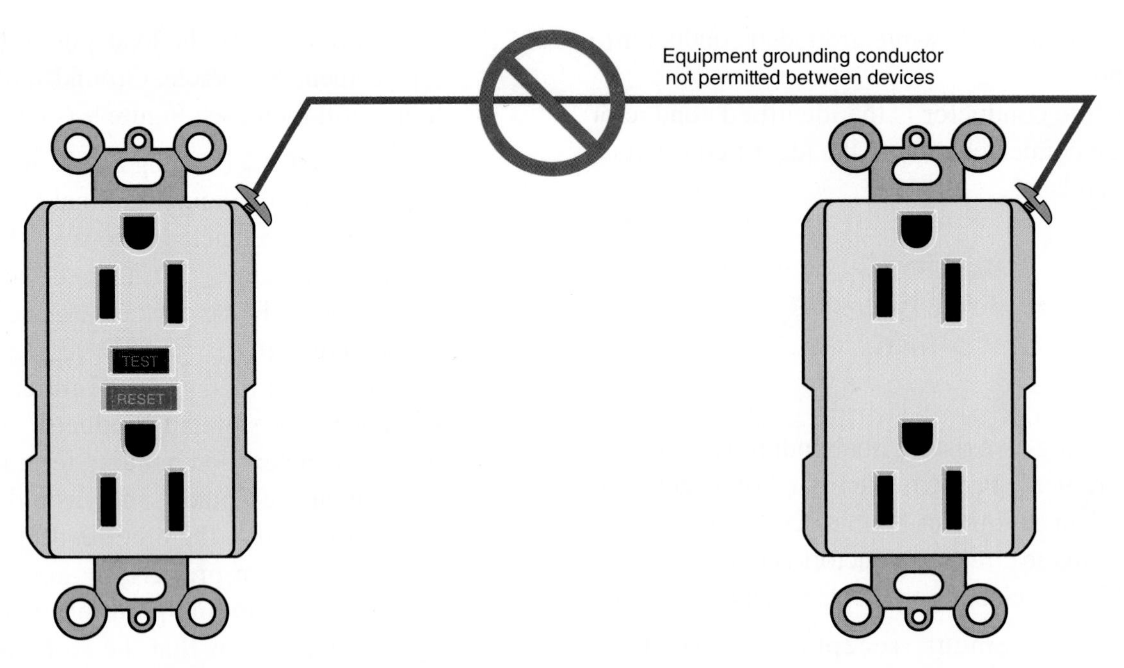

FIGURE 9-16 *Do not* connect a bonding conductor between the above receptacles. The text explains why this is not permitted, *Rule 26–700(9)*.

- The green hexagon equipment-bonding terminal of the replacement bonding-type receptacle (B) need not be connected to any grounding means. It may be left unconnected. The upstream feedthrough GFCI receptacle (C) trip mechanism will work properly when ground faults occur anywhere on its load side. Ground-fault protection is still there; see Figure 9-6.

4. The replacement receptacle may be a bonding type (B) if

   - A bonding conductor sized per *Table 16A* is run between the replacement receptacle's green hexagon terminal and a grounded water pipe or other effective bonding means, *Rules 26–700(8)* and *10–808*. The acceptable means of connecting the equipment-bonding conductor to the water pipe is described in *Rule 10–906*.

GROUND-FAULT PROTECTION FOR CONSTRUCTION SITES

Safety on construction sites is of paramount importance to workers and employers. Safety regulations such as provincial occupational health and safety acts and the *Canada Occupational Health and Safety Regulations* govern how certain aspects of a building's construction will be completed. Temporary lighting and power are both referenced in the acts and regulations. Along with the *CEC*, these determine the minimum requirements for the temporary wiring on a construction site.

Because of the nature of construction sites, the continual presence of shock hazard can lead to serious personal injury or death through electrocution. Workers often are standing in water, standing on damp or wet ground, or in contact with steel framing members. Electric cords and cables are lying on the ground, subject to severe mechanical abuse. All of these conditions spell danger.

GFCI receptacles provide protection for workers in the event of a ground fault. It is important that workers not attempt to bypass this protection. If the GFCI trips, it is an indication that there is a ground fault. The GFCI has just protected you from a ground fault. If you bypass the GFCI, you no longer have the protection offered by the GFCI. You may not get the chance to make the same mistake twice. Figure 9-17 illustrates some portable GFCI devices that provide ground-fault protection when plugged into non-GFCI receptacles.

The permanent site office and any temporary lighting and/or power panels must fully comply

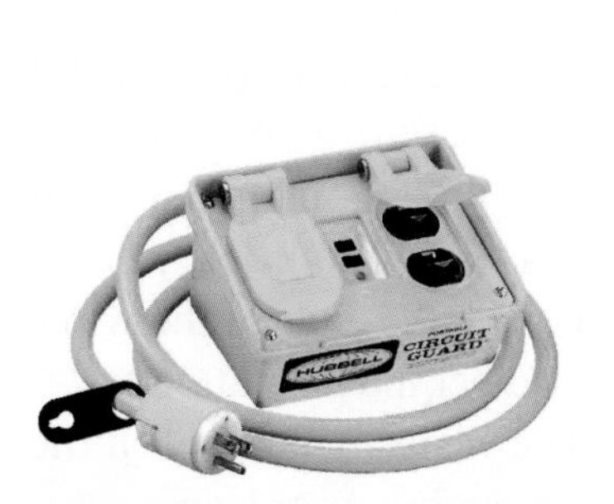

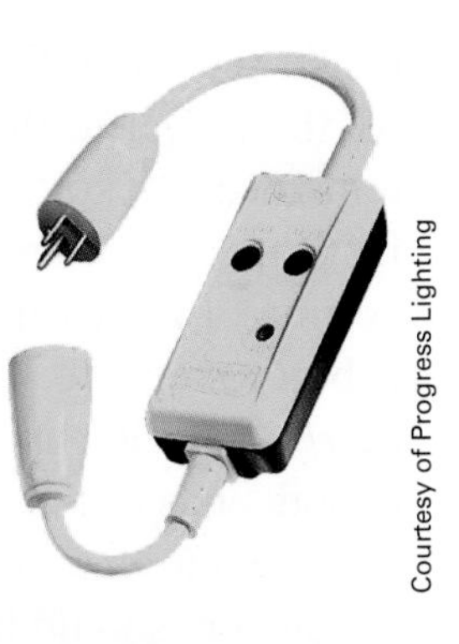

Courtesy of Progress Lighting

FIGURE 9-17 Plug- and cord-connected portable GFCI devices are easy to carry around and use anywhere when working with electrical tools and extension cords. These devices operate independently. These inexpensive portable GFCI in-line cords should always be used with portable electric tools.

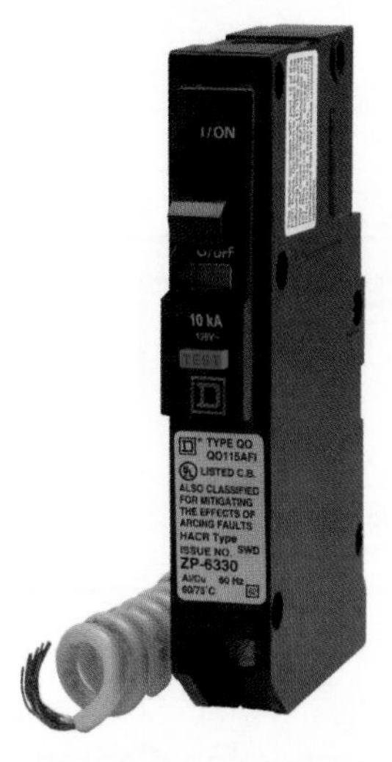

Courtesy of Schneider Electric

FIGURE 9-18 An arc-fault circuit-interrupter circuit breaker.

with the *CEC*. They must be inspected by the electrical inspection authority before connection can take place, *Rule 2–004 and Rule 76–002.*

ARC-FAULT CIRCUIT INTERRUPTERS (AFCIs)

About 10% of fires are electrical fires. One-third of electrical fires are caused by arcing at poor connections or between conductors due to a breakdown in electrical insulation. For example, if the strands in a conductor of the cord for an electric iron break, a few strands may still be touching, and there may be enough of a path to permit current to flow. When the iron draws current, arcing will occur at the point of the conductor break. The current drawn by the iron will not trip the circuit breaker or fuse protecting the circuit but can develop intense heat at the break in the wire. Arcing faults can result from such things as poor terminations during installation, equipment failure, and the failure of conductor insulation due to circuit overloading or improperly sized overcurrent protection. Figure 9-18 is an arc-fault circuit interrupter circuit breaker that would be installed in the main panel branch-circuit side.

Types of Arcs

A series arc is one that occurs when a current-carrying conductor breaks or there is a poor connection between the conductor and a terminal screw. The current from a series arc follows the normal circuit path.

Parallel arcing occurs between two conductors. The arcing may occur from line-to-line, line-to-neutral, or line-to-ground. Parallel arcing current does not follow the normal circuit path.

How AFCIs Work

An AFCI is designed to sense the fluctuations in voltage and current that may occur during an arcing sequence and de-energize the circuit if dangerous arcing occurs. AFCIs are designed to sense both series and parallel arcing and can distinguish between the arcing that occurs in the normal operation of an electrical system and the unwanted arcing that can lead to a fire. The arc that occurs when the filament of an incandescent lamps breaks, when a switch is turned on and off, or when appliances that are turned on are plugged into a receptacle are examples of things that produce arcs that are normally considered harmless.

Installing and Using AFCIs

AFCIs are required to protect receptacles except as per *CEC Rule 26–724(f)(i)(ii)*.Installing an AFCI is similar to installing a GFCI. With a single-pole AFCI, the white pigtail on the breaker is connected

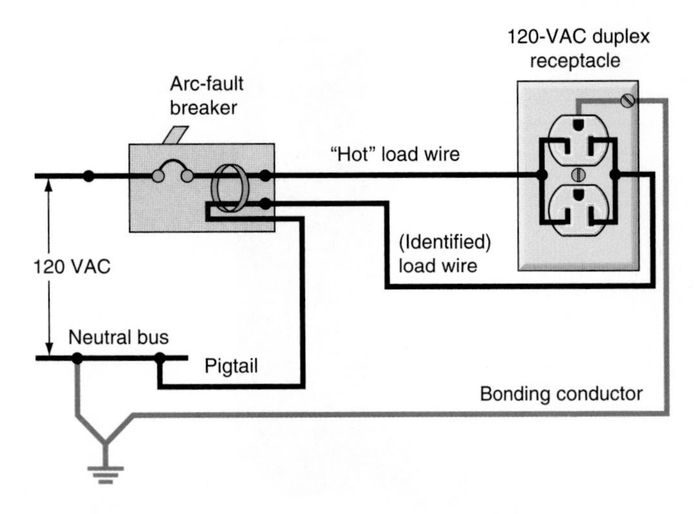

FIGURE 9-19 Connection diagram for single-pole AFCI circuit breaker.

to the neutral bus in the panel, and two circuit conductors, black (hot) and white (identified), are connected to the respective brass- and silver-coloured terminals of the breaker. See Figure 9-19.

Some manufacturers incorporate arc-fault protection and ground-fault protection on a single breaker.

If an AFCI detects a fault, unplug and/or disconnect all appliances connected to the circuit. Reset the AFCI. If the AFCI trips with the appliances removed from the circuit, this indicates a faulty AFCI breaker or a fault in the wiring system. Disconnect the wiring from the AFCI circuit breaker and retest the breaker. If the breaker still detects a fault, the breaker is faulty or has been damaged by the arc fault. Before replacing the breaker, perform an insulation test on the wiring system.

If the AFCI is reset and it doesn't trip, reconnect the appliances that were connected to the circuit, one at a time. If the AFCI still does not trip, it is possible that you have a loose connection or that the arc cleared the fault.

Performing an Insulation Test

An insulation test is normally performed with an insulation tester since it applies a voltage of around 500 volts to the system. If an insulation tester is not available, use your multimeter on its highest resistance scale. Follow all safety procedures. Turn off the power to the circuit. Isolate the wiring system from its source by disconnecting the wiring from the AFCI and disconnect all loads from the portion of the wiring system undergoing the test. (It is best to remove all the loads on the circuit, *particularly electronic loads.)* Isolate the ends of the wires at both ends of the run. Connect the insulation tester (megger) to the circuit. The resistance between the conductors should be one megohm or greater.

TRANSIENT VOLTAGE SURGE SUPPRESSION

In today's homes we find many electronic appliances (televisions, stereos, personal computers, videocassette recorders, digital stereo equipment, microwave ovens), all of which contain many sensitive, delicate electronic components (e.g., printed circuit boards, chips, microprocessors, transistors).

Voltage transients, called *surges* or *spikes,* can stress, degrade, and/or destroy these components; they can cause loss of memory in the equipment or "lock up" the microprocessor.

The increased complexity of electronic integrated circuits makes this equipment an easy target for "dirty" power that can and will affect the performance of the equipment.

Voltage transients cause abnormal current to flow through the sensitive electronic components. This energy is measured in joules. A *joule* is the energy of one ampere passing through a one-ohm resistance for one second. In other words, a joule is the amount of energy used by a one-watt load in one second. Hence, 1 watt-hour equals 3600 joules.

Line surges can be line-to-neutral, line-to-ground, and line-to-line.

Transients

Transients are generally grouped into two categories:

- *Ring wave* transients originate within the building and are caused by welders, elevators, photocopiers, motors, air conditioners, or other inductive loads.

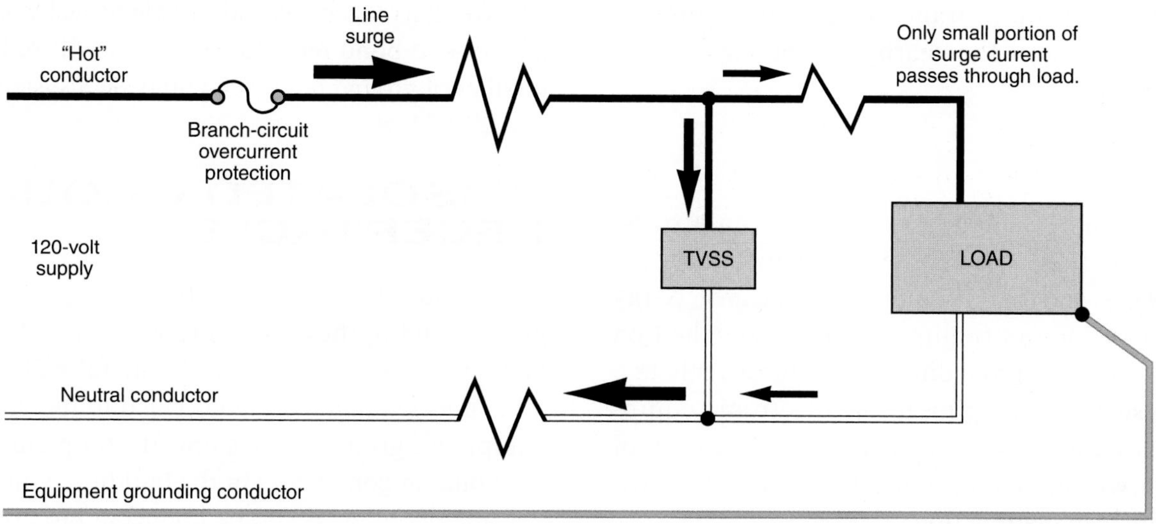

FIGURE 9-20 A transient voltage surge suppressor absorbs and bypasses (shunts) transient currents around the load. The MOV dissipates the surge in the form of heat.

- *Impulse* transients originate outside the building and are caused by utility company switching, lightning, and so on.

To minimize the damaging results of these transient line surges, service equipment, panels, load centres, feeders, branch circuits, and individual receptacle outlets can be protected with a *transient voltage surge suppressor* (TVSS), Figure 9-20.

A TVSS contains one or more metal-oxide varistors (MOVs) that clamp the transients by absorbing the major portion of the energy (joules) created by the surge, allowing only a small, safe amount of energy to enter the connected load.

The MOV clamps the transient in less than one nanosecond (one-billionth of a second) and limits the voltage spike passed through to the connected load to a maximum range of 400 to 500 peak volts.

Typical TVSS devices for homes are available as an integral part of receptacle outlets that mount in the same wall boxes as do regular receptacles. They look the same as a normal receptacle, and may have an audible alarm that sounds when an MOV has failed. The TVSS device may also have a visual indication, such as an LED (light-emitting diode), which glows continuously until an MOV fails. Then the LED starts to flash on and off. See Figure 9-21. TVSS devices are also available in plug-in strips; see Figure 9-22.

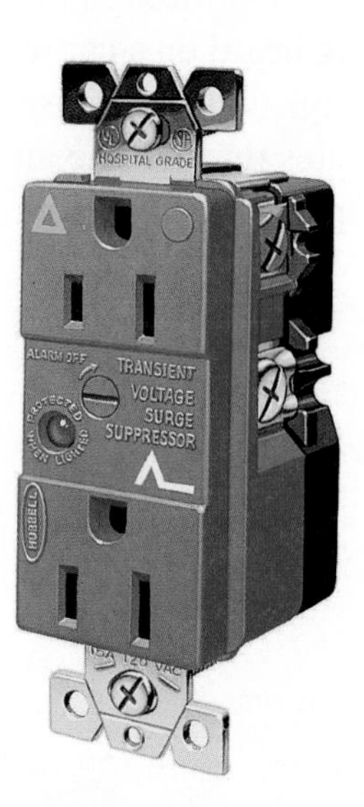

FIGURE 9-21 Surge suppressor.

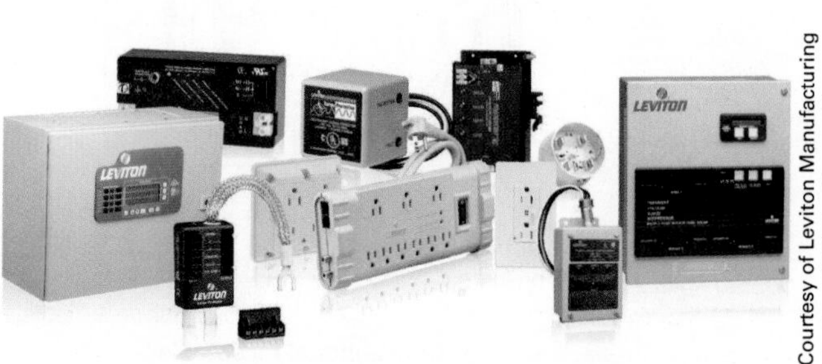

FIGURE 9-22 Surge suppressors.

One TVSS on a branch circuit will provide surge suppression for nearby receptacles on the same circuit.

Noise

Low-level non-damaging transients can also be present. These can be caused by fluorescent lamps and ballasts, electronic equipment in the area, X-ray equipment, motors running or being switched on and off, improper grounding, and so forth. Although not physically damaging to the electronic equipment, they can cause computers to lose memory or perform wrong calculations. Their intended programming can malfunction.

"Noise" comes from *electromagnetic interference* (EMI) and *radio frequency interference* (RFI). EMI is usually caused by ground currents of very low values from motors, utility switching loads, or lightning, and is transmitted through metal conduits. RFI causes the buzzing heard on a car radio when you are driving under a high-voltage transmission line. This interference radiates through the air from the source and is picked up by the grounding system of the building.

You can reduce this undesirable noise by installing an isolated bond receptacle, which will reduce the number of ground reference points on a system.

ISOLATED GROUND RECEPTACLE

In a standard receptacle outlet (Figure 9-23), the green bonding hexagon screw, the bonding contacts, the yoke (strap), and the metal wall box are all connected together and bonded to the building's equipment grounding system. If you picture all of the bonding conductors in the building, you can see that many outlets could be connected to ground by multiple paths.

In an isolated ground receptacle, the green bonding hexagon screw and the bonding contacts of the receptacle are isolated from the metal (strap) of the receptacle and also from the building's bonding system. A separate green insulated bonding conductor is then installed from the green hexagon screw on the receptacle all the way back to the nearest distribution panel, where it will be connected to the bond bus in the panel, *Rule 10–904(8)*. See Figure 9-23.

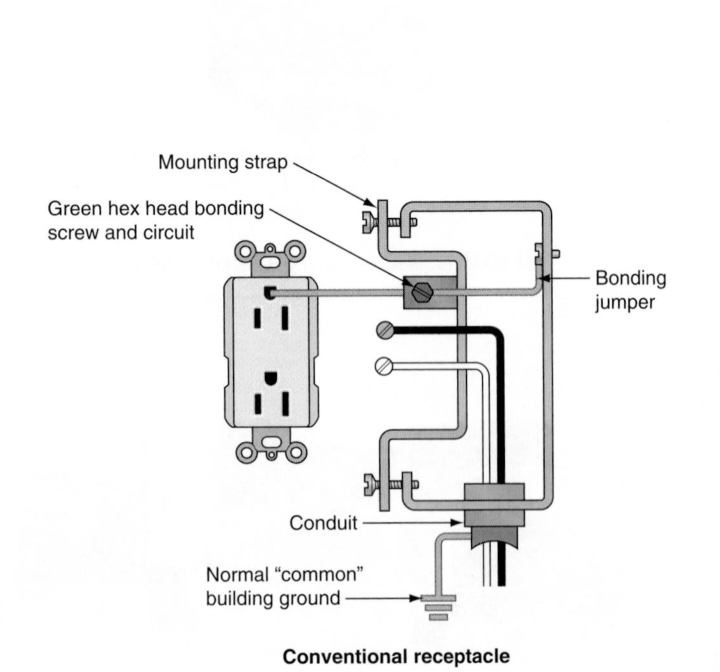

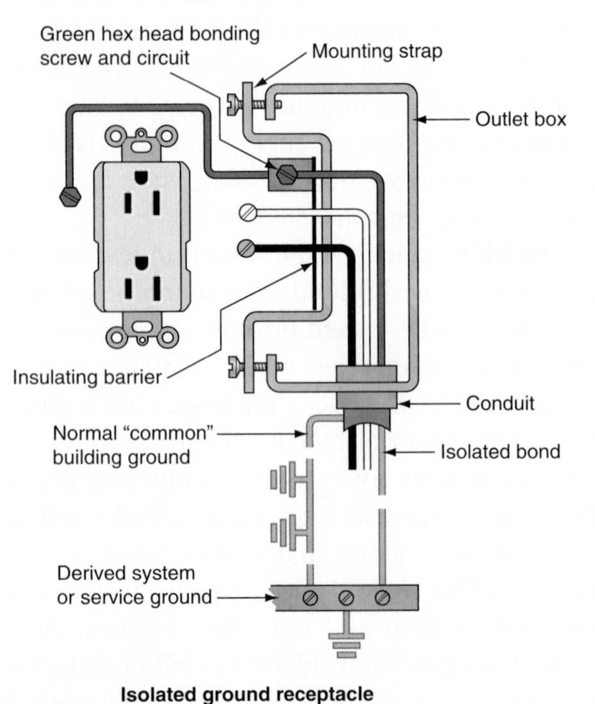

FIGURE 9-23 Standard conventional receptacle and isolated ground receptacle.

REVIEW

Note: Refer to the *CEC* or the blueprints provided with this textbook when necessary. Where applicable, responses should be written in complete sentences.

1. Explain the operation of a GFCI. Why are GFCI devices used? Where are GFCI receptacles required? ____________________

2. Residential GFCI devices are set to trip a ground-fault current of ____________________.

3. What are the residential applications of GFCI receptacles? ____________________

4. The *CEC* requires GFCI protection for certain receptacles in the bathroom. Explain where these are required. ____________________

5. Is it a *CEC* requirement to install GFCI receptacles in a fully carpeted, finished recreation room in the basement? Circle one. (Yes) (No)

6. A homeowner calls in an electrical contractor to install a separate circuit in an unfinished basement for a freezer. Is a GFCI receptacle required? ____________________

____

7. GFCI protection is available as (a) a branch-circuit breaker GFCI, (b) a feeder-circuit breaker GFCI, (c) an individual GFCI receptacle, (d) a feedthrough GFCI receptacle. What type would you install for residential use? Explain your choice.

____

8. Extremely long circuit runs connected to a GFCI branch-circuit breaker might result in [circle the letter of the correct answer]

   a. nuisance-tripping of the GFCI

   b. loss of protection

   c. the need to reduce the load on the circuit

9. If a person comes into contact with the hot and grounded circuit conductors of a two-wire branch circuit that is protected by a GFCI, will the GFCI trip? Why or why not?

____

____

____

10. What might happen if the line and load connections of a GFCI receptacle are reversed?

____

____

____

____

11. May a GFCI receptacle be installed on a replacement in an old installation where the two-wire circuit has no bonding conductor? Circle one. (Yes) (No)

12. What two types of receptacles may be used to replace a defective receptacle in an older home with knob-and-tube wiring where a bonding means does not exist in the box?

____

____

____

13. You are asked to replace a receptacle. Upon checking the wiring, you find that the wiring method is conduit and that the wall box is properly bonded. The receptacle is of the older style two-wire type that does not have a bonding terminal. You remove the old receptacle and replace it with [circle the letter of the correct answer]

a. the same type of receptacle as the type being removed
b. a receptacle of the bonding type
c. a GFCI receptacle

14. What colour are the terminals of a standard bonding-type receptacle? ______

15. What special shape are the bonding terminals of receptacle outlets and other devices?

16. Why are GFCI receptacles installed on construction sites? ______

17. In your own words, explain why the *CEC* does not require receptacle outlets in garages and basements to be GFCI-protected. ______

18. What do the letters TVSS stand for? ______

19. Transients (surges) on a line can cause spikes or surges of energy that can damage delicate electronic components. A TVSS device contains one or more ______ ______ that bypass and absorb the energy of the transient.

20. Undesirable noise on a circuit can cause computers to lock up or lose their memory, and/or can cause erratic performance of the computer. This noise does not damage the equipment. The two types of this noise are EMI and RFI. What do these letters mean?

21. Can TVSS receptacles be installed in standard device boxes? ______________________

____

22. When an isolated ground receptacle is installed, the *CEC* permits the separate equipment-bonding conductor to be carried back through the raceway or cable to the distribution panel. What section of the *CEC* references this topic? ______________

23. Briefly explain the difference between a GFCI breaker and an AFCI breaker.

____

____

____

UNIT **10**

Lighting Fixtures and Ballasts

OBJECTIVES

After studying this unit, you should be able to

- understand fixture terminology, such as Type IC, Type Non-IC, suspended ceiling fixtures, recessed fixtures, and surface-mounted fixtures
- connect recessed fixtures, both prewired and non-prewired types, according to *Canadian Electrical Code (CEC)* requirements
- specify insulation clearance requirements
- discuss the importance of temperature effects when planning recessed fixture installations
- describe thermal protection for recessed fixtures
- understand Class P ballasts
- discuss the *CEC* requirements for track lighting

TYPES OF LIGHTING FIXTURES

This unit begins with an in-depth coverage of recessed fixtures commonly installed in homes. It also considers other types of available lighting fixtures.

CEC Section 30 sets out the requirements for the installation of lighting fixtures. Although not involved in the actual manufacture of lighting fixtures, the electrician must "meet code" when installing fixtures, including mounting, supporting, grounding, live-parts exposure, insulation clearances, supply conductor types, and maximum lamp wattages.

The homeowner, interior designer, architect, or electrician has thousands of fixtures from which to choose. Specific needs, space requirements, and price considerations are among the factors to consider when selecting fixtures.

It is essential for the electrician to know early in the roughing-in stage of wiring a house what types of fixtures are to be installed, particularly for recessed-type fixtures. The electrician must work closely with the general building contractor, carpenter, plumber, heating contractor, and the other building tradespeople to ensure that requirements for location; proper and adequate support; and clearances from piping, ducts, combustible material, and insulation are complied with during construction.

The Canadian Standards Association (CSA) Group provides the safety standards to which a fixture manufacturer must conform. Common fixture types are as shown in the table below.

FLUORESCENT	INCANDESCENT
• Surface	• Surface
• Recessed	• Recessed
• Suspended ceiling	• Suspended ceiling

IMPORTANT: Always carefully read the label on the fixture. The information will help you to conform to *Section 30* of the *CEC* and complete a safe installation. The label provides such information as

- for wall mount only
- ceiling mount only
- maximum lamp wattage
- type of lamp
- access above ceiling required
- suitable for air handling use
- for chain or hook suspension only
- suitable for operation in ambient temperatures not exceeding ________ °C
- for line volt–amperes, multiply lamp wattage by 1.25
- suitable for use in suspended ceilings
- suitable for use in non-insulated ceilings
- suitable for use in insulated ceilings
- suitable for damp locations (such as bathrooms and under eaves)
- suitable for wet locations
- suitable for use as a raceway
- suitable for mounting on low-density cellulose fibreboard
- for supply connections, use wire rated for at least ________ °C
- not for use in dwellings
- thermally protected fixture
- Type IC
- Type Non-IC
- inherently protected

The *CEC, Part II,* and the fixture manufacturers' catalogues and literature are excellent sources of information. The CSA tests, lists, and labels fixtures for conformance to its standards and to the *CEC, Parts I and II.*

Recessed fixtures (Figures 10-1A and 10-1B) in particular have an inherent heat problem. Therefore, they must be suitable for the intended application and must be properly installed.

Courtesy of Cooper Industries

FIGURE 10-1A Typical recessed fixture.

FIGURE 10-1B Recessed fixture mounted in ceiling.

To protect against overheating, the CSA requires that some types of recessed fixtures be equipped with an integral thermal protector; see Figure 10-2. This device shuts the fixture off if it overheats, then resets once the fixture cools down.

Because of the potential fire hazard when mounting lighting fixtures near combustible materials, *Rule 30–200(1)* states that fixtures must be equipped with shades or guards to limit the temperature to which such materials may be exposed to a maximum of 90°C. If lights are installed over combustible material, each fixture must be controlled by an individual wall switch if it is located less than 2.5 m above the floor or if the fixture is not guarded so that the lamps cannot be readily removed or damaged, *Rule 30–200(3)*.

Figure 10-3 shows a fluorescent fixture mounted on low-density cellulose fibreboard treated with fire-retarding chemicals to meet specific standards.

All branch-circuit conductors used for wiring a ceiling outlet box must have insulation approved to 90°C (e.g., NMD 90), but their ampacity is de-rated to that of 60°C wire, *Table 2, Column 2, Rule 30–408(2)*. This applies to most residential applications.

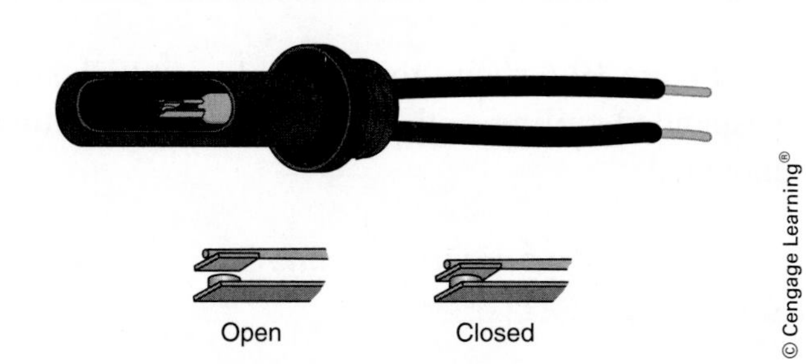

FIGURE 10-2 Recessed fixture thermal cutout.

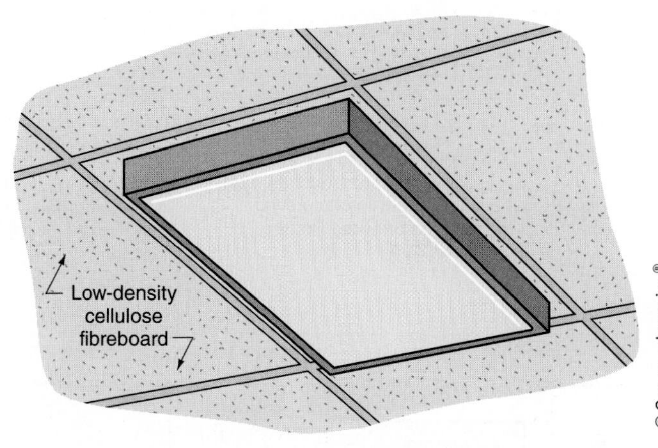

FIGURE 10-3 When mounted on low-density ceiling fibreboard, surface-mounted fluorescent fixtures must be installed so as to limit the temperature of the fibreboard to less than 90°C. See *Rule 30–200(1)*.

CEC REQUIREMENTS FOR INSTALLING RECESSED FIXTURES

CEC Rules 30–308, 30–408, and *30–900* through *30–912* list requirements for the installation and construction of recessed fixtures. Of particular importance are the restrictions on conductor temperature ratings, fixture clearances from combustible materials, and maximum lamp wattages. Recessed fixtures generate a great deal of heat within the enclosure, making them a fire hazard if they are not wired and installed properly. Figure 10-4 shows the roughing-in box of a recessed fixture with mounting brackets and junction box.

The branch-circuit conductors are run to the junction box for the recessed fixture. Here they are connected to conductors whose insulation can

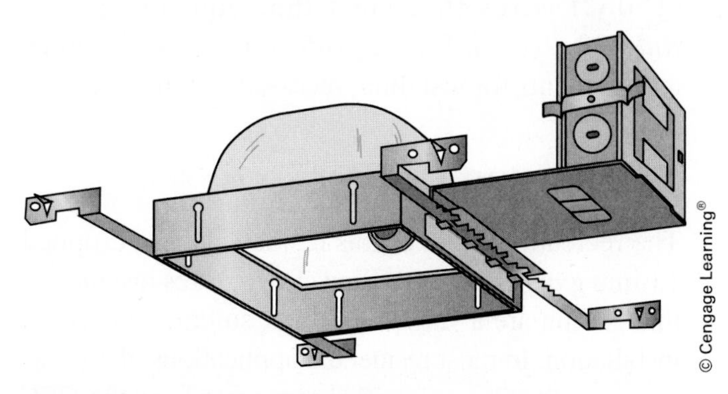

FIGURE 10-4 Roughing-in box of a recessed fixture with mounting brackets and junction box.

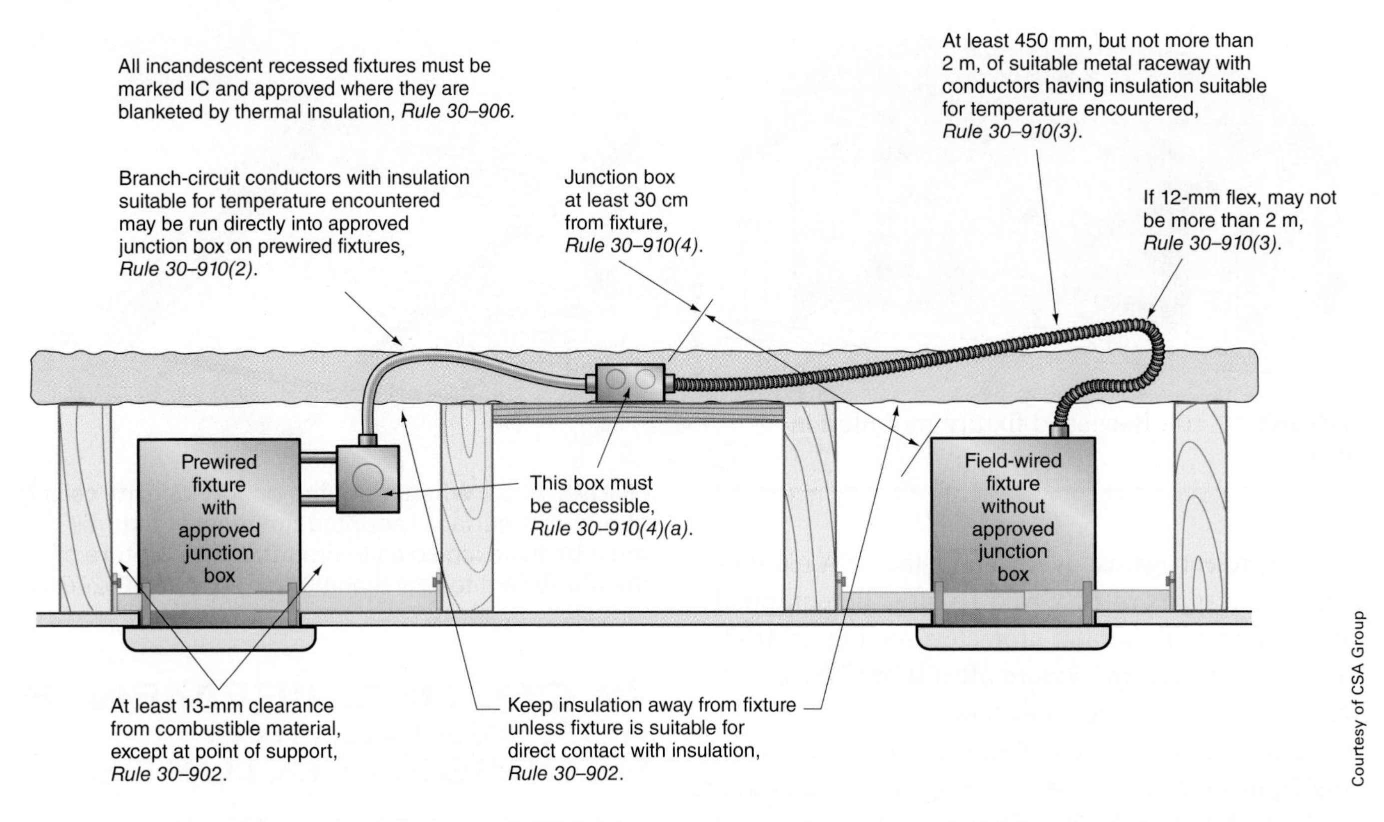

FIGURE 10-5 Clearance requirements for installing recessed lighting fixtures.

handle the temperature at the fixture lampholder. The junction box is placed at least 300 mm from the fixture, so heat radiated from the fixture cannot overheat the wires in the junction box. Conductors rated for higher temperatures must run in a metal raceway between the fixture and the junction box. This raceway must be 450 mm to 2 m long. Any heat conducted from the fixture along the metal raceway will be reduced before reaching the junction box. Many recessed fixtures are factory-equipped with a flexible metal raceway containing high-temperature (150°C) wires that meet the requirements of *Rule 30–910(1)*. Figure 10-5 illustrates clearance requirements for installing recessed lighting fixtures.

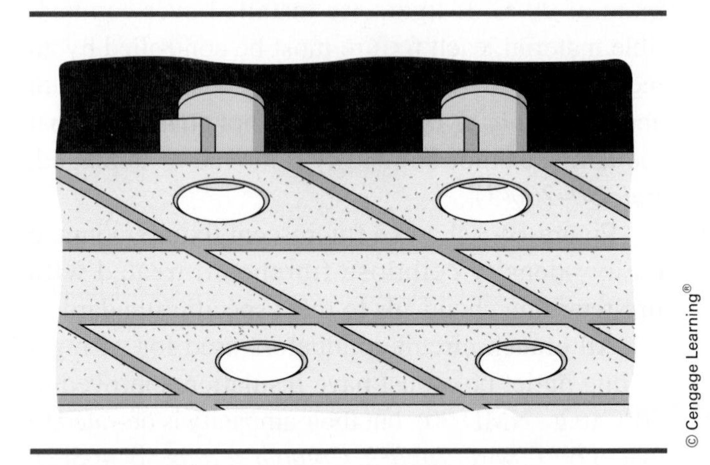

FIGURE 10-6 Suspended ceiling fixtures.

Suspended Ceiling Fixtures

The recreation room of this residence has a dropped 13-mm gypsum board ceiling. The fixtures installed in this ceiling are a "lay-in" fixture, suitable for recess installation. In most residential applications, these fixtures would still be classified as recessed and the *CEC* requirement of a minimum of 13-mm clearance from structural members must be maintained, *Rule 30–902*.

Figure 10-6 shows the typical layout look at the suspended ceiling with the recessed light fixtures mounted.

Support of Suspended Ceiling Fixtures

When framing members of a suspended ceiling grid are used to support lighting fixtures, the grid installer must securely fasten all members together and to the building structure. All suspended ceiling fixtures must be securely fastened to the ceiling grid members by bolts, screws, or rivets. Most fixture manufacturers supply special clips suitable for fastening the fixture to the grid. The fixture must NEVER be supported by the ceiling tile.

Some inspection authorities interpret the intent of *Rule 30–302* to mean that the recessed fixture must be supported independent of a suspended ceiling structure. This is usually accomplished by using a chain attached to the fixture and the building structure above the ceiling. Large fixtures such as fluorescents require at least two chain supports.

The logic behind all of the "securely fastening" requirements is that in the event of a major problem (e.g., earthquake or fire), the lighting fixtures will not fall down and injure someone.

Connecting Suspended Ceiling Fixtures

The most common way to connect suspended ceiling lay-in lighting fixtures is to complete all of the wiring above the suspended ceiling using normal wiring methods, EMT, non-metallic-sheathed cable, armoured cable, or whatever type of wiring method is acceptable by the governing electrical code, with outlet boxes strategically placed above the ceiling near the intended location of the fixtures. Then make a flexible connection between the outlet box and the fixture.

The flexible connection can be up to 2 m long and is usually made by installing either of the following items:

- Flexible metal conduit with conductors suitable for the temperature requirement as stated on the fixture label, usually 90°C, *Rule 30–910(1)*. *Rule 12–1010(3)* specifies that when flexibility is required, flexible metal conduit does not require a clip within 300 mm of the outlet box.
- Armoured cable containing conductors suitable for the temperature requirements as stated on the fixture label, usually 90°C. Be sure to staple the armoured cable within 300 mm of the outlet box. See *Rules 12–510(1)* and *12–618* and Figures 10-7A, 10-7B, and 10-7C.

It is necessary for the electrician who is installing recessed fixtures to work with the installer of the insulation to be sure that the clearances around the fixture are maintained. If the ceiling has insulation

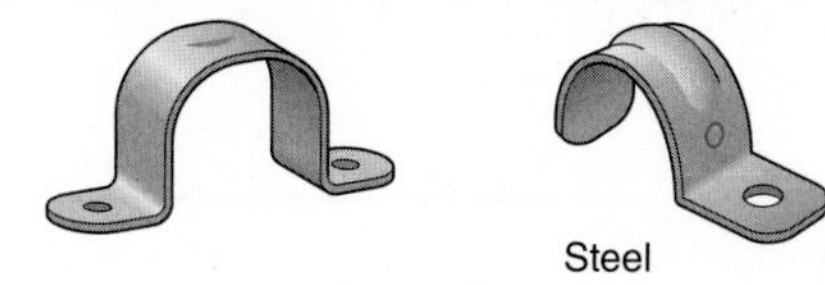

FIGURE 10-7A Clips for supporting 12-mm flex or AC90 cable.

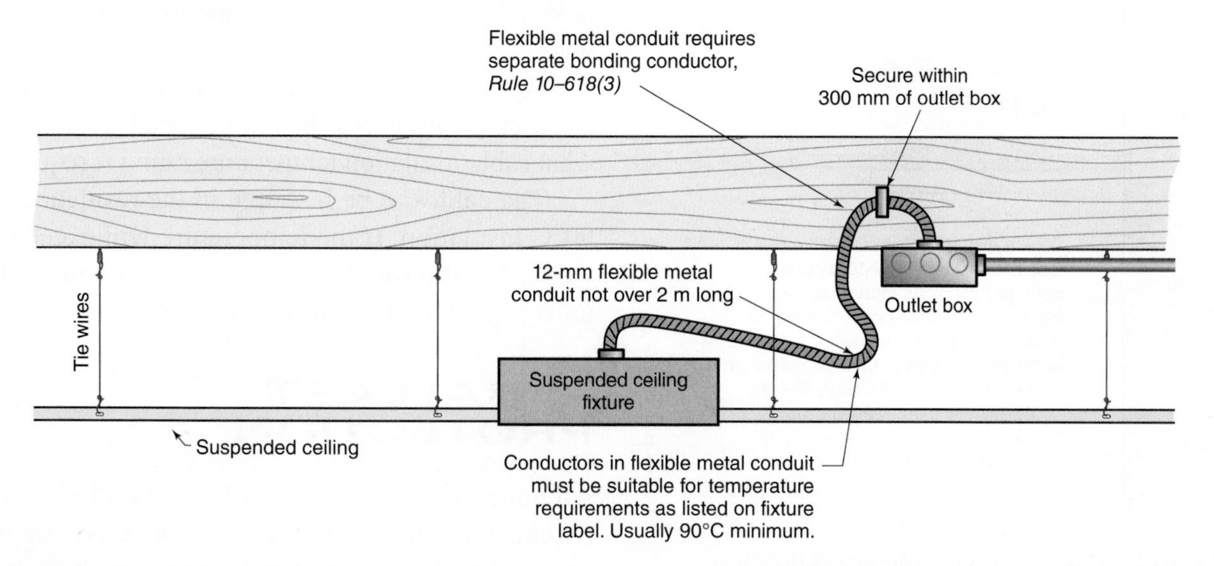

FIGURE 10-7B A suspended ceiling fixture supplied by not over 2 m of flexible metal conduit. *Rule 12–1010(3)* indicates that if the conduit is less than 900 mm long, it does not require clipping. Remember that 12-mm flexible conduit cannot be more than 2 m long, *Rule 30–910(3)*.

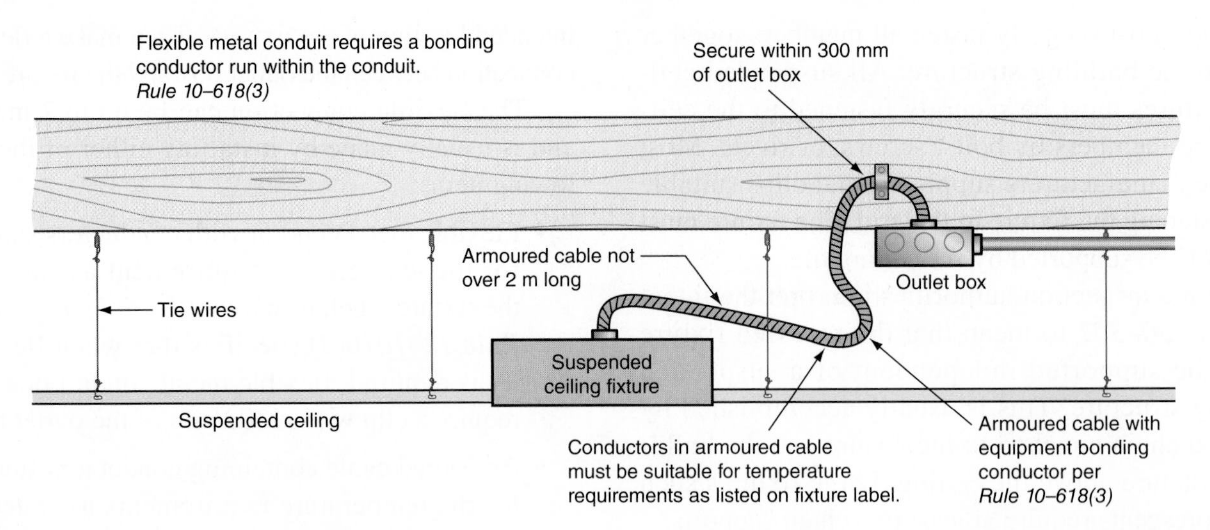

FIGURE 10-7C A suspended ceiling fixture supplied by armoured cable not over 2 m long. Cable must be secured (stapled) within 300 mm of the outlet box, *Rule 12–510.*

blanketing the fixture, it must be a Type IC fixture, *Rule 30–906* (Figure 10-8).

When the only access to the junction box is through the fixture opening, the opening must be at least 180 cm^2 with no dimension less than 150 mm unless the entire fixture and junction box can be removed from the ceiling, *Rule 30–910(5)*. Many small, low-voltage spotlights installed in today's homes do not meet this requirement. Therefore, an alternative access method would be required, such as through an attic, or a removable fixture must be installed.

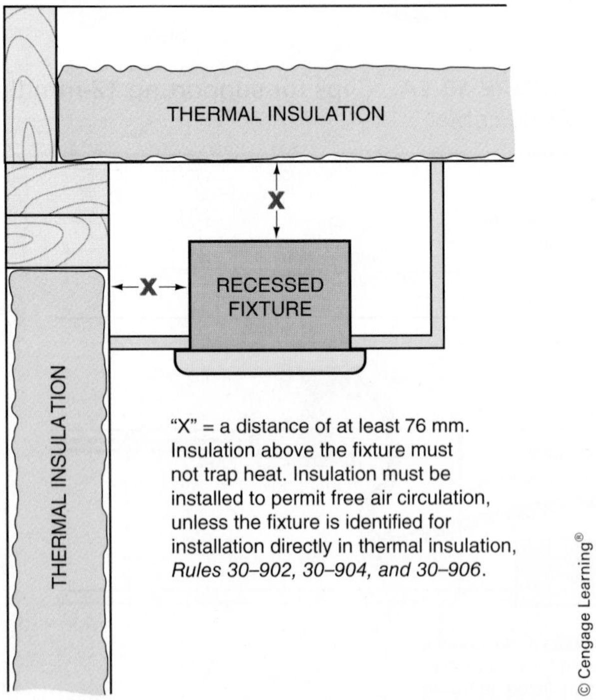

Prewired fixtures do not require additional wiring; see Figure 10-9. *Rule 30–910(8)* states that branch-circuit wiring shall not be passed through an outlet box that is an integral part of an incandescent fixture unless the fixture is approved and marked for through wiring.

For a recessed fixture that is not prewired, the electrician must check the fixture for a label indicating what cable insulation temperature rating is required.

The cables to be installed in the residence have 90°C insulation. If the temperature will exceed this value, conductors with other types of insulation must be installed, *Table 19* and *Table D1*.

THERMAL INSULATION

X

X

RECESSED FIXTURE

THERMAL INSULATION

"X" = a distance of at least 76 mm. Insulation above the fixture must not trap heat. Insulation must be installed to permit free air circulation, unless the fixture is identified for installation directly in thermal insulation, *Rules 30–902, 30–904, and 30–906.*

FIGURE 10-8 Clearances for recessed lighting fixtures installed near thermal insulation. Fixtures must be CSA-approved for use in thermal insulation, *Rule 30–906.*

BALLAST PROTECTION

All approved fluorescent ballasts installed indoors (except simple reactance-type ballasts) for both new and replacement installations must have built-in thermal protection from the manufacturer; see Figure 10-10. Ballasts provided with built-in thermal protection are listed by the CSA as Class P

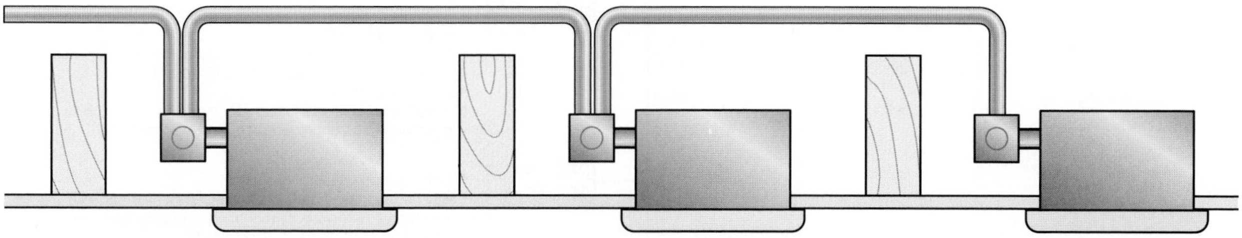

FIGURE 10-9 Installation permissible only with prewired recessed fixtures with approved junction box, *Rule 30–910(8)*.

ballasts. Under normal conditions, the Class P ballast has a case temperature not exceeding 90°C. The thermal protector must open within 2 hours when the case temperature reaches 110°C.

Some Class P ballasts also have a non-resetting fuse integral with the capacitor to protect against capacitor leakage and violent rupture. The Class P ballast's internal thermal protector will disconnect the ballast from the circuit in the event of a high temperature. Excessive temperatures can be caused by abnormal voltage and improper installation, such as being covered with insulation.

The reason for thermal protection is to reduce the fire hazard from an overheated ballast. Ballasts can overheat if they are shorted, grounded, covered with insulation, or lack air circulation.

LIGHTING FIXTURE VOLTAGE LIMITATIONS

The maximum voltage allowed for residential lighting fixtures is 150 volts between a conductor and ground, *Rules 2–104* and *30–102(1)*.

Rule 30–706 makes a further restriction in stating that for dwelling occupancies, no electrical discharge lighting equipment shall be used if it operates with an open-circuit voltage over 300 volts; see Figure 10-11. Fluorescent and neon lights are examples of discharge lighting. *Rule 30–706* addresses the increasing use of neon lighting for decorative purposes. Some important CSA and *CEC* requirements for installing recessed fixtures are summarized in Figure 10-12.

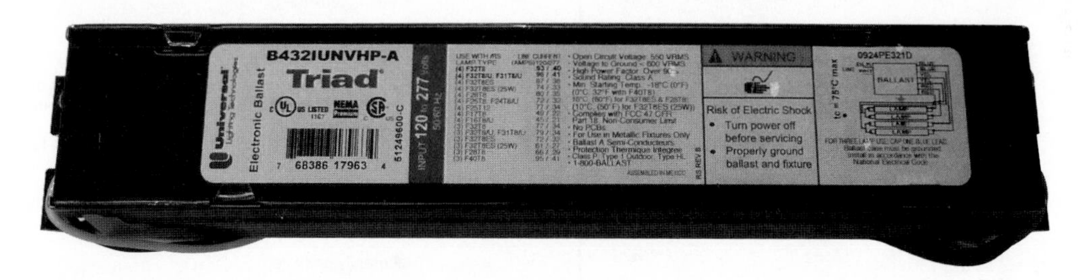

FIGURE 10-10 Fluorescent ballasts installed on new or replacement installations are required to have built-in thermal protection. These ballasts are called *Class P ballasts, CEC, Part II.*

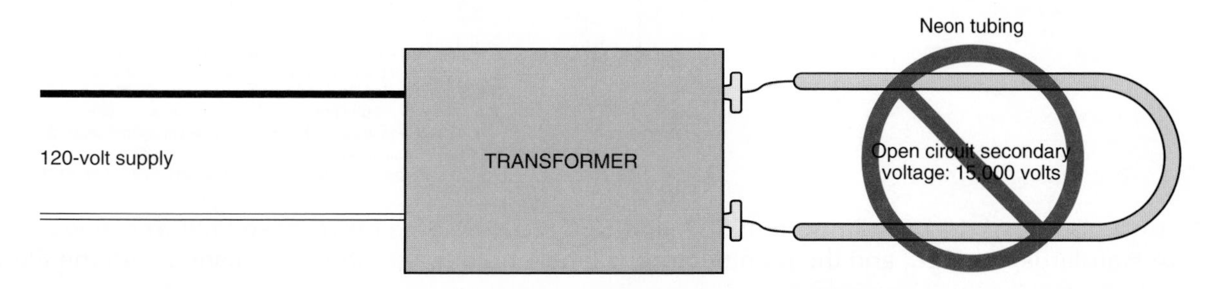

FIGURE 10-11 It is NOT permitted to use neon lighting fixtures in residences where the open circuit voltage is over 300 volts, unless there is no access to live parts when changing the lamps, *Rule 30–706*.

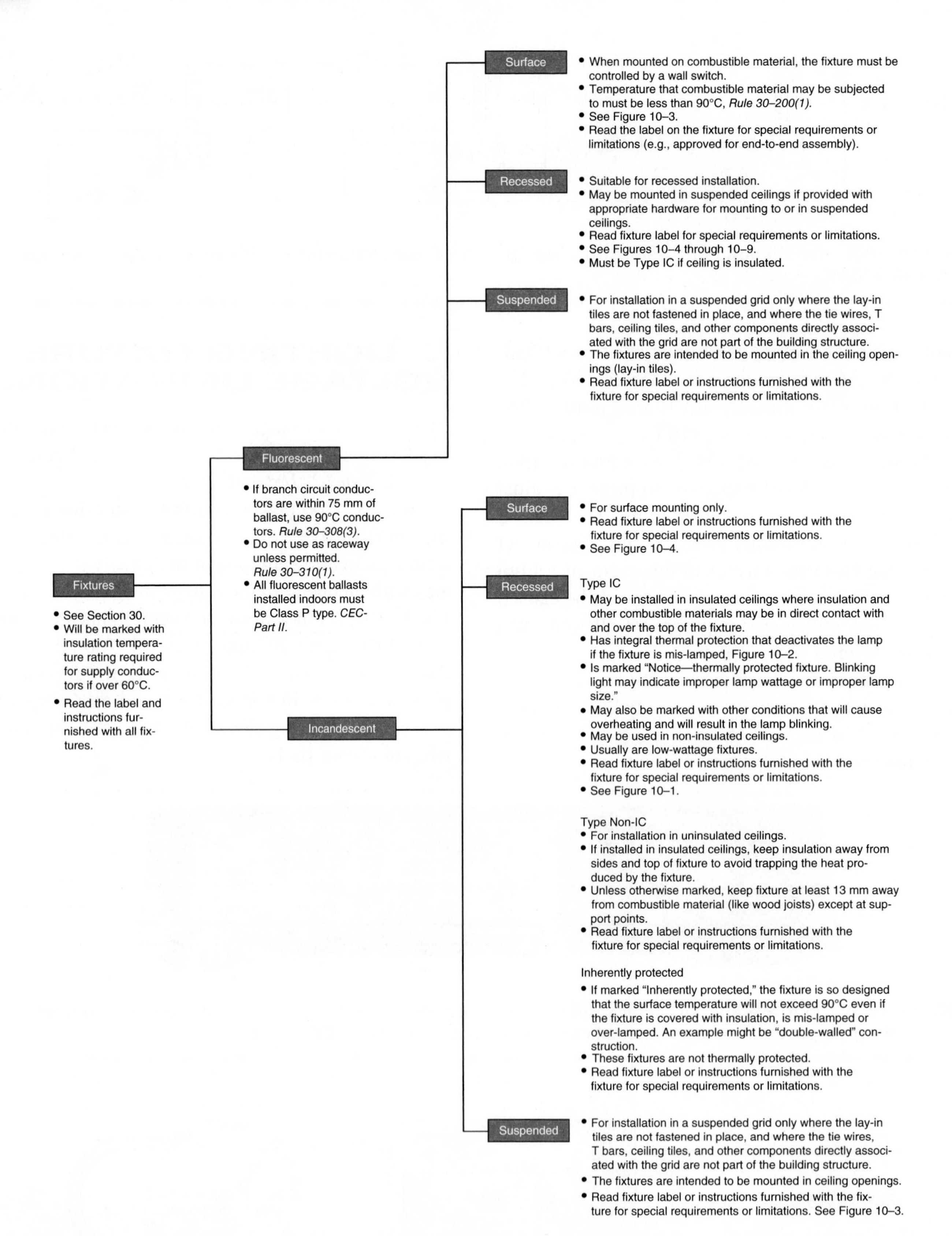

FIGURE 10-12 Some of the most important CSA and *CEC* requirements for recessed fixtures. Always refer to the CSA standards, the *CEC*, and the manufacturer's label and/or instructions furnished with the fixture.

REVIEW

Note: Refer to the *CEC* or the blueprints provided with this textbook when necessary. Where applicable, responses should be written in complete sentences.

1. Is it permissible to install a recessed fixture directly against wood ceiling joists? Explain why or why not. ______________________________

____

2. If a recessed fixture without an approved junction box is installed, what extra wiring must be provided? ______________________________

____

3. Recessed fixtures are available for installation in direct contact with thermal insulation. These fixtures bear the CSA mark "Type______________."

4. Unless specially designed, all recessed incandescent fixtures must be provided with factory-installed ______________________________.

5. Plans require the installation of a surface-mounted fluorescent fixture on the ceiling of a recreation room that is finished with low-density ceiling fibreboard. What sort of mark would you look for on the label of the fixture? ______________________________

____

6. If a recessed fixture bears no marking that it is listed for branch-circuit feedthrough wiring, is it permitted to run the circuit conductors from fixture to fixture? What section of the *CEC* covers this? ______________________________

____

7. Fluorescent ballasts for all indoor applications must be ____________ type. These ballasts contain internal ____________ protection to protect against overheating.

8. You are called on to install a number of lighting fixtures in a suspended ceiling. The ceiling will be dropped about 200 mm from the ceiling joists. Briefly explain how you might go about wiring these fixtures. ______________________________

____

____

____

9. The *CEC* places a maximum open circuit voltage on lighting equipment in homes. [Circle the correct answer.] The maximum voltage is (300) (600) (750) (1000). Where in the *CEC* is this voltage maximum referenced? ______________________________

____

____

UNIT **11**

Branch Circuits for the Bedrooms, Study, and Halls

OBJECTIVES

After studying this unit, you should be able to

- explain the factors that influence the grouping of outlets into circuits
- estimate loads for the outlets of a circuit
- draw a cable layout and a wiring diagram based on information given in the residence plans, specifications, and *Canadian Electrical Code (CEC)* requirements
- select the proper wall box for a particular installation
- explain how wall boxes are bonded to ground
- list the requirements for the installation of fixtures in clothes closets
- make connections for three-way switches

GROUPING OUTLETS

A branch circuit is all of the wiring from the final overcurrent device in the system to the point of utilization. The grouping of outlets into branch circuits must conform to *CEC* standards and good wiring practices. There are many possible combinations of groupings of outlets. In most residential installations, the electrician usually does the circuit planning. In larger, more costly residences, the architect may complete the circuit layout and include it on the plans. In either case, the electrician must ensure that the branch circuits conform with the *CEC.*

An electrician plans circuits that are economical without sacrificing the quality of the installation. See Figure 11-1. For example, some electricians prefer to have more than one circuit feed a room. Should one circuit have a problem, the second circuit would still supply power to the other outlets in that room; see Figure 11-2.

Some electricians consider it poor practice to include outlets on different floors on the same circuit. Here, too, the decision can be a matter of personal choice. Some local building codes limit this type of installation to lights at the head and foot of a stairway.

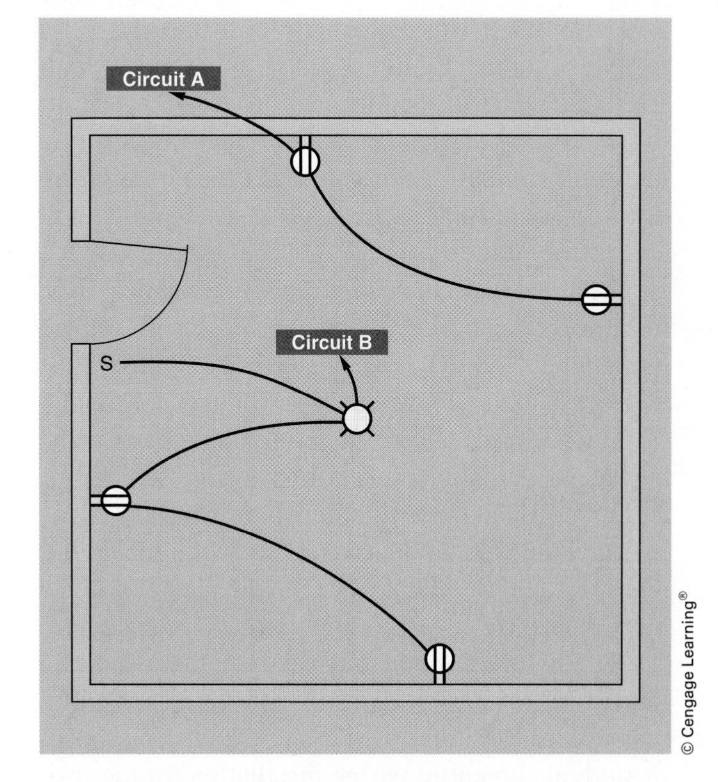

FIGURE 11-2 Wiring layout showing one room that is fed by two different circuits. If one of the circuits goes out, the other circuit still provides electricity to the room.

Residential Lighting

Residential lighting is a personal thing. The homeowner, builder, and electrical contractor must meet to decide on what types of lighting fixtures are to be installed in the residence. Many variables, including cost, personal preference, and construction obstacles, must be considered.

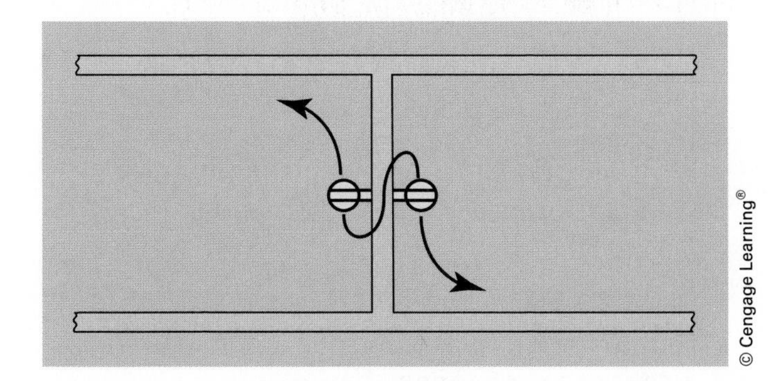

FIGURE 11-1 Receptacle outlets connected back to back can reduce the cost of the installation because of the short distance between the outlets.

Residential lighting can be divided into four groups: general, accent, task, and security lighting.

General Lighting. General lighting provides overall illumination for an area, such as the front entry hall, bedroom hall, and garage.

Accent Lighting. Accent lighting provides a "focus" or accent on an object or area in the home. This might be recessed spotlights over the fireplace or track lighting on the ceiling of the living room to accent a photo, painting, or sculpture.

Task Lighting. Task lighting provides proper lighting where tasks are performed. In the residence in our blueprints, the fluorescent fixtures over the workbench, the lighting inside the range hood in the kitchen, and the recessed fixture over the kitchen sink are all examples.

Security Lighting. Security lighting generally includes outdoor lighting (such as post lights, wall fixtures, walkway lighting) and all other lighting used for security and safety. Security lighting is often provided by the normal types of lighting fixtures found in the typical residence. In this residence, the outdoor bracket fixtures in front of the garage and next to the entry doors, and the post light might be considered to be security lighting, even though they add to the beauty of the residence.

A visit to the showroom of one of the many manufacturers or distributors of lighting fixtures offers the buyer an opportunity to select from hundreds—in some cases, thousands—of lighting fixtures. Most of the finer lighting showrooms are staffed with individuals highly qualified to make recommendations to homeowners so they get the most value for their money.

The lighting provided throughout this residence conforms to accepted trade practice. Certainly, many other variations are possible. For the purposes of studying an entire wiring installation for the residence, all load calculations, wiring diagrams, and so on use the fixture selection indicated on the plans and in the specifications.

CABLE RUNS

Studying the great number of wiring diagrams throughout this text, you will find many ways to run the cables and make up the circuit connections. Sizing of boxes that will conform to the *CEC* for the number of wires and devices is covered in great detail. When a circuit is run through a recessed fixture like the type shown in Figure 10-1B in Unit 10, that recessed fixture must be approved for "through wiring," a subject discussed in Unit 10. When you are unsure whether a fixture can accommodate the extra wiring for a run to another fixture or some other part of the circuit, it is best to design the circuit so that the wiring ends at the recessed fixture with just the conductors needed to connect it.

The grouping of outlets into circuits must satisfy the requirements of the *CEC*, local building codes, good wiring practices, and common sense. A good wiring practice is to

- divide all loads as evenly as possible among the circuits
- ensure that no more than 12 outlets are placed on a two-wire circuit
- keep cable runs as short as possible
- use boxes that have adequate space for conductors and devices.

See *Rules 8–104* and *8–304*.

ESTIMATING LOADS FOR OUTLETS

When calculating loads, remember that

$$\text{Volts} \times \text{Amperes} = \text{Volt–amperes}$$

Yet, many times we say that

$$\text{Volts} \times \text{Amperes} = \text{Watts}$$

What we really mean to say is that

$$\text{Volts} \times \text{Amperes} \times \text{Power factor} = \text{Watts (true power)}$$

In a pure resistive load, such as a simple light bulb, a toaster, an iron, or a resistance electric heating element, the power factor is 100%. Then

$$\text{Volts} \times \text{Amperes} \times 1 = \text{Watts (true power)}$$

With transformers, motors, ballasts, and other "inductive" loads, wattage is not necessarily the same as volt–amperes. Therefore, to be sure that branch-circuit wiring has adequate ampacity and to provide for feeder sizing and service calculations, the volt–amperes calculation would reflect the true current draw of the alternating current (AC) load.

EXAMPLE

Calculate (a) the wattage and (b) the volt–amperes of a 120-volt, 10-ampere resistive load.

SOLUTION

a. $120 \times 10 \times 1 = 1200$ watts true power

b. $120 \times 10 = 1200$ volt–amperes (apparent power)

EXAMPLE

Calculate (a) the wattage and (b) the volt–amperes of a 120-volt, 10-ampere motor load at 50% power factor.

SOLUTION

a. $120 \times 10 \times 0.5 = 600$ watts (true power)

b. $120 \times 10 = 1200$ volt–amperes (apparent power)

In most *Code* calculations, *watts* are used in calculating the demand on service or feeder conductors because the Canadian Standards Association (CSA) requires that all electrical equipment be marked with a *wattage* or *kilowatt* rating, *Rule 2–100(1)(e)*.

Most residential loads are resistive; therefore, the power factor of 1 gives the same value to the demand *watts* as *volt–amperes*, and kW ratings are equivalent to kVA values. Therefore, this text uses the terms *wattage* and *volt–amperes* when calculating and/or estimating loads.

Building plans typically do not specify the ratings in watts for the outlets shown. When planning circuits, the electrician must consider the types of fixtures that may be used at the various outlets, based on the general uses of the receptacle outlets in the typical dwelling.

The *CEC* specifies that the maximum number of receptacle and lighting outlets that may be connected to any 2-wire branch circuit (15-A or 20-A rated) in a residence is 12 outlets, *Rule 8–304(1)*.

However, *Rule 8–304(3)* could permit more outlets if there is a maximum loading of 80% on the overcurrent device and the load on the outlets is known.

The layout of the receptacles on the walls must satisfy *Rule 26–712*. This should virtually eliminate the use of extension cords, one of the highest reported causes of electrical fires. Rarely, if ever, would all of the outlets be fully loaded at the same time.

Circuit Loading Guidelines (*Rules 8–104(5) and 8–304*)

A good rule to follow in residential wiring is *never load the circuit to more than 80% of its capacity.*

A 15-ampere, 120-volt branch circuit would be calculated as

$$15 \times 0.80 = 12 \text{ amperes}$$

or

$$12 \text{ amperes} \times 120 \text{ volts} = 1440 \text{ volt-amperes}$$

Certain fixtures, such as recessed lights and fluorescent lights, are marked with their maximum lamp wattage; fluorescent fixtures also list their ballast current. Other fixtures, however, are not marked, and the electrician does not know the size of the lamps that will be installed in them. Furthermore, the electrician does not know the exact load that will be connected to the receptacle outlets as it is difficult to anticipate what the homeowner may do after the installation is complete. The room in which the outlets are located gives some indication of their possible uses; therefore, the electrician should plan the circuits accordingly.

Load Estimation

The electrician can estimate the lamp loads in the lighting fixtures. By determining the lamp wattages that will probably be needed to provide adequate lighting for an area, many lighting manufacturers publish useful information and recommendations in catalogues and on the Internet.

Estimating the Number of Outlets by Assigning an Amperage Value to Each

One method of determining the number of lighting and receptacle outlets to be included on one circuit is to assign a value of 1 ampere to each outlet, to a total of 12 amperes. Thus, a total of 12 outlets can be included in a 15-ampere circuit.

Although the *CEC* limits the number of outlets on one circuit to 12 outlets, local building codes may specify fewer. Therefore, before planning any circuits, the electrician must check the local building code requirements.

All outlets will not be required to deliver one ampere. For example, closet lights, night-lights, and clocks use only a small portion of the allowable current. A 60-watt closet light draws less than 1 ampere:

$$I = \frac{W}{E} = \frac{60}{120} = 0.5 \text{ amperes}$$

If low-wattage fixtures are connected to a circuit, it is quite possible that 15 or more lighting outlets could be connected to the circuit without a problem. On the other hand, if the load consists of high-wattage lamps, the number of outlets would be less. In this residence, estimated loads were calculated at 120 volt–amperes (one ampere) for those receptacle outlets intended for general use.

To estimate the number of 15-ampere lighting branch circuits desired for a new residence where the total count of lighting and receptacle outlets is, for example, 80:

$$\frac{80}{12} = 6.7 \text{ (seven 15-ampere branch circuits)}$$

We divide by 12 because each outlet counts as 1 ampere. Twelve amperes is 80%, or the maximum for a 15-ampere circuit, as $15 \times 80\% = 12$, *Rule 8–304(3)*.

In residential occupancies there is great diversity in the loading of lighting branch circuits.

CAUTION: This procedure for estimating the number of lighting and receptacle outlets is for branch circuits only. Where the branch circuits in the kitchen, laundry area, and other areas supply heavy concentrations of plug-in appliances, the 1-ampere-per-receptacle-outlet assumption is not applicable. These circuits are discussed in more detail in Units 14 through 16.

Review of How to Determine the Minimum Number of Lighting Circuits

1. Connect 12 outlets per circuit. The required number of outlets should be divided by 12; this will show the minimum number of branch circuits, *Rule 8–304(1)*.
2. Estimate the probable loading for each lighting and receptacle outlet (do not include the small-appliance receptacle outlets covered by *Rules 26–720* through *26–724*). Try to have a total load not over 1440 volt–amperes for each 15-ampere lighting circuit.
3. Alternatively, add up the actual loading of outlets in a circuit, verify that the overcurrent device is not loaded to more than 80% of its capacity, and the number may, in this instance, exceed 12, *Rule 8–304(3)*. Including outlets for smoke detectors is a good example of a situation in which the number could exceed 12. The wattage of a 120-volt smoke detector is approximately ½ watt.

Methods 1 and 3 are laid out in the *CEC*. Method 2 is only a guideline because the general lighting loads in homes are so varied. As the electrician lays out the circuits of a residence, these guidelines will generally provide adequate circuitry for both lighting outlets and receptacle outlets that are intended for general lighting, *not* appliance-circuit receptacle outlets. Be practical. Consider the diversity of uses.

Divide Loads Evenly

It is mandatory to divide loads evenly among the circuits to avoid overloading some while leaving others lightly loaded. At a main service panel, do not connect the branch circuits and feeders to result in, for instance, 120 amperes on Phase A and 40 amperes on Phase B. Look at the probable and/or calculated loads to attain a balance of loads on the A phase and the B phase.

SYMBOLS

The symbols used on the cable layouts in this textbook are the same as those found in the legend of symbols on the electrical plans for the residence.

Pictorial illustrations are used on all wiring diagrams in this text to make it easy for the reader to understand the diagrams; see Figure 11-3.

DRAWING THE WIRING DIAGRAM OF A LIGHTING CIRCUIT

The electrician must convert information from the building plans to the form most useful for planning an installation that is economical, conforms to all code regulations, and follows good wiring practice. To do this, the electrician makes a cable layout and a wiring diagram to clearly show each wire and all connections in the circuit.

After many years of experience, skilled electricians will not always prepare wiring diagrams for

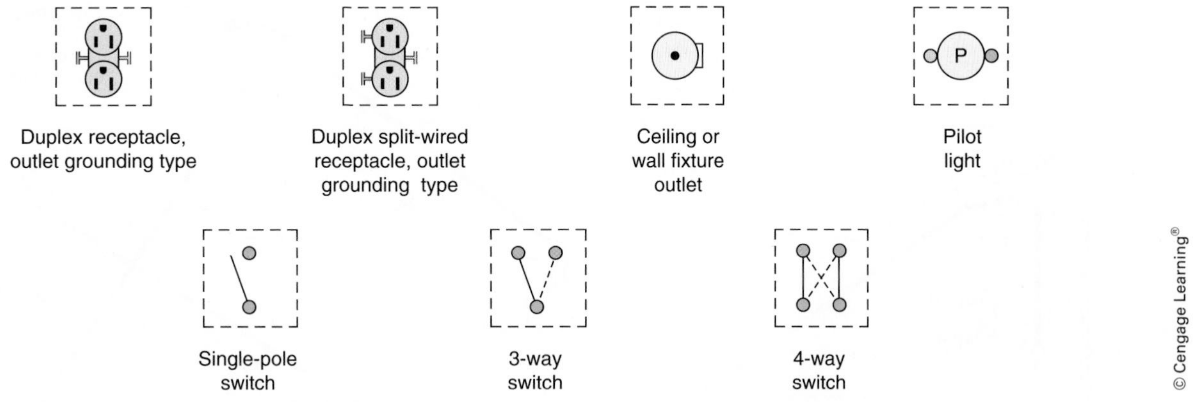

Duplex receptacle, outlet grounding type

Duplex split-wired receptacle, outlet grounding type

Ceiling or wall fixture outlet

Pilot light

Single-pole switch

3-way switch

4-way switch

FIGURE 11-3 Pictorial illustrations used on wiring diagrams in this text.

most residential circuitry because they have the ability to "think" the connections through. An unskilled individual should make detailed wiring diagrams.

The following steps will guide you in preparing wiring diagrams. (You will be required to draw wiring diagrams in later units of this textbook.) As you work on the wiring diagrams of the many circuits provided in this text, note that the receptacles are positioned exactly as though you were standing in the centre of the room. The *CEC* does not address receptacle orientation on the wall. Many electricians install receptacles with the ground pin on top. This provides for an added measure of safety. If a plug cap was not fully plugged in and a metal faceplate or other electrically conductive object should fall, it would hit the ground pin and not the live plug blades.

1. Refer to the plans and make a cable layout of all lighting and receptacle outlets; see Figure 11-4.
2. Draw a wiring diagram showing the traveller conductors for all three-way switches, if any; see Figure 11-5.
3. Draw a line between each switch and the outlet it controls. For three-way switches, do this for one switch only. This is the "switch leg."
4. Draw a line from the neutral bus bar on the lighting panel to each current-consuming outlet. This line may pass through switch boxes, but must not be connected to any switches. This is the white grounded circuit conductor, often called the *neutral* or *identified conductor*. It is not truly a neutral conductor unless it is part of a multiwire (three-wire) circuit.

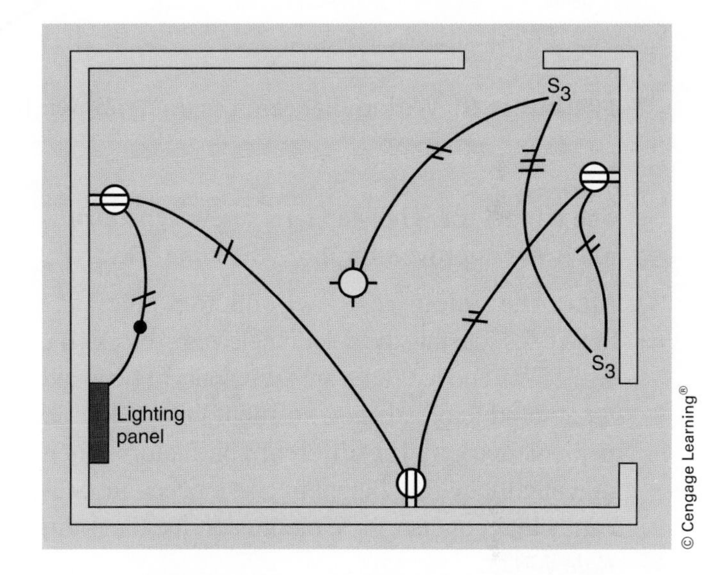

FIGURE 11-4 Typical cable layout.

Note: An exception to Step 4 may be made for double-pole switches. For these switches, all conductors of the circuit are opened simultaneously. They are rarely used in residential wiring.

5. Draw a line from the ungrounded ("hot") terminal on the panel to connect to each single-pole switch and unswitched outlet. Connect to common terminal of one three-way switch only.
6. Show splices as small dots where the various wires are to be connected. In the wiring diagram, the terminal of a switch or outlet may be used for the junction point of wires. However, the *CEC* does not permit more than one wire to be connected to a terminal unless the terminal is a type approved for more than

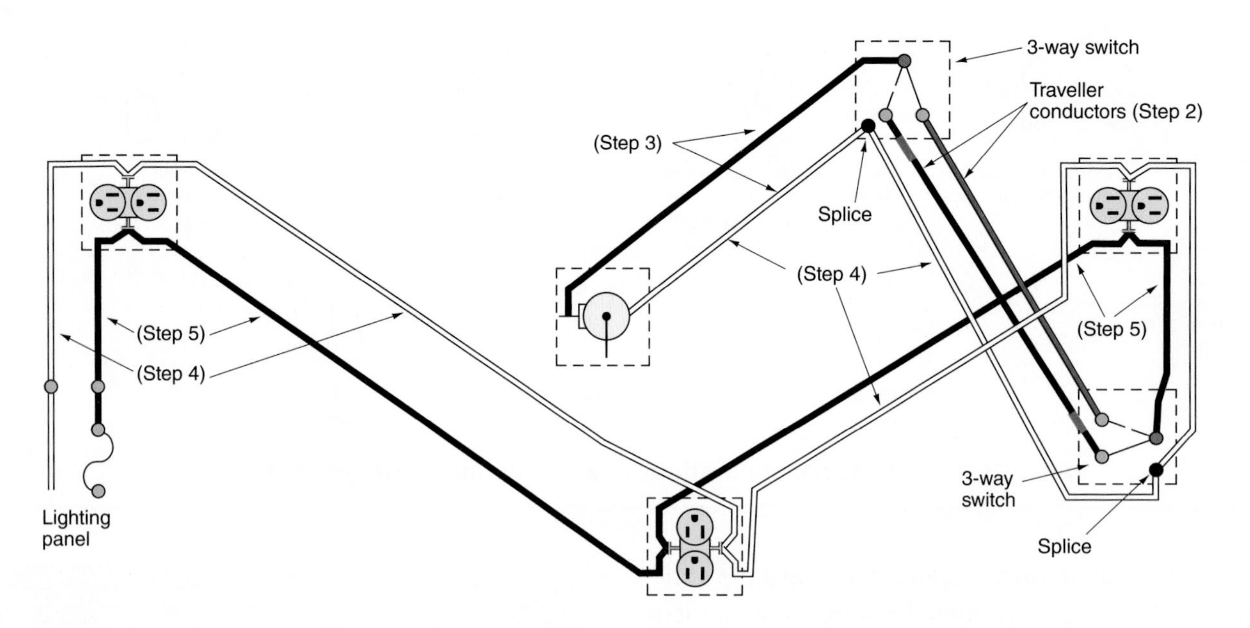

FIGURE 11-5 Wiring diagram of circuit shown in Figure 11-4. For simplicity, no bonding is shown.

one conductor. The standard screw-type terminal is *not* approved for more than one wire.

7. Mark the colour of the conductors; see Figure 11-5. Note that the colours selected—black (B), white (W), and red (R)—are the colours of two- and three-conductor cables. You should use coloured pens or markers for different conductors when drawing the wiring diagram. Use a pencil to indicate white conductors. Note this on the diagram, *Rule 4–028.*

Estimating Cable Lengths

The length of cable needed to complete an installation can be estimated roughly. It may be possible to run the cable in a straight line directly from one wall outlet to another. However, obstacles such as steel columns, sheet-metal ductwork, and plumbing may require the cable routing to follow a longer path.

To ensure that the estimate is not short, make all measurements "square." For example, measure from the ceiling outlet straight to the wall, and finally measure straight over to the wall outlet. Add 300 mm of cable at each cable termination. This allowance is for the free conductor needed to make connections in the junction and outlet boxes. Check the plans carefully to determine where the cables can be routed.

LIGHTING BRANCH-CIRCUIT A18 FOR FRONT BEDROOM AND STUDY

The receptacles in the bedrooms and living room are split-switched receptacles. Half the receptacle (normally the bottom half) is controlled by a wall switch in the room, and the other half is energized continuously. Receptacles in bedrooms, living rooms, and most other areas in a dwelling must be tamper-resistant, as per *Rule 26–712(g)*. These receptacles must be clearly marked "Tamper-Resistant" or "TR."

Radios, televisions, and other electrical items not intended to be controlled by the wall switch will be plugged into the unswitched ("hot" continuously) half of the duplex receptacle.

The recessed closet light is controlled by a single-pole switch outside and to the right of the closet door.

The outside weatherproof receptacle outside the study/bedroom is connected to Circuit A24. The front bedroom circuit is fed from A18 (Figure 11-6).

Checking the actual electrical plans, we find one television outlet and one telephone outlet are to be installed in each bedroom. Television and telephones are discussed in Unit 23. Table 11-1 summarizes the outlets in the front bedroom and the estimated load.

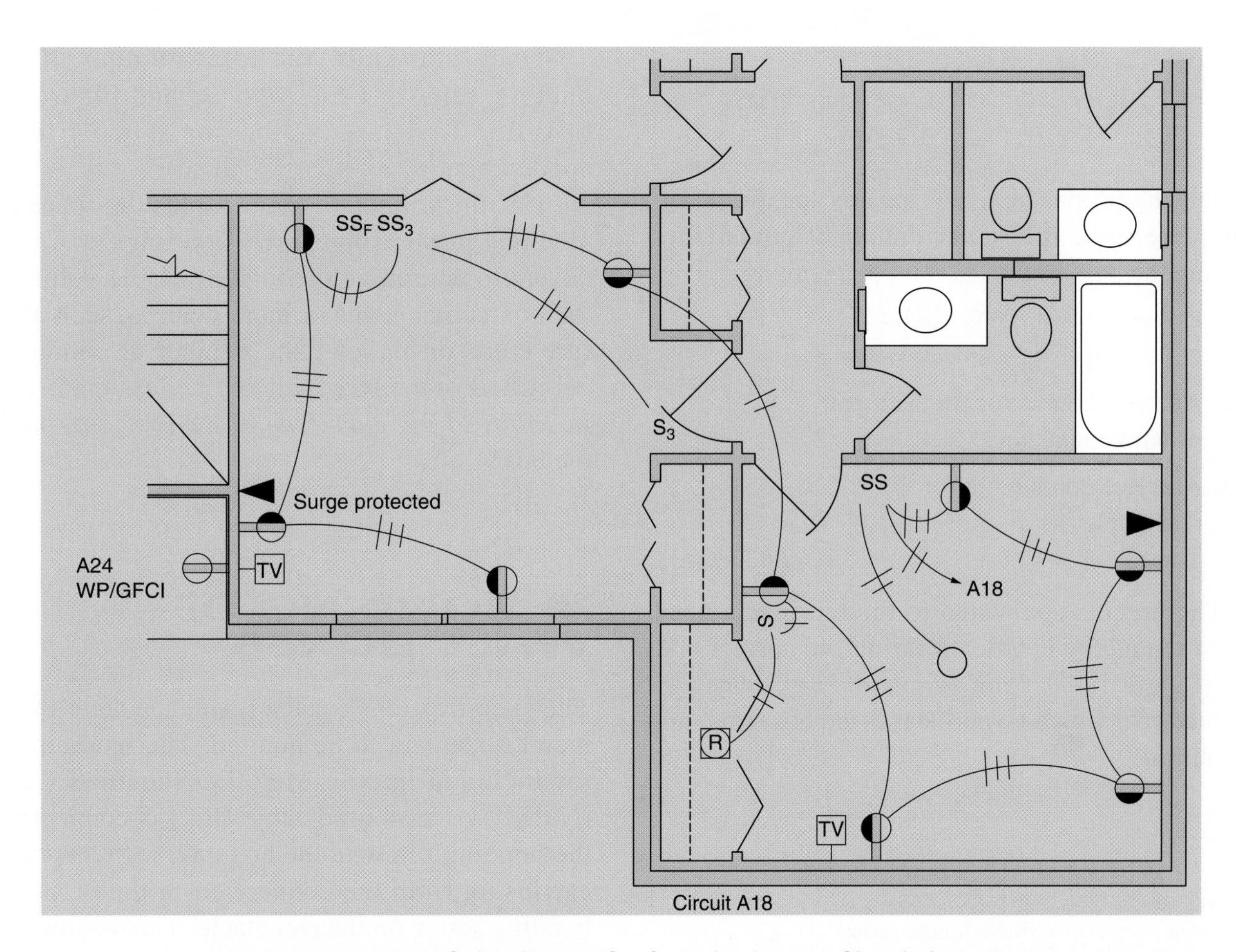

FIGURE 11-6 Cable layout for front bedroom, Circuit A18.

TABLE 11-1

Front bedroom and study/bedroom receptacles: outlet count and estimated load, Circuit A18.

DESCRIPTION	QUANTITY	WATTS	VOLT–AMPERES
Receptacles @ 120 watts each	10	1200	1200
Closed recessed fixture	1	60	60
Ceiling outlet	1	120	120
Totals	12	1380	1380

BRANCH-CIRCUIT A24 OUTSIDE RECEPTACLE

Branch-Circuit A24 is a 15-ampere circuit providing power to the outside receptacle outlet at the front of the house. The receptacle is Class A-type Ground-Fault Circuit Interrupter (GFCI)-protected with a weatherproof cover on the receptacle outlet box.

Rule 26–710(n) states that all outdoor receptacles 2.5 m or less above ground or grade level shall be Class A-type GFCI-protected and on a dedicated circuit, *Rule 26–726(a)*. Ordinary receptacles may be installed above 2.5 m, but must incorporate a weatherproof cover. Cover plates marked "Wet Location Only When Cover Closed" shall be installed facing downward at an angle of 45° or less from horizontal and located at least 1 m above finished grade and not in a wet location, *Rules 26–702(1) and (3)*.

DETERMINING THE WALL BOX SIZE

The electrician must consider several factors when determining the size of the wall box:

- the number of conductors entering the box
- the types of boxes available
- the space allowed for the installation of the box

Sizing Boxes According to the Number of Conductors in a Box *(Rule 12-3034)*

To find the proper box size for any location, you must determine the total number of conductors entering the box (Figure 11-7). For example:

1. Count the #14 AWG circuit conductors $2 + 3 + 3 = 8$
2. Add one conductor for each pair of wire connectors 2
3. Add two conductors for the receptacle 2

Total 12 conductors

The pigtails connected to the receptacle need *not* be counted when determining the correct box size (Figure 11-7). *Rule 12–3034(1)(c)* states, "A conductor, of which no part leaves the box, shall not be counted."

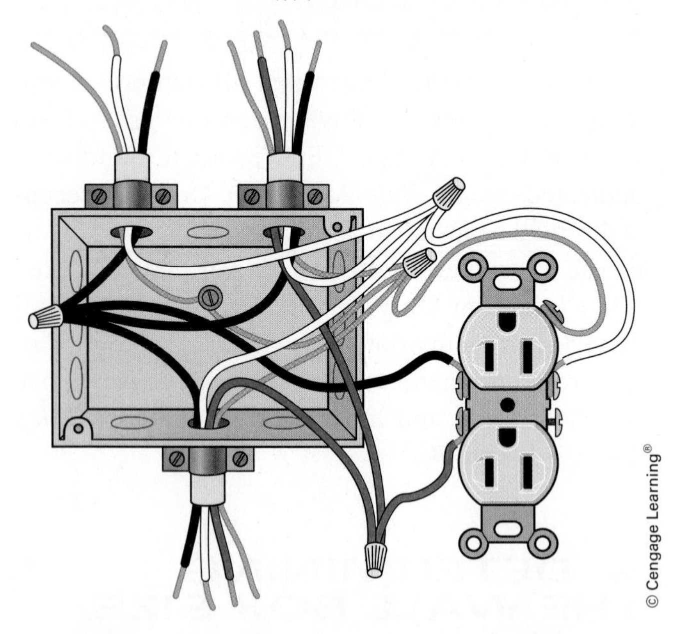

FIGURE 11-7 The tab must be broken between the red and black conductor terminals. The box size is determined according to the number of conductors. This box would be required to be 102 × 38 mm, complete with a plaster ring for the receptacle.

Once you know the total number of conductors, refer to *CEC Table 23* and Figure 5-16 in Unit 5 to determine the box suitable for the conductors.

Use the volume of the box plus the space provided by plaster rings, extension rings, and raised covers to determine the total available volume. If the box contains one or more devices, such as fixture studs or hickeys, the number of conductors permitted in the box shall be one less than shown in *Table 23* for *each type* of device contained in the box.

BONDING OF WALL BOXES

The specifications for the residence state that *all* metal boxes are to be bonded. The bare conductor included in Non-Metallic-Sheathed Cables (NMSC) or armoured cable (BX) is connected to the bonding screw in the box and, for receptacles, carries on from the connection in the box to the bonding screw on the receptacle. This conductor is used only to bond the metal box. This bare grounding conductor must *not* be used as a current-carrying conductor, since severe shocks can result.

According to *Diagram 1* and *Rule 26–700*, grounding-type receptacles must be installed on 15-ampere and 20-ampere branch circuits. Figure 11-7 shows how to attach the bonding conductor to the proper terminal on the receptacle.

POSITIONING OF SPLIT-CIRCUIT RECEPTACLES

The receptacle outlets shown in the front bedroom and study/bedroom are split-switched receptacles. The top portion of such a receptacle is hot at all times, and the bottom portion is controlled by the wall switch; see Figure 11-8. The electrician should wire the bottom of the receptacle as the switched section. Then, when the lamp is plugged into the switched bottom portion of the receptacle, the cord does not hang in front of the unswitched

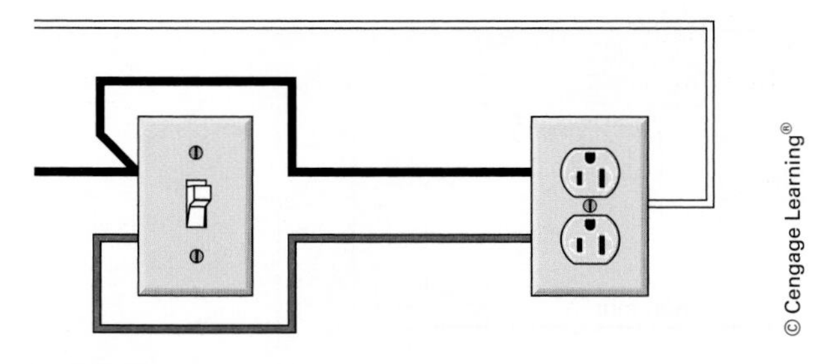

FIGURE 11-8 Split-switched wiring for receptacles in the bedroom.

section. This unswitched section can be used for a clock, vacuum cleaner, radio, stereo, CD player, personal computer, television, or some other appliance where a switch control is not necessary or desirable.

When mounting split-switched receptacles horizontally, locate the switched portion to the right.

POSITIONING OF RECEPTACLES NEAR ELECTRIC BASEBOARD HEATING

When laying out receptacles in residential applications, ensure that any cords plugged into the receptacles do not pass over heat-generating sources such as baseboard heaters, or hot air, hot water, or steam registers, *Rule 26–712(a)* and *Appendix B*.

Unit 17 discusses installing electric baseboard heating units and their relative positions below receptacle outlets.

FIXTURES IN CLOTHES CLOSETS

Clothing, boxes, and other material normally stored on closet shelves are potential fire hazards. These items may ignite on contact with the hot surface of exposed lamps. *Rule 30–204* gives very specific rules about the location and types of lighting fixtures permitted in clothes closets (Figure 11-9).

Light fixtures should not be installed above closet shelves. Fixtures may be installed on the ceiling in front of a shelf or above a closet door. See Figure 11-10 for suggested clearances.

The first-floor electrical drawing shows recessed lighting fixtures being used in the closets as well as other areas of the house. These fixtures should be Type IC, approved for insulated ceiling use, *Rule 30–906*. See Figure 11-11.

When installed, these fixtures will be covered with a vapour barrier and blanketed with insulation. Because heat is not dissipated from

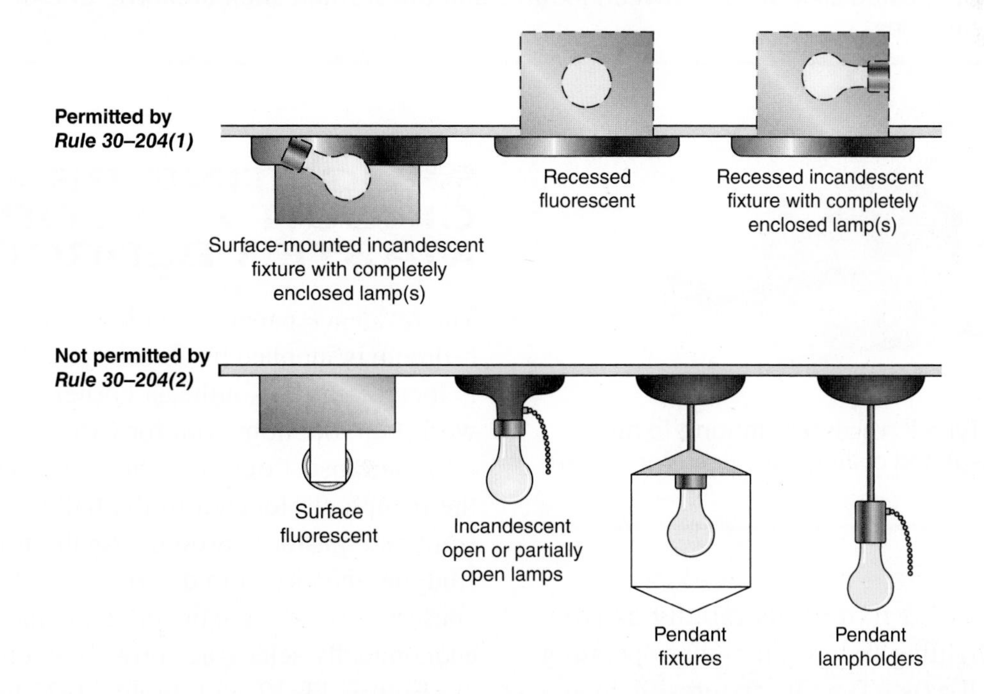

FIGURE 11-9 Types of lighting fixtures permitted in clothes closets. The *CEC* does not permit pendant fixtures, pendant lampholders, or luminaires with bare bulbs to be installed in clothes closets.

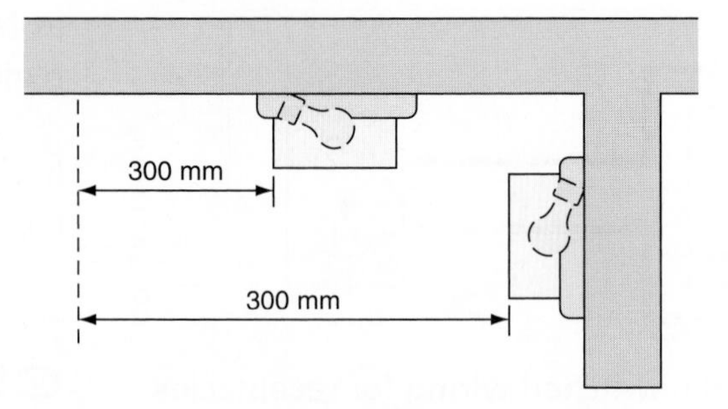

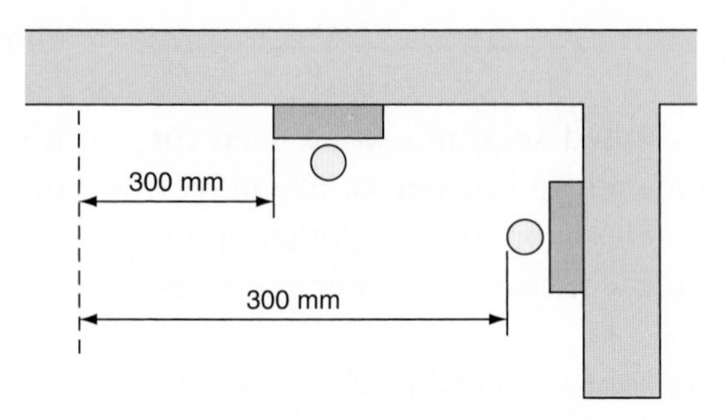

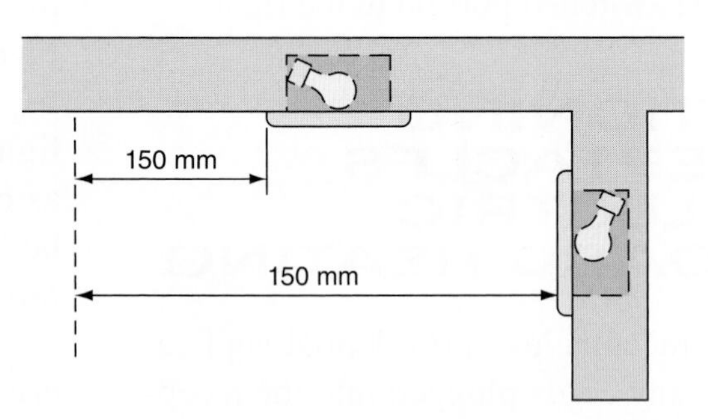

FIGURE 11-10 Suggested clearances between fixtures and the storage shelf area. The *CEC* does not specify exact clearances.

FIGURE 11-11 Type IC recessed lighting fixture approved for insulated ceiling use. Used for closet lighting.

insulation-blanketed fixtures as rapidly as from other types of lighting fixtures, a higher operating temperature will exist. Type IC fixtures include a thermal cutout device to prevent the maximum temperature from being exceeded.

LIGHTING BRANCH-CIRCUIT A14 FOR MASTER BEDROOM

The residence panel schedules show that the master bedroom is supplied by Circuit A14. Because Panel A is located in the southeast corner of the basement workshop, the home run for Circuit A14 is brought into the closest outlet to the panel. This would be the receptacle located in the hallway between the front and master bedrooms. Again, it is a matter of studying the circuit to determine the best choice for conservation of cable or conduit in these runs and to economically select the correct size of wall boxes.

Figure 11-12 and Table 11-2 show a sample device and wiring layout in which Circuit A14 has four split-switched receptacle outlets in the

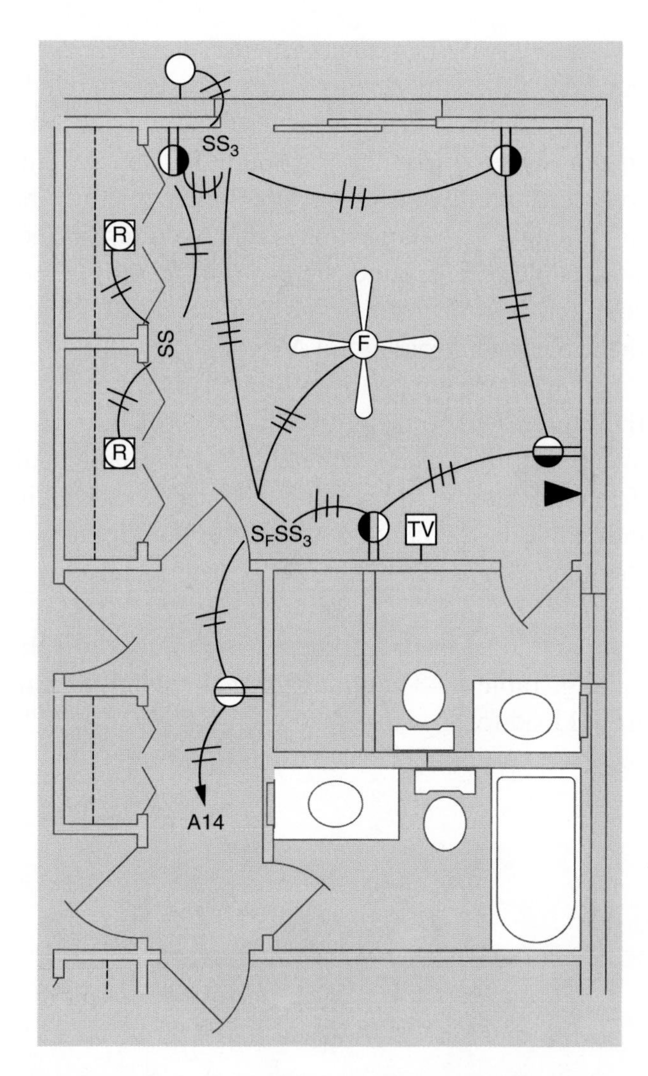

FIGURE 11-12 Cable layout for master bedroom, Circuit A14.

TABLE 11-2

Master bedroom: Outlet count and estimated load, Circuit A14.

DESCRIPTION	QUANTITY	WATTS	VOLT–AMPERES
Receptacles @ 120 watts each	5	600	600
Outdoor bracket fixture	1	100	100
Closet recessed fixtures			
One 60-W lamp each	2	120	120
Ceiling outlet	1	100	100
Totals	9	920	920

bedroom, one receptacle in the hall, one outdoor weatherproof light, two recessed closet fixtures each on a separate switch, one paddle fan/light fixture, one telephone outlet, and one television outlet.

In addition, an outdoor GFCI weatherproof receptacle is located adjacent to the sliding door and is connected to Circuit A24, which also includes the other outdoor receptacles.

Because there are two entrances to this room, the split-switched receptacle outlets are controlled by two three-way switches. One switch is located next to the sliding door. The use of split-switched receptacles offers the advantage of having switch control of one of the receptacles at a given outlet, while the other receptacle remains live at all times.

Next to the door to the hall is the other three-way switch plus the single-pole switch for the ceiling outlet.

SLIDING GLASS DOORS

CEC Rule 26–712(c) covers spacing receptacle outlets on walls where sliding glass doors are installed.

SELECTION OF BOXES

In the master bedroom, the electrician may decide to install a two-gang welded box or two sectional device boxes ganged together. The size of box depends on the number of conductors entering the box. The cable layout in Figure 11-12 shows that three cables enter the box next to the sliding glass door, for a total of seven conductors, each #14 AWG circuit conductors. *Rule 12–3034(2)(c)* stipulates that the number of conductors shall be reduced by two if the box contains one or more flush devices mounted on a single strap. Therefore, the number of conductors to be included in determining proper box size is seven circuit conductors plus two conductors for each switch, for a total of 11 wires.

1. Add the circuit conductors $2 + 2 + 3 = 7$
2. Add two for each switch $(2 + 2) = 4$

Total 11

Now look at the Quick-Check Box Selection Guide, Figure 5-16 in Unit 5, and select a box or a combination of gangable device boxes that are permitted to hold 11 conductors. For example, you can gang two 76 × 51 × 64-mm (3 × 2 × 2½-in.) device boxes together (8 × 2 = 16 #14 AWG conductors allowed).

LIGHTING CIRCUIT FOR STUDY AND MAIN BATHROOM

The study/bedroom circuit feeds the bedroom, which provides excellent space for a home office or den. Should it become necessary to have a third bedroom, the changeover is easily done. The circuit also supplies the light and receptacle in the main bathroom.

Circuit A16 originates at main Panel A. This circuit feeds the study/bedroom at the closet switch. From this point, the circuit feeds the closet recessed light and then on to the overhead light. Here it splits to feed the hallway light and smoke detector. The other leg goes to the switches by the door. A three-wire cable runs around the room to feed the other four split-switched from A18. Figure 11-13 shows the cable layout for Circuit A16.

There are four split-switched receptacles in the study/bedroom controlled by two three-way switches; one of the receptacles is surge-protected. The black and white conductors carry the live circuit, and the red conductor is the switch return for the control of the switched half of the split-circuit receptacles.

A paddle fan/light fixture is mounted on the ceiling. Table 11-3 summarizes the outlets and estimated load for the study/bedroom.

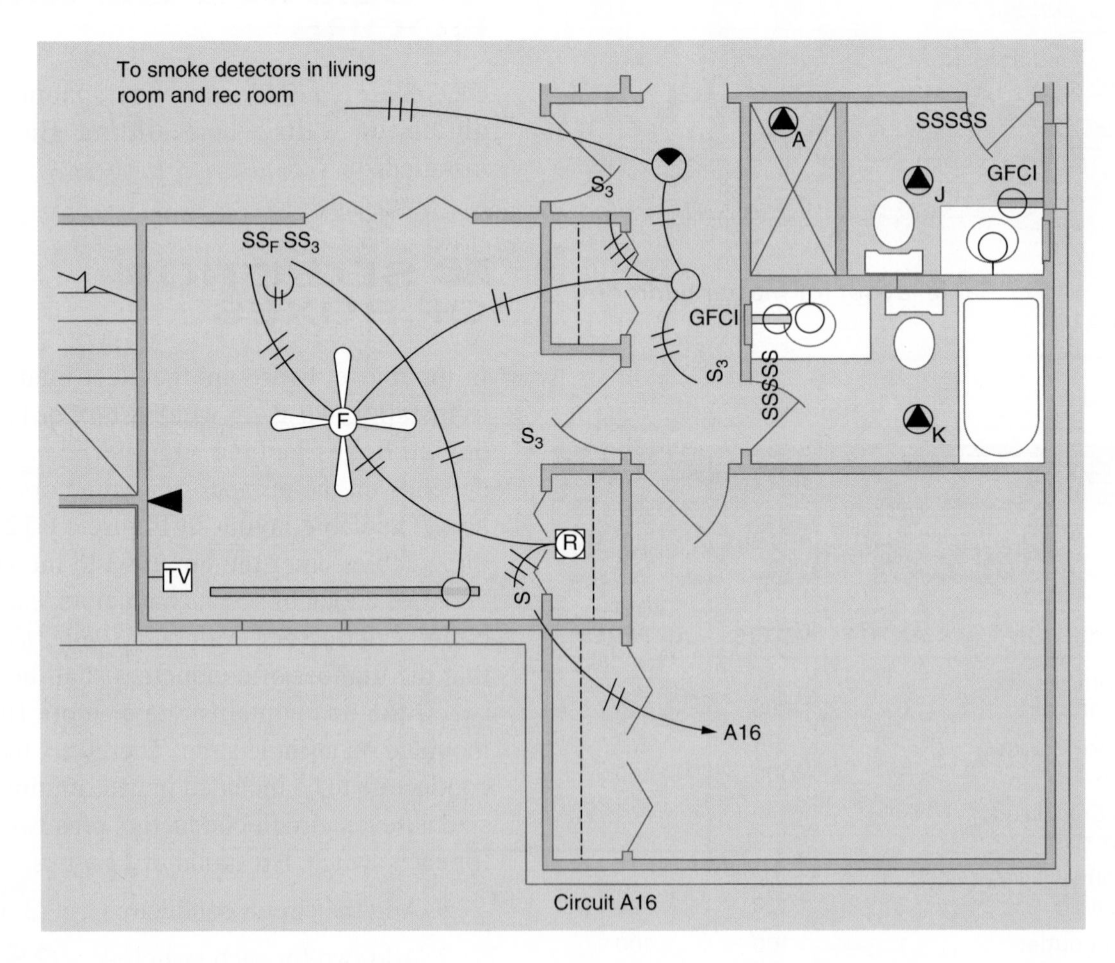

FIGURE 11-13 Cable layout for study/bedroom, Circuit A16. For receptacles in study, see Figure 11-6.

TABLE 11-3

Study/bedroom: Outlet count and estimated load, Circuit A16.

DESCRIPTION	QUANTITY	WATTS	VOLT–AMPERES
Receptacle @ 120 W	1	120	120
Closet recessed fixture	1	60	60
Paddle fan/light	1		
Three 50-W lamps		150	150
Fan motor (0.75 A @ 120 V)		80	90
Valance lighting	1		
Two 34-W fluorescent lamps		68	80
Hall light	1	100	100
Bathroom light	2	100	100
Totals	7	678	700

VALANCE LIGHTING

Indirect fluorescent lighting is provided above the windows. Figure 11-14 illustrates one method of installing fluorescent valance lighting.

Figure 11-15 shows one way to arrange the wiring for the paddle fan/light control, the valance lighting control, and the receptacle outlet control.

Since this room is intended to be used as a study or home office, a personal computer will probably be located here. Transient voltage surge suppressors are discussed in Unit 9. The wiring requirements for telephone and data systems are discussed in Unit 23.

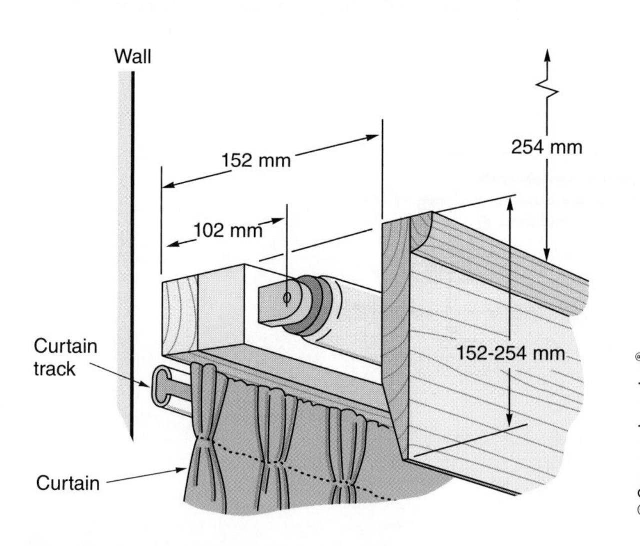

FIGURE 11-14 Fluorescent valance with draperies.

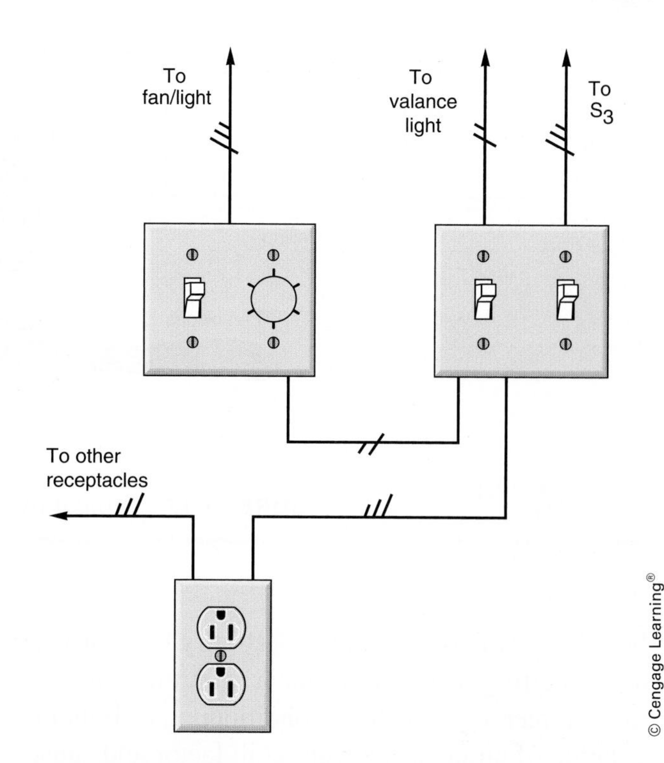

FIGURE 11-15 Conceptual view of how the study/bedroom switch arrangement is to be accomplished, Circuit A16.

PADDLE FANS

Paddle fans have become extremely popular in recent years; see Figure 11-16. These fans rotate slowly (60 to 250 rpm for home-type fans). The air currents they create can save energy during the

FIGURE 11-16 Paddle fan.

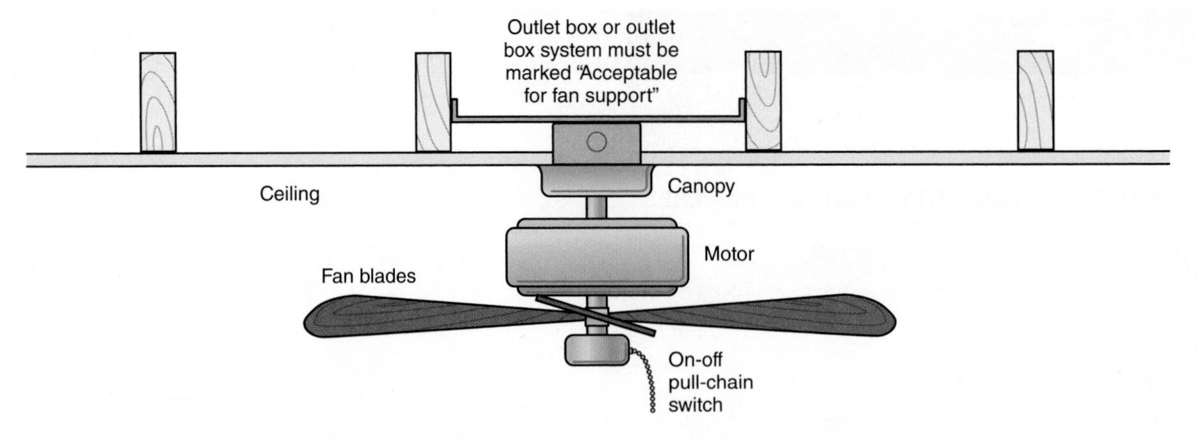

FIGURE 11-17 Typical mounting of a ceiling paddle fan.

heating season because they destratify the warm air at the ceiling, bringing it down to living levels. In the summer, even with air conditioning, a slight circulation of air creates a wind-chill factor and causes a person to feel cooler.

Residential paddle fans usually extend 305 mm or less below the ceiling; so, for 2.44 m ceilings, the fan blades are about 229 to 254 mm below the ceiling; see Figure 11-17.

Safely supporting a paddle fan involves

- the actual weight of the fan
- the twisting and turning motion when started
- vibration

To ensure the safe support of the fan, you should not use outlet boxes as the sole support of ceiling (paddle) fans unless the CSA approves and lists them for this purpose. The CSA also requires that the motor housing have a cable attached to an independent means of support to prevent the fan from falling if it vibrates loose from its mounting screws.

Always check the instructions furnished with the ceiling fan to be sure that the mounting and supporting methods will be safe. Boxes that are permitted to support a fan are marked "Acceptable for fan support."

Rule 12–3010(10)(8) requires that a pendant ceiling fan be mounted in an outlet box marked for fan support and independently of the outlet box if the fixture weighs more than 16 kg, *Rule 12–3010(9)*.

Figure 11-18 shows the wiring for a fan/light combination with the supply at the fan/light unit.

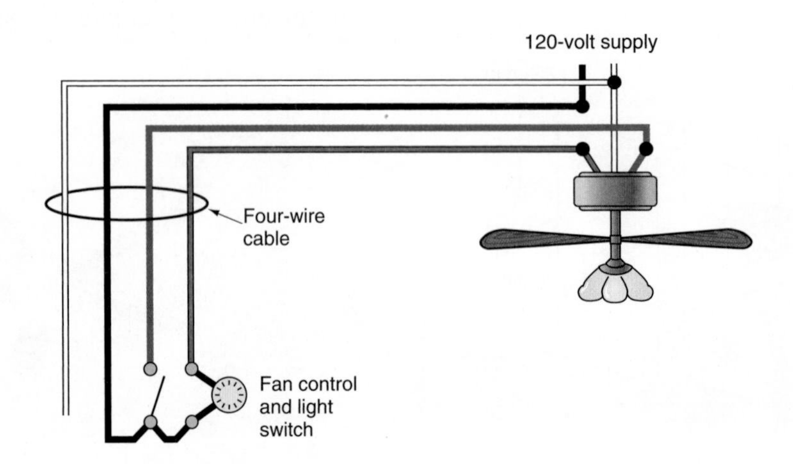

FIGURE 11-18 Fan/light combination with the supply at the fan/light unit.

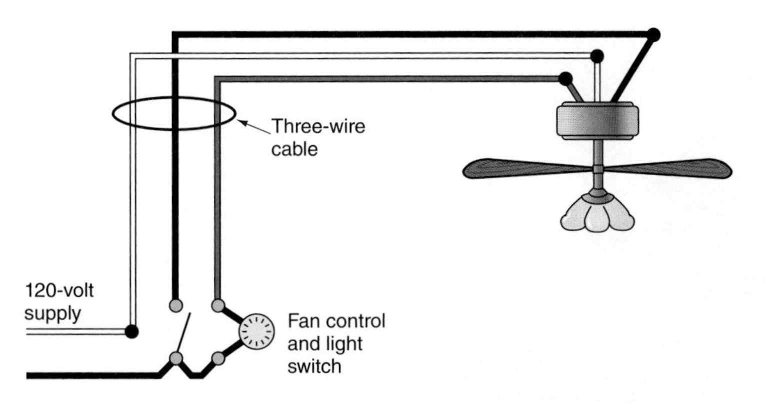

FIGURE 11-19 Fan/light combination with the supply at the switch.

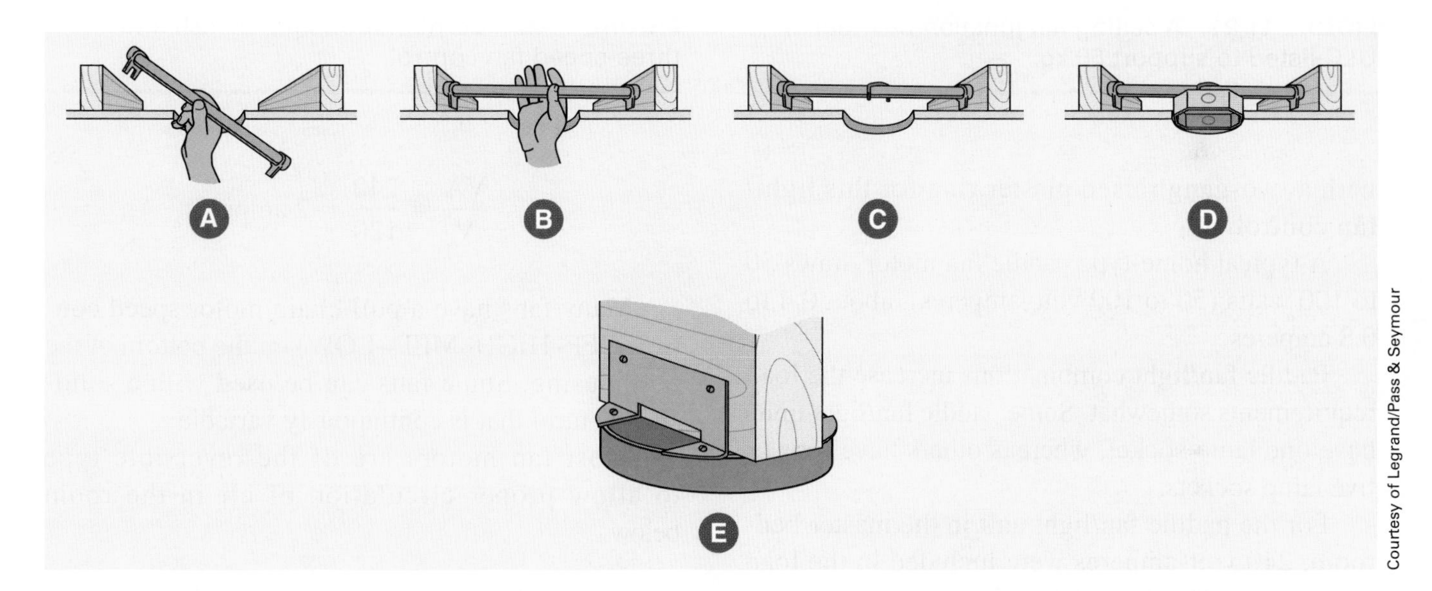

FIGURE 11-20 (A), (B), (C), and (D) show how a listed ceiling fan hanger and box assembly is installed through a properly sized and carefully cut hole in the ceiling. This is done for existing installations. Similar hanger/box assemblies are used for new work. The hanger adjusts for 400-mm and 600-mm joist spacing but can be cut shorter if necessary. (E) shows a type of box listed and identified for the purpose where the fan is supported from the joist, independent of the box, as required by *CEC Rules 12–3010(8)* and *12–3010(9)* for fans that weigh more than 16 kg.

Figure 11-19 shows the fan/light combination with the supply at the switch.

Figure 11-20 shows an adjustable-type ceiling fan hanger installed. This type of hanger is useful when there is no access to the ceiling cavity, since it can be installed through the hole where the outlet box will be mounted and adjusted to tighten against the joists.

The electrician must install a junction box where the fan or fan/light is to be located. Since the fan is quite heavy, the box must be solidly secured to the ceiling joist. One type of certified fan junction box is shown in Figure 11-21.

Figure 11-22 illustrates a typical combination light switch and three-speed fan control. The electrician must install a deep 102-mm box

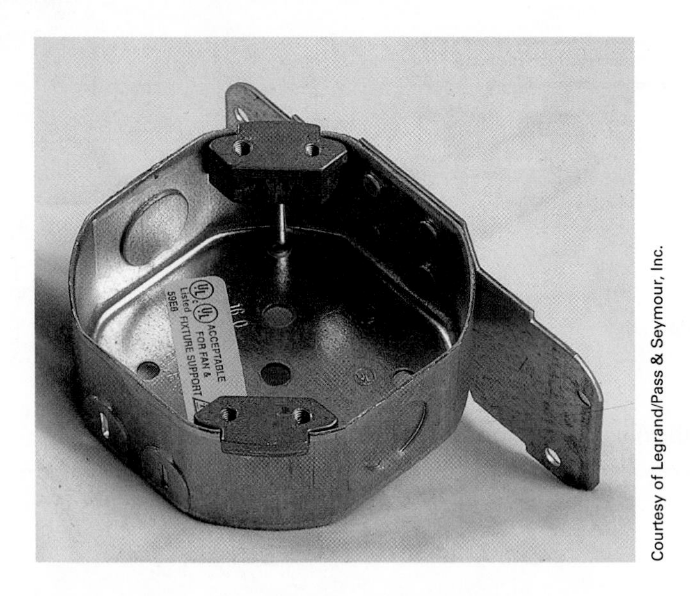

FIGURE 11-21 A ceiling fan junction box that is ULC-listed to support 60 kg.

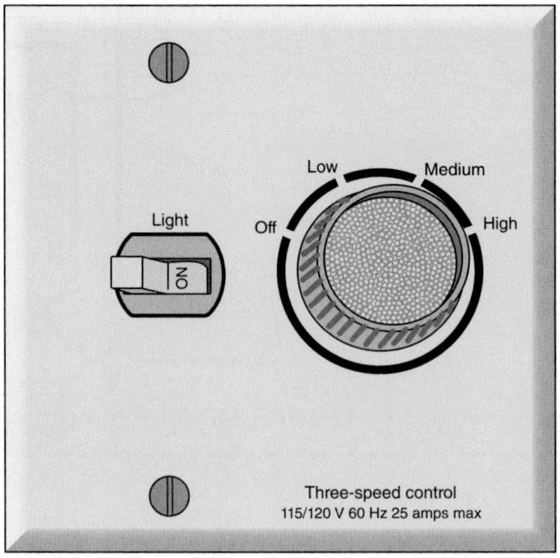

Light
ON
Off
Low
Medium
High
Three-speed control
115/120 V 60 Hz 25 amps max

FIGURE 11-22 Combination light switch and three-speed fan control.

with a two-gang raised plaster ring for this light/fan control.

A typical home-type paddle fan motor draws 50 to 100 watts (50 to 100 volt–amperes), about 0.4 to 0.8 amperes.

Paddle fan/light combinations increase the load requirements somewhat. Some paddle fan/light units have one lamp socket, whereas others have four or five lamp sockets.

For the paddle fan/light unit in the master bedroom, 240 volt-amperes were included in the load calculations. The current draw is

$$I = \frac{VA}{V} = \frac{240}{120} = 2 \text{ amperes}$$

Many fans have a pull-chain motor speed control (OFF–HIGH–MED–LOW) on the bottom of the fan housing. Some fans can be used with a solid-state control that is continuously variable.

Most fan motors are of the reversible type to allow proper circulation of air in the room below.

REVIEW

Note: Refer to the *CEC* or the blueprints provided with this textbook when necessary. Where applicable, responses should be written in complete sentences.

1. Why do some electricians prefer to have more than one circuit feeding a room? _____
____
____

2. Is it good practice to have outlets on different floors on the same circuit? Why?
____
____

3. What usually determines the grouping of outlets into a circuit? ____________________
____
____

4. The continuous load on a lighting branch circuit must not exceed ____________% of the branch-circuit rating.

5. To determine the maximum number of outlets in a circuit, ____________ amperes per outlet are allowed. For a 15-ampere circuit, this results in a maximum of ___________ outlets.

6. For this residence, what are the estimated wattages used in determining the loading of Branch-Circuit A16?

   Receptacles ____________ watts (volt–amperes)

   Closet recessed fixture ____________ watts (volts–amperes)

7. With respect to *Rule 8–104(1)*, the ampere rating of Circuit A16 is ______________.

8. What size of wire is used for the lighting circuit in the front bedroom? ____________

9. How many receptacles are connected to this circuit? ____________________
____

10. What main factor influences the choice of wall boxes? ____________________
____

11. How is a wall box grounded? ____________________
____

12. What is a split-switched receptacle? ____________________
____

13. Is the switched portion of an outlet mounted toward the top or the bottom? Why?
____
____

14. The following questions pertain to lighting fixtures in clothes closets.

    a. Does the *CEC* allow bare incandescent lamp fixtures such as porcelain keyless or porcelain pull-chain lampholders to be installed? ____________________
    ____

    b. Does the *CEC* allow bare fluorescent lamp fixtures to be installed? ____________
    ____

    c. Does the *CEC* permit pendant fixtures or pendant lampholders to be installed?
    ____

15. How many switches are in the front bedroom circuit, and of what type are they? ____
____
____

16. The following is a layout of the lighting circuit for the front bedroom. Using the cable layout shown in Figure 11-6, make a complete wiring diagram of this circuit. Indicate the colour of each conductor.

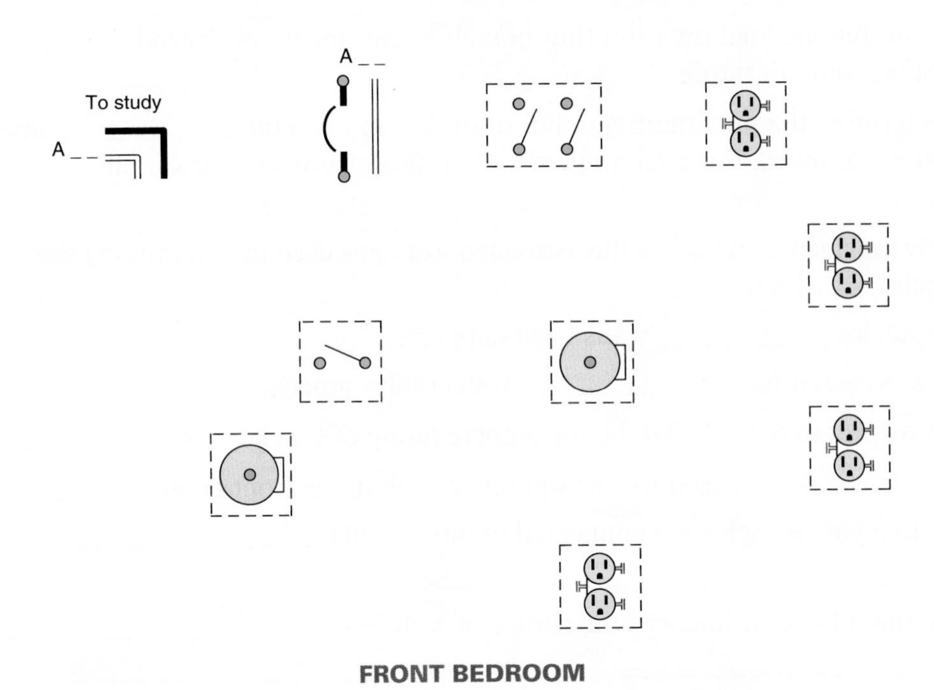

FRONT BEDROOM

17. When planning circuits, how should you divide loads? ______________________

____

____

18. The *CEC* uses the terms *watts, volt–amperes*, and kW. Explain their significance in calculating loads. ______________________

____

____

____

____

____

19. How many #14 AWG conductors are permitted in a device box that measures 76 × 51 × 64 mm (3 × 2 × 2½ in.)? ______________________

20. A 102 × 38 mm (4 × 1½ in.) octagon box has one cable clamp, one fixture stud, and three wire connectors. How many #14 AWG conductors are permitted? ______________________

____

____

21. What circuit supplies the master bedroom? ______________________

22. For the recessed closet fixtures, what wattage was used for calculating their contribution to the load on the circuit? ______________________

23. What is the estimated load in volt–amperes for the circuit supplying the master bedroom? ______________________

24. The sliding glass door in the master bedroom could affect the number of required receptacle outlets. Answer the following statements True or False.
    a. Sliding glass panels are considered to be wall space. ______________________
    b. Fixed panels of glass doors are considered to be wall space. ______________
25. What type of receptacle will be installed outdoors, just outside of the master bedroom?
    ____
    ____
26. Based on the total estimated load calculations, what is the current draw on the study/bedroom lighting circuit? ______________________________
27. The study/bedroom is connected to Circuit ______________________________
28. The conductor for this circuit is ______________________________
29. Is it necessary to install a receptacle in the wall space leading to the bedroom hallway?
    ____
    ____
    ____
30. Show the calculations needed to select a properly sized box for the receptacle outlet in Question 29. NMSC is the wiring method. Refer to the Quick-Check Box Selection Guide, Figure 5-16 in Unit 5.
    ____
    ____
    ____
    ____
31. Using the cable layout shown in Figure 11-13, make a complete wiring diagram of Circuit A16.

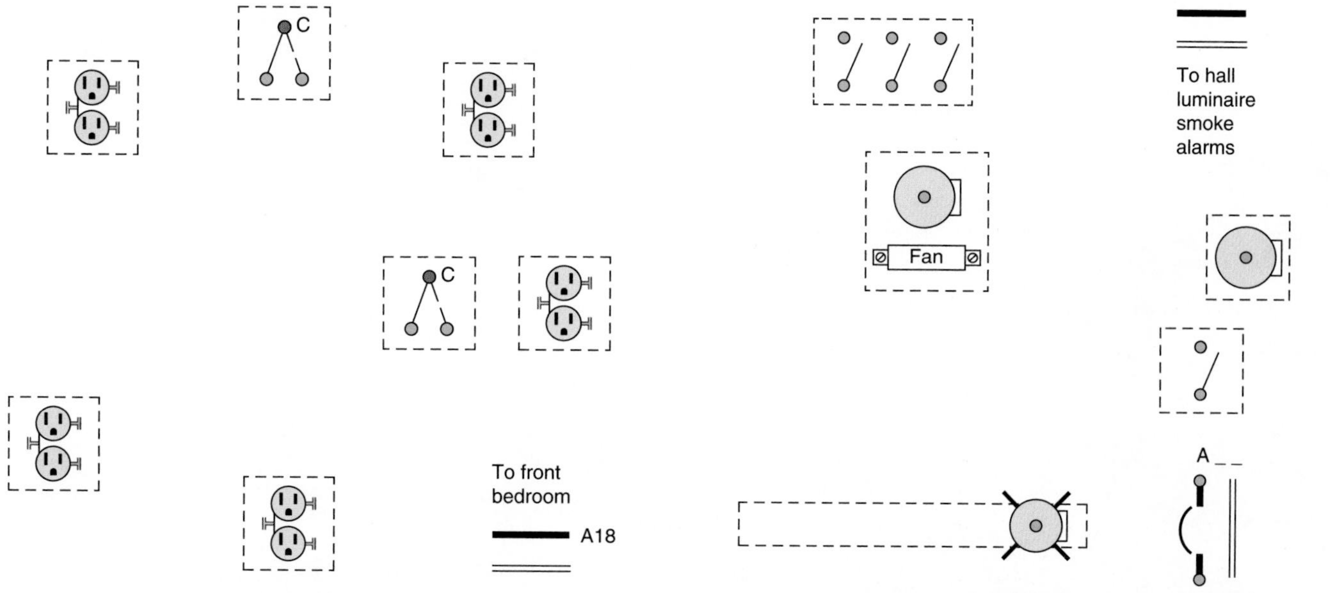

C
C
To front bedroom
A18

STUDY RECEPTACLE CIRCUIT

To hall luminaire smoke alarms
Fan
A

STUDY LIGHTING CIRCUIT

32. The following is a layout of the lighting circuit for the master bedroom. Using the cable layout shown in Figure 11-12, make a complete wiring diagram of this circuit. Indicate the colour of each conductor.

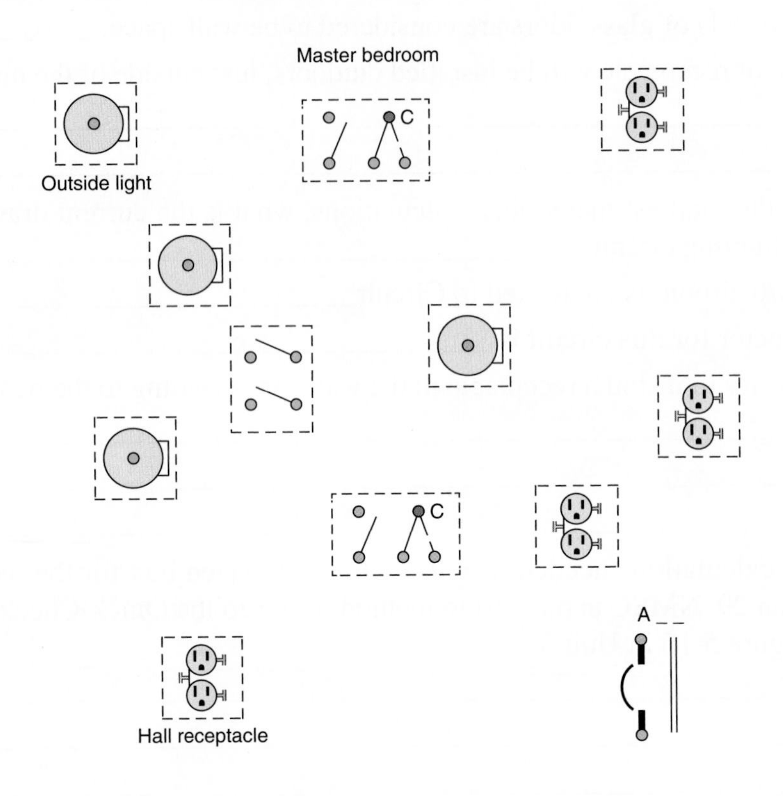

33. Using the electrical plans supplied with this textbook, lay out the devices and wiring in the master bedroom according to minimum *CEC* standards.

UNIT **12**

Branch Circuits for the Living Room and Front Entry

OBJECTIVES

After studying this unit, you should be able to

- understand various types of dimmer controls
- connect dimmers to incandescent lamp loads
- understand the phenomenon of incandescent lamp inrush current
- discuss Class P ballasts and ballast overcurrent protection
- discuss the basics of track lighting
- understand how to install a switch in a door jamb for automatic ON/OFF when the door is opened or closed
- discuss the type of fixtures recommended for porches and entries
- discuss the advantages of switching outdoor receptacles from indoors
- define *wet* and *damp* locations

LIGHTING CIRCUIT OVERVIEW

Figure 12-1 shows sample device and wiring layout for the living room area that is connected to Circuit B16, a 15-ampere circuit. The home run is brought from Panel B to the split-switched receptacle in the living room next to the sliding door.

A three-wire cable is then carried around the living room, feeding in and out of the eight split-switched receptacles. This three-wire cable carries the black and white circuit conductors plus the red wire, which is the switched conductor. Since these receptacles will provide most of the general lighting for the living room through floor and table lamps, it is a good idea to have some of these lamps controlled by the three switches (two three-way switches and one four-way switch). Only one-half of each receptacle is switched, *Rule 26–710(k)*.

The receptacle below the four-way switch is required because the wall space is more than 900 mm wide, *Rule 26–712(c)*, and there must be a receptacle within 1.8 m of any point along the wall line, *Rule 26–712(a)*. Because it is unlikely that a split-switched receptacle (Figure 12-2) will be needed in this location, a standard receptacle is used.

Accent lighting above the fireplace is in the form of a recessed fixture controlled by a single-pole dimmer switch. See Unit 10 for installation data and *Canadian Electrical Code (CEC)* requirements for recessed fixtures.

Table 12-1 summarizes the number of outlets and estimated load for the living room circuit. Unit 6 covers requirements for spacing receptacle outlets and locating lighting outlets, as well as spacing receptacles on exterior walls when sliding glass doors are involved, *Rule 26–712(c)*.

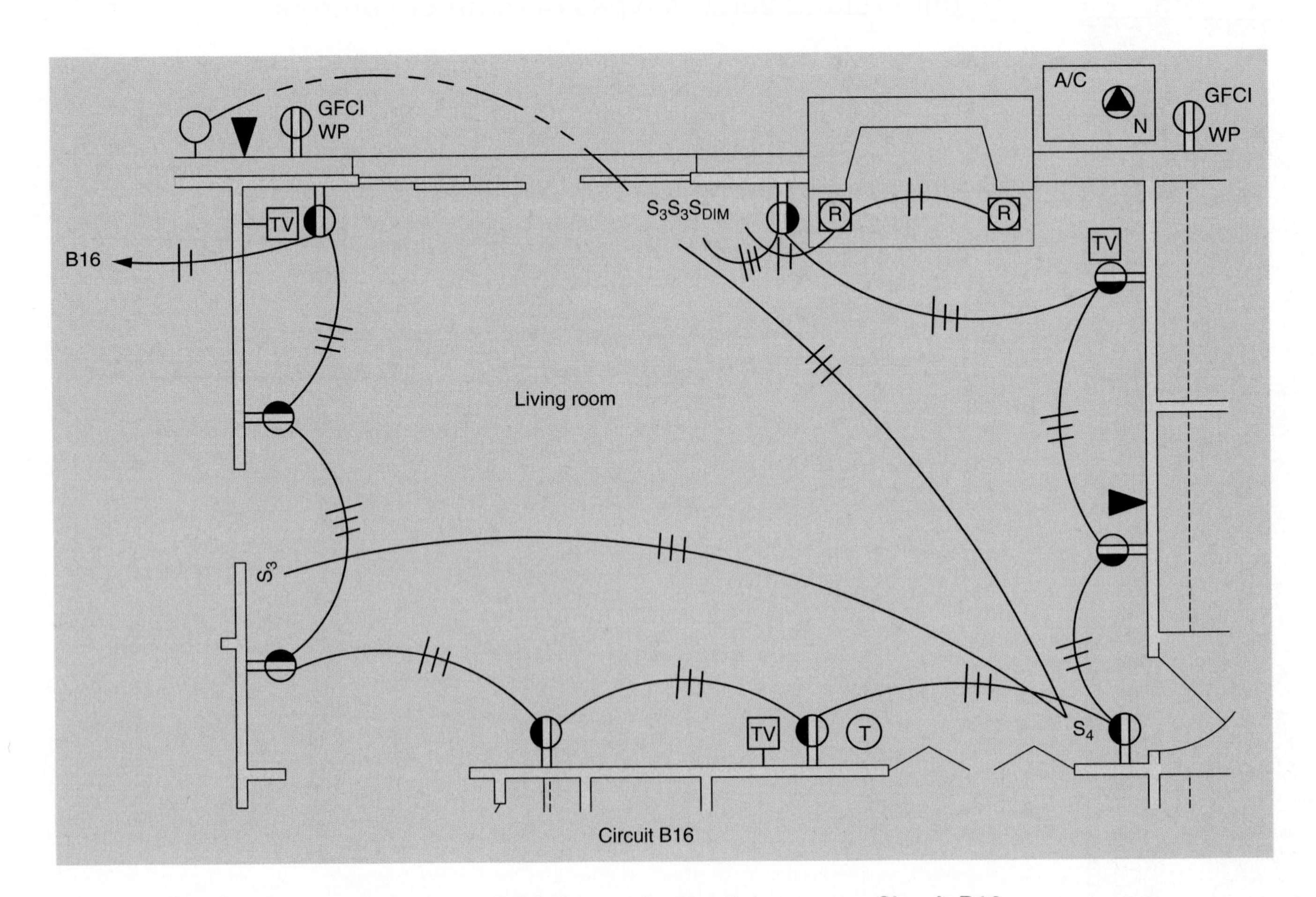

FIGURE 12-1 Cable layout for the living room, Circuit B16.

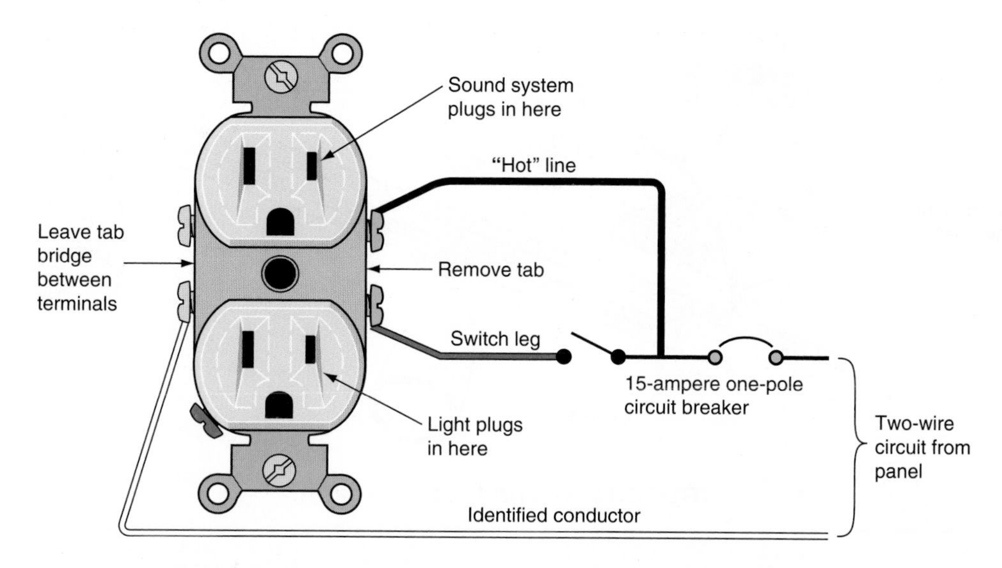

FIGURE 12-2 Split-switched receptacle wiring diagram.

TABLE 12-1

Living room: Outlet count and estimated load, Circuit B16.

DESCRIPTION	QUANTITY	WATTS	VOLT-AMPERES
Receptacles @ 120 W	9	1080	1080
Fireplace recessed fixtures @ 60 W each	1	60	60
Totals	10	1140	1140

TRACK LIGHTING

The plans show that track lighting is mounted on the ceiling of the living room on the wall opposite the fireplace. Track lighting provides accent lighting for a fireplace, a painting on the wall, or an item (sculpture or collection) that the homeowner wants to focus attention on. It is also used as task lighting for counters, game tables, or kitchen dining tables.

Unlike recessed fixtures and individual ceiling fixtures that are fixed in place, track lighting offers flexibility because the actual lampholders can be moved and relocated on the track as desired.

Lampholders, selected from hundreds of styles, are inserted into the extruded aluminum track at any point. The circuit conductors are in the track, and the plug-in connector on the lampholder completes the connection. The track is generally fastened to the ceiling at the outlet box and with toggle bolts or screws. Various track lighting installations are shown in Figures 12-3, 12-4, and 12-5.

On the residence plans provided with this textbook, the living room ceiling has wooden beams, a challenge when installing track lighting that is longer than the space between them. There are a number of things the electrician can do:

1. Install a pendant kit assembly that will allow the track to hang below the beam; see Figure 12-6. This fixture may be supported directly by the outlet box if the weight does not exceed 23 kg, *Rule 30–302(4)*.
2. Install conduit connector fittings on the ends of sections of the track, then drill a hole in the beam, and use 16-mm EMT to join the track sections. Running lighting track through walls or partitions is not recommended.
3. Run the branch-circuit wiring concealed above the ceiling to outlet boxes located at the points where the track is to be fed by the branch-circuit wiring; see Figure 12-7.

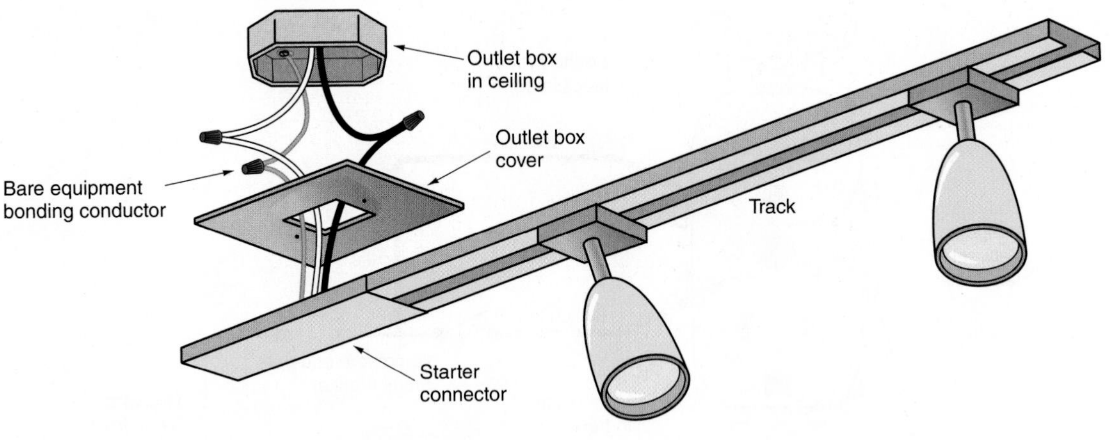

FIGURE 12-3 End-feed track light.

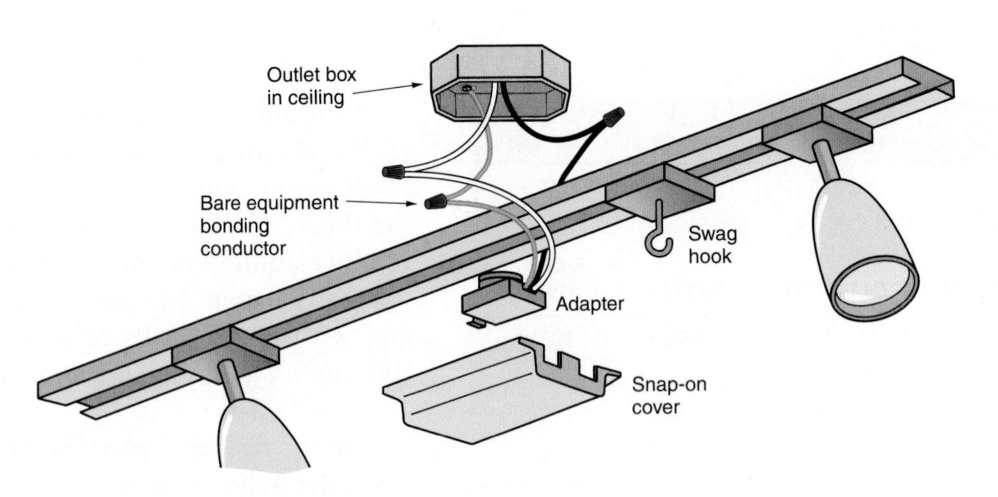

FIGURE 12-4 Centre-feed track light.

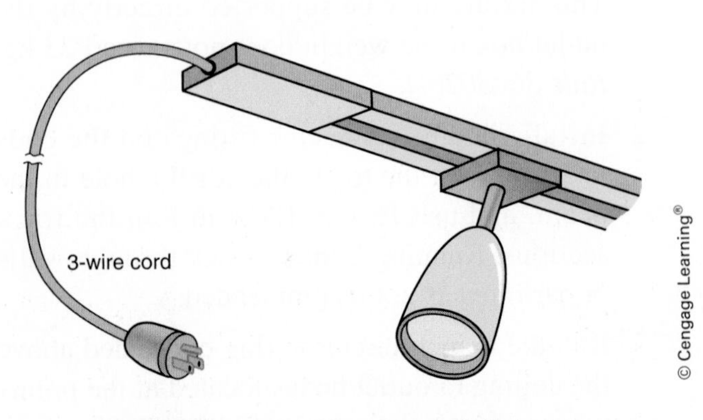

FIGURE 12-5 Plug-in track light.

Where to Mount Track Lighting

To install the lighting track at the proper distance from the wall, you must consider the ceiling height, the aiming angle, the type of lampholder, the type of lamp, and the type of lighting to be achieved.

For example, if you are designing an installation to illuminate an oil painting that is to be centred 1.65 m (average eye level) from the floor, where the ceiling height is 2.75 m, and the lampholder and lamp information tell us that the aiming angle should be 60°, you would find in the manufacturer's

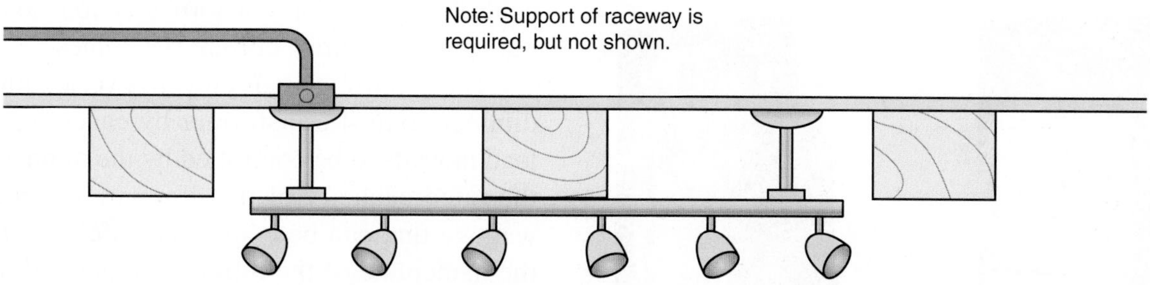

Note: Support of raceway is
required, but not shown.
© Cengage Learning®

FIGURE 12-6 A pendant kit assembly allows the track to hang below the beam.

Note: Support of raceway is
required, but not shown.
© Cengage Learning®

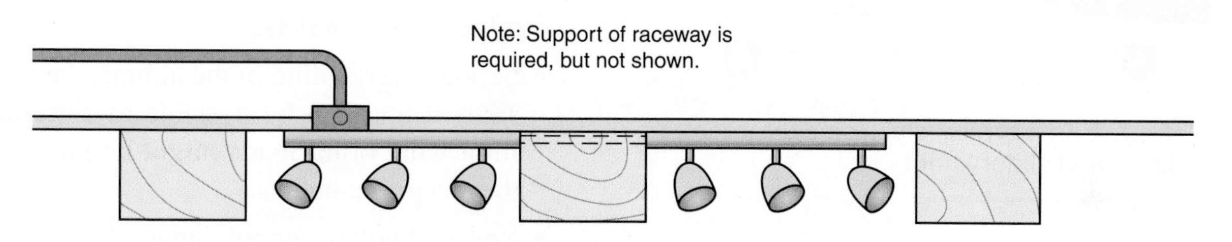

FIGURE 12-7 An outlet box feeds each section of track.

installation data that the track should be mounted 600 mm from the wall.

If the ceiling height in this example was 2.4 m, then the distance to the track from the wall would be 450 mm.

Since many variables affect the positioning of the track, it is necessary to consult the track lighting manufacturer's catalogue for recommendations that will achieve the desired results.

The track itself must be properly grounded to the outlet box. The plug-in connector on the movable fixture also has a ground contact to ensure that the body of the fixture is safely grounded.

The track lighting in the living room is turned on and off by a single-pole dimmer switch.

Supporting Track Lighting

Each lighting track must be fastened at a minimum of two places. This provides the secure support required by *Rule 30–302(1)*. If the track is extended, install at least one additional support for every 1.5 m or less of extension.

Load Calculations for Lighting Track

For track lighting installed in homes, it is not necessary to add additional loads for feeder or service entrance calculations. *Rule 8–200(1)(a)* includes lighting track load as part of the regular watts-per-square-metre computations.

For further information on track lighting, refer to lighting manufacturers' catalogues.

DIMMER CONTROLS

Electronic Solid-State Dimmer Controls

According to their Canadian Standards Association (CSA) listing, solid-state electronic dimmers of the type sold for home use "are intended only for the control of permanently installed incandescent fixtures"; see Figure 12-8. For residential installations,

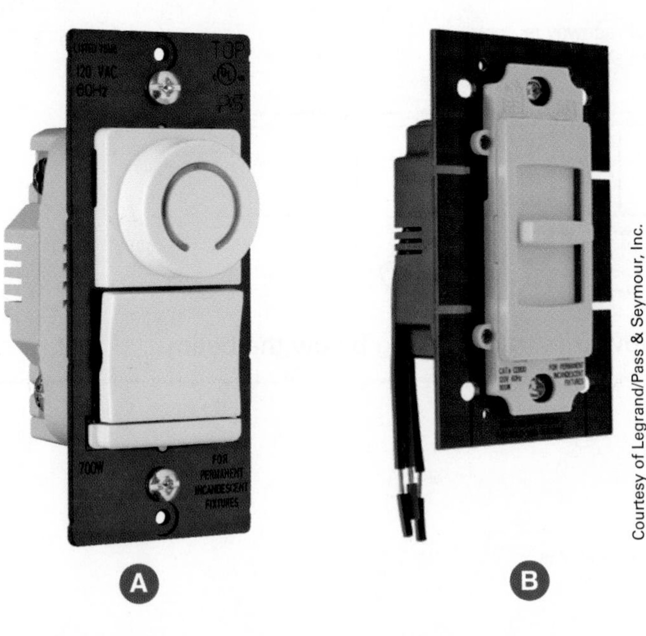

FIGURE 12-8 Some typical electronic dimmers; rotating knob and slider knob.

the dimmer commonly used is a 600-watt unit that fits into a standard single-gang wall box. Both single-pole and three-way dimmers are available.

Electronic dimmers use TRIACS to chop the alternating current (AC) sine wave, controlling the amount of voltage supplied to the load. Compact 1000-watt dimmer units are available for use in two-gang wall boxes.

These boxes should be No. 1004 boxes, which are 3 inches deep. Figure 12-9 shows how single-pole and three-way solid-state dimmer controls are used in circuits.

Autotransformer Dimmer Control

This type of dimmer control has a continuously adjustable autotransformer that changes the lighting intensity by varying the voltage to the lamps; see Figures 12-10 and 12-11. When the knob of the transformer is rotated, a brush contact moves over a bared portion of the transformer winding to vary the lighting intensity from complete darkness to full brightness. This type of control uses only the current required to produce the lighting desired.

Dimmer controls are used on 120-volt, 60-hertz, single-phase AC systems. The operation of such controls depends on transformer action, so they will not operate on direct current (DC) lines.

The electrician can prevent overloading of the dimmer control transformer by checking the total load in watts to be controlled by the dimmer against the rated capacity of the dimmer. The maximum wattage that can be connected safely is shown on the nameplate of the dimmer. In general, overcurrent protection is built into dimmer controls to prevent burnout due to an overload.

Instructions are furnished with most dimmer controls to help prevent such burnout. These instructions state that the dimmers are for control of incandescent lighting only and are not to be used to control receptacle outlets:

- Serious overloading of the dimmer could result, since it may not be possible to determine or limit what other loads might be plugged into the receptacle outlets.
- Reduced voltage supplying a television, radio, stereo components, computer, or motors of any kind, such as those in vacuum cleaners and food processors, can result in costly damage to the appliance.

Because they contain a bulky transformer, autotransformer dimmers are larger than solid-state dimmers. Autotransformers require a special box that is furnished with the dimmer. They are common in commercial establishments, whereas the smaller solid-state dimmer is most commonly used in residential installations.

INCANDESCENT LAMP LOAD INRUSH CURRENTS

An unusual action occurs when a circuit is energized to supply a tungsten filament lamp load. A tungsten filament lamp has a very low resistance when it is cold. When the lamp is connected to the proper voltage, the resistance of the filament increases very quickly (within 1/240 second, or one-quarter of a 60-Hz cycle) after the circuit is energized. During this period, there is an inrush current that is about 10 to 20 times greater than the normal operating current of the lamp.

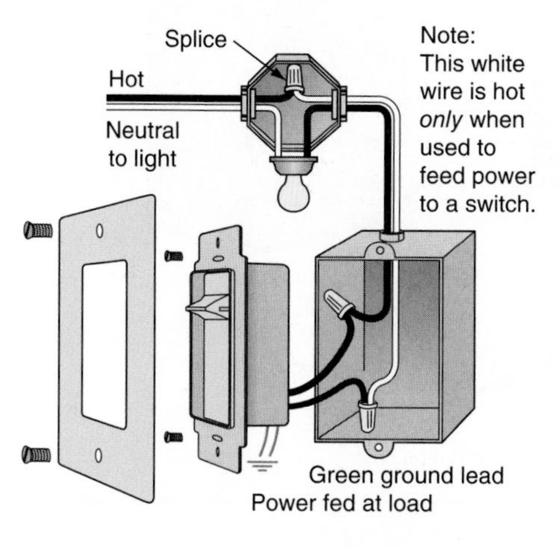

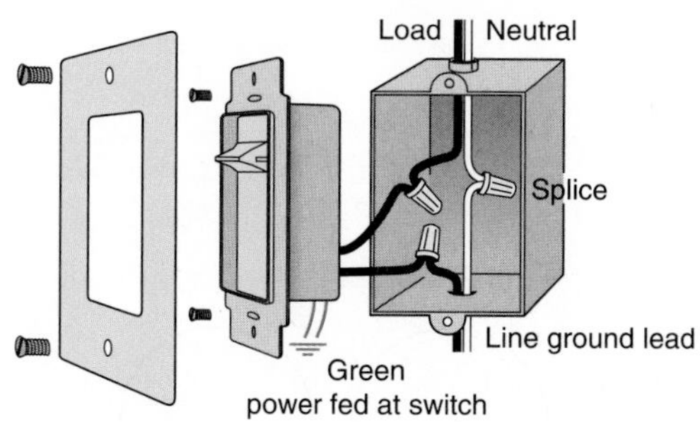

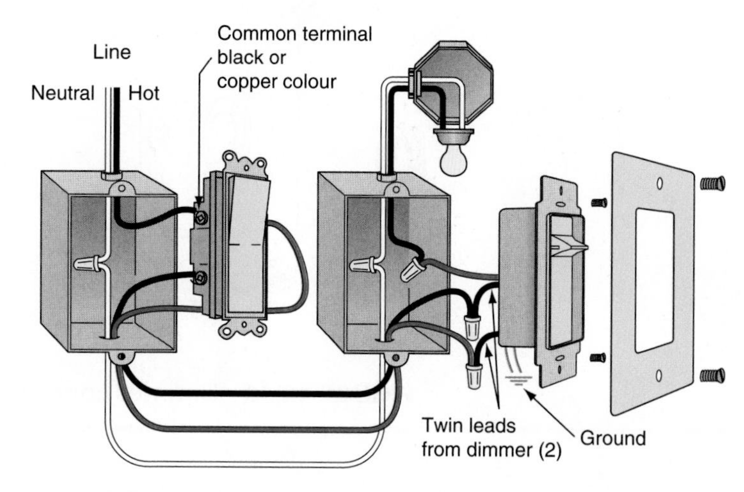

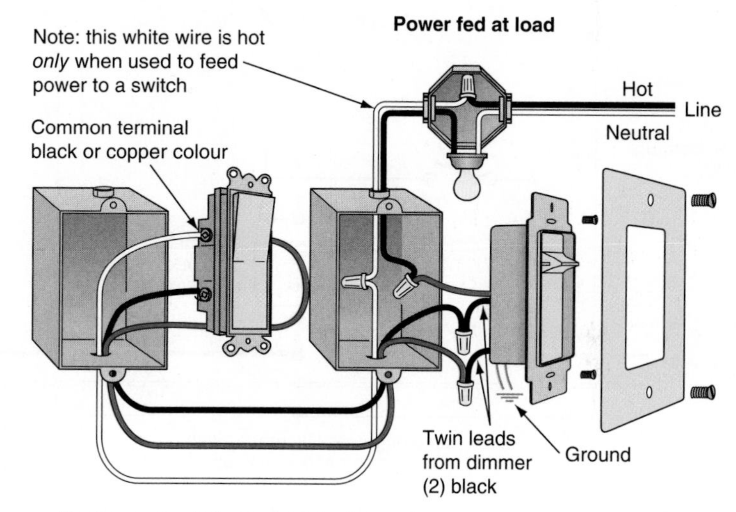

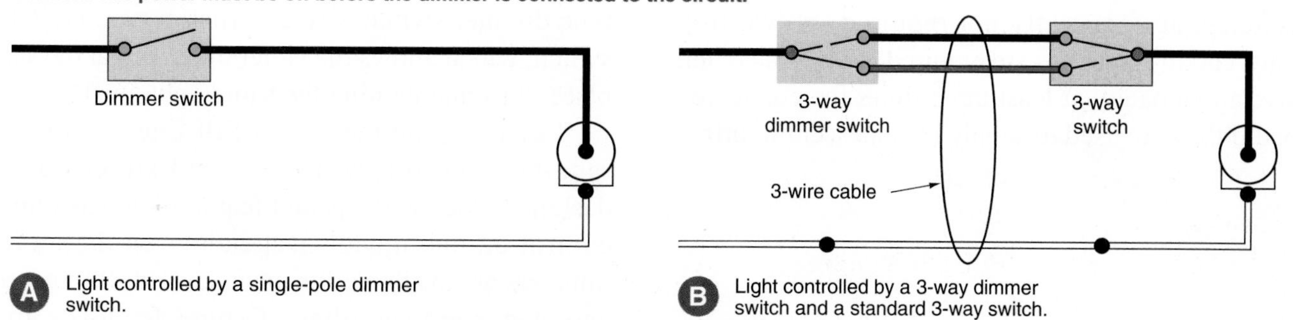

FIGURE 12-9 Single-pole and three-way solid-state dimmer control circuits.

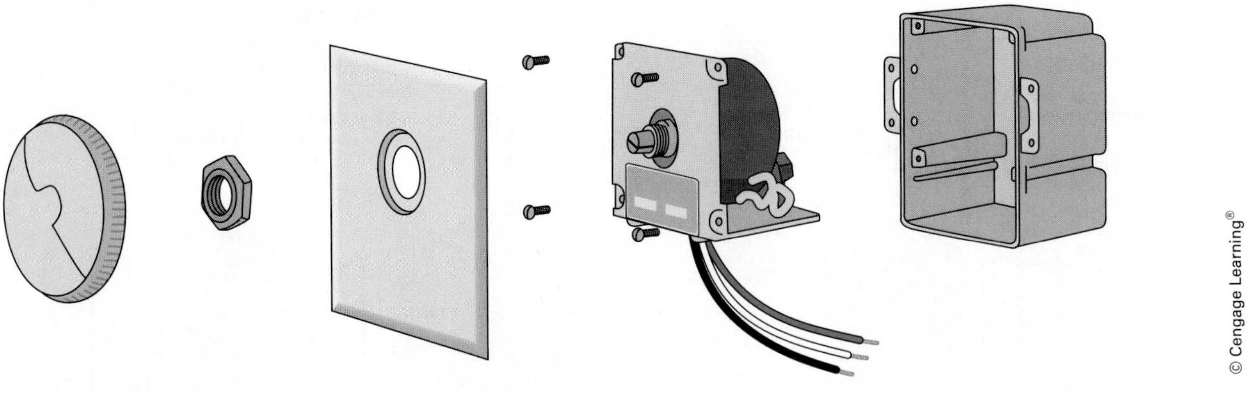

FIGURE 12-10 Dimmer control, autotransformer type.

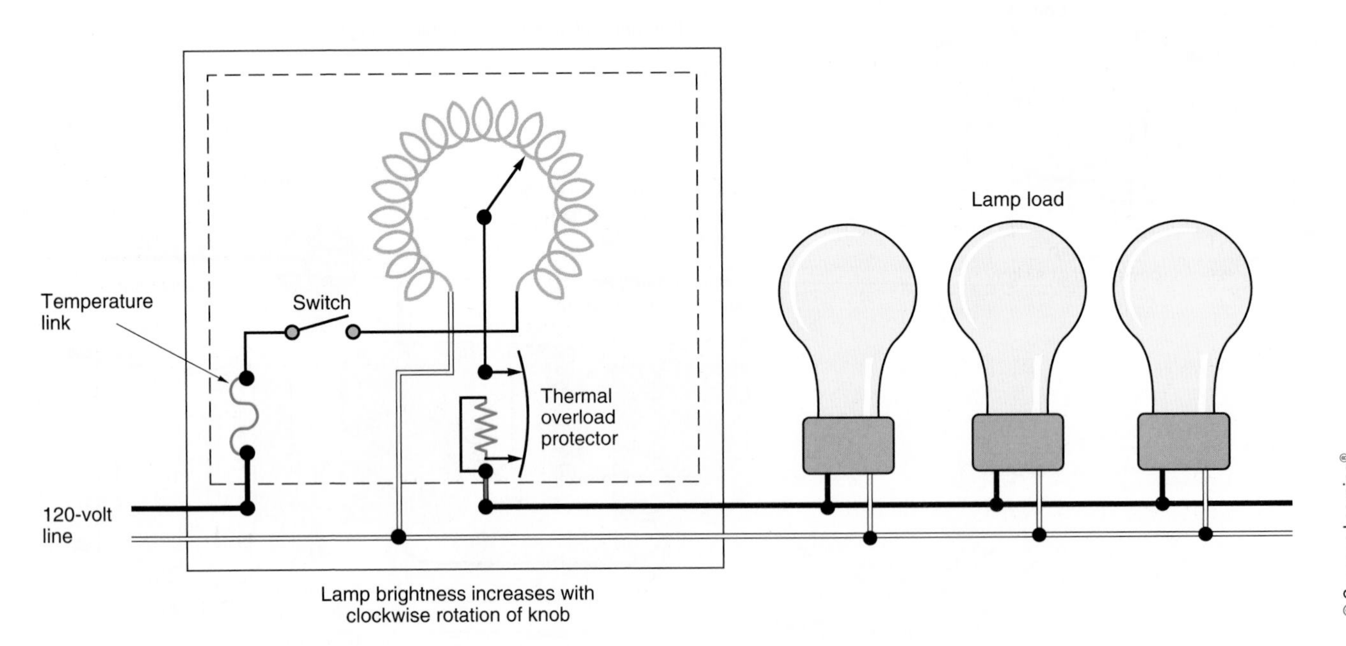

FIGURE 12-11 Dimmer control (autotransformer) wiring diagram for incandescent lamps.

T-rated switches are required for many applications of tungsten filament lamps. Although T-rated switches are generally not required in dwelling units, canopy switches are required to be T-rated and have an ampacity at least three times the connected load if they are used to supply incandescent lighting.

Dimmer Controls for Incandescent Lighting

The wiring diagram in Figure 12-9 shows the circuit used to control incandescent lamps. The entire load is cut off when the internal switch is in the OFF position. These dimmers are available in single-pole and three-way switches. Figure 12-12 shows slider-type dimmer switches with a rocker-type ON/OFF switch, which allows the slider to remain at the same place, thus maintaining the same light level.

Leviton manufactures a full line of dimmers ranging from rotary, toggle, and slider up to a designer series with special features such as remote control on one model. Depending on the model, dimmers are available for incandescent, fluorescent, and magnetic low-voltage fixtures for loads up to 600 or 1000 watts, with many models having matching paddle fan speed controls.

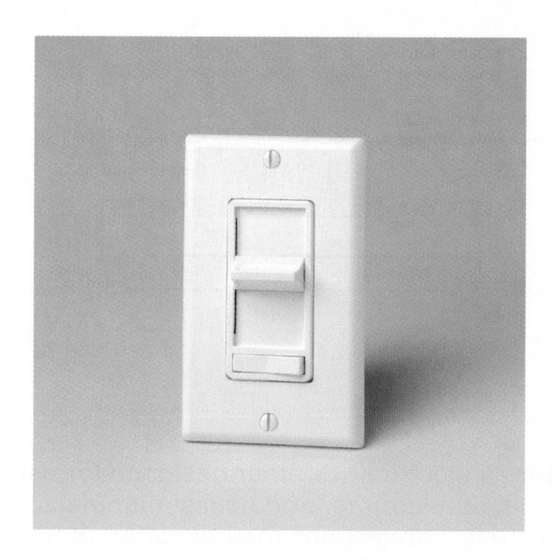

FIGURE 12-12 Solid-state slider dimmer control with rocker ON-OFF switch to maintain the same light level.

DIMMING FLUORESCENT LAMPS

To dim incandescent lamps, the voltage to the lamp filament is varied (raised or lowered) by the dimmer control. The lower the voltage, the less intense the light. The dimmer is simply connected "in series" with the incandescent lamp load.

In fluorescent lamps, the connection is not a simple "series" circuit. To dim fluorescent lamps, special dimming ballasts and dimmer switches are required. The dimmer control allows the dimming ballast to maintain a voltage to the cathodes that will maintain the cathodes' proper operating temperature and allow the dimming ballast to vary the current flowing in the arc. This varies the intensity of light coming from the fluorescent lamp.

However, some compact fluorescent lamps have an integrated circuit built right into the lamp, allowing it to be controlled by a standard incandescent wall dimmer. These dimmable compact fluorescent lamps can replace standard incandescent lamps of the medium screw shell base type. No additional wiring is necessary. They are designed for use with dimmers, photocells, occupancy sensors, and electronic timers that are marked "Incandescent Only."

Figure 12-13 illustrates the connections for an autotransformer-type dimmer for the control of fluorescent lamps using a special dimming ballast. The fluorescent lamps are F40T12 rapid-start lamps. With the newer electronic dimming ballasts, 32-watt, rapid-start T8 lamps are generally used.

Figure 12-14 shows an electronic dimmer for the control of fluorescent lamps using a special dimming ballast.

Fluorescent dimming ballasts and controls are available in many types, such as simple ON/OFF, 100% to 1% variable light output, 100% to 10% variable light output, and two-level 100%/50% light output. Wireless remote control is also an option.

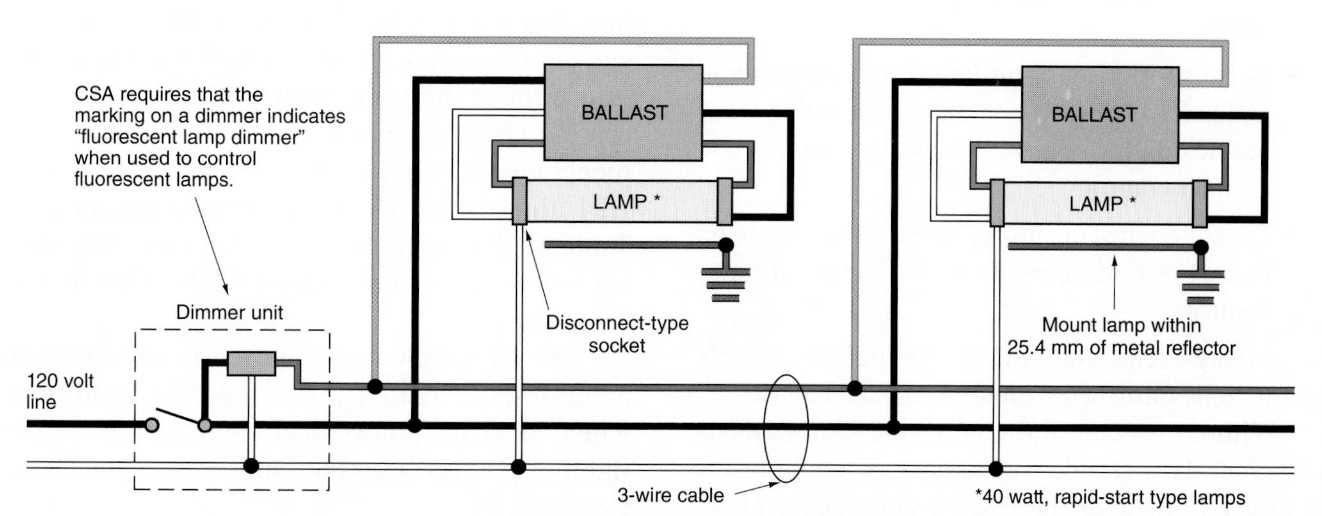

FIGURE 12-13 Dimmer control wiring diagram for fluorescent lamps.

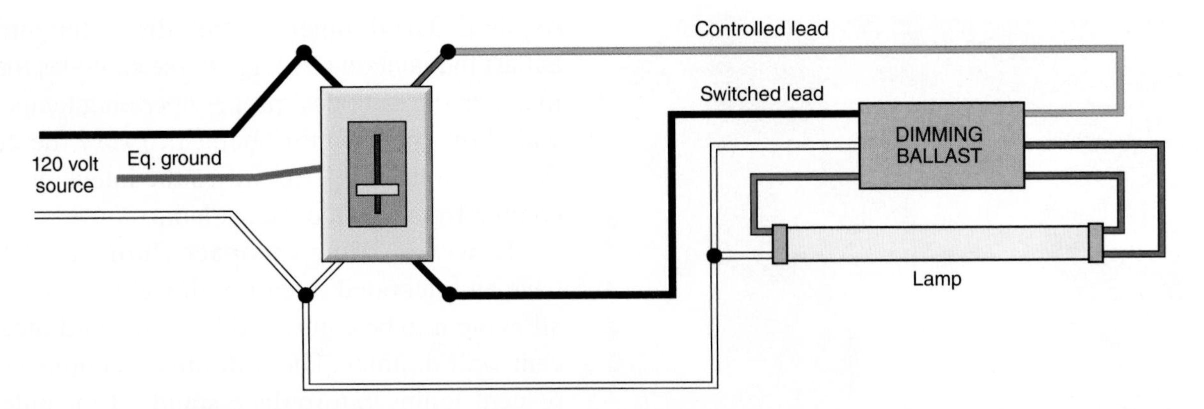

FIGURE 12-14 This wiring diagram shows the connections for an electronic dimmer designed for use with a special dimming ballast for rapid-start lamps. Always check the dimmer and ballast manufacturers' wiring diagrams, as the colour coding of the leads and the electrical connections may be different from this. The dimmer shown is a "slider" type, and offers wide-range (20% to 100%) adjustment of light output for the fluorescent lamp.

To ensure proper dimming performance of fluorescent lamps:

- Make sure that the lamp reflector and the ballast case are solidly grounded.
- Do not mix ballasts or lamps from different manufacturers on the same dimmer.
- Do not mix single- and two-lamp ballasts.
- "Age" the lamps for about 100 hours before dimming to allow the gas cathodes in the lamps to stabilize. Without "aging" (sometimes referred to as *seasoning*), it is possible to have unstable dimming, lamp striation (stripes), and/or unbalanced dimming characteristics between lamps.
- Because incandescent lamps and fluorescent lamps have different characteristics, they cannot be controlled simultaneously with a single dimmer control.
- Do not control electronic ballasts and magnetic ballasts simultaneously with a single dimmer control.

Always read and follow the manufacturer's instructions furnished with dimmers and ballasts.

Without special dimming fluorescent ballasts and controls, a simple way to have two light levels is to use four-lamp fluorescent luminaires. Then, control the ballast that supplies the centre two lamps with one wall switch, and control the ballast that supplies the outer two lamps with a second wall switch. This is sometimes referred to as an *inboard/outboard connection*.

BRANCH CIRCUIT FOR FRONT ENTRY, PORCH

Figure 12-15 shows a possible device and wiring layout for the front entry and porch area. This area is connected to Circuit A15. The home run enters the six-gang switch box next to the front door. From this box, the circuit spreads out, feeding the recessed closet light, the porch bracket fixture, the two bracket fixtures on the front of the garage, one receptacle outlet in the entry, and the post light in the front lawn. The outdoor weatherproof Ground-Fault Circuit Interrupter (GFCI) receptacle on the porch is connected to Circuit A24 along with the other outdoor receptacles. Table 12-2 summarizes the outlets and estimated load for the entry and outdoor lights at the front of the house.

The receptacle on the porch is controlled by a single-pole switch just inside the front door, which gives the homeowner control of decorative lighting plugged into the outdoor receptacle; see Figure 12-16.

Typical ceiling fixtures commonly installed in front entryways, where it is desirable to make

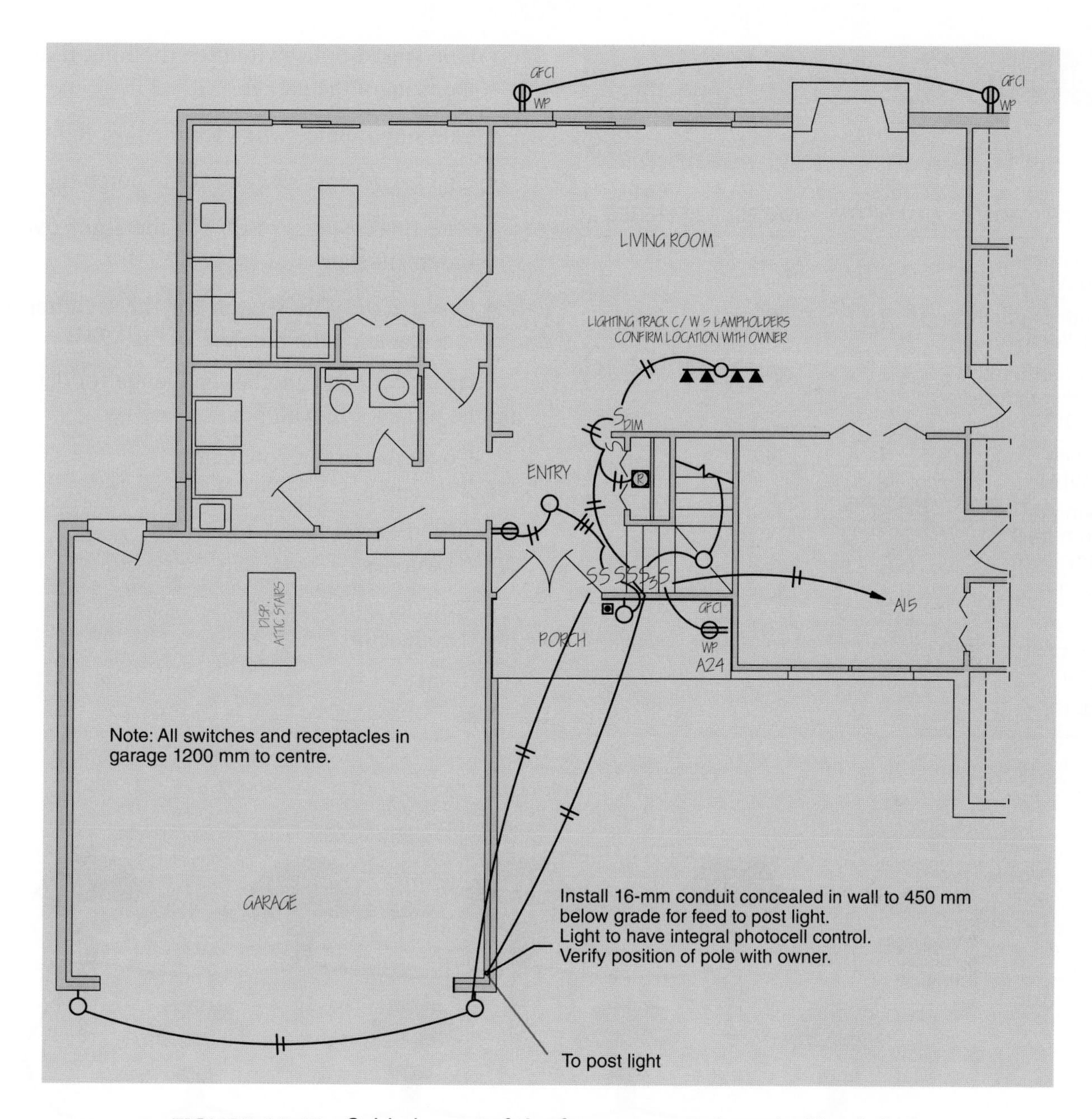

FIGURE 12-15 Cable layout of the front entry and porch, Circuit A15.

a good first impression on guests, are shown in Figure 12-17. Typical wall-mounted porch and entrance lighting fixtures are shown in Figure 12-18.

According to *Rule 30–318*, any wiring of light fixtures in damp or wet locations requires fixtures to be approved and marked for use in such locations. For example, the fixtures will be marked "Suitable for wet locations."

The *CEC* defines partially protected areas under roofs, canopies, or open porches as damp locations. Fixtures to be used in these locations must be marked "Suitable for damp locations." Many types of fixtures are available. Always check the label on the fixture for CSA approval or consult with your electrical supplier to determine the suitability of the fixture for a wet or damp location. See Figure 12-19.

Probably the most difficult part of the A15 circuit is planning the connections at the switch location just inside the front door, where there are six toggle switches:

- One single-pole switch for the front entrance ceiling fixture

TABLE 12-2

Front entry and porch: Outlet count and estimated load, Circuit A15.

DESCRIPTION	QUANTITY	WATTS	VOLT-AMPERES
Receptacles @ 120 W	1	120	120
Outdoor porch bracket fixture	1	100	100
Outdoor garage bracket fixtures @ 100 W each	2	200	200
Ceiling fixture	1	100	100
Living room track light	1	200	200
Post light	1	100	100
Closet recessed fixture	1	60	60
Totals	8	880	880

- One single-pole switch for the light fixtures at the front of the garage
- One single-pole switch for the post light
- One single-pole switch for the porch light
- One three-way switch for the light over the basement stairs
- One single-pole switch for the weatherproof receptacle on the porch (Circuit A24)

This location is another challenge for determining the proper size wall box:

1. Count the circuit conductors
   $2 + 2 + 2 + 2 + 2 + 3 + 3 = 16$
2. Add the switches
   6 switches $= 12$
3. Add the wire connectors
   2 connectors $= 1$

   Total 29

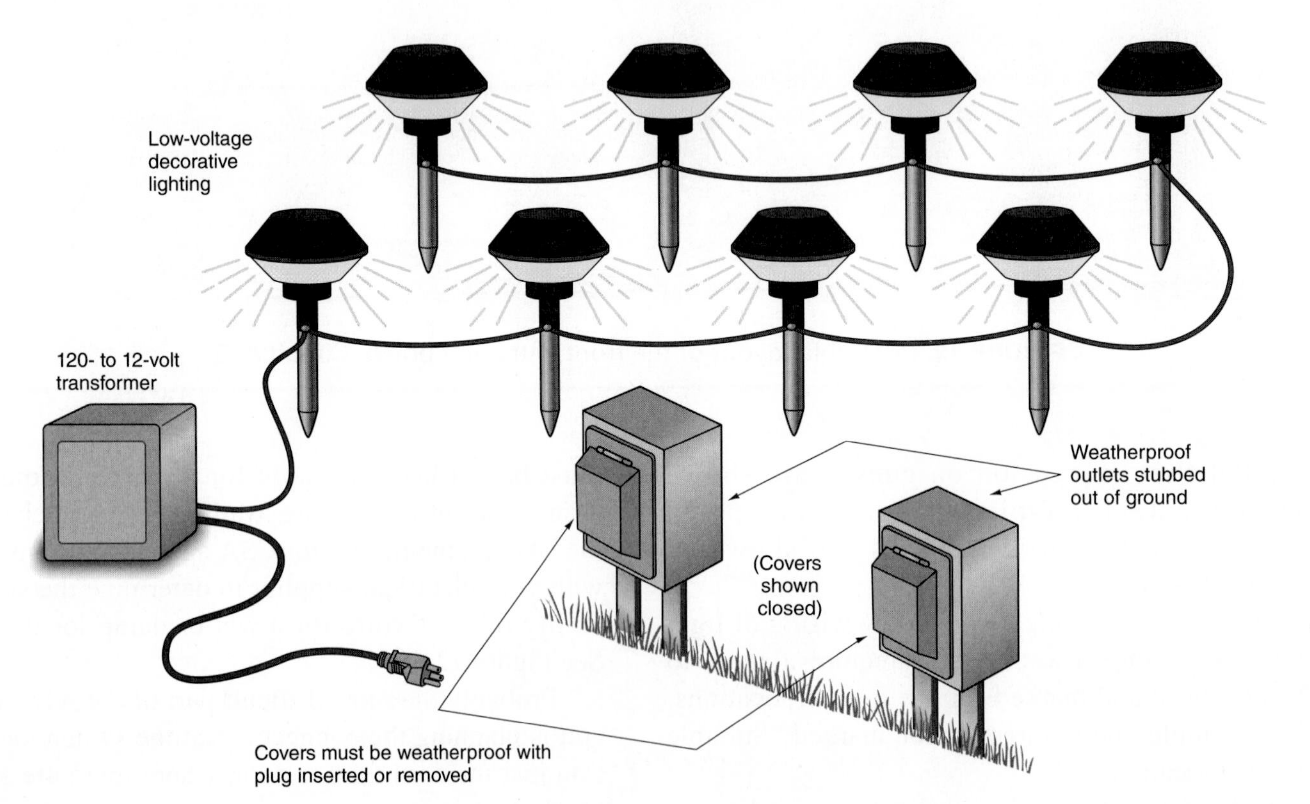

FIGURE 12-16 Typical low-voltage decorative lighting for use around shrubbery. Cord plugs into 120-volt receptacle and feeds transformer that reduces 120 volts to a safer 12 volts.

FIGURE 12-17 Hall luminaires: Ceiling mount and chain mount.

FIGURE 12-18 Outdoor porch and entrance luminaires: Wall bracket styles.

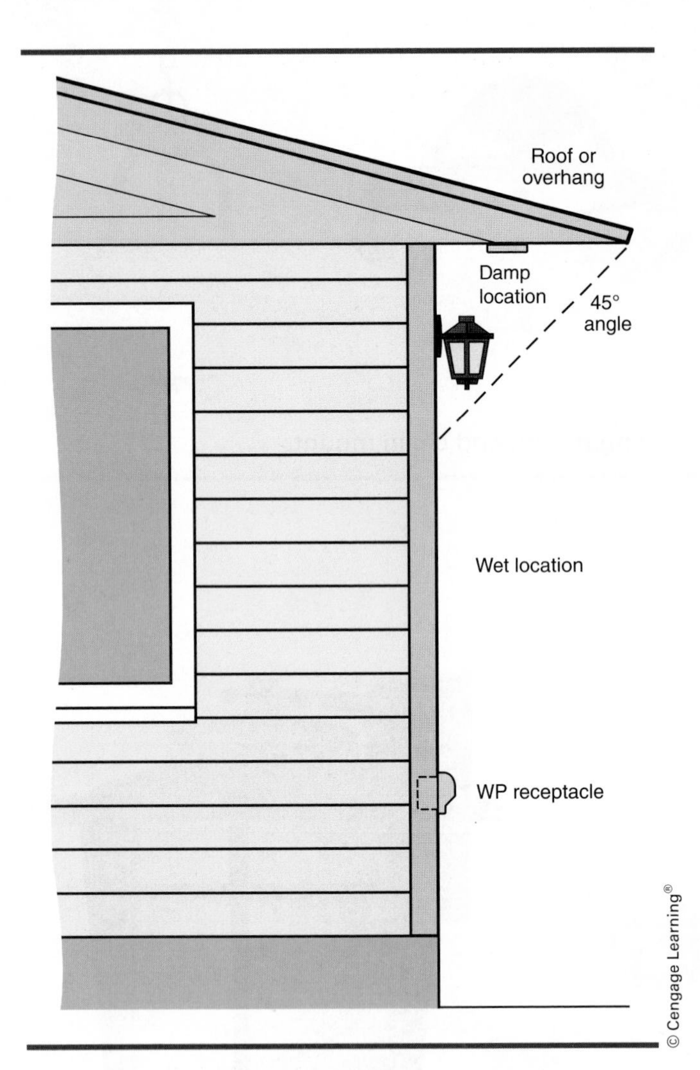

FIGURE 12-19 Outdoor areas considered to be *damp* or wet locations.

The Quick-Check Box Selection Guide (Figure 5-16) and Tables 5-1 and 5-2 give many possibilities. For example, two sets of three 76 × 51 × 64-mm (3 × 2 × 2½-in.) device boxes can be ganged together (6 × 8 = 48 conductors).

An interesting possibility presents itself for the front entry closet. Although the plans show that the recessed closet fixture is turned on and off by a standard single-pole switch to the left as you face the closet, a door jamb switch could have been installed. A door jamb switch is usually mounted about 1.83 m above the floor, on the inside of the 2 × 4 framing for the closet door. The electrician runs the cable to this point and lets it hang out until the carpenter can cut the proper size of opening into the door jamb for the box. After the finish woodwork is completed, the electrician installs the switch; see Figure 12-20. The plunger on the switch is pushed inward when the edge of the door pushes on it as it closes, shutting off the light. The plunger can be adjusted in or out to make the switch work properly. These door jamb switches come complete with their own special wall box.

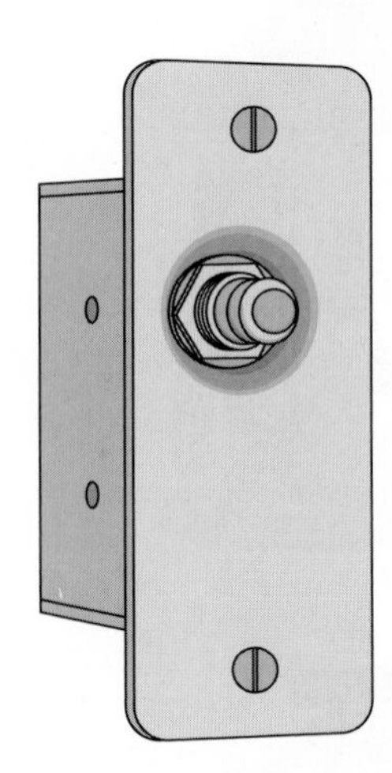

FIGURE 12-20 Door jamb switch.

REVIEW

Note: Refer to the *CEC* or the blueprints provided with this textbook when necessary. Where applicable, responses should be written in complete sentences.

1. How many convenience receptacles are connected to the living room circuit? _______

____

2. a. How many wires enter the switch box at the four-way switch location? ________

   b. What type and size of box may be installed at this location? ________

3. a. What is meant by incandescent lamp inrush current? ________

   b. Does the *CEC* require T-rated switches for incandescent lamps in dwelling units?

   ____

4. Complete the following wiring diagram for the dimmer and lamp.

120-volt supply

5. What is the difference between a split-switched receptacle and a split-wired (three-wire) receptacle? ________

6. How many branch circuits are required for

   a. a split-switched receptacle? ________

   b. a split-wired (three-wire) receptacle? ________

7. Explain why fluorescent lamps having the same wattage can draw different current values. ________

8. What is the total current consumption of the track lighting? ________

9. a. How many television outlets are provided in the living room? ________

   b. What symbol is used on the drawing to indicate the TV outlet? ________

10. a. Where is the telephone outlet located in the living room? ________

    b. What symbol is used on the drawing to indicate the telephone outlet? ________

11. A layout of the outlets, switches, dimmers, track lighting, and recessed fixtures is shown in the following diagram. Using the cable layout shown in Figure 12-1, make a complete wiring diagram of this circuit. Use coloured pencils or marking pens to indicate conductors.

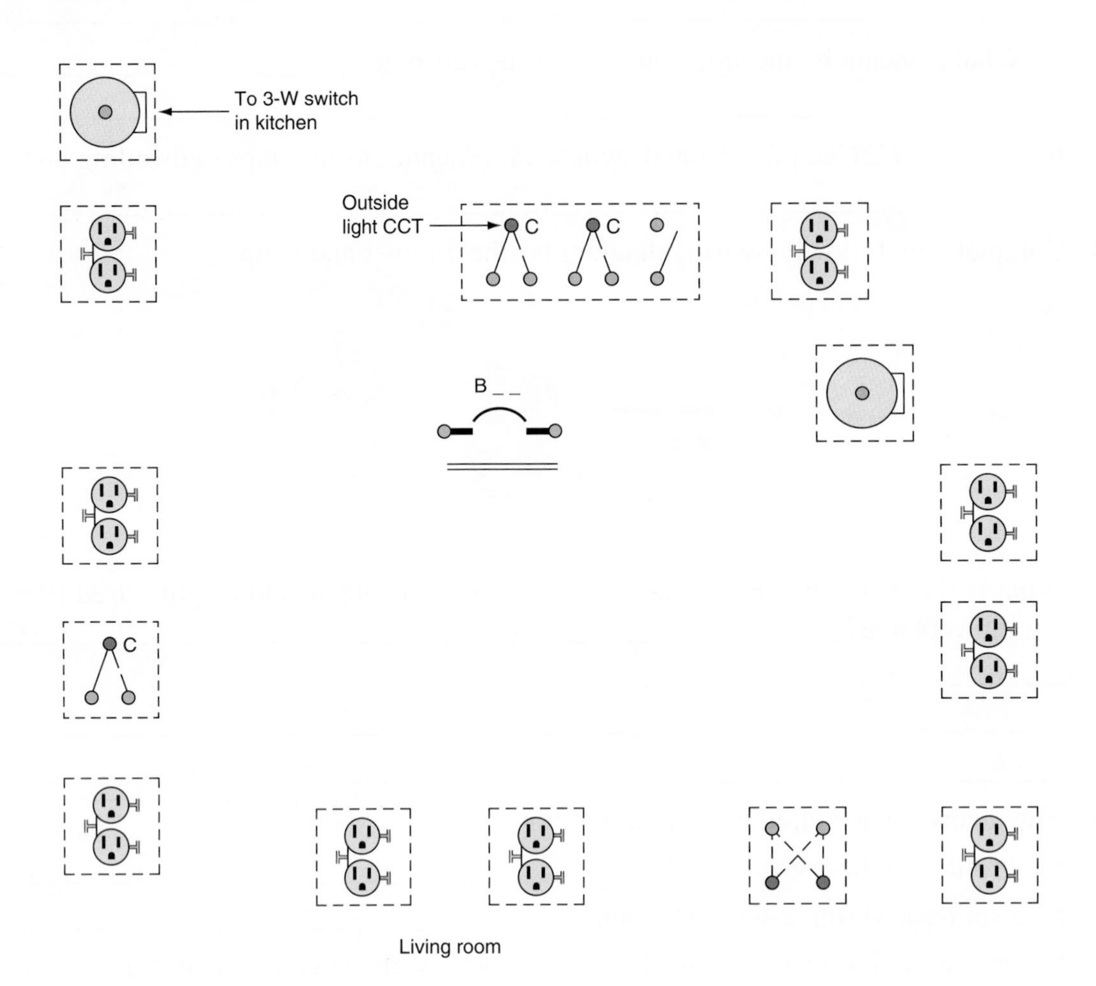

12. Must track lighting always be fed (connected) at one end of the track? ______________

____

13. If there is a 7-metre piece of track lighting, how many outlets is this considered to have?

____

____

14. Using the electrical plans provided, lay out devices and wiring in the living room area to *minimum CEC* standards.

15. a. How many circuit wires enter the front entry ceiling box? ______________

b. How many equipment bonding conductors enter the front entry ceiling box? _____

16. How many receptacle outlets and lighting outlets are supplied by Circuit A15?

____

17. Outdoor fixtures directly exposed to the weather must be marked as [circle the correct answer]

a. suitable for damp locations
b. suitable for dry locations only
c. suitable for wet locations

18. Make a materials list of all types of switches and receptacles connected to Circuit A15.

____

____

____

19. The following layout is for lighting Circuit A15, the front entry, porch, and front garage lights. Using the cable layout shown in Figure 12-15, make a complete wiring diagram of this circuit. Use coloured pencils or pens to indicate the colour of the conductors' insulation.

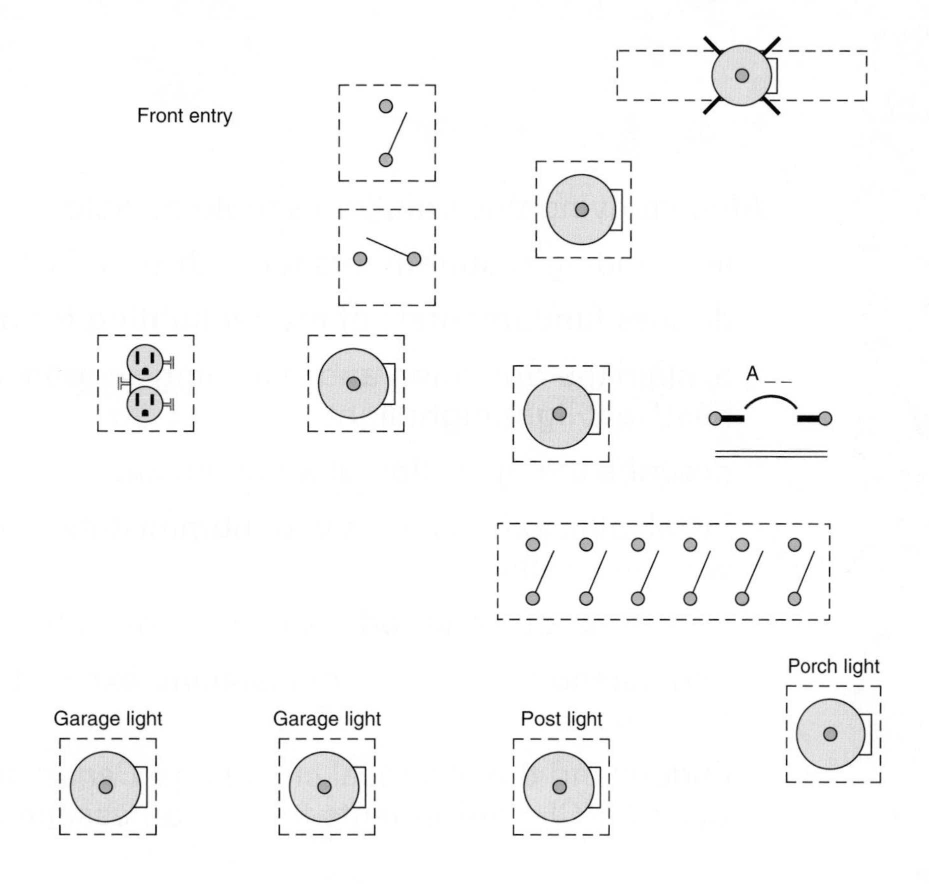

20. Using the electrical plans provided with this textbook, lay out the devices and wiring, including circuit numbers, for the front porch and entry to *minimum CEC* standards.

UNIT 13

Branch Circuits for Bathrooms

OBJECTIVES

After studying this unit, you should be able to

- list bonding requirements for bathroom installations
- discuss fundamentals of proper lighting for bathrooms
- explain the operation and switching sequence of the heat/vent/light/night light
- describe the operation of a humidistat
- install attic exhaust fans with humidistats, both with and without a relay
- list the various methods of controlling exhaust fans
- understand advantages of installing exhaust fans in residences
- understand the electrical circuit and *Canadian Electrical Code (CEC)* requirements for hydromassage bathtubs

Electrical equipment in bathrooms and washrooms requires special installation considerations. For example, in the residence used as an example in this textbook, the hydromassage bathtub in the ensuite bathroom requires ground-fault circuit protection. Switches must be properly located. There must be enough circuits to operate all of the equipment without the possibility of overloading circuits. It is also important to correctly locate luminaires to achieve proper lighting.

The vanity lighting for the two bathrooms is provided from the circuit that supplies the GFCI-protected receptacle in each bathroom (A20 and A22). Because the receptacles in the main bathroom and ensuite bath are likely to be used in the morning at the same time by people using hair dryers that draw as much as a toaster or kettle (1200 W), we have put the receptacles on their own circuit. This is not a *CEC* requirement.

Each bathroom has a combination ceiling heater/light/night light/fan unit that is connected to separate circuits shown as (▲)$_J$ and (▲)$_K$. A hydromassage tub, located in the bathroom serving the master bedroom, is connected to a separate circuit (▲)$_A$.

Figure 13-1 and the electrical plans for this area of the home show that each bathroom has a fixture

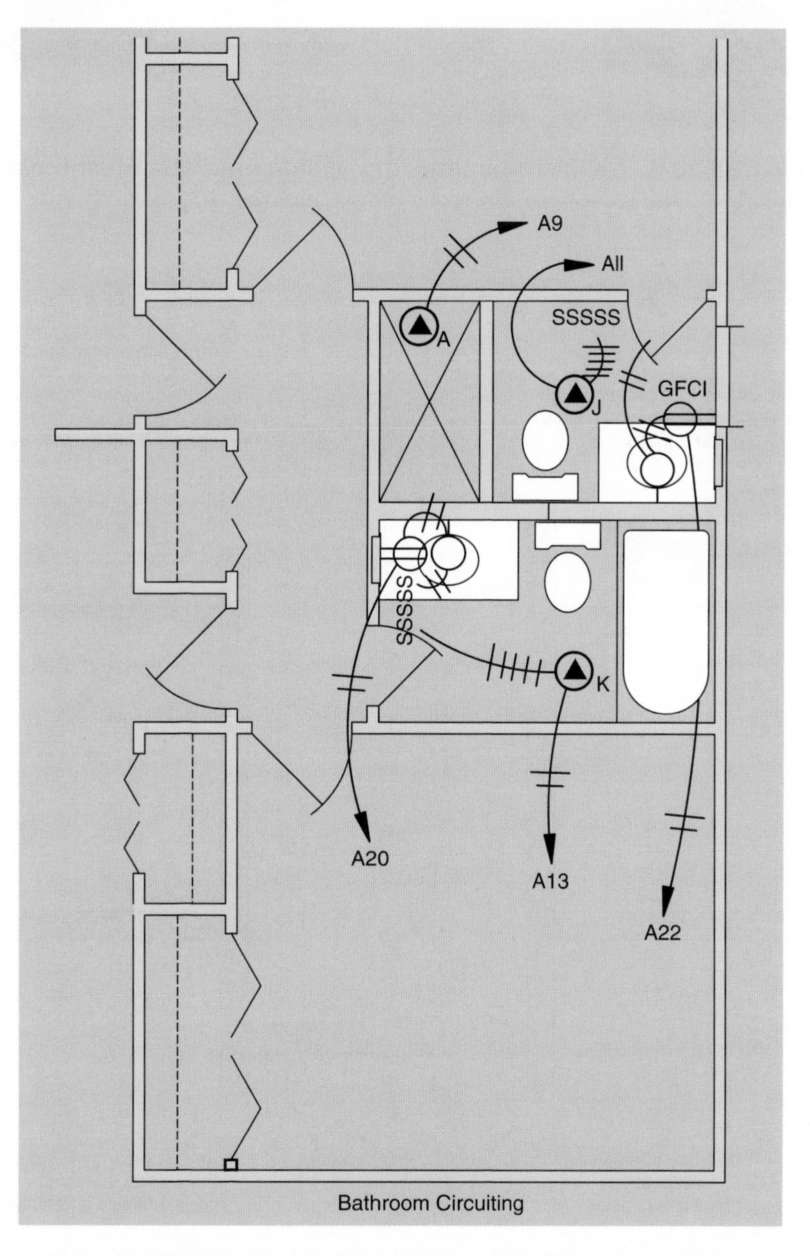

FIGURE 13-1 Cable layout for bathrooms.

above the vanity mirror: Some typical fixtures are shown in Figure 13-2. The homeowner might decide to purchase a medicine cabinet complete with a self-contained lighting fixture. Figure 13-3 illustrates how to rough-in the wiring for medicine cabinets with or without such a lighting fixture. These fixtures are controlled by single-pole switches at the doors.

All photos: Courtesy of Progress Lighting

FIGURE 13-2 Vanity luminaires of the side bracket and strip types.

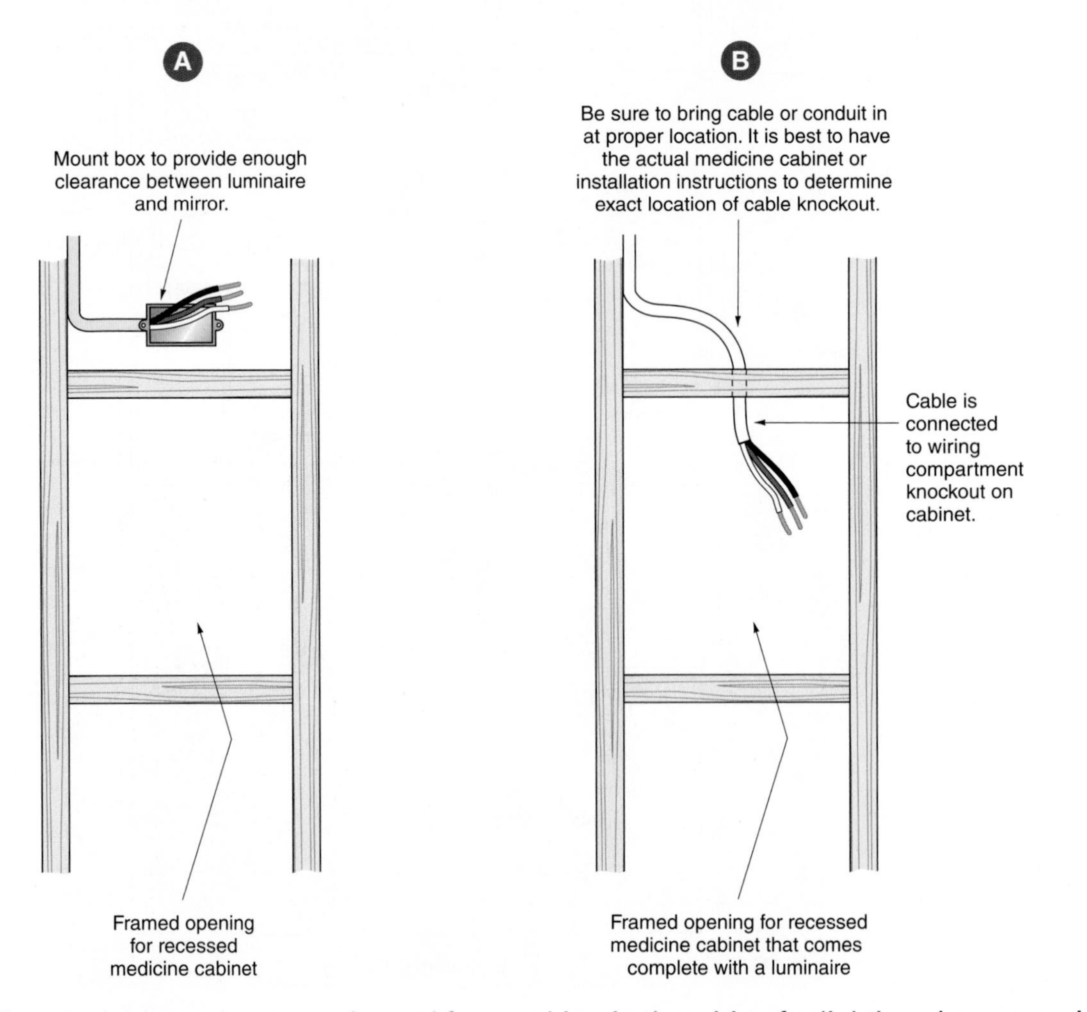

FIGURE 13-3 Two methods most commonly used for roughing-in the wiring for lighting above a vanity.

A bathroom or powder room should have proper lighting for shaving, combing hair, putting on makeup, and so on. If the face is poorly lit and in shadow, that is what will be reflected in the mirror.

A lighting fixture directly overhead will light the top of one's head, but will cause shadows on the face. Mirror lighting and/or adequate lighting above and forward of the standing position at the vanity can provide excellent lighting in the bathroom. Figures 13-4 to 13-6 show pictorial and section views of typical soffit lighting above a bathroom vanity.

The receptacles in each bathroom are located adjacent to the vanity and within 1 m of the wash basin, *Rule 26–710(f)*. Bathroom receptacles must be at least 1 m from the bathtub or shower when practicable, *Rule 26–710(g)*. Switches controlling lights must not be within 1 m of the shower or bathtub, *Rule 30–320(3)*. As required by *Rule 26–700(11)* and as discussed in Unit 9, these must be Ground-Fault Circuit Interrupter (GFCI) receptacles. *Section 0* defines a bathroom as "a room containing a wash basin together with bathing or showering facilities"; see Figure 13-7.

GENERAL COMMENTS ON LAMPS AND COLOUR

Incandescent lamps (light bulbs) provide pleasant colour tones, bringing out warm flesh tones similar to the way that natural light does.

With fluorescent lighting, you should use fluorescent lamps that bring out the warm flesh tones, such as warm white (WW) and warm white deluxe (WWX), which simulate incandescent lamps, or cool white deluxe (CWX), which simulate outdoor daylight on a cloudy day.

FIGURE 13-4 Positioning bathroom lighting fixtures.

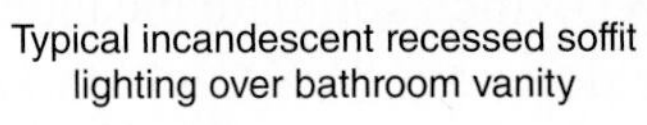

Typical incandescent recessed soffit lighting over bathroom vanity

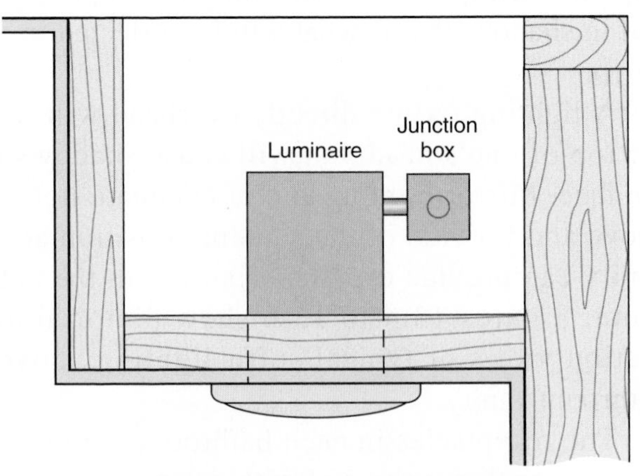

End cutaway of soffit showing recessed incandescent luminaires in typical soffit above bathroom vanity. Two or three luminaires generally installed to provide proper lighting.

FIGURE 13-5 Incandescent soffit lighting.

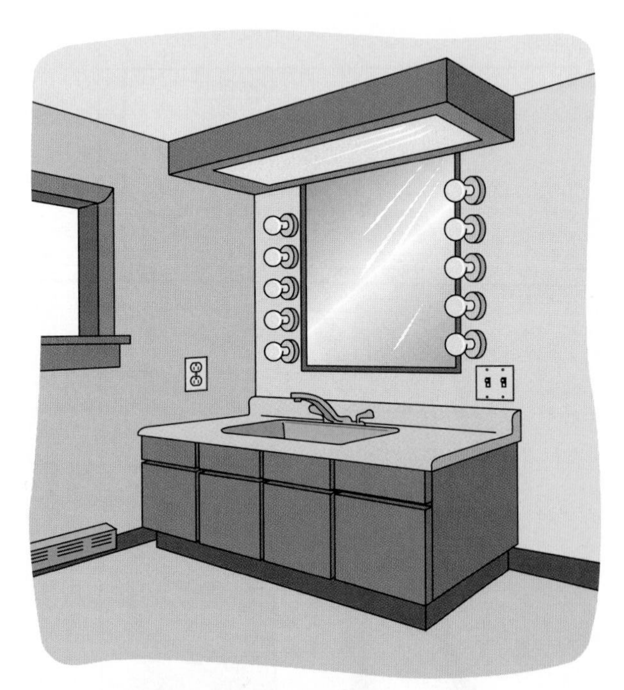

Typical fluorescent recessed soffit lighting over bathroom vanity. Note additional incandescent "side-of-mirror" lighting.

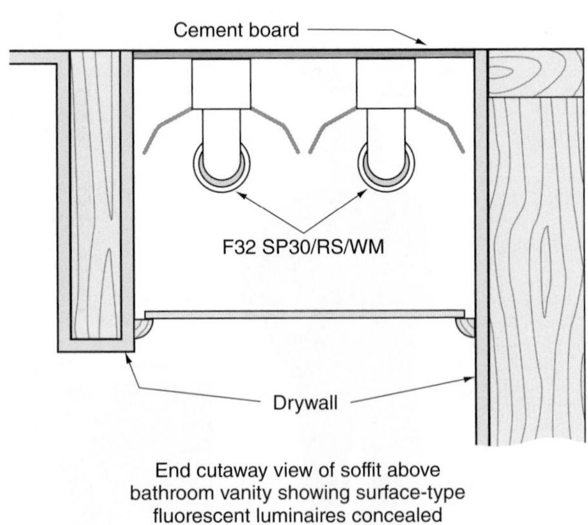

End cutaway view of soffit above bathroom vanity showing surface-type fluorescent luminaires concealed above translucent acrylic lens

FIGURE 13-6 Combination fluorescent and incandescent bathroom lighting.

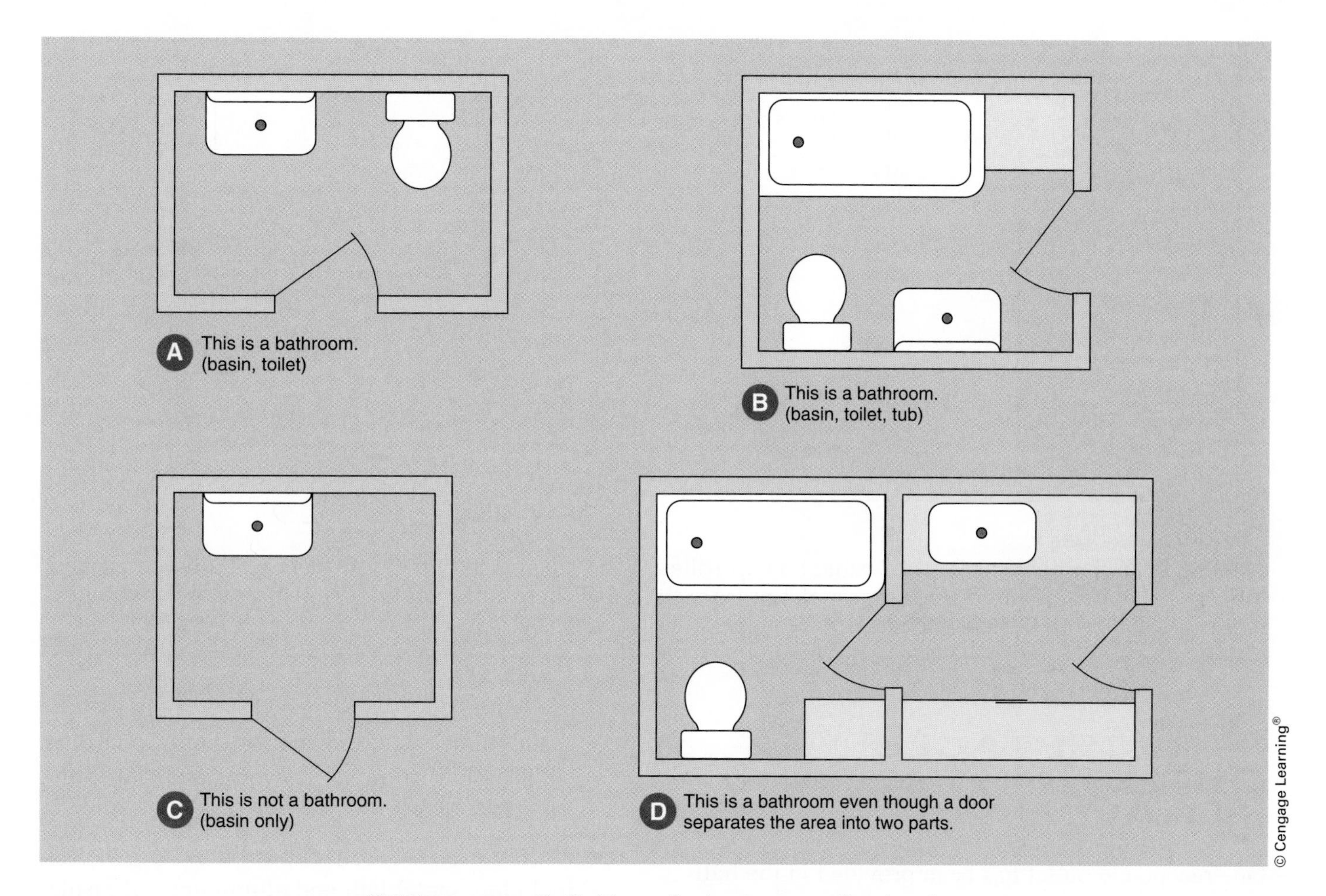

FIGURE 13-7 Definition of a bathroom, *Section 0.*

White (W) or cool white (CW) fluorescents downplay flesh tones and give the skin a pale appearance. Compact fluorescent lamps provide excellent colour rendition.

LIGHTING FIXTURES IN BATHROOMS

Rule 30–320(1) requires any light fixture (luminaire) or lampholder that is located within 2.5 m vertically or 1.5 m horizontally of bathtubs, laundry tubs, or other grounded surfaces to be controlled by a wall switch. The switch must be located at least 1 m from a tub or shower so that a person standing in the tub or shower will not be able to reach the switch. In installations where this distance is impracticable, the *CEC* does allow us to reduce this distance to 500 mm. However, the circuit must be ground fault-protected. See Figures 13-8 and 13-9.

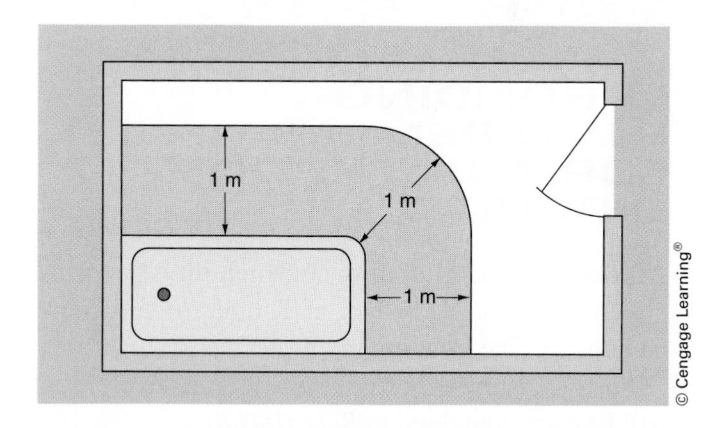

FIGURE 13-8 The horizontal area around a tub in which luminaires must be controlled by a wall switch, *Rule 30–320.*

HALLWAY LIGHTING

The hallway lighting is provided by one ceiling fixture controlled with two three-way switches located at either end of the hall and supplied from Circuit A16.

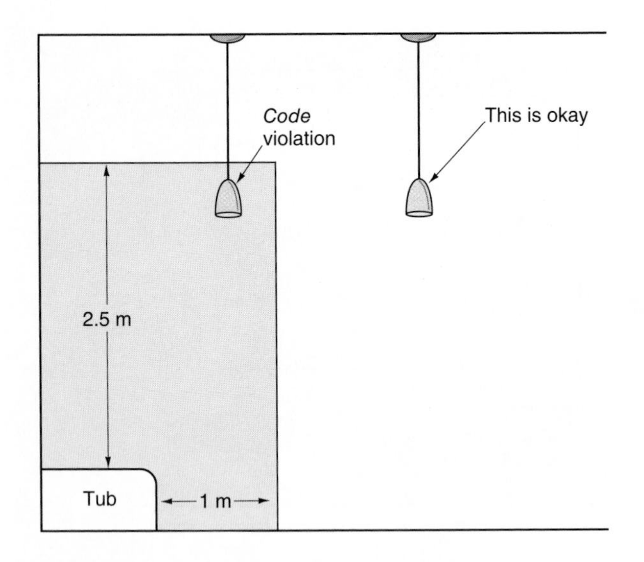

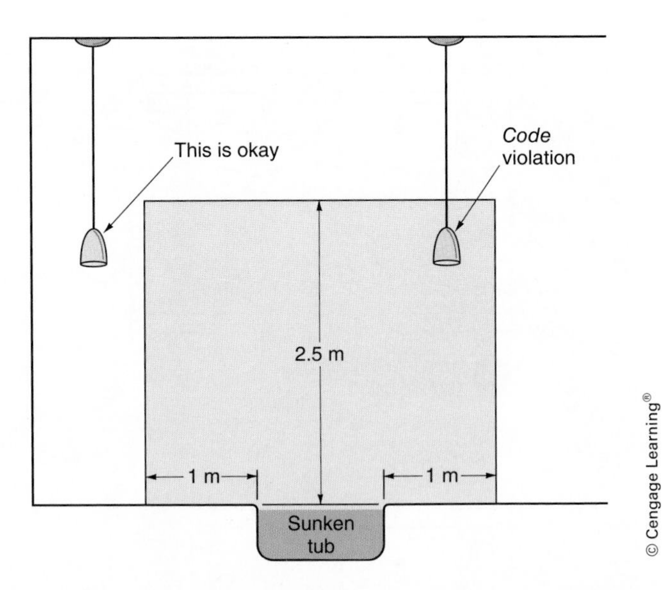

FIGURE 13-9 Luminaires B and C must be controlled by a wall switch outside the reach of a person in the tub. A wall switch must be a minimum of 1 m from the tub. Where this is not possible, the spacing may be reduced to 500 mm if the switch is protected by a GFCI, *Rule 30–320*.

RECEPTACLE OUTLETS IN HALLWAYS

One receptacle outlet has been provided in the hallway, supplied from Circuit A14. *Rule 26–712(f)* requires that in a dwelling unit, no point in a hallway shall be more than 4.5 m from a duplex receptacle.

BONDING REQUIREMENTS FOR A BATHROOM CIRCUIT

All exposed metal equipment, including fixtures, electric heaters, faceplates, and similar items, must be bonded. The bonding requirements for bathroom circuits and any other equipment in a residential building are contained in *Rule 10–400*.

In general, all exposed non-current-carrying metal parts of electrical equipment must be bonded together and connected to ground

- if wiring supplying the equipment contains a bonding conductor, *Rule 10–400(b)*
- if the ground or a grounded surface and the electrical equipment can be touched at the same time, *Rule 10–400(d, e)*
- if they are located in wet or damp locations, such as bathrooms, showers, and outdoors, *Rule 10–400(c)*
- if they are in electrical contact with metal, including metal lath and aluminum foil insulation, *Rule 10–400(g)*

Equipment is considered bonded to ground when it is properly and permanently connected to a metal raceway, a separate bonding conductor, or the bonding conductor in non-metallic-sheathed cable or armoured cable. The bonding conductor must itself be properly grounded.

According to *Rule 10–408*, residential cord-connected electrical appliances must be grounded. There are two exceptions to this basic requirement. *Rule 10–408(3)* accepts "double insulating" or equivalent in lieu of grounding. *Rule 10–408(4)* provides an exemption from the bonding requirements of *Subrule (1)* provided the equipment is used only in a location where reliable grounding cannot be obtained and the equipment is supplied from a double-insulated portable Class A GFCI.

Double-insulated appliances must be clearly marked ⧈, or with the words "Double insulated," and do not require a separate bonding conductor. These appliances are furnished with a two-wire cord that is designed for a polarized receptacle.

The short blade must be connected to the "hot" conductor and the long blade to the neutral or identified conductor, *Rule 10–408(3)* and *Appendix B*.

Residential electrical appliances that must be grounded, unless double insulated, are

- facial saunas
- freezers and refrigerators
- clothes washers and dryers
- dishwashers
- aquarium equipment
- hand-held motor-operated tools
- electric motor-operated hedge trimmers, lawn mowers, and snow blowers
- air conditioners
- sump pumps

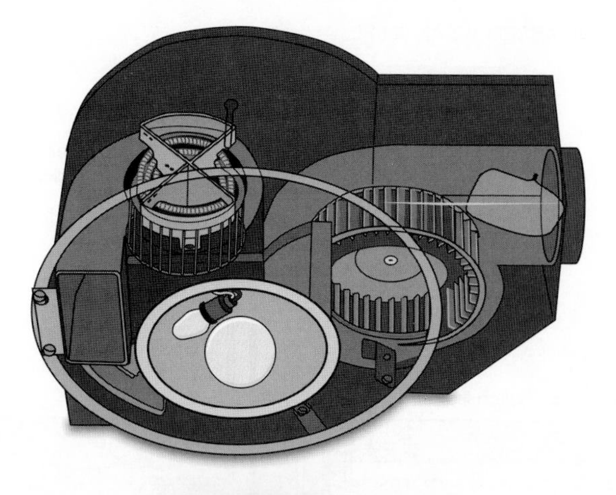

Courtesy of Broan-NuTone Canada Inc.

FIGURE 13-10 Combination heater, ventilation fan, light, and night light.

BATHROOM CEILING HEATER CIRCUITS Ⓐ$_K$ Ⓐ$_J$

In the ceilings of both bathrooms is a combination heater, light, night light, and exhaust fan, similar to the one shown in Figure 13-10. The symbols Ⓐ$_K$ and Ⓐ$_J$ represent the outlets for these units.

The bathroom ventilation units are heat/vent/light/night light combination units. Each unit consists of a 1500-watt heater, separate heater and ventilation motors rated at 0.9 ampere each, a 7-watt night light, and either a 100-watt incandescent or 26-watt compact fluorescent main lamp. The unit is IC-rated for installation in ceilings insulated to R40 and has a sound rating of 3.5 sones. The unit is not designed for mounting over a tub or shower stall. The unit has been tested and certified by CSA Standard C22.2 for electric air heaters.

According to the manufacturer's specifications and installation instructions, the minimum circuit requirements for the unit are 120 volts, 18.6 amperes, 60 Hz. (The actual full-load current based on a 1500-watt heater, two 0.9 FLA motors, a 60-watt main light, and a 7-watt night light would be 14.9 amperes.) It must be wired on a separate 20-ampere circuit. Install #12 AWG copper NMD 90 cable from the panel to a two-gang switch box and from the switch box to the bathroom heat/light/ventilation combination unit shown in Figures 13-10 and 13-11. The switches must be located so they cannot be reached from the tub or shower. A single gang box is used for the vanity light.

Operation of the Heat/Vent/Light

The heat/vent/light is controlled by four switches:

1. one switch for the heater
2. one switch for the light
3. one switch for the exhaust fan
4. one switch for the night light

When the heater switch is turned on, the heater blower motor starts and the heating element begins to give off heat. The heat activates a bimetallic coil attached to a damper section in the housing of the unit. The heat-sensitive coil expands until the damper closes the discharge opening of the exhaust

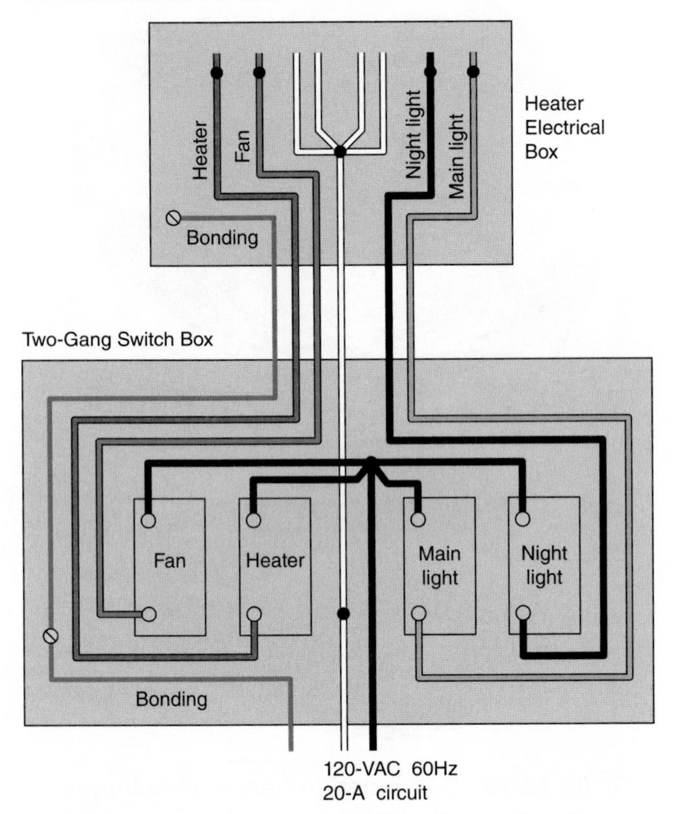

FIGURE 13-11 Wiring for bathroom combination units. The heater identified here is composed of a fan/heating element, internally wired. The fan identified here is a ventilation fan motor.

fan. Air is taken in through the outer grille of the unit. The air is blown downward over the heating element. The blower wheel circulates the heated air back into the room.

When the heater is turned off and the exhaust fan switch is on, only the exhaust fan is on. The air is pulled into the unit and is exhausted to the outside of the house by the blower wheel.

Bonding the Combination Heater

To bond the unit, connect the bare bonding conductor of the non-metallic-sheathed cable (NMSC) to the bonding terminal or lug on the unit.

When connecting exhaust ducting to the unit, it is critical to vent the unit to the outdoors. If a unit is vented to the attic, the moisture in the warm exhaust air will condense in the attic during the winter, which can cause a build-up of ice on the inside of the roof, resulting in significant damage to the structure of the building during freeze–melt cycles. It also reduces the effectiveness of the ceiling insulation.

HYDROMASSAGE TUB CIRCUIT Ⓐ A

The master bedroom is equipped with a hydromassage bathtub, sometimes referred to as a *whirlpool bath*. Pools, tubs, and spas are covered by *CEC Section 68*.

Rule 68–050 defines a hydromassage bathtub as a permanently installed bathtub having an integral or remote water pump or air blower, and having a fill and drain water system.

In other words, fill—use—drain!

The significant difference between a hydromassage bathtub and a regular bathtub is the recirculating piping system and the electric pump that circulates the water. Both types of tubs are drained completely after each use.

Spas and hot tubs are intended to be filled and then used. They are *not* drained after each use because they have a filtering and heating system.

Electrical Connections

The hydromassage tub is fed with Circuit A9, a separate 15-ampere, 120-volt circuit using #14 AWG NMD 90.

Table A–1, Schedule of Special-Purpose Outlets, in this book's Appendix A, indicates that the hydromassage bathtub in this residence has a ½-hp motor that draws 10 amperes.

Rule 68–302 requires that the circuit supplying a hydromassage bathtub has Class A GFCI protection, as discussed in Unit 9.

Rule 68–304(1) requires that the hydromassage bathtub be controlled by an on/off device. An automatic shut-off timer is recommended; see Figure 13-12.

Rule 68–304(2) requires that any associated electrical controls be located behind a barrier not less than 1 m horizontally from the wall of the hydromassage bathtub. Pneumatic (air) controls are often used instead of electrical controls. These controls can be mounted on the tub. The 1-m rule does not apply.

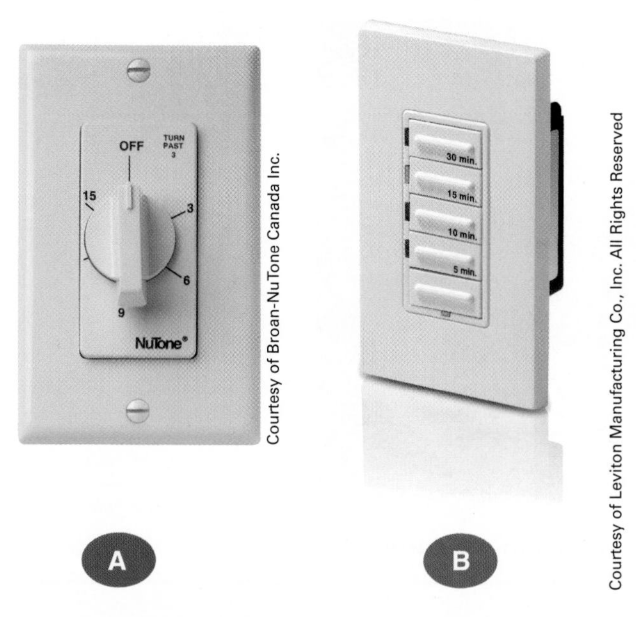

FIGURE 13-12 (A) A 15-minute mechanical timer and (B) a 30-minute electronic timer for control of a hydromassage bathtub.

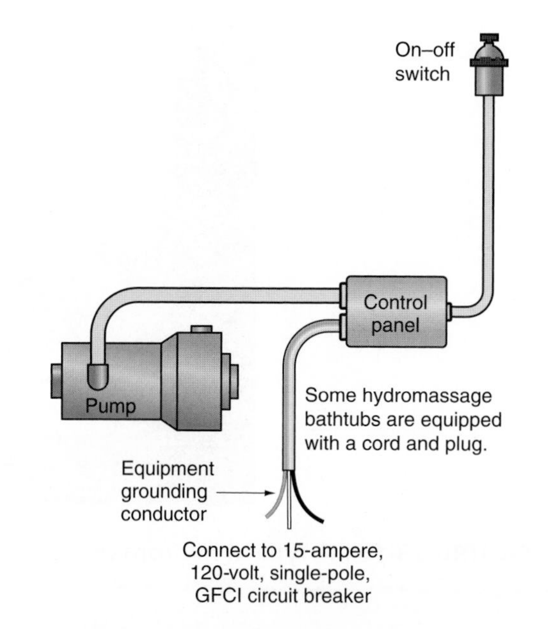

FIGURE 13-13 Typical wiring diagram of a hydromassage tub showing the motor, power panel, and electrical supply leads, *Rule 68–300.*

Spas, hot tubs, and swimming pools introduce additional electrical shock hazards; these issues are covered in Unit 22.

The hydromassage bathtub's electrical control is prewired by the manufacturer. The electrician just has to run the separate 15-ampere, 120-volt GFCI-protected circuit to the end of the tub where the pump and control are located. Generally, the manufacturer will supply a 900-mm length of watertight flexible conduit that contains one black, one white, and one green equipment bonding conductor; see Figure 13-13. Make sure that the equipment-bonding conductor of the circuit is properly connected to the green bonding conductor of the hydromassage tub.

Do not connect the green bonding conductor to the grounded (white) circuit conductor.

Make proper electrical connections in the junction box where the branch-circuit wiring is to be connected to the hydromassage wires. Because it contains splices, this junction box must be accessible, usually by means of a removable cover on the side of the hydromassage tub.

The pump and power panel may also need servicing. Access may be from underneath or the end, whichever is convenient for the installation; see Figure 13-14.

Figure 13-15 shows a typical hydromassage tub.

FIGURE 13-14 The basic roughing-in of a hydromassage bathtub is similar to that of a regular bathtub. The electrician runs a separate 15-ampere, 120-volt GFCI-protected circuit to the area where the pump and control are located. Check the manufacturer's specifications for this data. An access panel from the end or from below is necessary to service the wiring, the pump, and the power panel after the installation is complete.

Robert Nolan/Shutterstock.com

FIGURE 13-15 A typical hydromassage bathtub, sometimes referred to as a *whirlpool bathtub.*

REVIEW

Note: Refer to the *CEC* or the blueprints provided with this textbook when necessary. Where applicable, responses should be written in complete sentences.

1. Most appliances of the type commonly used in bathrooms, such as hair dryers, electric shavers, and curling irons, have two-wire cords. These appliances are ______________ insulated.
2. The *CEC* requires that all receptacles in bathrooms be ______________ protected.
3. Draw the symbol or other markings used to identify double-insulated equipment.

____

____

____

BATHROOM CEILING HEATER CIRCUITS ⓐK ⓐJ

1. What is the wattage rating of the heat/vent/light/night light? ______________
2. To what circuits are the heat/vent/lights connected? ______________
3. What type of box would you use for the rough-in of the switch assembly for the heat/vent/light? ______________

____

Why? ____

____

____

4. a. How many wires are required to connect the control switch and the heat/vent/light/night light? ______
   b. What size of wires are used? ______
5. Can the heating element be energized when the fan is not operating? ______
6. Can the fan be turned on without the heating element? ______
7. What device can be used to provide automatic control of the heating element and the fan of the heat/vent/light? ______
8. Where does the air enter the heat/vent/light? ______
9. Where does the air leave this unit? ______
10. Who is to furnish the heat/vent/light? ______
11. For a ceiling heater rated 1200 watts at 240 volts, what is the current draw? ______

HYDROMASSAGE BATHTUB CIRCUIT Ⓐ A

1. What circuit supplies the hydromassage bathtub? ______
2. Should the circuit supplying the hydromassage bathtub have GFCI protection? ______
3. What conductor size feeds the hydromassage tub? ______
4. What is the fundamental difference between a hydromassage bathtub and a spa? ______
5. What rules of the *CEC* refer to hydromassage bathtubs? ______
6. Must the metal parts of the pump and power panel of the hydromassage tub be grounded? ______
7. Where should the electric controls be located for the hydromassage tub? ______
8. Complete the following wiring diagram.

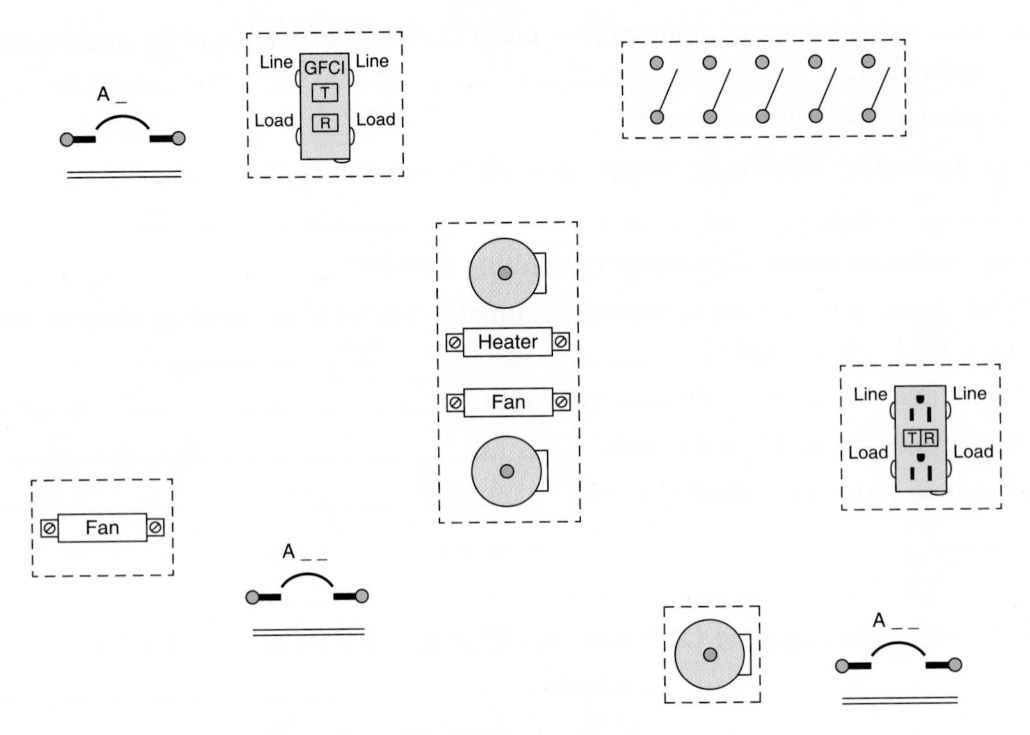

MASTER BATHROOM CCTS A9, A11, A22

UNIT 14

Lighting Branch Circuit and Small-Appliance Circuits for the Kitchen

OBJECTIVES

After studying this unit, you should be able to

- explain the *Canadian Electrical Code (CEC)* requirements for small-appliance circuits in kitchens
- understand the *CEC* requirements for receptacles that serve countertops
- discuss split-circuit receptacles
- discuss multiwire circuits
- discuss exhaust fan noise ratings
- discuss typical kitchen, dining area, and undercabinet lighting

LIGHTING CIRCUIT B7

Circuit B7 supplies the lighting in the kitchen and service entrance as well as the outside light between the sliding doors in the kitchen and living room; see Figure 14-1. There are eight outlets on Circuit B7. See Table 14-1.

KITCHEN LIGHTING

The general lighting for the kitchen is provided by two track lights controlled by three-way switches located next to the doorway leading to the living room and at the rear entrance. All fixtures must be Canadian Standards Association (CSA)-approved.

An adjustable swag fixture (Figure 14-2A) is mounted on a track to provide lighting above the

TABLE 14-1

Kitchen lighting: Outlet count and estimated load, Circuit B7.

DESCRIPTION	QUANTITY	WATTS	VOLT-AMPS
Track light	2	400	400
Recessed fixture over sink	1	60	60
Outdoor rear bracket	1	100	100
Range hood fan Lamps (two × 60 W) Motor (120 volt, 1.33 amps)	1	 120 160	 120 180
Recessed lights in service entrance	2	120	120
Receptacle in service entrance	1	120	120
Total	8	1080	1100

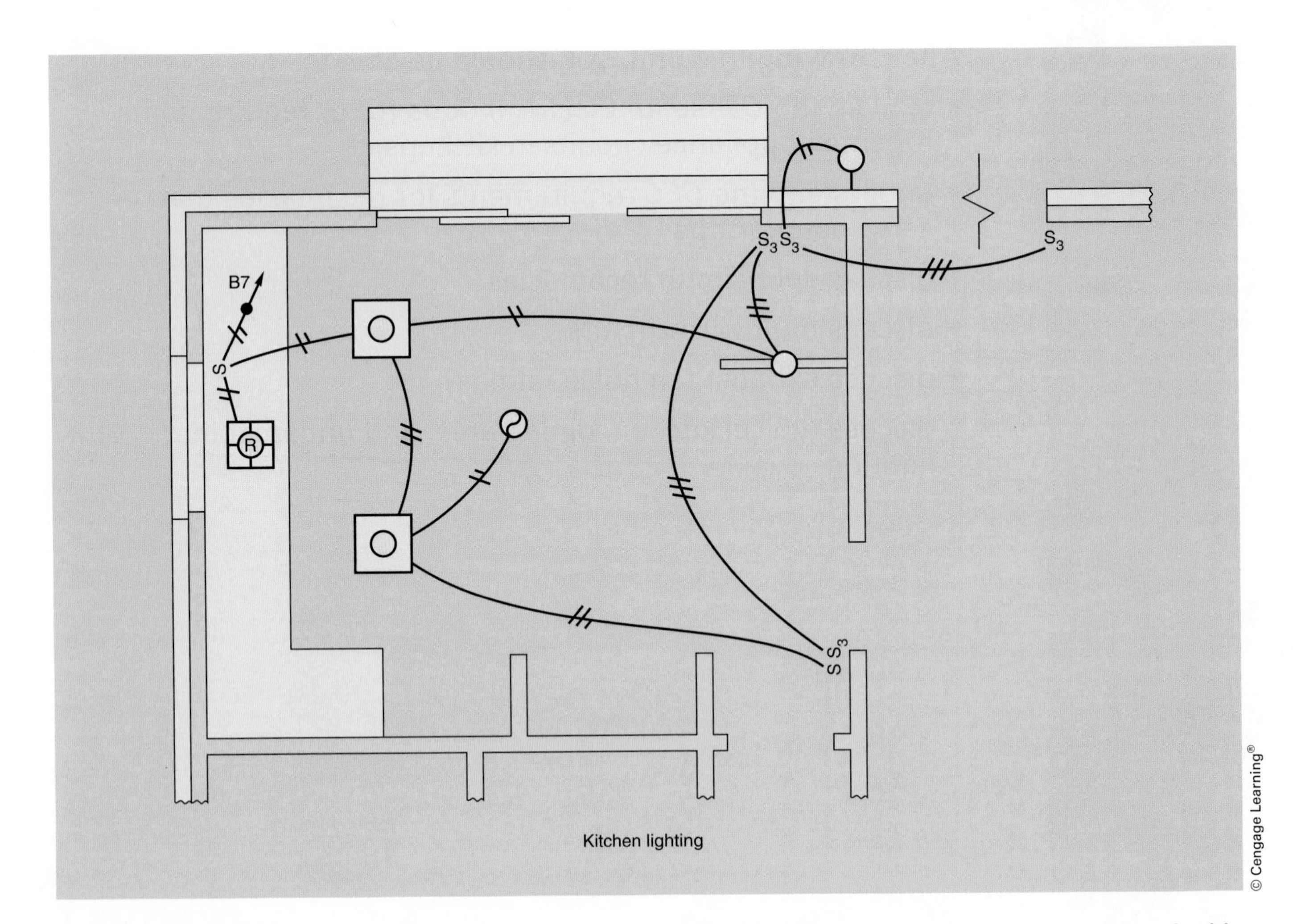

FIGURE 14-1 Cable layout for the kitchen lighting, Circuit B7. The kitchen receptacles are not shown in this lighting circuit layout.

breakfast nook table. This fixture is controlled by three-way switches: one located adjacent to the sliding door, the other at the doorway leading to the living room. Many fixtures of this type are provided with a dimmer switch to permit selection of the lighting level. Figures 14-2A through 14-2C show other types of ceiling-mounted lighting fixtures used in kitchens, dining areas, and dining rooms.

One goal of good lighting is to reduce shadows in work areas, which is particularly important in the kitchen. The two ceiling fixtures, the fixture above the sink, and the fixture in the range hood will provide excellent lighting in the work area of the kitchen. Track lighting in the nook offers good lighting for that area.

Undercabinet Lighting

In homes with little natural light, the architect might specify strip fluorescent fixtures under the kitchen cabinets. Several methods may be used to install undercabinet lighting fixtures: They can be fastened to the wall just under the upper cabinets (Part A in Figure 14-3), installed in a recess that is part of the upper cabinets so that the lights are hidden from view (Part B in Figure 14-3), or fastened under and to the front of the upper cabinets (Part C in Figure 14-3). All three require close coordination with the cabinet installer to be sure that the wiring is brought out of the wall at the proper location to connect to the undercabinet fixtures. The installation of

All photos: Courtesy of Progress Lighting

FIGURE 14-2A Fixtures that can be hung either singly or on a track over dinette tables.

Courtesy of Progress Lighting

FIGURE 14-2B Typical kitchen ceiling fixture, which might have three 13-watt compact fluorescent lamps or two 34-watt U-shaped lamps.

FIGURE 14-2C Types of fixtures used over dining room tables.

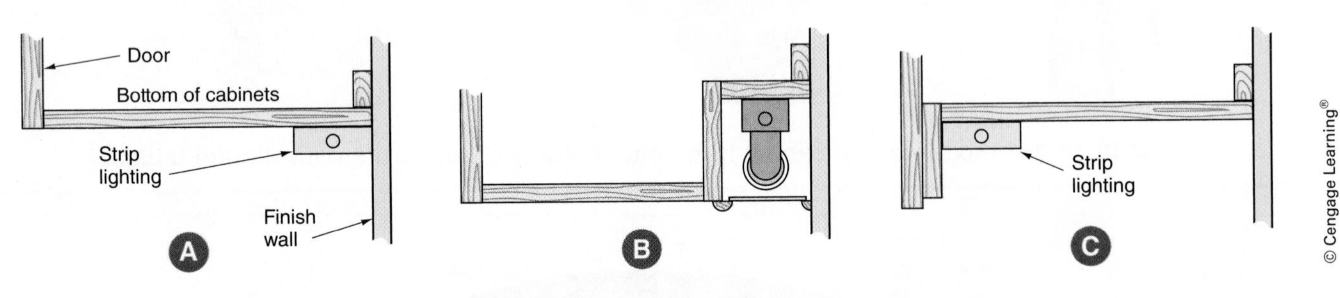

FIGURE 14-3 Methods of installing undercabinet lighting fixtures.

extra-low-voltage cabinet lighting systems is covered by *Rule 30–1208*.

Lamp Type

Good colour rendition in a kitchen area is achieved with incandescent lamps or fluorescent fixtures with compact lamps or standard warm white deluxe (WWX), warm white (WW), cool white deluxe (CXW), or Ultralume™ 3000 lamps.

Fluorescent lamps use up to 80% less energy than incandescent ones and last 10 to 20 times longer, saving on the cost of replacement. Fluorescent lighting also produces less heat, thus saving on air conditioning.

For example, two F40 34-watt fluorescent lamps (78 watts including ballast) provide about the same illumination as five 60-watt incandescent lamps (300 watts). When used 5 hours a day at $0.08/kWH, a luminaire with two fluorescent lamps will save $32.41 annually. Still even more saving can be achieved through the use of light-emitting diode (LED) systems, which offer even greater efficiency and longevity but have the drawback of being initially more expensive.

Fan Outlet

The fan outlet is connected to lighting branch Circuit B7. Fans can be installed in a kitchen either to exhaust the air to the outside of the building via ducts or vents, or to recirculate the air through various types of filters (e.g., charcoal or foam). Both types of fans can be used in a residence. The ductless hood fan does not exhaust air to the outside and does not remove humidity from the air. Figure 14-4 shows a typical range hood exhaust fan.

Homes heated electrically with resistance-type baseboard units usually have excess humidity. These heating units neither add nor remove moisture from the air. The proper use of weather stripping, storm doors, storm windows, and a vapour barrier of polyethylene plastic film tends to retain humidity.

Vapour barriers are installed to keep moisture out of the insulation. This protection normally is installed on the warm side of ceilings, walls, and floors; under concrete slabs poured directly on the ground; and as a ground cover in crawl spaces. Insulation must be kept dry to maintain its effectiveness. For example, a 1% increase of moisture in insulating material can reduce its efficiency by 5%. (Humidity control using a humidistat connected to an exhaust fan is covered in Unit 16.) Building inspectors require the vapour barrier to be intact around all outlet boxes on the outside walls of the house. It is the electrician's responsibility to maintain the integrity of the vapour barrier, *Rule 12–3000* and *Appendix B.*

An exhaust fan simultaneously removes odours and moisture from the air and exhausts heated air; see Figure 14-4. The electrician, builder, and/or architect may make suggestions to guide the owner in making the choice of type of fan to install.

For the residence in the blueprints that accompany this book, a separate wall switch is not required because the speed switch, light, and light switch are integral parts of the fan.

Michael Higginson/Shutterstock.com

FIGURE 14-4 Typical range hood exhaust fan. Speed control and light are integral parts of the unit.

Fan Noise

The fan motor and the air movement through an exhaust fan generate noise. The unit used to define fan noise is the *sone* (rhymes with *tone*), a subjective unit equal to the loudness of a 1000-hertz tone at 40 decibels as perceived by a person with normal hearing. One sone is about the noise level of an average refrigerator. The lower the sone rating, the quieter the fan. All manufacturers of exhaust fans provide this information in their descriptive literature.

Clock Outlets

A deep sectional box is recommended because a recessed clock outlet takes up considerable room in the box; see Figure 14-5.

Some decorative clocks have the entire motor recessed so that only the hands and numbers are exposed on the surface of the soffit or wall. Some of these clocks require special outlet boxes, usually furnished with the clock. Other clocks require a standard 4-inch square outlet box. Accurate measurements must be taken to centre the clock between the ceiling and the bottom edge of the soffit. If the clock is available when the electrician is roughing in

FIGURE 14-5 Recessed clock receptacle.

FIGURE 14-6 Single and duplex receptacles.

the wiring, its dimensions can be checked to help in locating the outlet.

If an electrical clock outlet has not been provided, battery-operated clocks are available.

SMALL-APPLIANCE AND BRANCH CIRCUITS FOR CONVENIENCE

Receptacles in Kitchen

Figure 14-6 shows single and duplex receptacles. The *CEC* requirements for small-appliance circuits and the spacing of receptacles in the kitchens of dwellings are covered in *Rules 26–712, 26–722, and 26–724*. This was discussed in detail in Unit 6.

Highlights are

- A microwave gets its own circuit and so does a refrigerator.
- These small-appliance circuits shall be assigned a maximum load of 12 amperes each when calculating branch-circuit connectors.
- Receptacles may not be either directly behind or in front of the sink.
- The clock outlet may be connected to the refrigerator circuit only when this outlet is to supply a recessed clock receptacle intended for use with an electric clock, *Rule 26–722(a)* (Figure 14-7). In the residence in the blueprints, the clock outlet is connected to the lighting circuit.

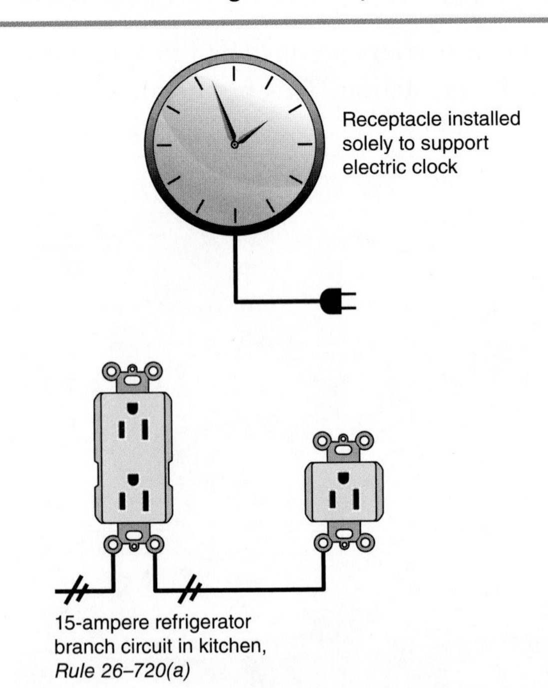

FIGURE 14-7 A receptacle installed solely to supply and support an electric clock may be connected to a refrigerator receptacle, *Rule 26–722(a)*.

- Countertop receptacles must be supplied by at least two 15-ampere, three-wire branch circuits. As an alternative, 5–20 RA receptacles may be used in accordance with *Rule 26–724(b)*; see Figure 14-8.
- There must be a receptacle within 900 mm of any point along the wall behind all counters except those less than 300 mm long; see Figure 14-9.

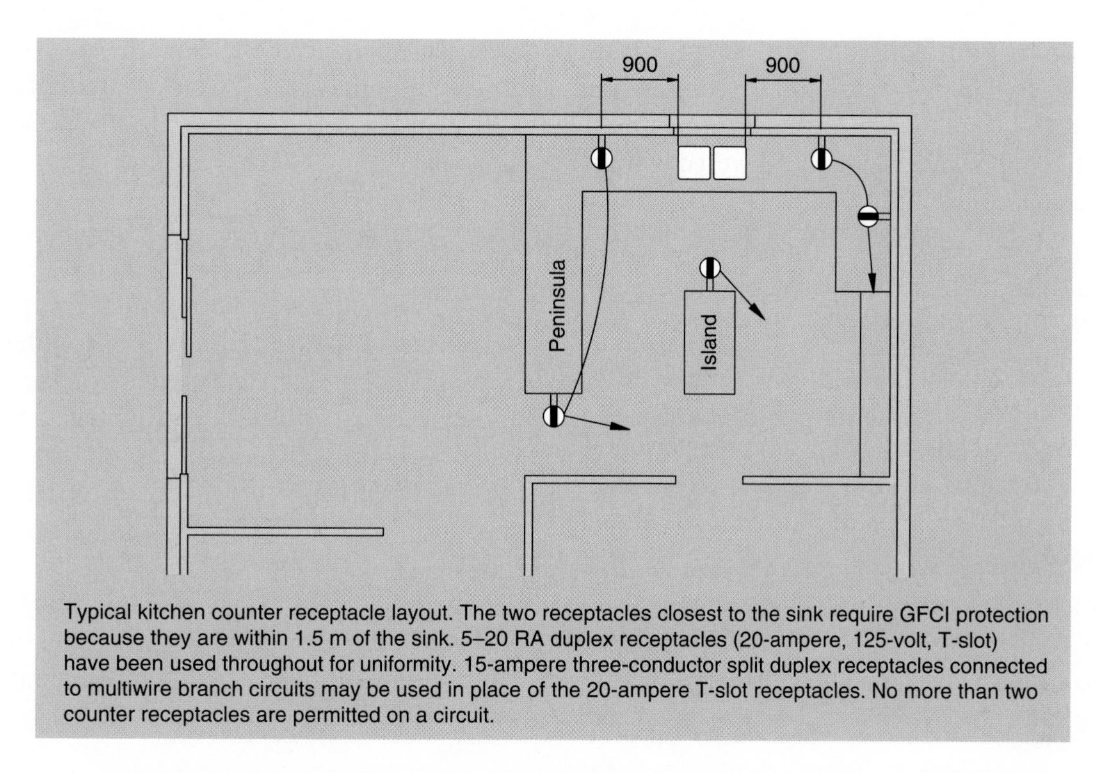

FIGURE 14-8 Layout of receptacles behind the counter in a kitchen. If two or more countertop receptacles are required by *Rule 26–712*, at least two 20-ampere branch circuits or two 15-ampere, three-wire branch circuits are required, *Rule 26–724(b)*.

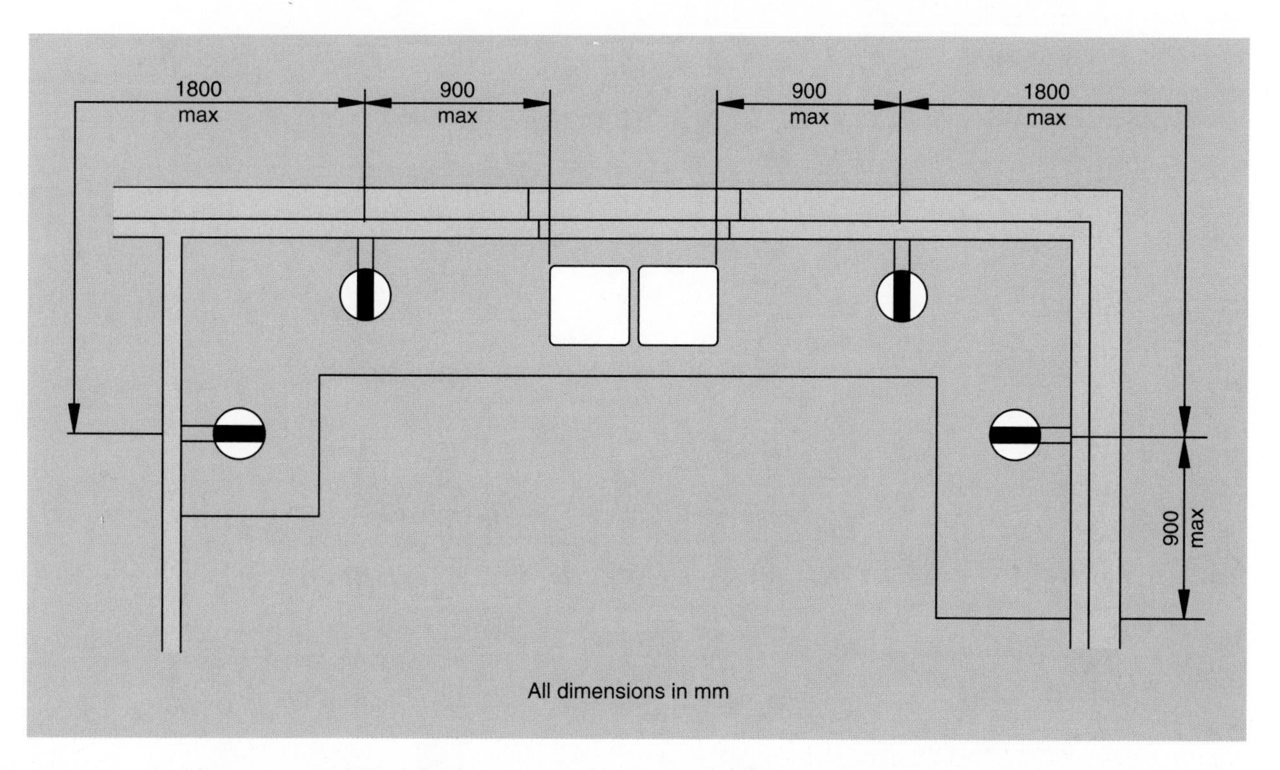

FIGURE 14-9 In kitchens, all 125-volt, 15-ampere duplex receptacles located along the wall of counter work surfaces must be of the split-wired type (three-wire circuit) or 20-ampere 125-volt T-slot receptacles must be used. Receptacles must be located so that no point along the wall line is more than 900 mm from a receptacle, *Rule 26–712(d)(iii)*.

- Microwave enclosures must be supplied with a receptacle on a separate circuit, *Rule 26–722(d)*. If no cupboard or enclosure is provided, the microwave would be powered by one of the countertop split receptacles.
- All outdoor receptacles within 2.5 m of the finished grade must be on a separate GFCI-protected circuit, *Rule 26–710(n)*.
- Counter receptacles may be three-conductor split-circuit or 20-ampere T-slot receptacles. Counter receptacles prevent circuit overloading when small appliances such as kettles and toasters are plugged in at the same time, *Rule 26–724(b)*.

In the kitchen, split-circuit receptacles and 20-ampere T-slot receptacles prevent circuit overloading from the heavy concentration and use of electrical appliances. For example, one type of cord-connected microwave oven is rated 1440 watts at 120 volts. The current required by this appliance alone is

$$I = \frac{W}{E} = \frac{1440}{120} = 12.0 \text{ amperes}$$

A maximum of two counter receptacles are permitted on one circuit. This provides the best availability of electrical power for the variety of appliances that will be used in the heavily concentrated work areas in the kitchen.

Figure 14-10 shows the circuiting for the special-purpose outlets and counter receptacles in the kitchen.

A receptacle outlet installed below the sink for easy plug-in connection of a food waste disposal is *not* included in the *CEC* requirements for split

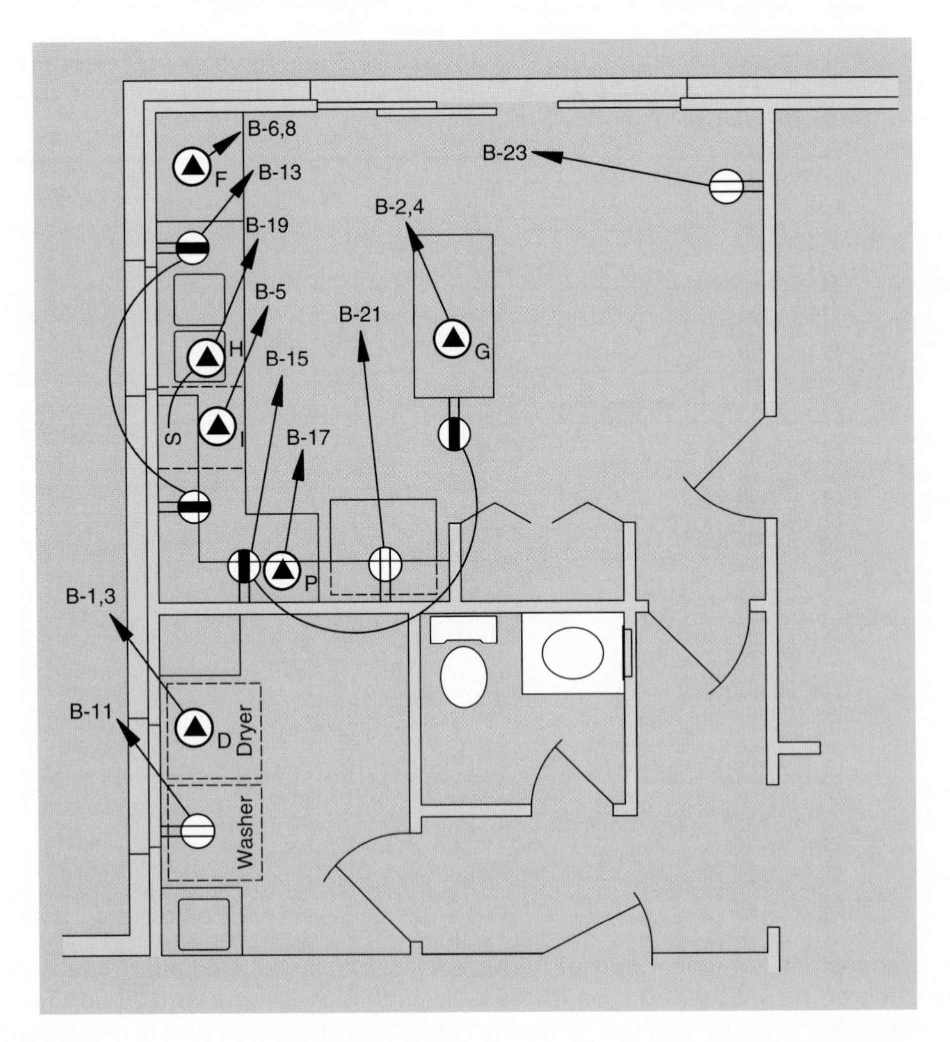

FIGURE 14-10 Circuiting for special-purpose and counter receptacles in the kitchen and laundry.

circuits and is not required to be GFCI-protected or tamper-resistant.

SPLIT-CIRCUIT RECEPTACLES AND MULTIWIRE CIRCUITS

Split-wired receptacles may be installed where a heavy concentration of plug-in load is anticipated, in which case, each half of the duplex receptacle is connected to a separate circuit, as in Figures 14-11 and 14-12. Examples of two-pole 15-ampere breakers that feed three-wire branch circuits for use with kitchen split receptacles are found in Figure 14-13.

Split or 20-ampere T-slot receptacles are mandatory along the countertop in the kitchen, *Rule 26–712*. These must be arranged so that no more than two receptacles are on a circuit. Any receptacle within 1.5 m of the sink must be a GFCI receptacle or GFCI-protected.

When a multiwire branch circuit is used to supply fixed lighting loads and non-split duplex receptacles, the continuity of the identified neutral conductor must be independent of the device

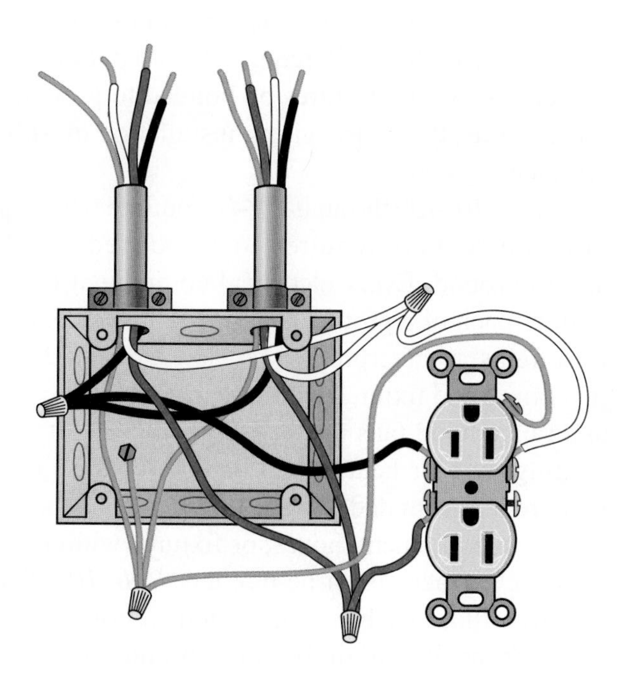

Ⓐ Proper way to connect grounded neutral conductors in a multiwire branch circuit. The receptacle can be removed without disrupting the circuit.

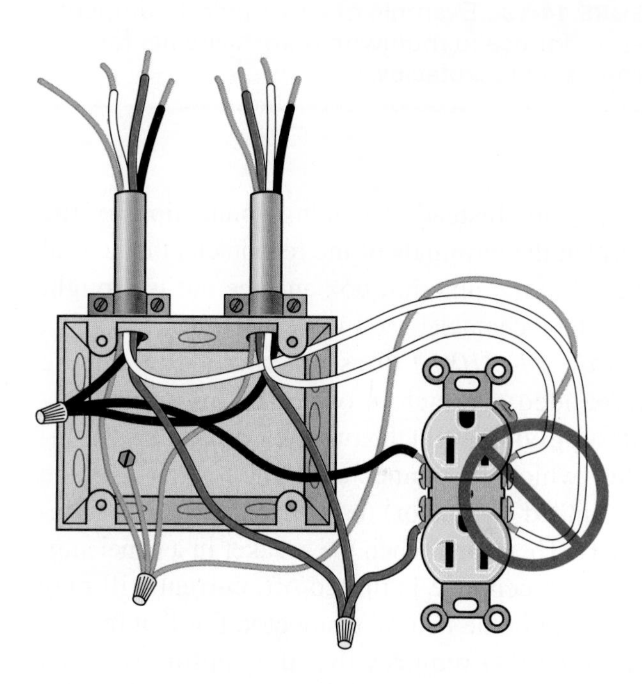

Ⓑ Improper way to connect grounded neutral conductors in a multiwire branch circuit. Not permitted. Removing the receptacle will disrupt the circuit because the neutral bar on the receptacle is part of the circuit.

FIGURE 14-11 Connecting the neutral in a three-wire (multiwire) circuit, *Rule 4–026(a)*.

120/240 volt 3-wire circuit (multiwire circuit)

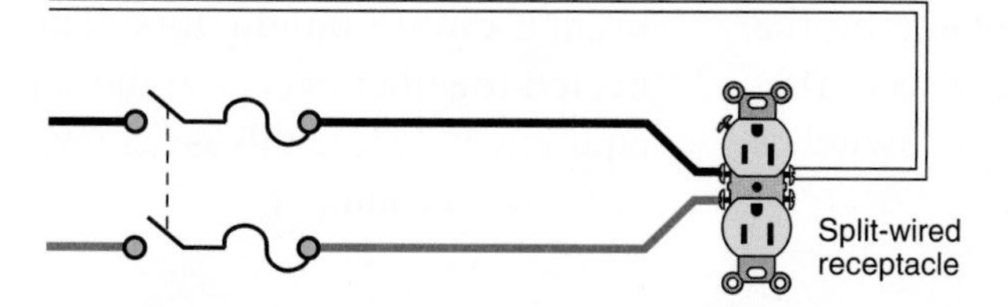

Rule 26–710 requires that a means must be provided to simultaneously disconnect both ungrounded conductors at the panelboard where the branch circuit originates. This could be a two-pole switch with fuses, a two-pole circuit breaker, or two single-pole circuit breakers with a listed tie.

FIGURE 14-12 Split receptacle connected to a multiwire circuit.

Courtesy of Natalie Barrington

FIGURE 14-13 Example of a two-pole 15-ampere breaker for use in multiwire branch circuits for kitchen split receptacles.

connections. Instead of making connections of the neutral at the terminals of the receptacle, the neutral is tailed in the junction box and the tail is brought out to the receptacle; see Figure 14-11.

Rule 14–010(b) does not require all of the ungrounded conductors of multiwire branch circuit supplying light fixtures or non-split receptacles (which are connected to the neutral and one ungrounded conductor) to be disconnected simultaneously. Therefore, when the breaker in a panel supplying a receptacle is turned off, current still may be flowing in the neutral conductor. For that reason, *Rule 4–036(4)* requires that the continuity of an identified conductor be independent of device connections to luminaires and receptacles. The hazards of open neutrals are covered in Unit 19.

When multiwire branch circuits supply more than one receptacle on the same yoke, a means must be provided to simultaneously disconnect all of the hot conductors at the panelboard where the branch circuit originates. Figure 14-12 shows that although 240 volts are present on the wiring device, only 120 volts are connected to each receptacle on the wiring device. The "simultaneous disconnect" rule applies to any receptacles, lampholders, or switches mounted on one yoke, *Rule 14–010(b)*.

In the case of the three-conductor split-duplex receptacles in the kitchen connected to multiwire branch circuits, the two-pole breaker in the panel will disconnect all ungrounded conductors of the circuit simultaneously. The local inspection authority may permit you to make connections of the neutral at the terminals of the receptacle when looping between the counter receptacles in the kitchen.

GENERAL BONDING CONSIDERATIONS

The specifications for the residence in the blueprints provided with this book require that all outlet boxes, switch boxes, and fixtures be bonded to ground. In other words, the entire wiring installation must be a grounded system.

Rules 10–400 through *10–414* outline what types of equipment are required to be bonded together and to ground. Every electrical component located within reach of a grounded surface must be bonded. Post lights, weatherproof receptacles, basement wiring, boxes and fixtures within reach of sinks, ranges, and range hood fans must all be bonded. Hot water heating, hot air heating, and steam heating registers are all grounded surfaces. This means that any electrical equipment, boxes, or fixtures within reach of these surfaces must be bonded. *Rule 10–400(g)* requires that any boxes installed in contact with metal or metal lath, tinfoil, or aluminum insulation also be bonded.

Bonding is accomplished through the proper use of non-metallic-sheathed cable, armoured cable, and metal conduit.

When using non-metallic-sheathed cable (NMSC), do not confuse the bare *bonding* conductor with the *grounded* circuit conductor. The grounding or bonding conductor is used to ground equipment, whereas the white grounded conductor is one of the branch-circuit conductors. They may not be connected together except at the main service entrance equipment, where this is achieved through the bonding screw or jumper.

REVIEW

Note: Refer to the *CEC* or the blueprints provided with this textbook when necessary. Where applicable, responses should be written in complete sentences.

1. What provision does the *CEC* require for microwave ovens in enclosures?

   ____

   What is the relevant rule number? ______________________________

2. Which circuit feeds the kitchen lighting? What size of conductors is used? ____________

   ____

   ____

3. How many lighting fixtures are connected to the kitchen lighting circuit? ____________

   ____

4. Which colour of fluorescent lamps are recommended for residential installations? ____

   ____

   ____

5. a. What is the minimum number of 15-ampere split circuits or 20-ampere circuits required for a kitchen according to the *CEC*? ______________________________

   b. How many are there in this kitchen? ______________________________

6. How many duplex receptacle outlets are provided in the kitchen? ____________

7. What is meant by the term *split-circuit receptacle*? ______________________________

   ____

   ____

   ____

   ____

   ____

8. A single receptacle connected to a circuit must have a rating ____________ the ampere rating of the circuit.

9. In kitchens, receptacles along the countertop must be installed no farther than ________ mm apart.

10. A fundamental rule regarding the grounding of metal boxes, fixtures, and so on is that they must be grounded when "in reach of ______________________________."

11. How many circuit conductors enter the box

    a. in the track lighting ceiling box over the eating area? ____________

    b. at the two-gang switch located to the right of the sliding door? ____________

12. How much space is there between the countertop and upper cabinets? ____________

13. Where is the speed control for the exhaust fan located? ______________________

14. Who furnishes the range hood? ______________________

15. List the appliances in the kitchen that must be connected. ______________________

____

____

____

____

16. Complete the wiring diagram below by connecting feedthrough GFCI 1 to protect receptacle 1, both to be supplied by Circuit A1. Connect feedthrough GFCI 2 to also protect receptacle 2, both to be supplied by Circuit A2. Use coloured pencils or pens to show insulation colours. Assume that the wiring method is EMT, where more freedom in the choice of insulation colours is possible.

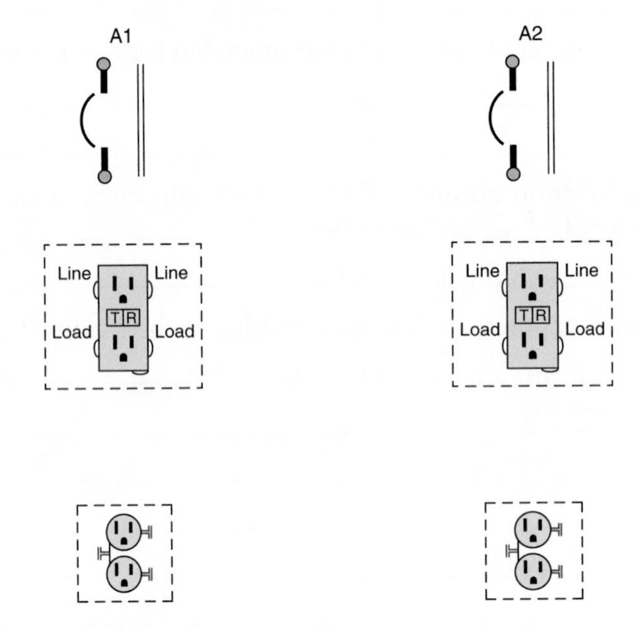

17. Each 15-ampere appliance circuit load demand shall be determined at [circle one] (1440 watts) (1500 watts) (1800 watts).

18. It is permitted to connect an outlet supplying a clock receptacle to the branch circuit dedicated to the refrigerator circuit. [Circle one.] (True) (False)

19. a. It is permitted to connect a receptacle in the dining area of a kitchen to one of the kitchen split circuits. [Circle one.] (True) (False)

   Which *CEC* rule applies to this situation? ______________________

   b. Receptacles located above the countertops in the kitchen must be supplied by at least ______________ 15-ampere, three-wire circuits.

20. The following is a layout for the lighting circuit for the kitchen. Complete the wiring diagram using coloured pens or pencils to show the conductors' insulation colour.

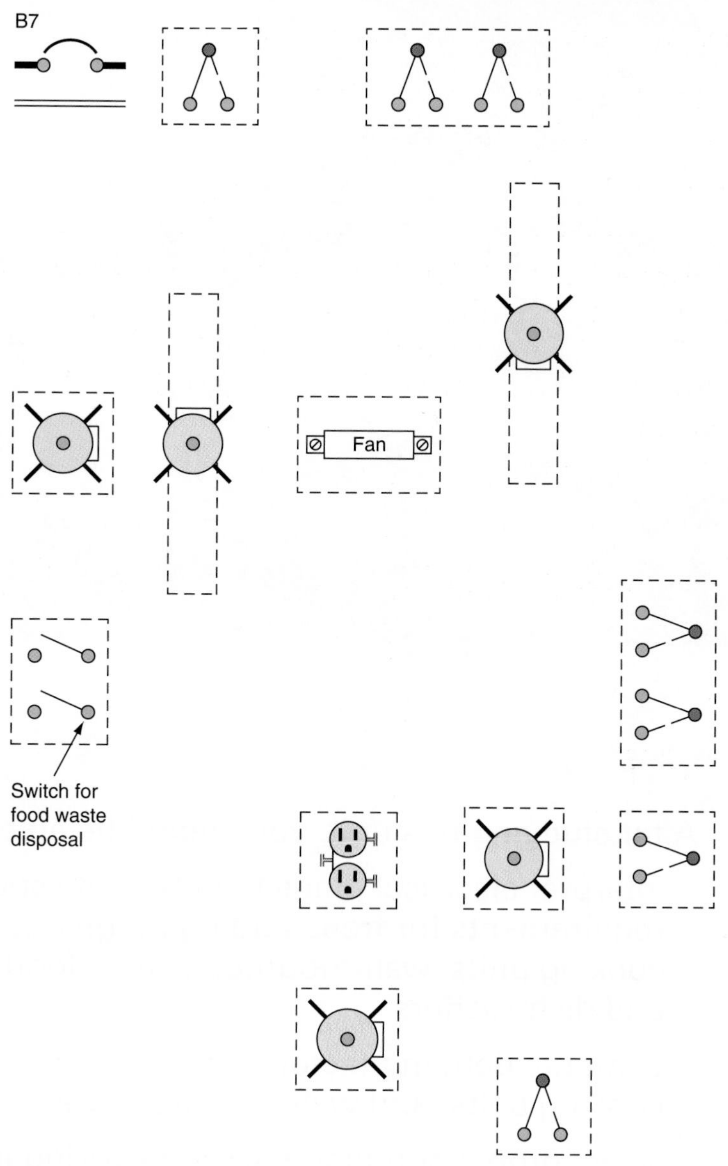

21. a. According to *Rule 26–712(a)*, no point along the floor line shall be more than _____ m from a receptacle outlet.

    b. According to *Rule 26–712(d)*, a receptacle must be installed in any wall space _____ m wide or greater.

22. The *CEC* states that in multiwire circuits, the screw terminals of a receptacle must *not* be used to splice the neutral conductors. Why? Quote the rule number.

____

____

____

23. Exhaust fans produce noise. It is possible to compare the noise levels of different fans before installation by comparing their _______________ ratings.

24. Is it permitted to connect the white grounded circuit conductor to the grounding terminal of a receptacle? _______________________________

UNIT 15

Special-Purpose Outlets for Ranges, Counter-Mounted Cooking Units ⓖG, Wall-Mounted Ovens ⓖF, Food Waste Disposals ⓖH, and Dishwashers ⓖI

OBJECTIVES

After studying this unit, you should be able to

- interpret electrical plans to determine special installation requirements for freestanding ranges, counter-mounted cooking units, wall-mounted ovens, food waste disposals, and dishwashers
- compute demand factors for ranges, counter-mounted cooking units, and wall-mounted ovens
- select proper conductor sizes for wiring installations based on the ratings of appliances
- ground all appliances properly regardless of the wiring method used
- understand how infinite heat temperature controls operate
- supply counter-mounted cooking units and wall-mounted ovens by connecting them to one feeder using the "tap" rule
- understand how to install a feeder to a load centre, and divide the feeder to individual circuits to supply the appliances
- install circuits for dishwashers and food waste disposals

The residence used as an example in this textbook has one built-in oven and one built-in cooktop. Checking Table A-1, Schedule of Special-Purpose Outlets, in Appendix A at the back of the book, we find that the built-in wall-mounted oven is supplied by Circuit B6/B8, and the built-in cooktop is supplied by Circuit B2/B4.

Running separate circuits to individual appliances is probably the most common method used to connect appliances. However, according to *Rule 26–746(3)*, two or more separate built-in cooking units shall be counted as one appliance. A single circuit is run from the main panel to a junction box located close to both the oven and the cooktop. From this junction box, "taps" of smaller conductors are made to feed each appliance. These taps should be less than 7.5 m long. Which method to use is governed by local rules and consideration of the cost of installation (time and material).

This is how to use a single circuit for a separate cooktop and oven, and for a freestanding range.

COUNTER-MOUNTED COOKING UNIT CIRCUIT Ⓐ$_G$

Counter-mounted cooking units are available in many styles. Units may have two, three, or four surface heating elements. The surface heating elements of some cooking units are completely covered by a sheet of high-temperature ceramic. These units have the same controls as standard units and are wired according to the general *Canadian Electrical Code (CEC)* guidelines for cooking units. When the elements of these ceramic-covered units are turned on, the ceramic above each element changes colour slightly. Most of these cooking units have a faint design in the ceramic to locate the heating elements. The ceramic is rugged, attractive, and easy to clean.

In the residence in the plans, a separate circuit is provided for a standard, exposed-element, counter-mounted cooking unit. This unit is rated at 7450 watts and 120/240 volts. The ampere rating is

$$I = \frac{W}{E} = \frac{7450}{240} \text{ amperes} = 31.04 \text{ amperes}$$

The cooking unit outlet is indicated by the symbol Ⓐ$_G$.

Rule 8–300 discusses how to calculate the demand on the conductors feeding the range top and cooking unit.

Installing the Cooking Unit

CEC Table 2 shows that #10 AWG NMD 90 may be used as the circuit conductors for the 31.04-ampere cooking unit installation. These circuit conductors are connected to Circuit B2/B4, a two-pole, 40-ampere circuit in Panel B.

The electrician must obtain the roughing-in dimensions for the counter-mounted cooking unit so that the cable can be brought out of the wall at the proper height. In this way, the cable is hidden from view. To simplify both servicing and installation, a plug and receptacle combination may be used in the supply line.

Most cooktop manufacturers provide a junction box on their units: The electrician can terminate the circuit conductors of the appliance in this box and make the proper connections to the residence wiring. Or the manufacturer may connect a short length of flexible conduit to the appliance: This conduit contains the conductors to which the electrician will connect the supply conductors in a junction box furnished by either the electrician or the manufacturer.

In the residence in the blueprints, the cooktop range is installed in the island. (See the plans for details.) The feed will be a three-conductor nonmetallic-sheathed cable with three #10 AWG copper conductors, plus an equipment-bonding conductor. The cable runs out of the top of Panel B, through the ceiling joists in the recreation room, then upward into the base of the island cabinet. Because the recreation room has a drywall ceiling, the cable is run to a point just below the intended location of the kitchen island, leaving ample cable length hanging in the basement. Before the gypsum board in the basement is installed, the cable must be run up into the kitchen.

Later when the kitchen cabinets and the island are installed and other finishing touches are being done in the kitchen, the built-in appliances are set into place. Following this, the electrician completes the final hookup of the built-in appliances.

The plans show that a receptacle outlet is to be installed on the side of the island containing the cooktop. This outlet is supplied by Circuit B15; see Figure 14-10 in Unit 14.

Temperature Effects on Conductors

Before beginning the installation, always check the appliance instructions provided by the manufacturer. The supply conductors may require an insulation rated for temperatures higher than the standard rating of 60°C.

If such information is not provided, it is safe to assume that the conductors used may have insulation rated for 60°C. *Table 19* gives the conductor insulation temperature limitations and the accepted applications for conductors.

In addition, it is important to remember that the conductors are hooked up to lugs or breakers that have a temperature rating of their own and this imposes a limitation on the conductors. If a 90°C-rated conductor is hooked up to a 75°C-rated breaker, the conductor may not carry more current than is proscribed by the 75°C column of *CEC Table 2, Rule 4–006(1)*. All circuit breakers are marked with a maximum temperature of either 60°C or 75°C.

General Wiring Methods

You may use any of the standard wiring methods to connect counter-mounted cooking units and wall-mounted ovens. Freestanding ranges and clothes dryers must be connected using receptacles and cord-connected plugs, according to *Rule 26–744*. Otherwise, armoured or non-metallic-sheathed cable (NMSC) may be used.

Verify any local restrictions on the use of the wiring methods listed before proceeding.

Bonding Requirements

CEC Section 10 requires electric ranges, oven cooktop units, and electric clothes dryers to be bonded to ground, with a separate bonding conductor sized according to *Table 16A*.

Surface Heating Elements for Cooking Units

The heating elements used in surface-type cooking units are manufactured in several steps. First, spiral-wound nichrome resistance wire is impacted in magnesium oxide (a white, chalklike powder). The wire is then encased in a nickel–steel alloy sheath, flattened under very high pressure, and formed into coils. The flattened surface of the coil makes good contact with the bottoms of cooking utensils. Thus, efficient heat transfer takes place between the elements and the utensils.

Cooktops with a smooth ceramic surface offer the user ease in cleaning and a neater, more modern appearance. The electric heating elements are in direct contact with the underside of the ceramic top.

Figures 15-1 and 15-2 illustrate typical electric range surface units.

FIGURE 15-1 Typical 240-volt electric range surface heating element of the type used with infinite heat controls.

FIGURE 15-2 Typical electric range surface unit of the type used with seven-position controls, as discussed in this textbook.

TEMPERATURE CONTROL

Infinite-Position Temperature Controls

Modern electric ranges are equipped with infinite-position temperature knobs (controls) or programmable touch screens; see Figure 15-3. Older-style heat controls connect the heating elements of a surface unit to 120 volts or 240 volts in series or in parallel to attain a certain number of specific heat levels.

Some control knobs on ranges have a rotating cam that provides an infinite number of heat positions. The contacts open and close as a result of a heater-bimetal device that is an integral part of the rotary control. When the control is turned on, a separate set of pilot light contacts closes and remains closed, while the contacts for the heating element open and close to attain and maintain the desired temperature of the surface heating unit; see Figure 15-4.

FIGURE 15-3 Infinite-heat control switch.

Inside these infinite-position temperature controls, contacts open and close repetitively. The temperature of the surface heating element is regulated by the ratio of the time the contacts are in the closed position versus the open position. This is termed *input percentage.*

The contacts in a temperature control of this type have an expected life of over 250 000 automatic cycles. The knob rotation expected life exceeds 30 000 operations.

Heat Generation by Surface Heating Elements

The surface heating elements generate a large amount of radiant heat. For example, a 1000-watt heating unit generates about 3415 Btu of heat per hour. A *British thermal unit* (Btu) is the amount of heat required to raise the temperature of 1 pound (0.45 kg) of water by 1°F (0.556°C).

WALL-MOUNTED OVEN CIRCUIT Ⓕ

A separate circuit is provided for the wall-mounted oven. The circuit is connected to Circuit B6/B8, a two-pole, 40-ampere circuit in Panel B and wired using three #10 AWG NMD 90. The receptacle for the oven is indicated by the symbol Ⓕ.

The wall-mounted oven installed in the residence is rated at 6.6 kW or 6600 watts at 120/240 volts. The current rating is

$$I = \frac{W}{E} \text{ amperes} = \frac{6600}{240} \text{ amperes} = 27.5 \text{ amperes}$$

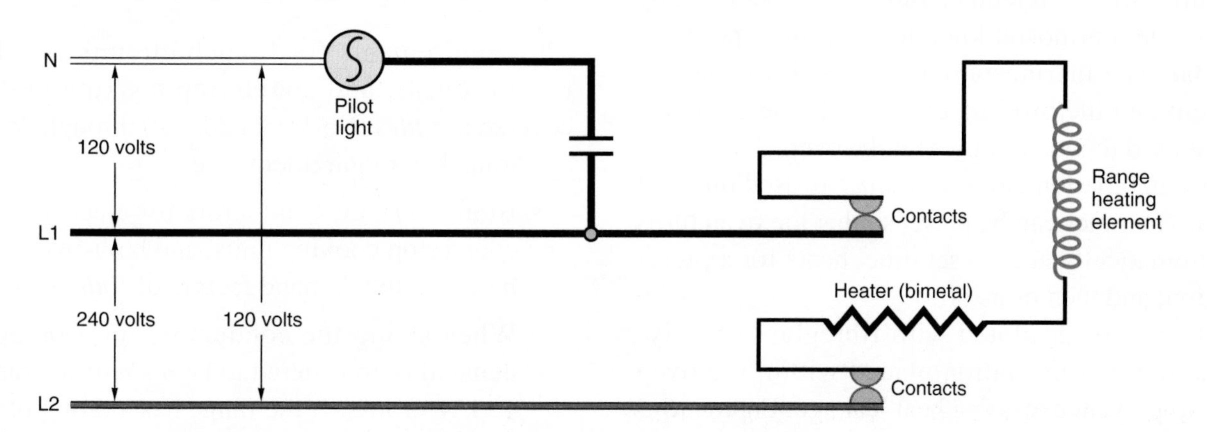

FIGURE 15-4 Typical internal wiring of an infinite-heat surface control.

After determining the current rating (I = W/E), refer to *CEC Table 2*. You find #12 AWG copper NMD 90 cable may be used to supply this 27.5- ampere wall-mounted oven. This 27.5-ampere load will exceed 80% of the rating of the overcurrent device. Therefore, a 40-ampere circuit breaker and #10 AWG NMD 90 will be used. See branch-circuit requirements for ranges, ovens, and countertop cooking units later in this unit.

Self-Cleaning Oven

Self-cleaning ovens are popular appliances because they automatically remove cooking spills from the inside of the oven. Lined with high-temperature material, a self-cleaning oven can be set for a cleaning temperature that is much higher than baking and broiling temperatures. When the self-clean control is turned on, oven temperatures will approximate 427°C. At this temperature, all residue in the oven, such as grease and drippings, is burned until it is fine ash that can be removed easily. During the self-clean cycle, for safety reasons, a special latching mechanism makes it impossible to open the oven door until a predetermined safe temperature is reached. Self-cleaning ovens do not require special wiring and are electrically connected in the same manner as standard electric ovens.

Operation of the Oven

The oven heating elements are similar to the surface elements. The oven elements are mounted in removable frames. A thermostat controls their temperature. Any oven temperature can be obtained by setting the thermostat knob to the proper point on the dial. The thermostat controls both the baking element and the broiling element. These elements can be used together to preheat the oven.

A combination clock and timer is used on most ovens. The timer can be preset so that the oven turns on automatically at a preset time, heats for a preset duration, and then turns off.

Ovens are insulated with fibreglass or polyurethane foam insulation placed within the oven walls to prevent excessive heat leakage. Figure 15-5 shows a typical wiring diagram for a standard oven unit.

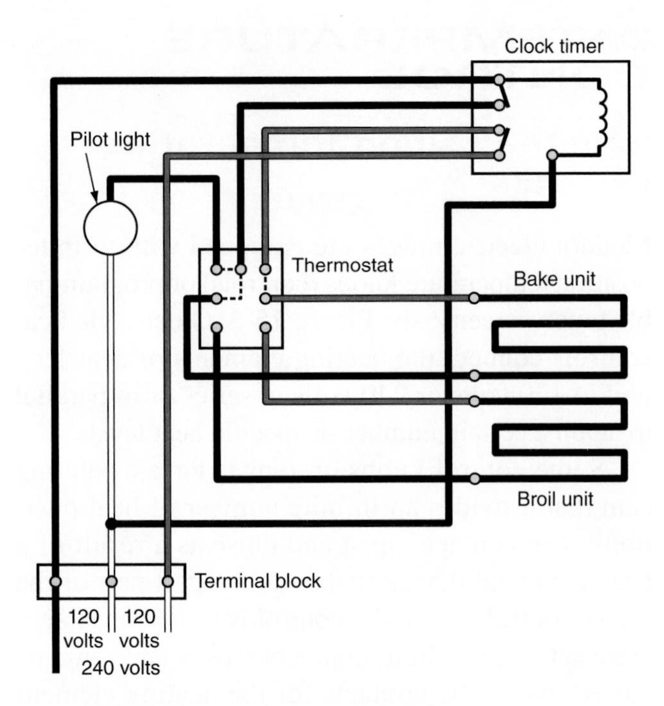

FIGURE 15-5 Wiring diagram for an oven unit.

Microwave Oven

The electrician's primary concern is how to calculate the microwave oven load and how to install the circuit for microwave appliances to conform to *Rule 26–722*. The same rules apply for both electrical and electronic microwave cooking appliances.

BRANCH-CIRCUIT REQUIREMENTS FOR RANGES, OVENS, AND COUNTERTOP COOKING UNITS*

The requirements for branch circuits supplying ranges, ovens, and countertop cooking units are covered in *Rules 8–300* and *26–740* through *26–746*.

Some key requirements are

- Branch-circuit conductors for electric ranges, countertop cooking units, and built-in ovens are based on the demand factors of *Rule 8–300.*
- When sizing the conductors for a range, the demand is considered to be 8 kW if the range is 12 kW or less. If the range has a rating of more

*Courtesy of CSA Group

than 12 kW, the demand is considered to be 8kW + 40% of the amount by which the range exceeds 12 kW, *Rule 8–300.*

- Two or more separate built-in units may be considered as a single range, *Rule 8–300.*
- When a single branch circuit is used to supply separate built-in units, tap conductors to an individual unit must have an ampacity equal to the rating of the unit they supply, *Rule 26–742.*
- The demand factors of *Rule 8–300* may not be applied to cord-connected appliances such as hot plates and rangettes.
- When a gas range is installed, a receptacle will be required in the wall behind the range. The receptacle should be mounted as close to the centreline of the range as possible and not more than 130 mm above the finished floor, *Rule 26–712(d)(ii).*

FREESTANDING RANGE

The demand calculations for a single freestanding range are simple. For example, a single electric range to be installed has a rating of 14 050 watts (exactly the same as the combined built-in units):

1. 14 050 watts − 12 000 watts = 2050 watts or 2 kW
2. According to *Rule 8–300(1)(b),* the conductors of a branch circuit supplying a range shall be considered to have a demand of 8 kW plus 40% of the amount that the rating of the range exceeds 12 kW.

   For example,

   2050 watts × 40% = 820 watts or 0.8 kW
3. Calculated load = 8 kW + 0.8 kW = 8.8 kW
4. In amperes, the calculated load is

$$I = \frac{W}{E} = \frac{8820}{240} \text{ amperes} = 36.75 \text{ amperes}$$

This range requires that #8 NMD 90 conductors be used for the branch circuit to the range. The size of the neutral is the same as that of the line conductors even though the load is minimal for the element pilot lights and clock. If a range is rated at not over 12 kW, the maximum power demand is based on *Rule 8–300(1)(a).* For example, if the range is rated at 11.4 kW, the maximum demand is based on 8 kW:

$$I = \frac{W}{E} = \frac{8000}{240} \text{ amperes} = 33.3 \text{ amperes}$$

Therefore, #10 AWG three-wire NMD 90 cable can be installed for this range.

The disconnecting means for a freestanding range having a demand of 50 amperes or less shall be a cord and plug; see Figure 15-6. For electric ranges, these cords and plugs are rated at 50 amperes and must have an attachment plug of Canadian Standards Association (CSA) configuration 14-50R. Figure 15-7 shows the correct wiring to the terminals on the receptacle. The receptacle for this plug must

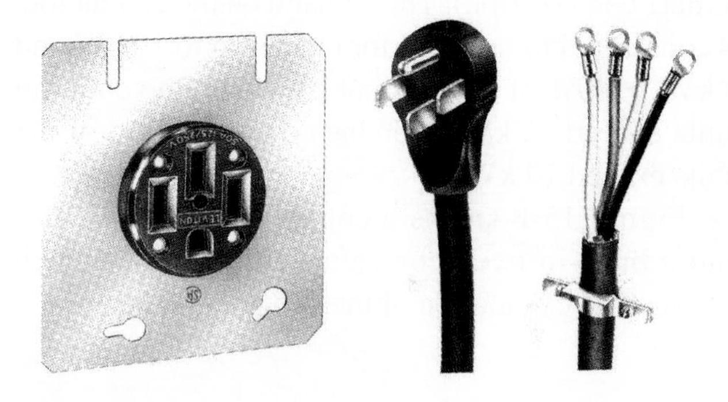

FIGURE 15-6 Range receptacle and cord set rated at 50 amperes, 125/250 volts. Be sure to mount the 119-mm box so that the ground pin will be to either side.

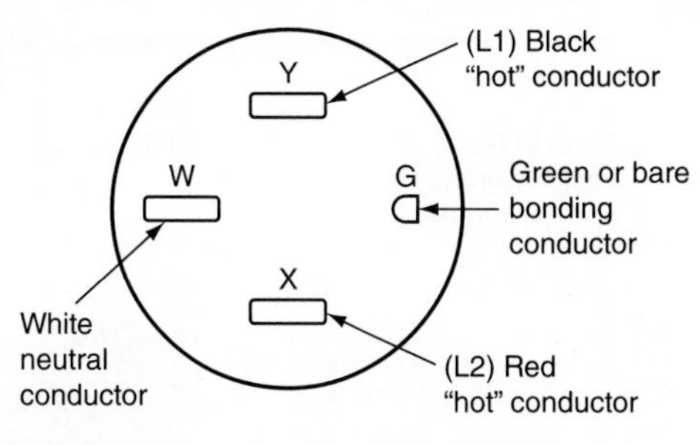

FIGURE 15-7 Correct wiring connections for the terminals on the range receptacle rated at 50 amperes, 125/250 volts, CSA configuration 14–50R, *Diagram 1.*

be flush-mounted, *Rule 26–744(9),* and be installed according to *Rule 26–744(6)*:

- centre of the receptacle not more than 130 mm above the finished floor (AFF)
- as near as possible to the centreline of the space the range will occupy
- flush with the wall when practical
- with the U-ground oriented to either side

BRANCH CIRCUITS SUPPLYING SEPARATE BUILT-IN COOKING UNITS

The branch circuit that supplies a separate countertop cooking unit and a built-in oven may be treated the same as a circuit for a single range. (Remember to consult your local inspection authority before you initiate construction). The demand on the conductors is considered to be 8 kW for ranges up to 12 kW and 8 kW + 40% of the amount by which the cooking units exceed 12 kW when the combined load of the units exceed 12 kW.

Figure 15-8 shows a countertop cooking unit and a built-in oven connected to a single branch circuit. The calculation of the demand load is

Counter-mounted cooking unit	7 450 watts
Built-in oven	6 600 watts
Total connected load	14 050 watts

The demand on the conductors will be considered to be

8 kW + 40% of the amount the units exceed 12 000 watts. The units exceed 12 000 watts by 14 050 watts − 12 000 = 2050 watts

The demand on the conductors is then

(8000 + 40% of 2050) watts = 8820 watts
8820/240 amperes = 36.8 amperes

Therefore, #8 AWG copper NMD 90 cable may be used as the branch-circuit conductors for the cooking unit and built-in oven. The wire can handle up to 90°C, but it is hooked up to a breaker and breakers have a limitation of either 60°C or 75°C. After breakers pass that temperature, they tend to trip prematurely.

The tap conductors to the cooking unit and built-in oven must have a current-carrying capacity that is equal to the ratings of the units they supply.

Counter-mounted cooking unit:

7450/240 amperes = 31.04 amperes
#10 AWG copper NMD 90 *(Table 2)*

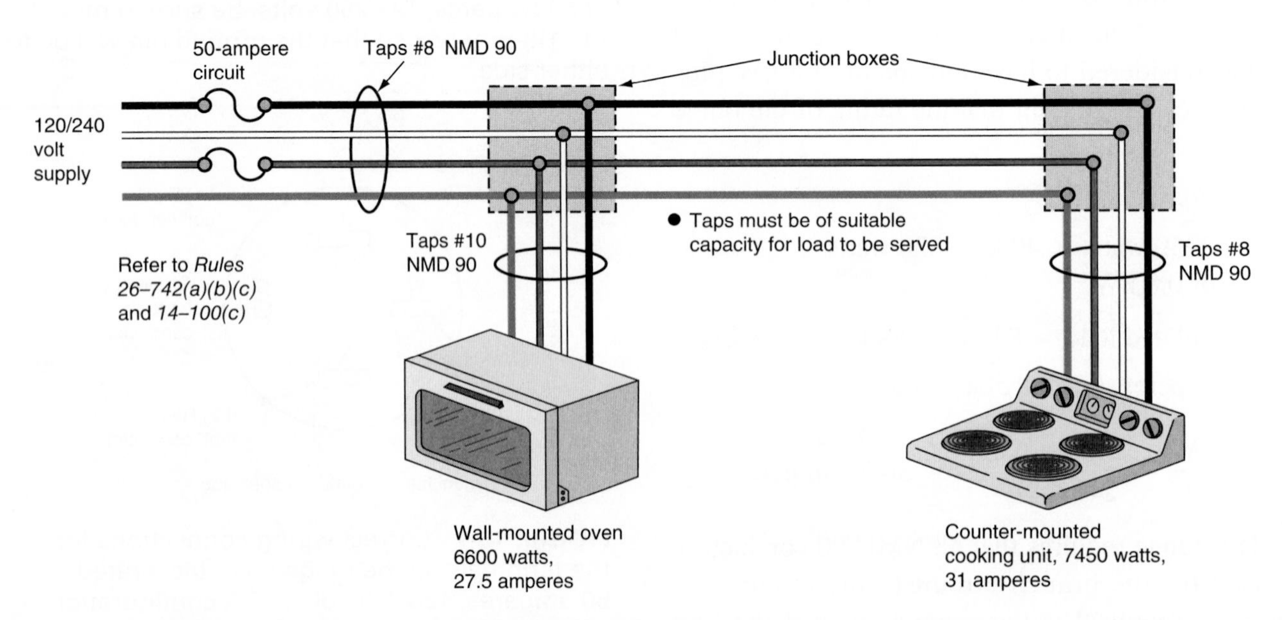

FIGURE 15-8 Counter-mounted cooktop and wall-mounted oven connected to one circuit.

Built-in oven:

6600/240 amperes = 27.50 amperes
#10 AWG copper NMD 90 *(Table 2)*

The tap conductors should not be longer than necessary. If you are in doubt about the length of these conductors, consult with your electrical inspector; otherwise, consider 7.5 m to be the limit.

At one time, a separate panel was required to provide overcurrent protection for a countertop cooking unit and built-in oven. This requirement has been deleted from the *CEC*.

FOOD WASTE DISPOSAL Ⓐ H

The food waste disposal outlet is represented on the plans by the symbol Ⓐ H. The food waste disposal is rated at 7.2 amperes. This appliance is connected to Circuit B19, a separate 15-ampere, 120-volt branch circuit in Panel B. The #14 NMD 90 conductors are used to supply the unit. The conductors terminate in a junction box provided on the disposal.

Overcurrent Protection

The food waste disposal is a motor-operated appliance. Food waste disposals normally are driven by a ¼- or ⅓-hp, split-phase, 120-volt motor. Therefore, running overcurrent protection must be provided. Most food waste disposal manufacturers install a built-in thermal protector to meet *CEC* requirements. Either a manual reset or an automatic reset thermal protector may be used.

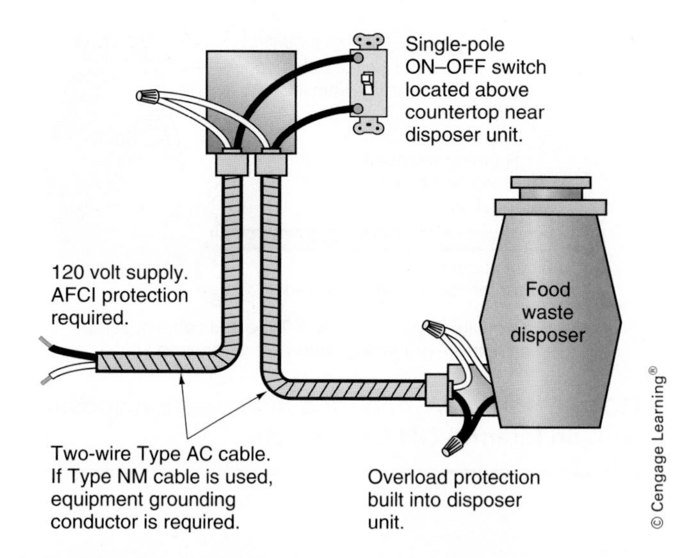

FIGURE 15-9 Wiring for a food waste disposal operated by a separate switch located above the countertop near the sink.

Disconnecting Means

All electrical appliances must be provided with some means of disconnecting them, *Rule 14–010*. The disconnecting means may be a separate ON/OFF switch (see Figure 15-9), a cord connection (Figure 15-10), or a branch-circuit switch or circuit breaker, such as that supplying the installation in Figure 15-11.

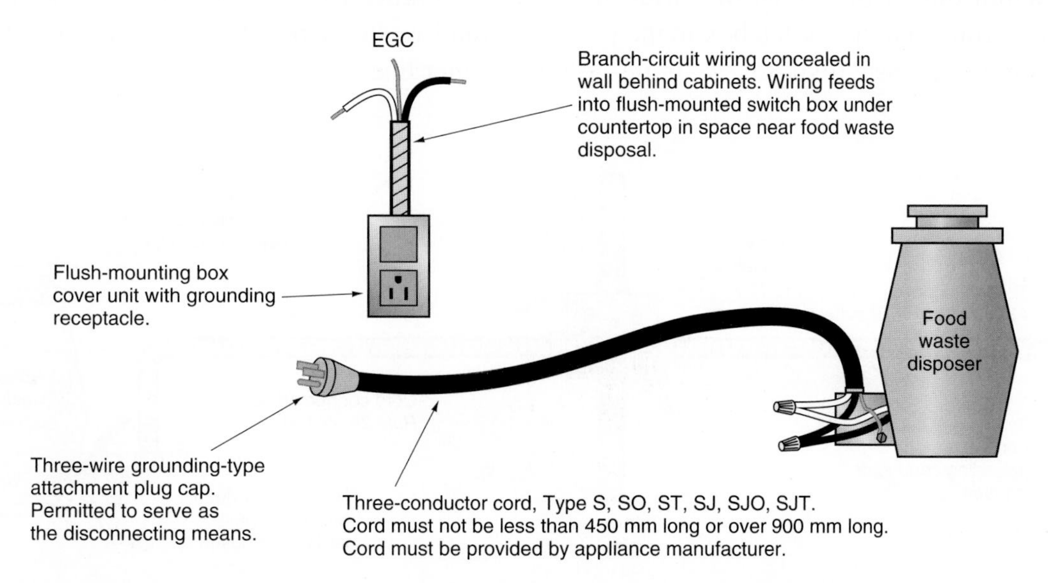

FIGURE 15-10 Typical cord connection for a food waste disposal, *Rule 26–744(8)*.

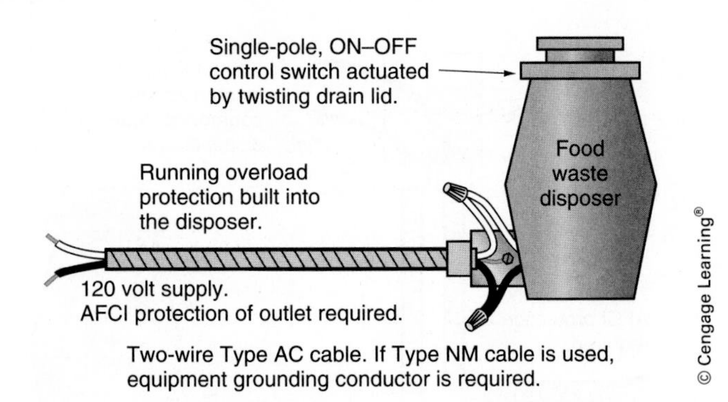

FIGURE 15-11 Wiring for a food waste disposal with an integral ON/OFF switch.

Some local codes may require that food waste disposals, dishwashers, and trash compactors be cord-connected (Figure 15-10) to make it easier to disconnect the unit, replace it, service it, and reduce noise and vibration. The appliance shall be intended or identified for flexible cord connection.

Turning the Food Waste Disposal On and Off

Separate ON/OFF Switch. When a separate ON/OFF switch is used, a simple circuit arrangement can be made by running a two-wire supply cable to a switch box located next to the sink at a convenient location above the countertop (Figure 15-9). Not only is this convenient, but it positions the switch out of the reach of children. A second cable is run from the switch box to the junction box of the food waste disposal. A single-pole switch in the switch box provides the ON/OFF control of the disposal.

In the kitchen of this residence, the food waste disposal is controlled by a single-pole switch located to the left of the kitchen sink above the countertop.

Another possibility, but not as convenient, is to install the switch for the disposal inside the cabinet space directly under the sink near the food waste disposal.

Integral ON/OFF Switch. A food waste disposal may be equipped with an integral prewired control switch. This integral control starts and stops the disposal when the user twists the drain cover into place. An extra ON/OFF switch is not required. To connect the disposal, the electrician runs the supply conductors directly to the junction box on the disposal and makes the proper connections; see Figure 15-11.

Disposal with Flow Switch in Cold Water Line. Some manufacturers of food waste disposals recommend that a *flow switch* be installed in the cold water line under the sink; see Figure 15-12. This switch is connected in series with the disposal motor. The flow switch prevents the disposal from operating until a predetermined quantity of water is flowing through it. The cold water helps prevent clogged drains by solidifying any grease in the disposal. Thus, the addition of a flow switch means that the disposal cannot be operated without water.

Figure 15-12 shows one method of installing a food waste disposal with a flow switch in the cold water line.

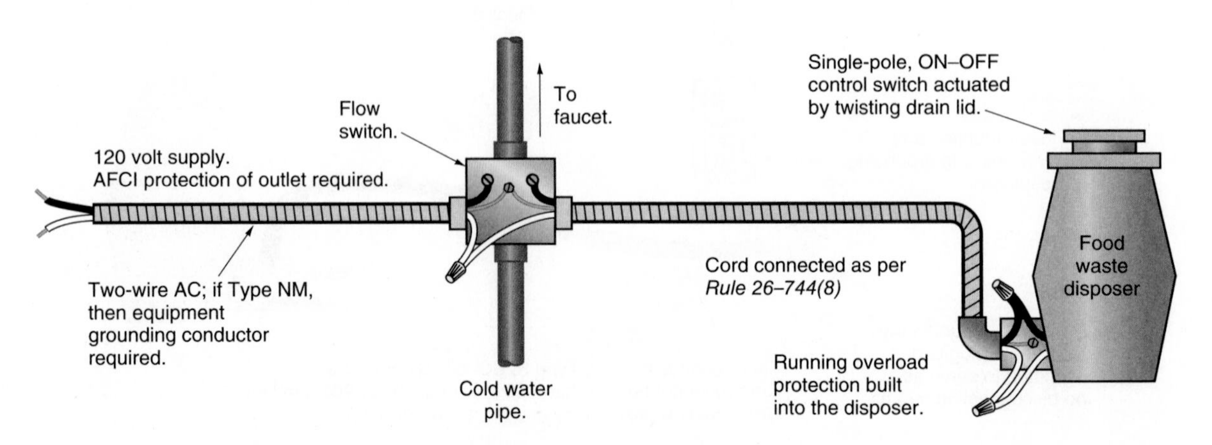

FIGURE 15-12 Disposal with a flow switch in the cold water line.

BONDING REQUIREMENTS

Rule 10–400 states that all exposed, non-current-carrying parts are to be bonded to ground. The presence of any of the conditions outlined in *Rules 10–400* through *10–404* requires that all electrical appliances in the dwelling be grounded. Food waste disposals can be grounded using any of the methods covered in the previous units, such as the separate bonding conductor of non-metallic-sheathed cable.

DISHWASHER Ⓐ$_{I}$

The dishwasher is supplied by a separate 15-ampere circuit connected to Circuit B5. The #14 NMD 90 conductors are used to connect the appliance. The dishwasher has a ¼-hp motor rated at 5.8 amperes and 120 volts. The outlet for the dishwasher is shown by the symbol Ⓐ$_{I}$. During the drying cycle of the dishwasher, a thermostatically controlled, 750-watt electric heating element turns on.

ACTUAL CONNECTED LOAD		CIRCUIT CALCULATION
Motor	5.8 A	5.8 A × 1.25 = 7.25 A
Heater	6.25 A	6.25 A
Total	12.05 A	13.5 A

Use a 15-ampere circuit.

For most dishwashers, the motor does not run during the drying cycle. Thus, the actual maximum demand on the branch circuit would be only the larger of the two loads—the 750-watt heating element. In some dishwashers, there is a fan or blower to speed up the drying time. In either case, for the dryer example shown, a 15-ampere circuit is more than adequate.

Energy-saving dishwashers are equipped with a built-in booster water heater that allows the homeowner to adjust the regular water heater temperature to 50°C or lower. The booster heater of the dishwasher then raises the water temperature from 50°C to a temperature not less than 60°C for the wash cycle, and 80°C for the rinse cycle, the temperature considered necessary to sanitize dishes.

Wiring the Dishwasher

The dishwasher manufacturer normally supplies a terminal or junction box on the appliance. The electrician connects the supply conductors in this box. The electrician must verify the dimensions of the dishwasher so that the supply conductors are brought through the wall at the proper point.

The dishwasher is a motor-operated appliance and requires running overcurrent protection. Such protection prevents the motor from burning out if it becomes overloaded or stalled. Normally, the required protection is supplied by the manufacturer as an integral part of the dishwasher motor. If the dishwasher does not have integral protection, the electrician must provide it as part of the installation of the unit.

Grounding

The dishwasher must be bonded to ground. The grounding of appliances is described fully in previous units and in the discussion of the food waste disposal.

PORTABLE DISHWASHERS

Portable dishwashers have one hose that is attached to the water faucet and a water drainage hose that hangs in the sink. The obvious location for a portable dishwasher is near the sink, so the dishwasher probably will be plugged into the outlet nearest the sink.

Portable dishwashers are supplied with a three-wire cord containing two circuit conductors and one bonding conductor and a three-wire grounding-type attachment plug cap. If the three-wire plug cap is plugged into the three-wire grounding-type receptacle, the dishwasher is adequately grounded. **Whenever there is a chance that the user may touch the appliance and a grounded surface such as the water pipe or faucet at the same time, the equipment must be bonded to ground to reduce the shock hazard.** The exceptions to *Rule 10–408* are *Subrules (3)* and *(4)*, which permit the double insulation of appliances and portable tools instead of grounding. Double insulation means that the appliance or tool has two separate insulations between the hot conductor and the person using the device. Although refrigerators, cooking units, water heaters, and other large appliances are not double-insulated, many portable hand-held appliances and tools have double insulation.

All double-insulated tools and appliances must be marked by the manufacturer to indicate this feature. These tools and appliances must be marked with the words "Double Insulated" or the symbol ⧈ as specified in CSA Standard C22.2 No. 0.1–M1985.

CORD CONNECTION OF FIXED APPLIANCES

As with the previously discussed food waste disposal, residential built-in dishwashers are permitted to be cord-connected (Figure 15-10).

Some electrical inspectors feel that using the cord-connection method to connect these appliances is better than direct connection (sometimes referred to as "hard-wired") because it allows them to be easily removed for maintenance and repairs.

REVIEW

Note: Refer to the *CEC* or the blueprints provided with this textbook when necessary. Where applicable, responses should be written in complete sentences.

COUNTER-MOUNTED COOKING UNIT CIRCUIT Ⓖ

1. a. What circuit supplies the counter-mounted cooking unit in this residence? ________
   ____

   b. What is the rating of this circuit? ________________________

2. What methods may be used to connect counter-mounted cooking units? ____________
   ____
   ____
   ____
   ____
   ____
   ____
   ____

3. Is it permissible to use standard 90°C insulated conductors to connect all counter-mounted cooking units? Why? ________________________
   ____
   ____
   ____

4. Indicate the maximum operating temperature (in degrees Celsius) for the following.
   a. Type TW conductors ________________
   b. Type TWN75 conductors ________________
   c. Type R90 conductors ________________

5. May NMWU cable be used to connect counter-mounted cooking units? ___________

____

6. a. What *CEC* rule applies to bonding a counter-mounted cooking unit?

____

b. How is the counter-mounted cooking unit bonded to ground? ___________

____

____

____

____

7. a. What is the mounting height of the receptacle supplying a range? ___________

b. In which direction should the U-ground slot be oriented? ___________

____

c. Give the *CEC* section number and rule. ___________

____

____

8. When the voltage to an element is doubled, the wattage [circle one]

a. increases. b. decreases.

9. One kilowatt equals ___________ Btu per hour.

10. How much heat will a 1000-watt heating element produce if operated continuously for 1 hour? ___________

WALL-MOUNTED OVEN CIRCUIT Ⓐ F

1. To what circuit is the wall-mounted oven connected? ___________

2. An oven is rated at 7.5 kW. This is equal to

a. ________ watts. b. ________ amperes at 240 volts.

3. a. What section of the CEC governs the bonding of a wall-mounted oven?

____

b. How are wall-mounted ovens bonded to ground? ___________

____

____

____

4. What is the type and rating of the overcurrent device protecting the wall-mounted oven?

____

____

____

5. How many metres of cable are required to connect the oven in the residence?

____

6. When connecting a wall-mounted oven and a counter-mounted cooking unit to one feeder, how long are the taps to the individual appliances? ____________________

____

____

____

7. A 6-kW counter-mounted cooking unit and a 4-kW wall-mounted oven are to be installed in a residence. Calculate the demand according to *Rule 8–300(1)*. Show all calculations.

8. What size conductors will feed

   a. from the panel to the junction box? ____________________

   b. each of the individual units in Question 7? ____________ and ____________

9. a. A freestanding range is rated at 11.8 kW, 240 volts. According to *Rule 8–300(1)*, what is the demand? ____________________

   b. What size wire (Type NMD 90) is required? ____________________

   c. What type of receptacle is required for the range? ____________________

FOOD WASTE DISPOSAL CIRCUIT DISPOSALS ▲H

1. How many amperes does the food waste disposal draw? ____________________

2. a. To what circuit is the food waste disposal connected? ____________________

   b. What size wire is used to connect the food waste disposal? ____________________

3. Means must be provided to disconnect the food waste disposal. The homeowner need not be involved in electrical connections when servicing the disposal if the disconnecting means is ____________________

____

4. How is running overcurrent protection provided in most waste disposal units?

____

____

5. Why are flow switches sometimes installed on food waste disposals? ____________________

____

____

6. What *CEC* sections relate to the grounding of appliances? ____________________

____

7. Do the plans show a wall switch for controlling the food waste disposal? ____________________

____

8. A separate circuit supplies the food waste disposal in this residence. How many metres of cable will be required to connect the disposal? ____________________

____

DISHWASHER CIRCUIT Ⓐ

1. a. To what circuit is the dishwasher in this residence connected? ______________

   b. What size wire is used to connect the dishwasher? ______________

2. The motor on the dishwasher is [circle one]

   a. ¼ hp   b. ⅓ hp   c. ½ hp

3. The heating element is rated at [circle one]

   a. 750 watts   b. 1000 watts   c. 1250 watts

4. How many amperes at 120 volts do the following heating elements draw?

   a. 750 watts ______________

   b. 1000 watts ______________

   c. 1250 watts ______________

5. How is the dishwasher in this residence grounded? ______________

6. What type of cord is used on most portable dishwashers? ______________

7. How is a portable dishwasher grounded? ______________

8. a. What is double insulation? ______________

   b. What is the symbol for double insulation? ______________

9. Who is to furnish the dishwasher? ______________

10. Does the *CEC* require a separate circuit for the dishwasher?

____

UNIT 16

Branch Circuits for the Laundry, Washroom, and Attic

OBJECTIVES

After studying this unit, you should be able to

- make the proper wiring and grounding connections for large appliances, using various wiring methods
- understand the *Canadian Electrical Code (CEC)* requirements governing the receptacle outlet(s) for laundry areas
- discuss the *CEC* rules pertaining to wiring methods in attics
- demonstrate the proper way to connect pilot lights and pilot light switches

DRYER CIRCUIT Ⓓ

A separate circuit is provided in the laundry room for the electric clothes dryer (Figure 16-1). This appliance demands a large amount of power. The special circuit provided for the dryer is indicated on the plans by the symbol $\blacktriangle_D$. The dryer is connected to Circuit B1/B3.

Clothes Dryer

Clothes dryer manufacturers make many dryer models with different wiring arrangements and connection provisions. All electric dryers have electric heating units and a motor-operated drum, which tumbles the clothes as the heat evaporates the moisture. Dryers also have thermostats to regulate the temperature of the air inside the dryer. Timers regulate the lengths of the various drying cycles. Drying time, determined by the type of fabric, can be set to a maximum of about 85 minutes. Dampness controls are often provided to stop the drying process at a preselected stage of dampness so that it is easier to iron the clothes.

Clothes dryers create a lot of humid air. Therefore, be sure the dryer is vented to the outside. This residence also has an exhaust fan located in the laundry room.

Table 16-1 shows the estimated loads for the various receptacles and fixtures in the laundry, washroom, and attic.

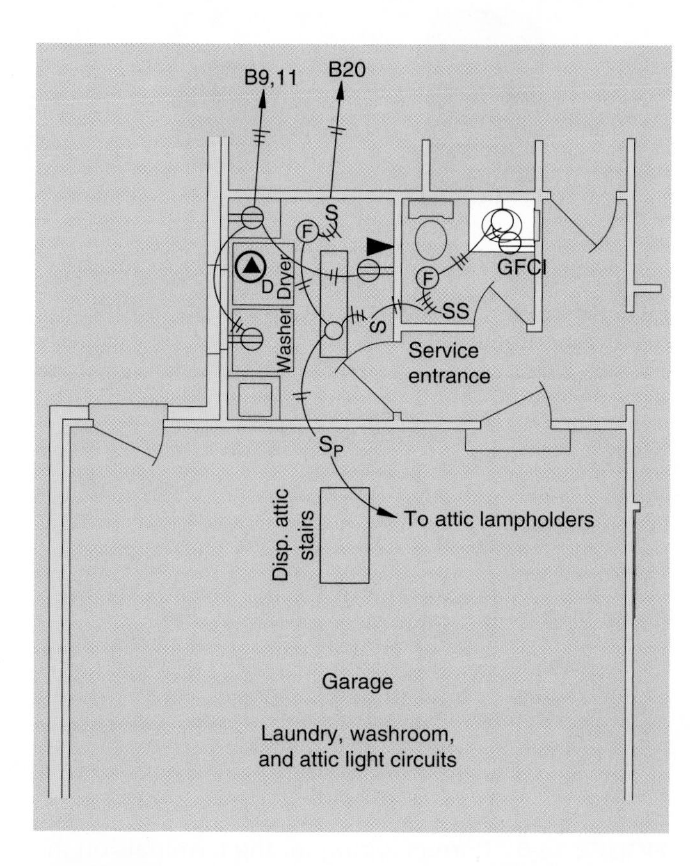

FIGURE 16-1 Cable layout for Circuits B9, B11, and B20.

Connection of Electric Clothes Dryer

The connection of electric clothes dryers is governed by *Rule 26–744(2, 3)*. Figure 16-2 shows the internal components and wiring of a typical electric clothes dryer.

A separate receptacle, as shown in CSA configuration 14–30R, *Diagram 1*, should be installed on the wall adjacent to the dryer. *Rule 26–744(3)* requires that a cord set as shown in CSA configuration 14–30P be used to connect the dryer to the receptacle; see Figure 16-3. A disconnecting means is not required for this appliance, since the attachment plug and receptacle may be used instead of a switch, *Rule 26–746(2)(a)*.

The receptacle can be wired with NMD 90 cable if concealed wiring is desired. If the NMD 90 is

TABLE 16-1

Laundry–washroom–attic, Circuit B20.

DESCRIPTION	QUANTITY	WATTS	VOLT–AMPERES
Washroom receptacles (GFCI)	1	120	120
Washroom vanity fixture	1	200	200
Laundry room fluorescent fixtures @ 32 W each	2	64	69
Attic lights @ 60 W each	4	240	240
Laundry exhaust fan	1	80	90
Washroom exhaust fan	1	80	90
Totals	10	784	809

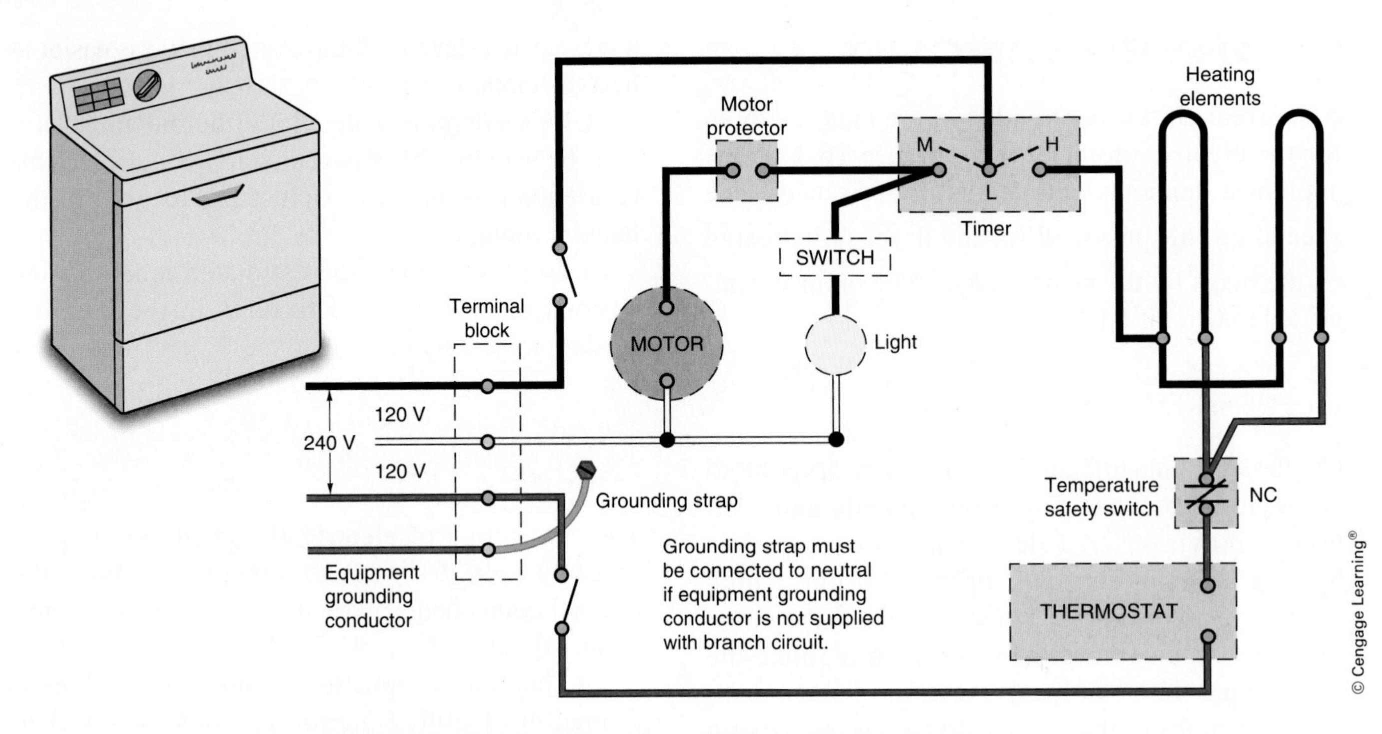

FIGURE 16-2 Laundry dryer: Wiring and components.

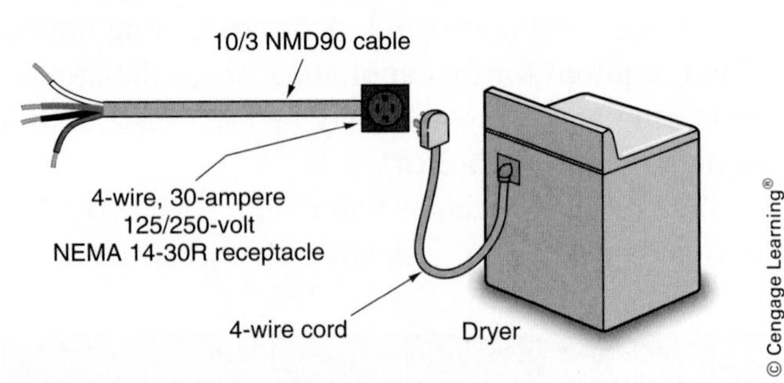

FIGURE 16-3 Dryer connection using a cord set.

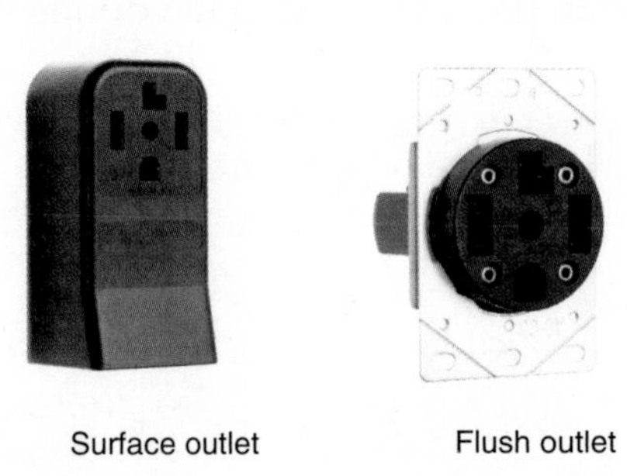

FIGURE 16-4 Surface mount and flush mount dryer receptacle

run exposed, it must be protected from mechanical injury. In the residence, the wiring method for the dryer is non-metallic-sheathed cable (NMSC), which runs from Panel B to a dryer receptacle on the wall behind the dryer. A cord set is attached to the dryer and is plugged into this receptacle. The receptacle and cord have a current-carrying capacity that is rated at 125% greater than the dryer rating.

A standard dryer receptacle and cord set, as shown in Figure 16-4, are usually rated at 30 amperes, 125/250 volts. Figure 16-5 shows the correct wiring connections to the terminals on the receptacle. Cords and receptacles rated for a higher current are available. The L-shaped neutral slot of the outlet does not accept a 50-ampere cord set.

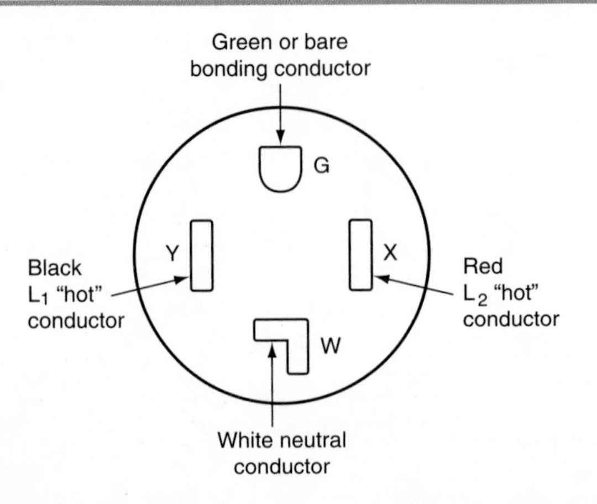

FIGURE 16-5 Correct wiring for the terminals on the dryer receptacle rated at 30 amperes, 125/250 volts, CSA configuration 14–30R. Refer to *Diagram 1*.

The receptacle must be flush-mounted, according to *Rule 26–744(9)*. The outlet will fit a 119- × 54-mm square box. Make sure that the box is mounted in the proper orientation to ensure that the appliance cord hangs down and reduces strain on the receptacle.

In this residence, the dryer is installed in the laundry room. The schedule of special-purpose outlets in the specifications shows that the dryer is rated at 5700 watts and 120/240 volts. The schematic wiring diagram, Figure 16-2, shows that the heating elements are connected to the 240-volt terminals of the terminal block. The dryer motor and light are connected between the hot wire and the neutral terminal, 120 volts.

The motor of the dryer has integral thermal protection. This prevents the motor from reaching dangerous temperatures as a result of an overload or failure to start. *Rule 14–610* specifies the overcurrent protection requirements for electric clothes dryers and states that, when fuses are used, they must be time-delay (TD) or low melting point fuses (P-rated).

To determine the feeder rating for the house, the *CEC* requires that the load included for a dryer be 25% of the nameplate rating, assuming that an electric range has been included.

The branch-circuit ampacity is based on the nameplate rating of the dryer. If it is not clear what the load of an appliance is, always base it on the nameplate rating.

Conductor Size

Rule 26–746(1) states that any appliance rated at more than 1500 watts shall be supplied from a branch circuit used solely for one appliance.

The electric clothes dryer in this residence has a nameplate rating of 5700 watts. Thus,

$$I = \frac{W}{E} = \frac{5700}{240} = 23.75 \text{ amperes}$$

If we apply *Rule 8–104*, that is, to multiply the current rating times 125% to find the required circuit ampacity, we find that

$$23.75 \text{ amperes} \times 1.25 = 29.7\ (30) \text{ amperes}$$

Accordingly, the dryer Circuit B1/B3 is a three-wire, 30-ampere, 240-volt circuit. Consulting *CEC Table 2*, it can be seen that the conductors in the non-metallic-sheathed cable that feed the dryer could be #12 AWG wire if 90°C-rated and #10 AWG wire if 75°C-rated. Since all devices in this house are to be hooked up to circuit breakers and since circuit breakers have either a 60°C or 75°C rating, the #10 non-metallic-sheathed cable will have to be used.

By *CEC* definition, a branch circuit refers to the circuit conductors between the final overcurrent device and the outlet. Therefore, the conductors between the overcurrent device in Panel B and the dryer outlet constitute the dryer's branch circuit.

Overcurrent Protection

Rule 14–104 states that the branch-circuit overcurrent device rating is not to exceed the allowable ampacity of the conductors. The size of the fuse or circuit breaker may be obtained from *Table 13*. Here, again, the manufacturer of the appliance will mark the circuit overcurrent device rating. If the appliance is not marked, the overcurrent protection shall be in accordance with *Rule 4–004* for the conductors—30 amperes for our example, as discussed above.

Bonding Frames of Ranges and Dryers

Rule 10–408(1) requires that the exposed non-current-carrying metal parts of portable equipment such as clothes dryers or ranges be bonded to ground.

The branch-circuit conductors for the dryer are connected to Panel B, Circuit B1/B3. This is a double-pole circuit that breaks only the two ungrounded conductors. The neutral conductor remains solid and unbroken.

RECEPTACLE OUTLETS—LAUNDRY

Rule 26–710 requires that at least two receptacles be installed in a laundry room—one for a washing machine and at least one more. *Rule 26–722* requires a minimum of one branch circuit to be installed solely for receptacles in the laundry room.

Wall type

Ceiling type

FIGURE 16-6 Exhaust fans.

In the laundry room in this residence, the 15-ampere Circuit B11 supplies the receptacle for the washer. Circuit B11 serves no other outlets. The receptacle is not required to be Ground-Fault Circuit Interrupter (GFCI)-protected. Figure 16-1 shows the cable layout for the laundry room.

Circuit B9 connects to the two receptacles in the laundry, which will serve an iron, a sewing machine, and other small portable appliances.

Circuit B20 has already been included in the watts per square metre calculations for service entrance and feeder conductor sizing, *Rule 8–200(1)(a)(i,ii)*. See the complete calculations in Unit 3.

LIGHTING CIRCUIT

The general lighting circuit for the laundry and washroom is supplied by Circuit B20.

The types of vanity lights, ceiling fixtures, and standard wall receptacle outlets, as well as the circuitry and switching arrangements, are similar to those discussed in other units of this text.

Exhaust Fans

Ceiling exhaust fans are installed in the laundry and washroom and connected to Circuit B20.

The exhaust fan in the laundry will remove excess moisture resulting from the use of the clothes washer.

Exhaust fans may be installed in walls or ceilings; see Figure 16-6. The wall-mounted fan can be adjusted to fit the thickness of the wall. If a ceiling-mounted fan is used, sheet-metal duct is recommended between the fan unit and the outside of the house.

Any exhaust ducting located in an attic space must be insulated. During the cold season, any moisture that is removed by an exhaust fan will condense in an uninsulated duct. This will lead to premature rusting of the exhaust duct and fan. The moisture can also leak out of the fan or duct, causing serious water damage.

The fan unit terminates in a metal hood or grille on the exterior of the house. The fan has a shutter that opens as the fan starts up and closes as the fan stops. The fan may have an integral pull-chain switch for starting and stopping, or it may be used with a separate wall switch. In either case, single-speed or multispeed control is available. The fan in use has a very small power demand, 90 volt–amperes.

HUMIDITY CONTROL

An electrically heated dwelling can experience a problem with excess humidity due to the "tightness" of the house, because of the care taken in the installation of the insulation and vapour barriers. High humidity is uncomfortable. It promotes the growth of mould and the deterioration of fabrics and floor coverings. In addition, the framing members, wall panels, and plaster or drywall of a dwelling may deteriorate because of the humidity. Insulation must be kept dry, or its efficiency decreases. A low humidity level can be maintained by automatically controlling the exhaust fan.

One type of automatic control is the *humidistat.* This device starts the exhaust fan when the relative humidity reaches a high level. The fan exhausts air until the relative humidity drops to a comfortable level. The comfort level is about 50% relative humidity. Adjustable settings are from 0 to 90% relative humidity.

The electrician must check the maximum current and voltage ratings of the humidistat before the device is installed. Some humidistats are extra-low-voltage devices and require a relay. Other humidistats are rated at line voltage and can be used to switch the motor directly (a relay is not required). However, a relay must be installed on the line-voltage humidistat if the connected load exceeds the maximum allowable current rating of the humidistat.

The switching mechanism of a humidistat is controlled by a nylon element (see Figure 16-7) that is very sensitive to changes in humidity. A bimetallic element cannot be used because it reacts to temperature changes only.

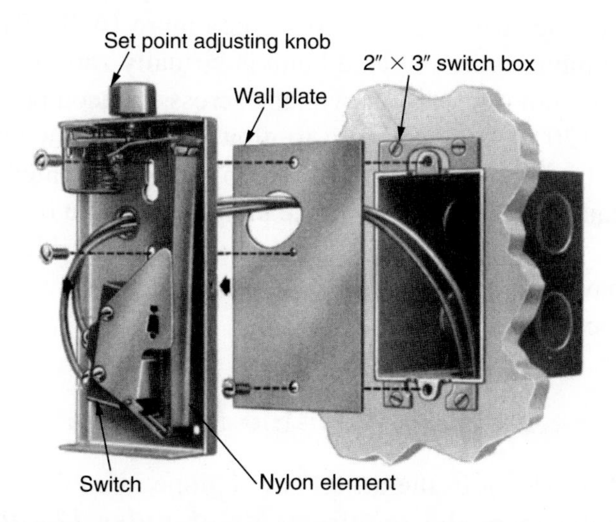

FIGURE 16-7 Details of a humidity control used with an exhaust fan.

Wiring

The humidistat and relay in the dwelling are installed using #14 NMD 90. The cable runs from Circuit A10 to a 4-in. square, $1^1/_2$-in. deep outlet box located in the attic near the fan; see Figure 16-8, Part B. A motor circuit switch is mounted on this outlet box. This switch unit serves as a disconnecting means within sight of the motor, as required by *Rules 28–600* and *28–602(3)(e)*. Motor overload protection is provided by the heater unit installed in the motor-rated switch. As an alternative, an AC general-purpose switch rated at 125% of the full load amperes (FLA) of the motor may be used as the disconnecting means.

Mounted next to the switch unit is a relay that can carry the full-load current of the motor. Extra-low-voltage thermostat wire runs from this relay to the humidistat. (The humidistat is located in the hall between the bedrooms.) The thermostat cable is a two-conductor cable since the humidistat has a single-pole, single-throw switching action. The line-voltage side of the relay provides single-pole switching to the motor. Normally, the white grounded conductor is not broken.

Although the example in this residence utilizes a relay to use extra-low-voltage wiring, it is also proper to install line-voltage wiring and use line-voltage thermostats and humidity controls. These are becoming very popular, with the switches actually being solid-state speed controllers, permitting operation of the fan at an infinite number of speeds between high and low settings.

ATTIC LIGHTING AND PILOT LIGHT SWITCHES

It is always good practice to install at least one lighting outlet in an attic and to control it by a switch located near the access point of the attic. The light should have a guard to protect it from injury or a short flexible drop light may be used, *Rule 30–314(2)*. Many attics are used for storage. They may also contain equipment that might need to be serviced, so install a lighting outlet near it. Ambient temperatures in attics can exceed the ratings of some equipment. Any circuit

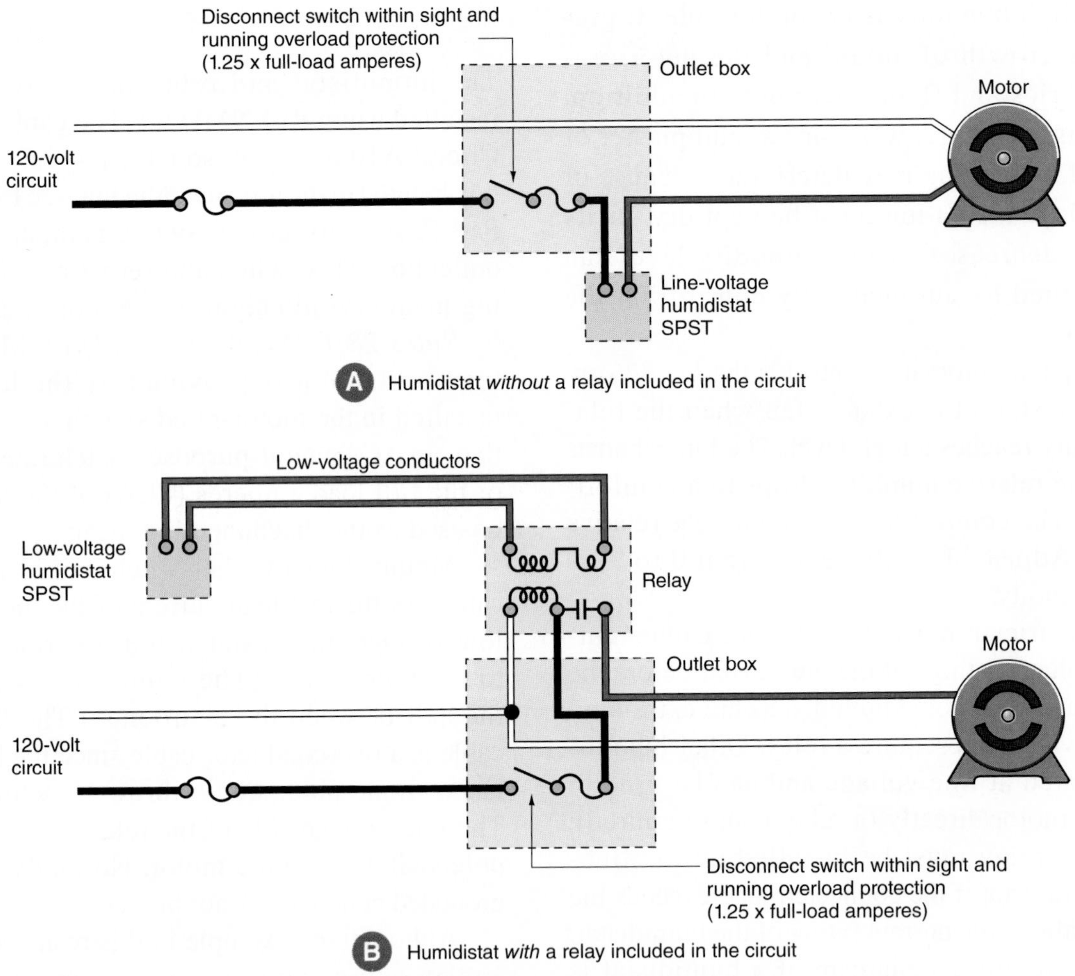

Disconnect switch within sight and running overload protection (1.25 x full-load amperes)
Outlet box
Motor
120-volt circuit
Line-voltage humidistat SPST

A Humidistat *without* a relay included in the circuit

Low-voltage conductors
Low-voltage humidistat SPST
Relay
Outlet box
Motor
120-volt circuit
Disconnect switch within sight and running overload protection (1.25 x full-load amperes)

B Humidistat *with* a relay included in the circuit

FIGURE. 16-8 Wiring for humidity control.

and equipment wiring calculations and requirements must take this into account, *Rule 4–004(8)*. Table 16-2, which reproduces *CEC Table 5A*, shows the de-rating factors for ambient temperatures greater than 30°C. These must be applied for wiring in attics where the ambient temperature could easily exceed 30°C.

The four keyless lampholders in the attic are turned on and off by a single-pole switch on the garage wall close to the attic storable stairway. Associated with this single-pole switch is a pilot light. The pilot light may be located in the handle of the switch or separately mounted. Figure 16-9 shows how pilot lamps are connected in circuits containing either single-pole or three-way switches.

If a neon pilot lamp in the handle (toggle) of a switch does not have a separate grounded conductor connection, it will glow only when the switch is in the OFF position, because the neon lamp will then be in series with the lamp load; see Figure 16-10. The voltage across the load lamp is virtually zero, so it does not light, and the voltage across the neon lamp is 120 volts, allowing it to glow. When the switch is turned on, the neon lamp is bypassed (shunted), causing it to turn off and the lamp load to turn on.

Use this type of switch when it is desirable to have a switch glow in the dark to make it easy to locate.

Installation of Cable in Attics

The wiring in the attic is to be done in cable and must meet the requirements of *Rules 12–500* through *12–526* for non-metallic-sheathed cable. These rules describe how the cable is to be

TABLE 16-2

CEC Table 5A.

CEC TABLE 5A
CORRECTION FACTORS APPLYING TO *TABLES 1, 2, 3,* AND *4*
(AMPACITY CORRECTION FACTORS FOR AMBIENT TEMPERATURES ABOVE 30°C)
(Rules 4–004(8), 12-2210,* and *12–2260* and *Tables 1, 2, 3, 4, 57,* and *58)

Ambient temperature °C	CORRECTION FACTOR								
	Insulation Temperature Rating, °C								
	60	75	90	105*	110*	125*	150*	200*	250*
35	0.91	0.94	0.96	0.97	0.97	0.97	0.98	0.99	0.99
40	0.82	0.88	0.91	0.93	0.94	0.95	0.96	0.97	0.98
45	0.71	0.82	0.87	0.89	0.90	0.92	0.94	0.95	0.97
50	0.58	0.75	0.82	0.86	0.87	0.89	0.91	0.94	0.95
55	0.41	0.67	0.76	0.82	0.83	0.86	0.89	0.92	0.94
60	—	0.58	0.71	0.77	0.79	0.83	0.87	0.91	0.93
65		0.47	0.65	0.73	0.75	0.79	0.84	0.89	0.92
70	—	0.33	0.58	0.68	0.71	0.76	0.82	0.87	0.90
75	—	—	0.50	0.63	0.66	0.73	0.79	0.86	0.89
80	—	—	0.41	0.58	0.61	0.69	0.76	0.84	0.88
90	—	—	—	0.45	0.50	0.61	0.71	0.80	0.85
100	—	—	—	0.26	0.35	0.51	0.65	0.77	0.83
110						0.40	0.58	0.73	0.80
120	—	—	—		—	0.23	0.50	0.69	0.77
130							0.41	0.64	0.74
140	—	—	—		—	—	0.29	0.59	0.71
Col. 1	Col. 2	Col. 3	Col. 4	Col. 5	Col. 6	Col. 7	Col. 8	Col. 9	Col. 10

* These ampacities are applicable only under special circumstances where the use of insulated conductors having this temperature rating is acceptable.

Notes:

1. These correction factors apply to *Tables 1, 2, 3,* and *4.* The correction factors in Column 2 also apply to *Table 57.*
2. The ampacity of a given conductor type at these higher ambient temperatures is obtained by multiplying the appropriate value from *Tables 1, 2, 3,* or *4* by the correction factor for that higher temperature.

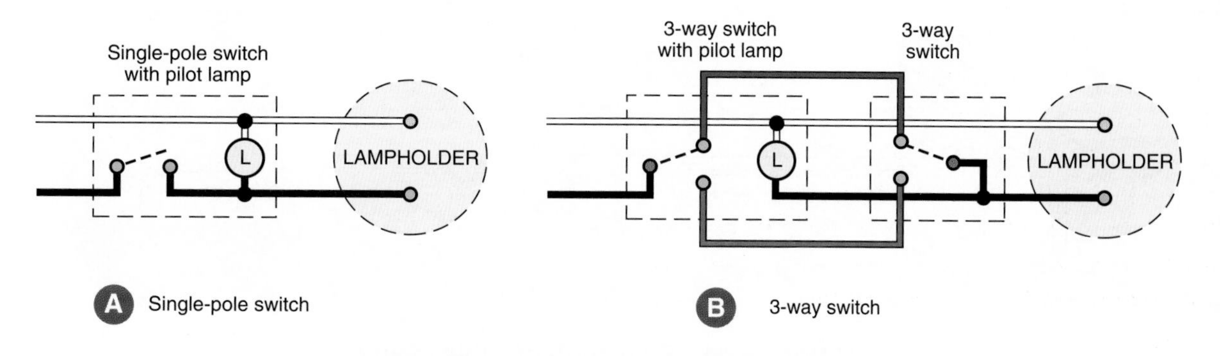

FIGURE 16-9 Pilot lamp connections.

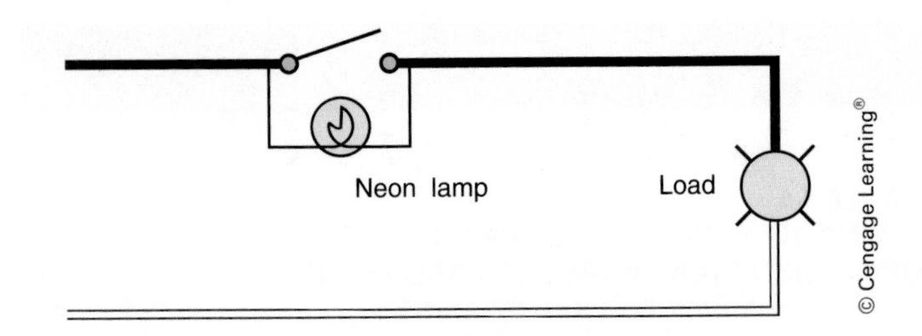

FIGURE 16-10 Neon pilot lamp in the handle of the toggle switch.

protected; see Figure 16-11. In accessible attics (see Figure 16-11), cables must *not* be

- run across the top of floor or ceiling joists ①
- run across the face of studs ② or rafters ③ that have a vertical distance between the floor or floor joists exceeding 1 m.

Guard strips are not required if the cable is run along the sides of rafters, studs, or floor joists ④, provided *Rule 12–516(1, 2)* is met.

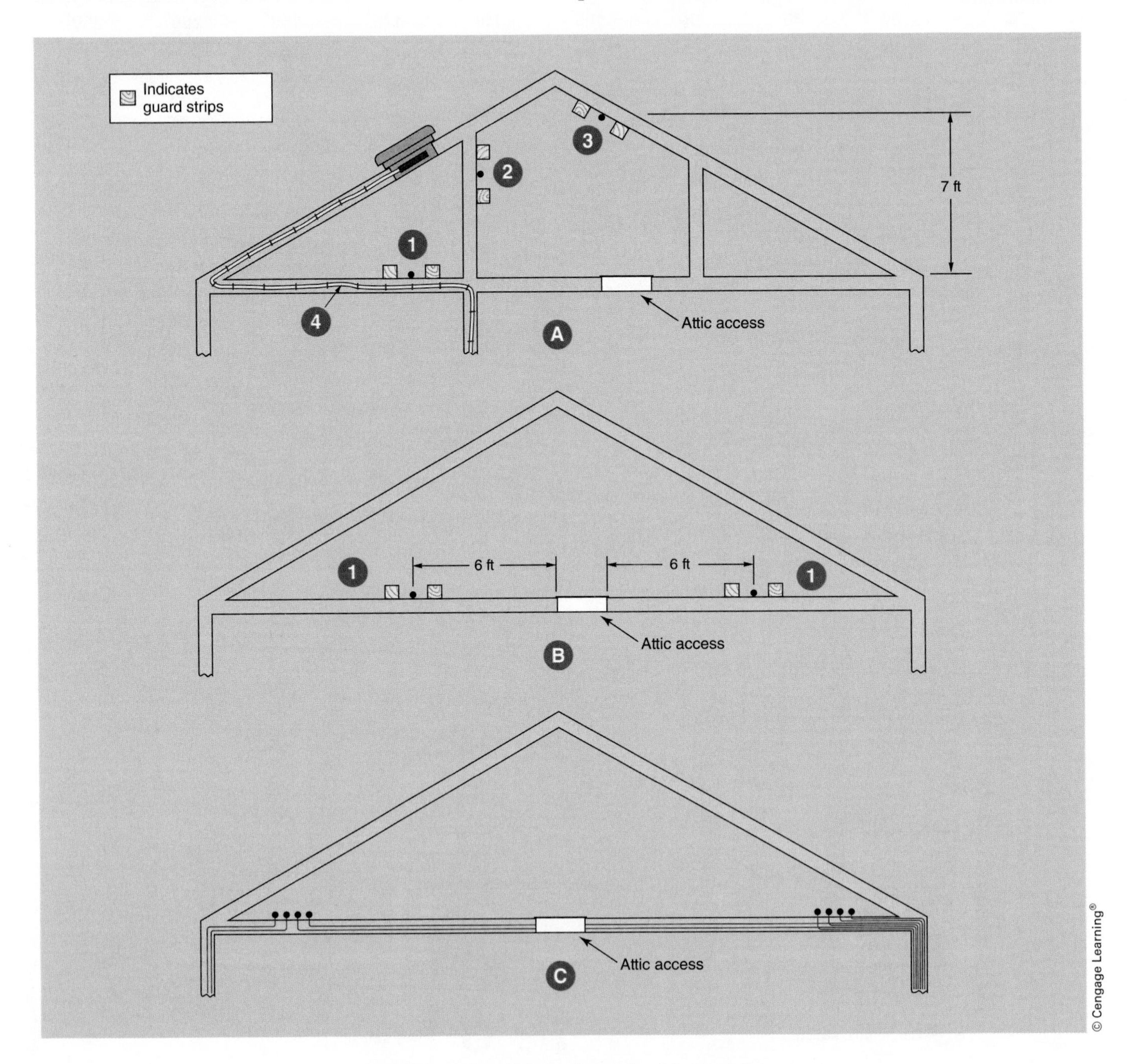

FIGURE 16-11 Protection of cable in attic.

Figure 16-11C illustrates a cable installation that most electrical inspectors consider to be safe. Because the cables are installed close to the point where the ceiling joists and the roof rafters meet, they are protected from physical damage. It would be very difficult for a person to crawl into this space or to store cartons in an area with a clearance of less than 1 m; see Figure 16-11(B).

Although the plans for this residence do not show a 600-mm-wide catwalk in the attic, it is installed. In the future the owner may decide to install additional flooring in the attic to obtain more storage space. Because of the large number of cables required to complete the circuits, it would interfere with flooring to install guard strips wherever the cables run across the tops of the joists; see Figure 16-12. However, the cables can be run through holes bored in the joists and along the sides of the joists and rafters. In this way, the cables do not interfere with the flooring.

Most building codes will not allow manufactured roof systems to be drilled. This includes roof trusses.

When running cables through framing members, be careful to maintain at least 32 mm between the cable and the edge of the framing member. This minimizes the possibility of driving nails into the cable.

ATTIC EXHAUST FAN CIRCUIT Ⓐ$_L$

An exhaust fan is mounted in the roof of the house; see Figure 16-13A. When running, this fan removes hot, stagnant, or humid air from the attic. It will also draw in fresh air through louvres located in the end gables. Such an attic exhaust fan can be used to lower the indoor temperature of the house by as much as 5°C to 10°C.

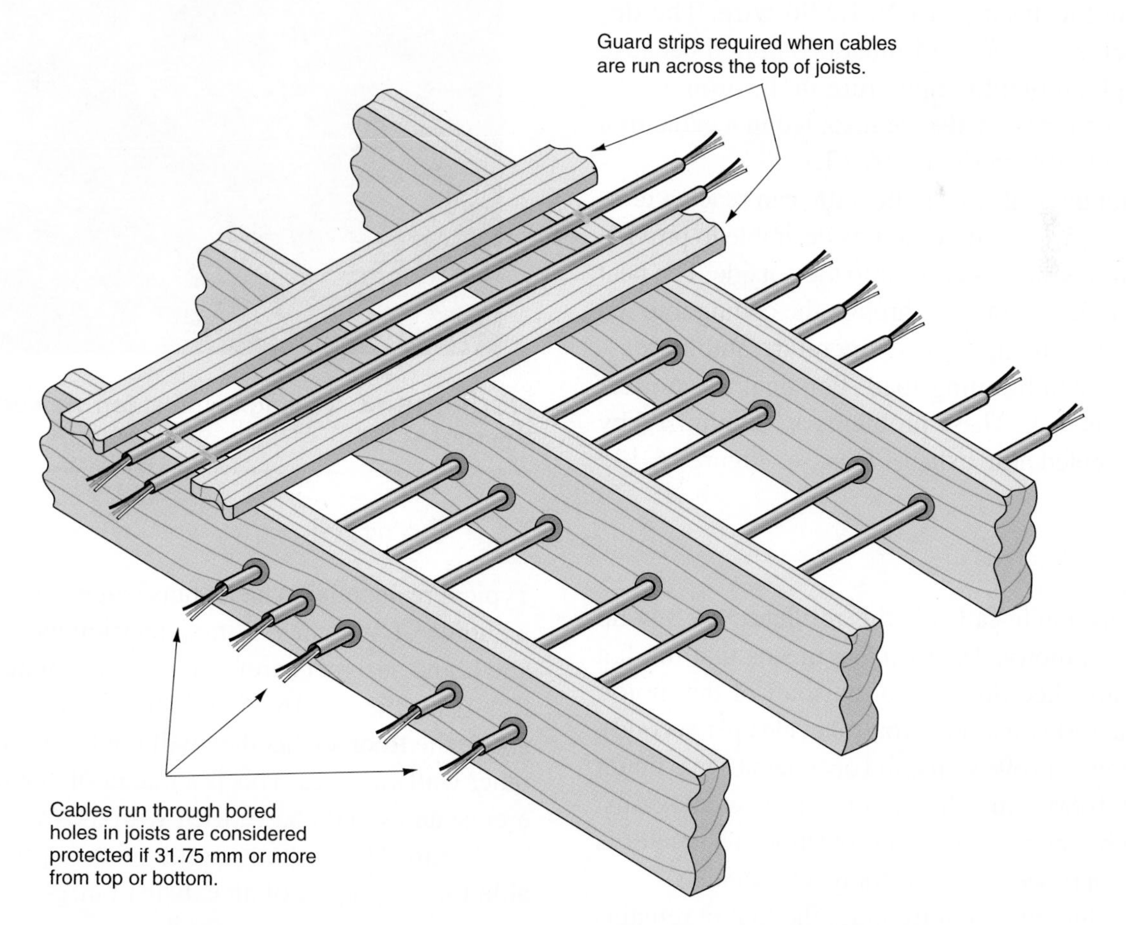

FIGURE 16-12 Methods of protecting cable installations in attics.

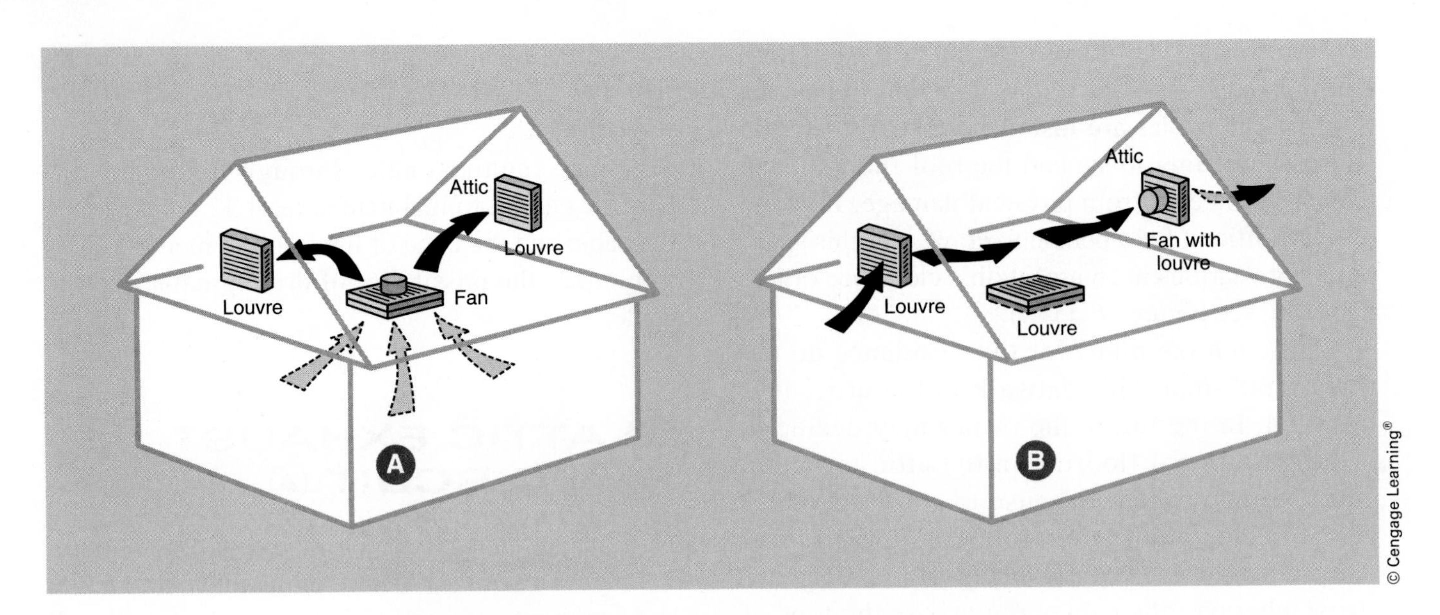

FIGURE 16-13 Exhaust fan installation in an attic.

The exhaust fan in this residence is connected to Circuit A10, a 120-volt, 15-ampere circuit in the main panel utilizing #14 NMD 90 wire. The derating factors in *Table 5A* must be applied because of the high ambient temperature in the attic.

Exhaust fans can also be installed in a gable of a house; see Part B in Figure 16-13.

On hot days, the air in the attic can reach a temperature of 65°C or more, so it is desirable to provide a means of taking this attic air to the outside. The heat from the attic can radiate through the ceiling into the living areas, raising their temperature and increasing the air-conditioning load. Personal discomfort increases as well. These problems are minimized by properly vented attic exhaust fans; see Figure 16-14.

FIGURE 16-14 Ceiling exhaust fan viewed from the roof.

Fan Operation

The exhaust fan installed in this residence has a ¼-hp direct-drive motor. *Direct-drive* means that the fan blade is attached directly to the shaft of the motor. It is rated 120 volts, 60 hertz (60 cycles per second), 5.8 amperes, and 696 watts, and operates at a maximum speed of 1050 rpm. The motor is protected against overload by internal thermal protection. The motor is mounted on rubber cushions for quiet operation.

When the fan is not running, the louvre remains in the closed position. When the fan is turned on, the louvre opens. When the fan is turned off, the louvre automatically closes to prevent the escape of air.

Fan Control

Typical residential exhaust fans can be controlled by a choice of switches. Some electricians, architects, and home designers prefer to have fan switches (controls) (see Figures 16-15 and 16-16) mounted 1.83 m above the floor so that they will not be confused with other wall switches. This is a matter of personal preference and should be verified with the owner.

Figure 16-17 shows some of the options available for the control of an exhaust fan:

a. A simple ON/OFF switch (A).

b. A speed control switch that allows multiple and/or an infinite number of speeds, (B). (The

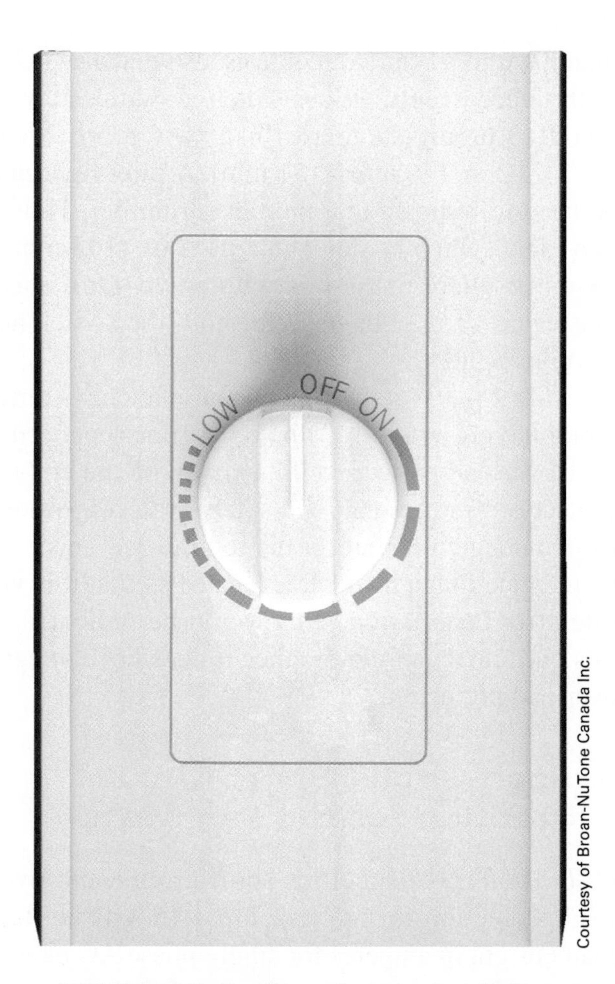

FIGURE 16-15 Speed control switch.

FIGURE 16-16 Timer switch with speed control.

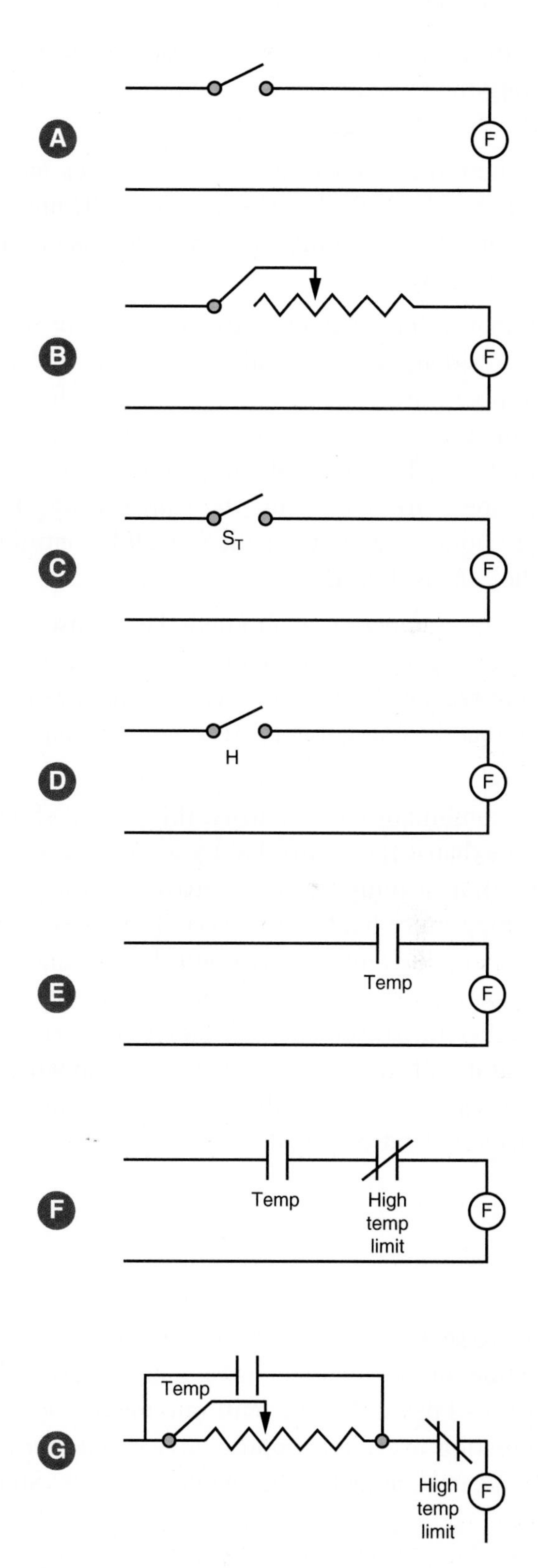

FIGURE 16-17 Options for the control of exhaust fans.

exhaust fan in this residence has this type of control.) See also Figure 16-15.

c. A timer switch that allows the user to select how long the exhaust fan is to run, up to 12 hours (C). See also Figure 16-16.

d. A humidity control switch that senses moisture build-up (D). See the section on "Humidity Control" in this unit for further details on this type of switch.

e. Exhaust fans mounted into end gables or roofs of residences are available with an adjustable temperature control on the frame of the fan. This thermostat generally has a start range of 21°C to 54°C, and will automatically stop at a temperature 5°C below the start setting, thus providing totally automatic ON/OFF control of the exhaust fan (E).

f. A high-temperature automatic heat sensor that will shut off the fan motor when the temperature reaches 93°C (F). This is a safety feature so that the fan will not spread a fire. Connect in series with other control switches.

g. A combination of controls, this circuit shows an exhaust fan controlled by an infinite-speed switch, a temperature control, and a high-temperature heat sensor (G). The speed control and the temperature control are connected in parallel so that either can start the fan. The high-temperature heat sensor is in series so that it will shut off the power to the fan when it senses 93°C, even if the speed control or temperature control is in the ON position.

Overload Protection for the Fan Motor

There are several ways to provide running overload protection for the ¼-hp attic fan motor. For example, an overload device can be built into the motor, or a combination overload tripping device and thermal element can be added to the switch assembly. Some manufacturers of motor controls also make switches with overcurrent devices. Such a switch may be installed in any standard flush switch box opening 3 × 2 in. (76 mm × 51 mm). A pilot light may be used to indicate that the fan is running. Thus, a two-gang switch box or 4-in. × 1½-in. (102-mm × 38-mm) square outlet box with a two-gang raised plaster cover may be used for both the switch and the pilot light.

Rule 28–308 states that an automatically started motor having a rating of 1 hp or less does not need to have overload protection if the motor on the unit has protection features that will safely protect the motor from damage if it stalls or the rotor locks. This must be indicated on a nameplate located so that it is visible after installation. This would be indicated on this particular fan motor since it does have integral overload protection.

Branch-Circuit Short-Circuit Protection for the Fan Motor

CEC Table D16 specifies short-circuit and overload protection for motors; *Table 45* lists the full-load current in amperes for single-phase AC motors. For a full-load current rating of 5.8 amperes, *Table D16* shows that the rating of the branch-circuit overcurrent device shall not exceed 15 amperes if time-delay or nontime-delay fuses are used and 15 amperes if standard thermal-magnetic circuit breakers are used.

The exhaust fan is connected to Circuit A10, which is a 15-ampere, 120-volt circuit. This circuit does not feed any other loads. The circuit supplies the exhaust fan only.

Time-delay fuses sized at 115% to 125% of the full-load ampere rating of the motor will provide both running overload protection and branch-circuit short-circuit protection. If the motor is equipped with inherent, built-in overload protection, the time-delay fuse in the circuit provides secondary backup protection, providing double protection against possible motor burnout.

REVIEW

Note: Refer to the *CEC* or the blueprints provided with this textbook when necessary. Where applicable, responses should be written in complete sentences.

DRYER CIRCUIT AND LAUNDRY ROOM CIRCUITS ▲D

1. List the switches, receptacles, and other wiring devices that are connected to Circuit B20.________________________________

2. What *CEC* rule(s) refers (refer) to the receptacle required for the laundry equipment? Briefly state the requirements. ________________________________

3. The receptacle outlets are connected to branch circuits that have been calculated to have a maximum load of ___ watts per circuit. Circle one: 1200, 1440, 1500 watts.
   a. What regulates the temperature in the dryer? ________________________________
   b. What regulates the drying time? ________________________________

4. What is the method of connection to an electric clothes dryer? ________________________________

5. What is the unique shape of the neutral blade of a 30-ampere dryer cord set?

6. a. What is the minimum number of receptacle outlets that must be installed in a laundry room? ________________________________
   b. Is a separate branch circuit required? ________________________________
   c. Quote the *CEC* rule numbers. ________________________________

7. What is the maximum permitted current rating of a portable appliance on a 30-ampere branch circuit? ________________________________

8. What is the current draw of the exhaust fan in the laundry? ______________________
____
____
____
____

9. What special type of switch controls the attic lights? ______________________
____

10. If an attic exhaust fan draws 12 amperes at 120 volts and the ambient temperature reaches 50°C, what size and type of wire will be required to feed the fan? Refer to *CEC, Table 5A*. ______________________
____

11. When installing cables in an attic along the top of the floor joists, ______-m clearance must not be exceeded.

12. The total estimated watts for a circuit has been calculated to be 1300 volt-amperes. How many amperes is this at 120 volts? ______________________

13. In what circumstances would lighting and/or power be provided in an attic? ______________________
____

14. The following is a layout of the lighting circuit for the laundry, washroom, rear entrance hall, and attic. Complete the wiring diagram using coloured pens or pencils to indicate conductor insulation colour.

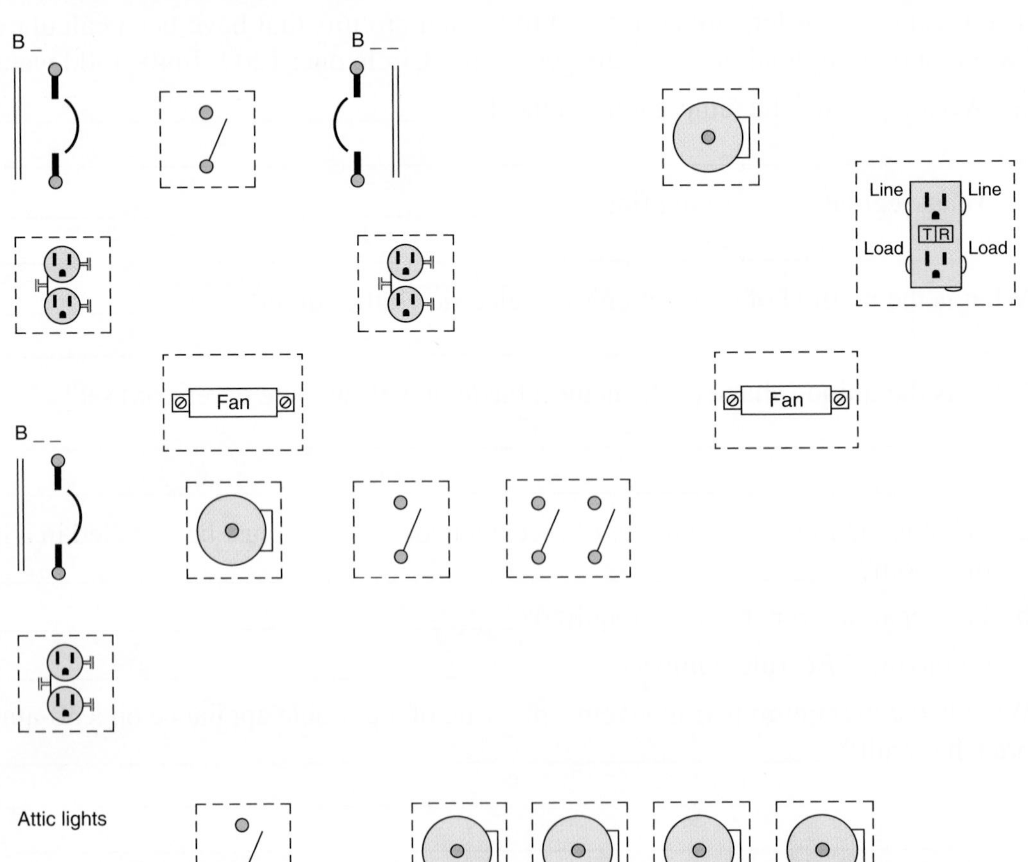

ATTIC EXHAUST FAN CIRCUIT (▲)L

1. What is the purpose of the attic exhaust fan? ____________________
2. At what voltage does the fan operate? ____________________
3. What is the horsepower rating of the fan motor? ____________________
4. Is the fan direct-driven or belt-driven? ____________________
5. How is the fan controlled? ____________________
6. What is the setting of the running overload device? ____________________
7. What is the rating of the running overload protection if the motor is rated at 10 amperes?
   ____
8. What is the basic difference between a thermostat and a humidistat? ____________________
9. What size of conductors are to be used for this circuit? ____________________
10. How many metres of cable are required to complete the wiring for the attic exhaust fan circuit? ____________________
11. May the metal frame of the fan be grounded to the grounded circuit conductor?
    ____
12. What *CEC* rule prohibits grounding equipment to a grounded circuit conductor?
    ____

UNIT 17

Electric Heating and Air Conditioning

OBJECTIVES

After studying this unit, you should be able to

- list the advantages of electric heating
- describe the components and operation of electric heating systems (baseboard heater, heating cable, furnace)
- describe thermostat control systems for electric heating units
- install electric heaters with appropriate temperature control according to *Canadian Electrical Code (CEC)* rules
- discuss the *CEC* requirements and electrical connections for air conditioners and heat pumps
- explain how heating and cooling may be connected to the same circuit

GENERAL DISCUSSION

Many types of electric heat are available for heating homes—unit heaters, boilers, electric furnaces, heat pumps, duct heaters, baseboard heaters, and radiant heating panels. The residence discussed in this textbook is heated by an electric furnace located in the workshop.

Detailed requirements are found in *CEC Section 62*, which covers fixed electric space and surface heating systems.

This textbook cannot cover in detail the methods used to calculate heat loss and the wattage required to provide a comfortable level of heat in the building. Many variables must be taken into account when doing heat loss calculations. Software is available for doing these calculations, both on the Internet and from manufacturers. The Heating, Refrigeration and Air Conditioning Institute of Canada offers courses for calculating heat losses for residential and commercial/industrial buildings.

Some of the variables that must be taken into account include construction of the building envelope, type and thickness of the insulation, size of windows, and whether the siding is of brick or wood. Another variable would be in what direction most windows and doors face. One of the most important variables is the location of a house within the country. Each region has a different number of degree days, which affects the heat requirements to maintain a comfortable living temperature. For this residence, the total estimated wattage is 13 000 watts (13 kW).

Compared with other types of heating systems, electric heating is not widely used. However, it does have a number of advantages over other systems. Electric heating is flexible when baseboard heating is used, because each room can have its own thermostat. Thus, one room can be kept cool while an adjoining room is warm. This type of zone control for an electric, gas-fired, or oil-fired central heating system is more complex and more expensive.

Electric heating is safer than heating with fuels. The system does not require storage space, tanks, or chimneys. Electric heating is quiet and does not add to or remove anything from the air. As a result, electric heat is cleaner. This type of heating is considered to be healthier than fuel heating systems, which remove oxygen from the air. The only moving part of an electric baseboard heating system is the thermostat. This means that there is a minimum of maintenance. The biggest disadvantage to electric heat is cost. Generally, the cost of electricity per British thermal unit (Btu) is greater than that of other fuels.

If an electric heating system is used, adequate insulation must be provided to minimize electric bills. Insulation also helps to keep the residence cool during the hot summer months. The cost of extra insulation is offset through the years by the decreased burden on the air-conditioning and heating equipment. Energy conservation measures require the installation of proper and adequate insulation as well as a good-quality vapour barrier.

TYPES OF ELECTRIC HEATING SYSTEMS

Electric heating units are available in baseboard, wall-mounted, and floor-mounted styles. These units may have built-in thermal overload protection. The type of unit to be installed depends on structural conditions and the purpose for which the room is to be used.

Another method of providing electric heating is to embed resistance-type cables in the plaster of ceilings or between two layers of drywall on the ceiling, *Rules 62–214* through *62–222*. Resistance-type heating cables may also be used for melting snow and ice on roof overhangs, eavestroughs, walkways, and driveways, *Rule 62–300*.

Electric heat can also be supplied by an electric furnace and a duct system similar to the type used on conventional hot-air central heating systems. The heat is supplied by electric heating elements rather than the burning of fuel. Air conditioning, humidity control, air circulation, electronic air cleaning, and zone control can be provided on an electric furnace system.

Many homes are now built or renovated with ceramic tile flooring. Often, heating cable is installed under the tile to eliminate cold floors. It is very important to follow the manufacturer's installation instructions. Although some heat will radiate from the tile into the room, underfloor heating cable sets are not intended to be the sole source of heat for the room. The wattage of the heating cable set is usually quite low, as it is meant to heat only the floor.

SEPARATE CIRCUIT REQUIRED

Rules 62–110(1)(a) and *26–806(1)* require that central heating equipment be supplied by a separate branch circuit. This includes electric, gas, and oil central furnaces and heat pumps. The exception to *Rules 62–110(1)* and *26–806(1)* is equipment directly associated with central heating equipment, such as humidifiers and electrostatic air cleaners, which are permitted to be connected to any branch circuit, *Rule 26–806(1)*.

The "separate circuit" rule is due to the possibility that if the central heating equipment was connected to another circuit, such as a lighting circuit, a fault on that circuit would shut off the power to the heating system. This could cause freezing of water pipes. Refer to Figure 17-1.

A disconnecting means must be provided to open all ungrounded conductors to the central heating unit. This disconnect must be within the sight of and no farther than 9 metres from the unit, *Rule 62–206(5)*.

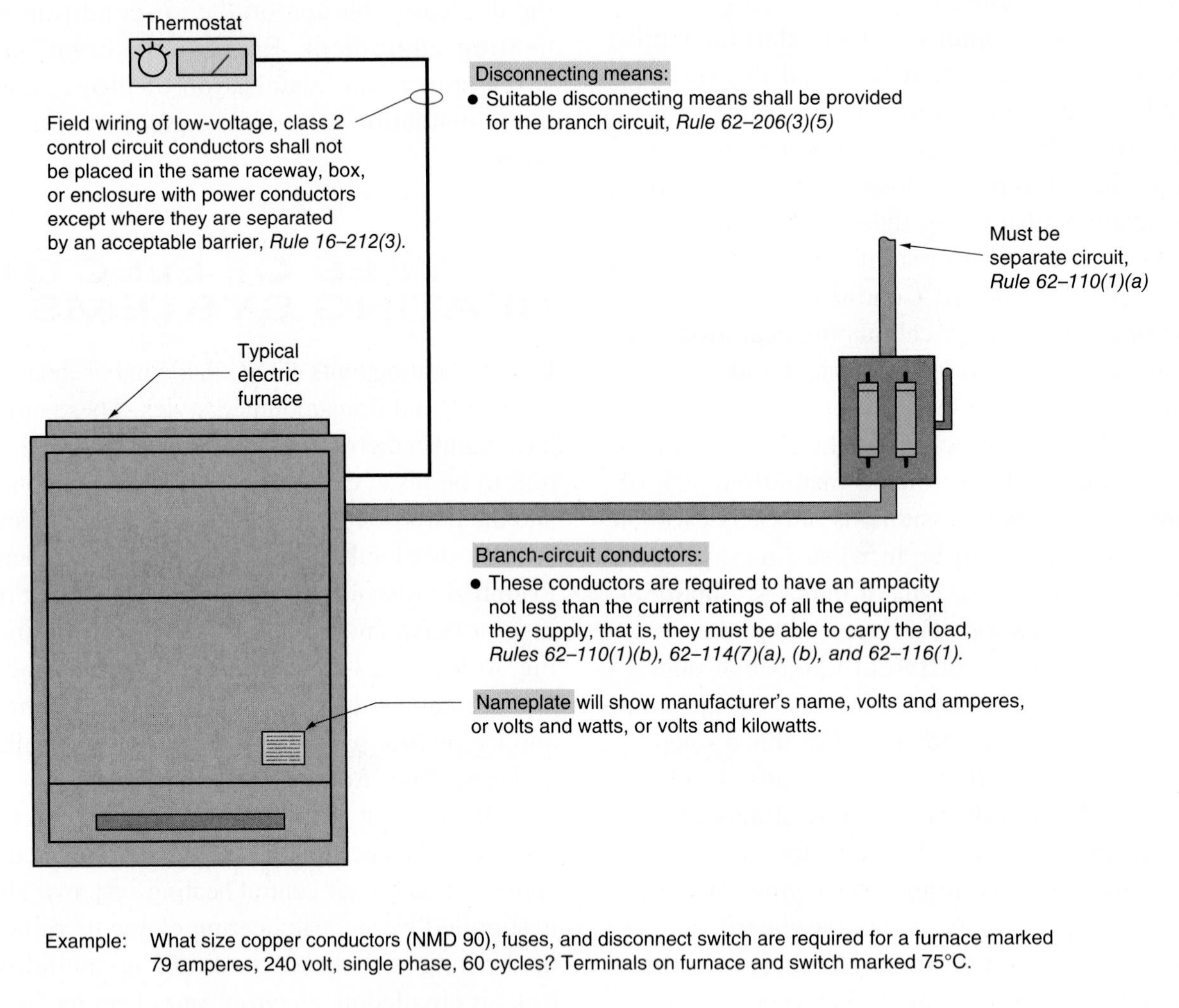

Example: What size copper conductors (NMD 90), fuses, and disconnect switch are required for a furnace marked 79 amperes, 240 volt, single phase, 60 cycles? Terminals on furnace and switch marked 75°C.

Answer:

Conductor size:	From *CEC Table 2*, select No. 4 NMD 90. [85 amperes at 75°C, *Rule 4–006(1)*] *Rule 62–116(1)*.
Size of overcurrent device:	Install 100-ampere fuse or circuit breaker. *Rule 62–114(7)(b)*
Switch:	100-ampere switch $79 \times 1.25 = 98.75$ ampere fuses.

FIGURE 17-1 Basic *CEC* requirements for an electric furnace. These "package" units have all of the internal components, such as limit switches, relays, motors, and contactors, prewired. The electrician provides both the branch-circuit and the control-circuit wiring, together with the thermostat wiring. See *CEC Section 62* for additional information.

Rule 26–804 and *Appendix B* give excerpts from CSA Standard C22.2 No. 3 entitled "Electrical Features of Fuel-Burning Equipment." *Rule 4.8.8* of this standard states that a safety control shall interrupt the current in the ungrounded conductor of the circuit between the overcurrent protection and the load. This must be provided for in the control circuit connected to the Class 2 transformer.

CONTROL OF ELECTRIC HEATING SYSTEMS

Line-voltage thermostats can be used to control the heat load for most baseboard electric heating systems; see Figure 17-2. Common wattage limits for line-voltage thermostats are 2500 watts, 3000 watts, and 5000 watts. (Additional ratings are available.) The electrician must check the nameplate ratings of the heating unit and the thermostat to ensure that the total connected load does not exceed the thermostat rating.

The wattage limit of a thermostat is found using the amperage formula. For example, if a thermostat has a rating of 20 amperes at 125 or 250 volts, the wattage value is

$$W = E \times I = 125 \times 20 \text{ watts } 5 \text{ } 2500 \text{ watts}$$

or

$$W = 250 \times 20 \text{ watts} = 5000 \text{ watts}$$

The total connected load for this thermostat must not exceed 20 amperes.

When the connected load is larger than the rating of a line-voltage thermostat, as in the case of an electric furnace, low-voltage thermostats may be used. In this case, a relay must be connected between the thermostat and load. The relay is an integral component of the furnace. The low-voltage contacts of the relay are connected to the thermostat. The line-voltage contacts of the relay are used to switch the actual heater load. Low-voltage thermostat cable is

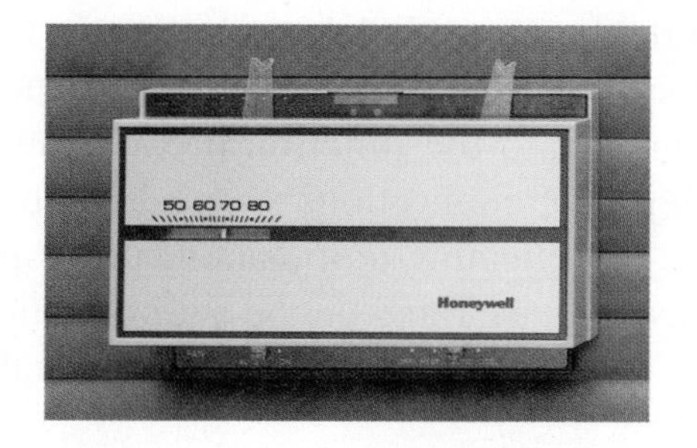

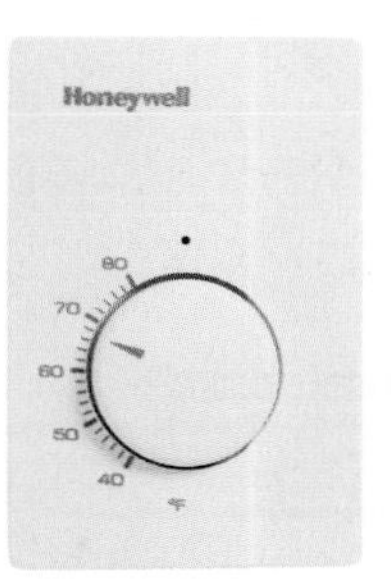

Courtesy of Honeywell Inc.

FIGURE 17-2 Thermostats for electric heating systems.

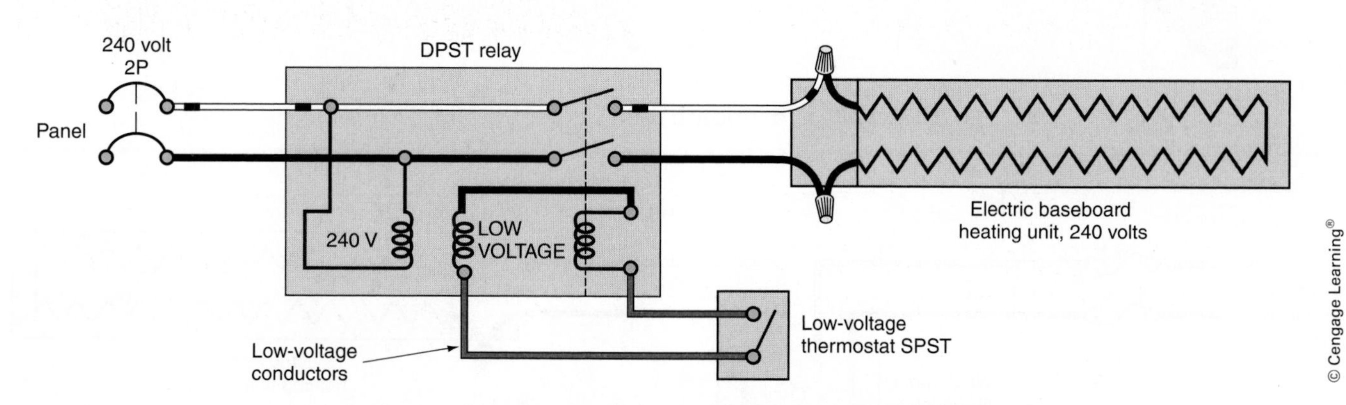

© Cengage Learning®

FIGURE 17-3 Wiring for a baseboard electric heating unit having a low-voltage thermostat and relay. *Note:* If using black and white wires in a two-wire non-metallic-sheathed cable (NMSC) or AC cable, tape or paint the white wire red, *Rule 4–036(2).* When using NMSC, many inspection authorities require the use of special cable that contains one black and one red conductor.

run between the low-voltage terminals of the relay and the thermostat; see Figure 17-3. In general, contacts and relays serve the function in the electrical industry of using a low voltage or low current to control a much higher one. This is done for safety and to lower expense, as the smaller control wires of a relay or contactor are much cheaper to run than the large conductors that bring power to heating or cooling devices.

When the connected load is larger than the maximum current rating of a thermostat and relay combination, the thermostat can be used to control a heavy-duty relay or contactor; see Figure 17-4. An example of such a relay is the magnetic switch used for motor controls. The load side of the magnetic switch feeds a distribution panel containing as many 15- or 20-ampere circuits as needed for the connected load. A 40-ampere load may be divided into three 15-ampere circuits and still be controlled by one thermostat. The proper overcurrent protection is obtained by dividing the 40-ampere load into several circuits having lower ratings.

CIRCUIT REQUIREMENTS FOR BASEBOARD UNITS

Figure 17-5 shows typical baseboard electric heating units. These units are rated at 240 volts, but are also available in 120-volt ratings.

The wiring for an individual baseboard heating unit or group of units is shown in Figures 17-6 and 17-7. A two-wire cable (armoured cable or nonmetallic-sheathed cable with ground) would be run from a 240-volt, two-pole circuit in the main panel to the outlet box or switch box installed at the thermostat location. A second two-wire cable runs from the thermostat to the junction box on the heater unit. The proper connections are made in this junction box. Most heating unit manufacturers provide knockouts at the rear and on the bottom of the junction box. The supply conductors can be run through

Courtesy of Honeywell Inc.

FIGURE 17-4 Relay for controlling an electric heater that can be used in baseboards.

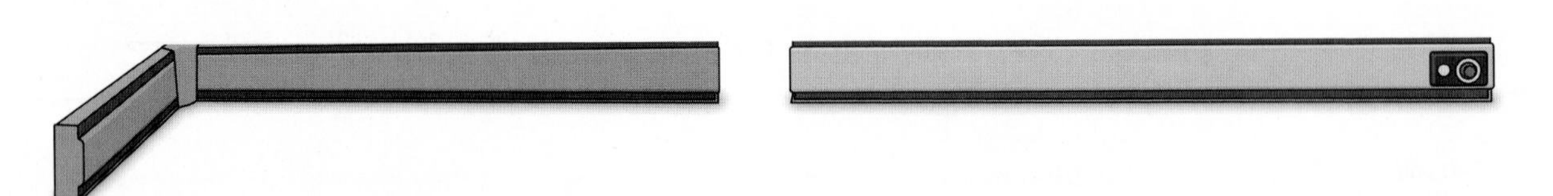

© Cengage Learning®

FIGURE 17-5 Baseboard electric heating systems.

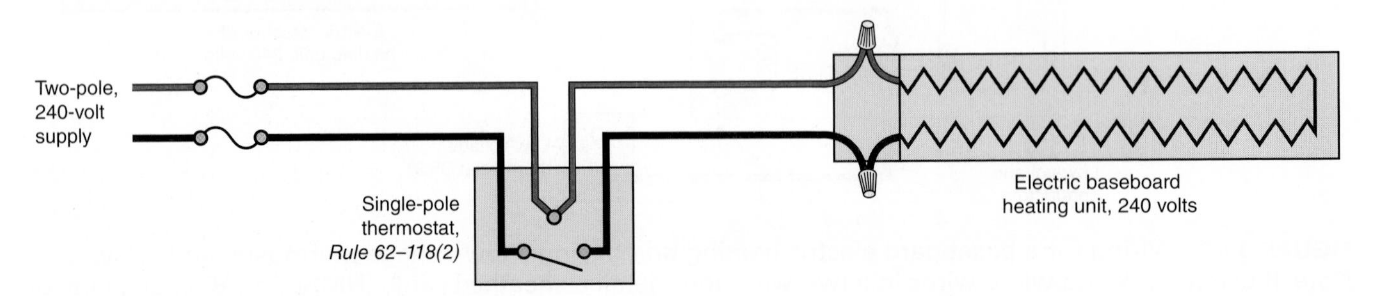

FIGURE 17-6 Wiring for a single baseboard electric heating unit. See the note in the caption of Figure 17-3.

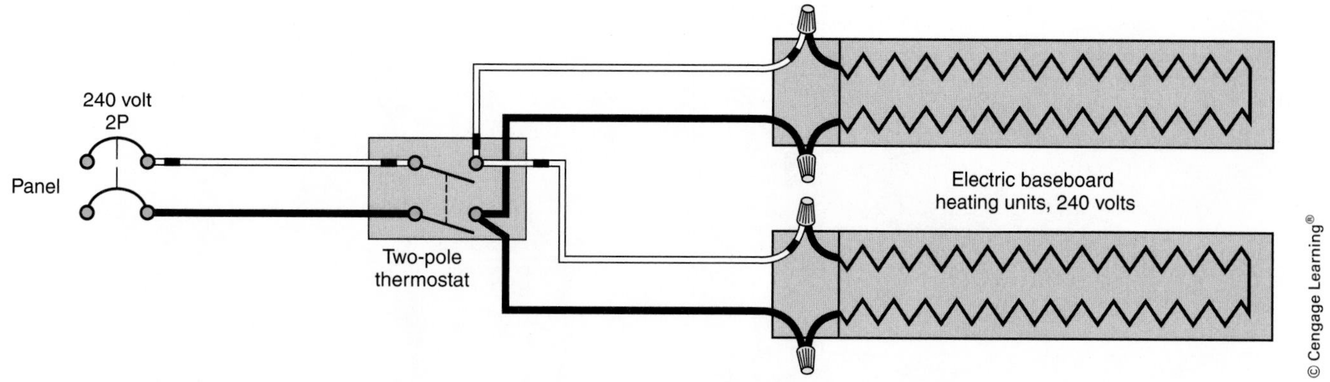

FIGURE 17-7 Wiring for two baseboard electric heating units at different locations in a room. See the note in the caption of Figure 17-3.

these knockouts, using proper connectors for their wiring method.

Most baseboard units also have a channel or wiring space running the full length of the unit, usually at the bottom. When two or more heating units are joined together, the conductors are run in this wiring channel. Most manufacturers indicate the type of wire required for these units because of conductor temperature limitations.

A variety of fittings, such as internal and external elbows (for turning corners) and blank sections, are available from the manufacturer.

Most wall and baseboard heating units are available with built-in thermostats (see Figure 17-5). The supply cable for such a unit runs from the main panel to the junction box on the unit.

Manufacturers of electric baseboard heaters list "blank" sections ranging in length from 0.5 m to 2 m. These blank sections give the installer the flexibility needed to spread out the heater sections and to install these blanks where wall receptacle outlets are encountered; see Figure 17-8.

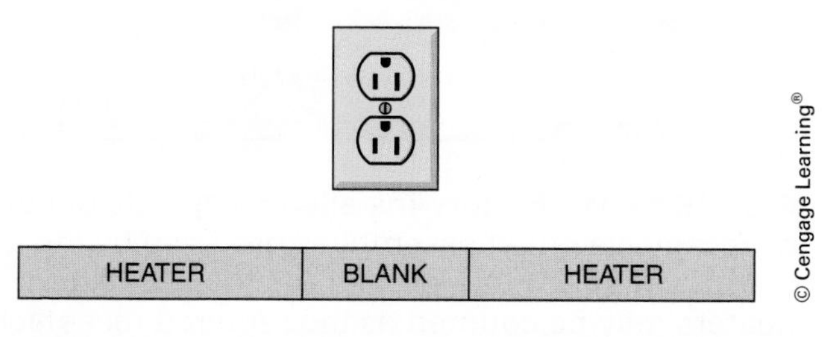

FIGURE 17-8 Use of blank baseboard heating sections.

LOCATION OF ELECTRIC BASEBOARD HEATERS IN RELATION TO RECEPTACLE OUTLETS

Approved electric baseboard heaters shall not be installed below wall receptacle outlets unless the instructions furnished with the heaters indicate that they may be installed there, *Rule 26–712(a)* and *Appendix B*. The reason for this is the possible fire and shock hazard if a cord hangs over and touches the heated electric baseboard unit. The insulation on the cord can melt from the heat. See Figures 17-9 and 17-10.

The electrician must pay close attention to the location of receptacle outlets and the electric

FIGURE 17-9 Code violation. Unless the instructions furnished with the electric baseboard heater specifically state that the unit is listed for installation below a receptacle outlet, this installation is a violation of *Rule 26–712(a)* and *Appendix B*. In this type of installation, cords attached to the receptacle outlet would hang over the heater, creating a fire hazard and possible shock hazard if the insulation on the cord melted.

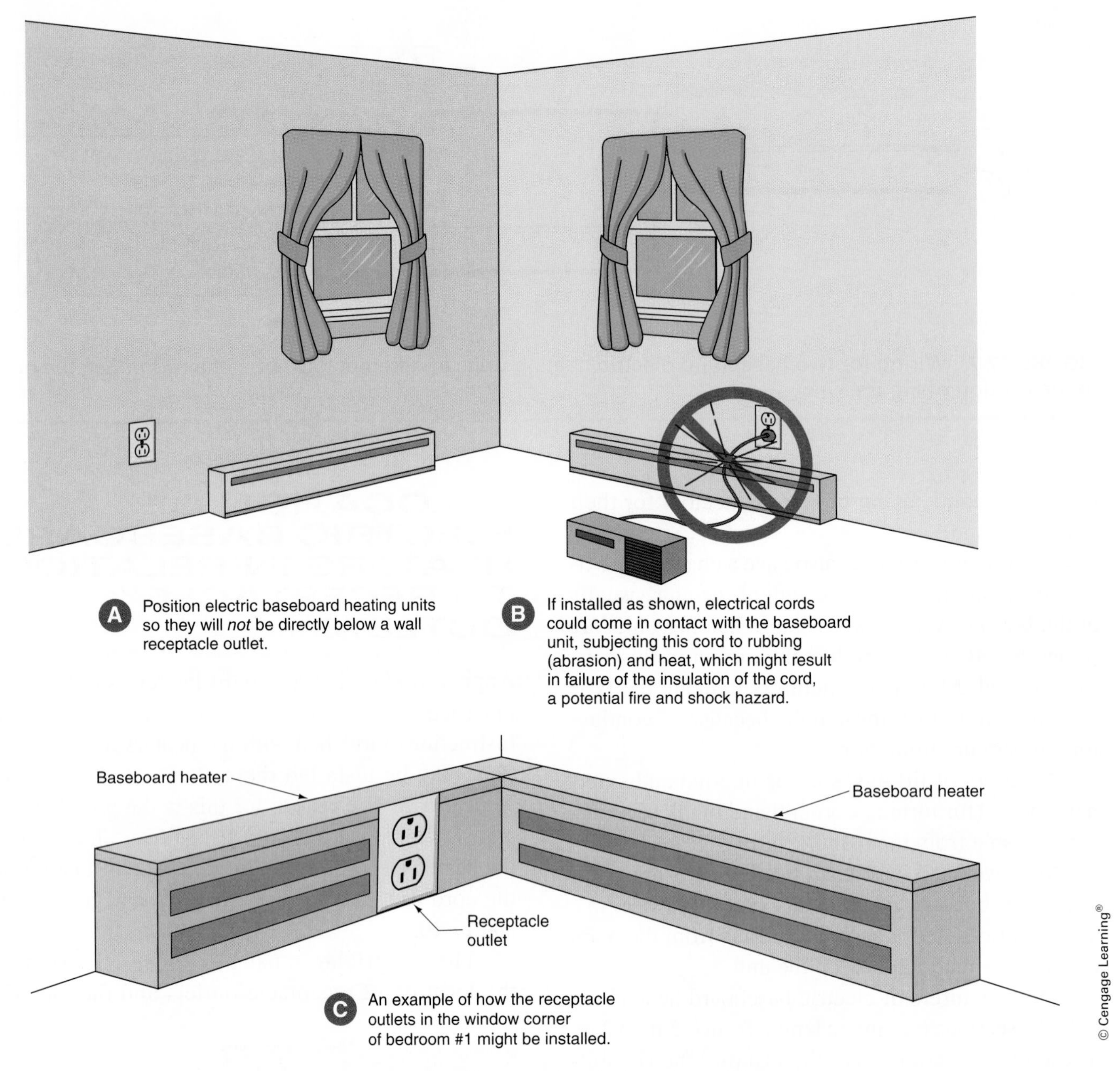

FIGURE 17-10 Position of electric baseboard heaters.

baseboard heaters in the plans and specifications. Blank spacer sections may be installed, following the *CEC* requirements for spacing receptacle outlets and just how much (wattage and length) baseboard heating is required.

The Canadian Standards Association (CSA) approves the use of factory-installed or factory-furnished receptacle outlet assemblies as part of permanently installed electric baseboard heaters. These receptacle outlets shall not be connected to the heater circuit; see Figure 17-11.

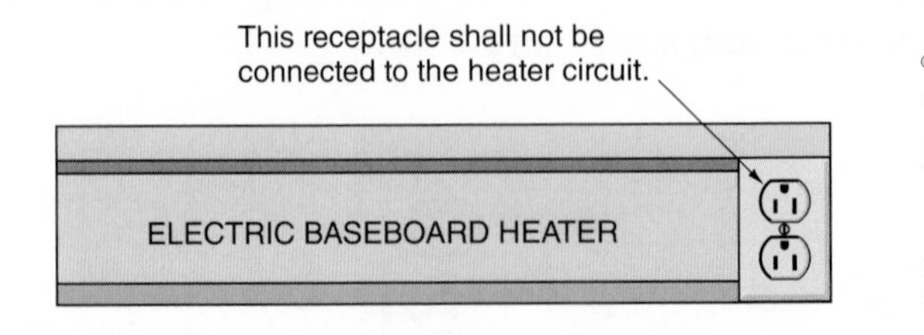

FIGURE 17-11 Factory-installed receptacle outlets or receptacle outlet assemblies provided by the manufacturer for use with its electric baseboard heaters may be counted as the required receptacle outlet for the space occupied by a permanently installed heater, *Rule 26–712(a)*.

CIRCUIT REQUIREMENTS FOR ELECTRIC FURNACES

In an electric furnace, resistance heating element(s) are the source of heat. The blower motor assembly, filter section, condensate coil, refrigerant line connection, fan motor speed control, high-temperature limit controls, relays, and similar components are similar to those found in gas furnaces.

In most installations, the supply conductors must be copper conductors having a minimum 90°C temperature rating, such as Type T90. The instructions furnished with the unit will have information relative to supply conductor size and type.

Since proper supply voltage is critical to ensure full output of the heating elements, voltage must be maintained at *not less* than 98% of the unit's rated voltage to provide its rated wattage. Therefore, the voltage drop should not be allowed to exceed 2% of the unit's rated voltage to achieve its rated output. The *CEC* allows a maximum voltage drop of 3% on a branch circuit, *Rule 8–102(1)(a).*

Many electric furnaces furnish the wattage (or kilowatt) output for more than one voltage level. For every 1% drop in voltage, there will be an almost 2% drop in wattage output. For instance, if a 240-volt-rated heater is connected to 208 volts, its actual output of heat is less than its rated watts; see Table 17-1.

TABLE 17-1

Comparison of wattages and currents for electric heating elements at 240 and 208 volts.

WATTS @ 240 VOLTS	AMPERES @ 240 VOLTS	WATTS @ 208 VOLTS	AMPERES @ 208 VOLTS
500	2.08	375.6	1.81
750	3.13	563.3	2.71
1000	4.17	751.1	3.61
1250	5.21	938.9	4.51
1500	6.25	1126.7	5.42
1750	7.29	1314.4	6.32
2000	8.33	1502.2	7.22
2500	10.42	1877.8	9.03

The actual voltage at the supply terminals of the electric furnace or baseboard heating unit determines the true wattage output of the heating elements. The effect that different voltages have on electric heating elements was covered in Unit 15.

The demand factors for service conductors used to supply heating equipment are set out in *Rule 62–118*, which requires service conductors to have an ampacity not less than the current ratings of all of the equipment supplied. In the case of an electric furnace, this would be 100% of the total connected load.

If baseboard heating is being used, *Rule 62–118(3)* allows the ampacity of the main service conductors to be reduced. If each room or heated area is thermostatically controlled, the demand factor is 75% for each kW above 10 kW. So, if the total connected load was 20 kW for baseboard heating with thermostats in each room, the demand load must be calculated as

1st 10 kW @ 100%	=	10.0 kW
2nd 10 kW @ 75%	=	7.5 kW
Total		17.5 kW
Total connected load	=	20.0 kW
Total demand load	=	17.5 kW

These demand watts would be applied to the service conductors.

Two or more heating fixtures, such as baseboard heaters, may be connected on a single heating branch circuit, provided that the overcurrent device does not exceed 30 amperes. Therefore, the maximum load on each branch circuit is 80% or 24 amperes or 5760 watts at 240 volts, *Rule 62–114(2)(6).*

The electric furnace must be supplied by an individual branch circuit, *Rule 62–110.* The "separate circuit" requirement pertains to any type of central heating, including oil, gas, electric, and heat pump systems, *Rule 26–806.*

HEAT PUMPS

A heat pump cools and dehumidifies on hot days and heats on cool days. To provide heat, the pump picks up heat from the outside air, which it adds to the heat created by the compression of the refrigerant. This heat is used to warm the air flowing through the unit, which is then delivered to the inside of the building.

The heat pump can also be used to cool the building when the outside air temperature is warm. To achieve this, the direction of flow of the refrigerant is reversed. Warm air is then delivered to the outside, and the house is cooled.

Heat pumps are available as self-contained packages that provide both cooling and heating. In the Canadian climate, the heat pump alone may not provide enough heat. Therefore, the heat pump may use supplementary heating elements contained in the air-handling unit, or be used in combination with a forced-air gas furnace.

BONDING

Bonding of electrical baseboard heating units and other electrical equipment and appliances has been covered previously in this textbook. Refer to *Rule 10–400.*

MARKING THE CONDUCTORS OF CABLES

Two-wire cable contains one white conductor, one black conductor, and one bonding conductor. Using two-wire cable to supply 240-volt electric baseboard heaters appears to violate the *CEC*. However, according to *Rules 4–030(4)* and *4–036(1)*, two-wire cable may be used for the 240-volt heaters if the white conductor is permanently marked as ungrounded by paint, coloured tape, or other effective means. This is necessary, since people may assume that an unmarked white conductor is a grounded circuit conductor having no voltage to ground. However, the white wire in the heater circuit is connected to a "hot" phase and has 120 volts to ground. Therefore, a person can be subject to a harmful shock by touching this wire and the grounded baseboard heater (or any other grounded object) at the same time.

Most electrical inspectors accept black paint or tape as a means of changing the colour of the white conductor. The white conductor must be made permanently unidentifiable at every accessible point, such as at heater and service panels. A more commonly used method is to install two-wire cable with black and red conductors, *Rule 4–038.* One such cable is Heatex™, a red-sheathed NMSC cable manufactured by Nexans Inc., specifically designed for heating circuits. Unit 8 also addresses wire colour coding.

ROOM AIR CONDITIONERS

For homes without central air conditioning, window or through-the-wall air conditioners may be installed. These room air conditioners are available in both 120-volt and 240-volt ratings. Because room air conditioners are plug-and-cord connected, the receptacle outlet and the circuit capacity must be selected and installed according to the *CEC.*

The basic *CEC* rules for installing these units and their receptacle outlets are

- the air conditioners must be grounded
- the air conditioners must be connected using a cord and attachment plug
- the rating of the branch-circuit overcurrent device must not exceed the branch-circuit conductor rating or the receptacle rating, whichever is less
- the air-conditioner load shall not exceed 80% of the branch-circuit ampacity if no other loads are served
- the attachment plug cap may serve as the disconnecting means, *Rule 28–602(3)(c)*

RECEPTACLES FOR AIR CONDITIONERS

Figure 17-12 shows three types of receptacles that can be used for air-conditioning installations: (A) shows a 20-ampere, 250-volt receptacle for use on a 240-volt installation; (B) a 20-ampere, 125-volt receptacle for use with a 120-volt air-conditioner circuit; and (C) a 15-ampere, 250-volt receptacle for use with a 240-volt installation. *Diagram 1* illustrates the many different voltage/ampere receptacle configurations.

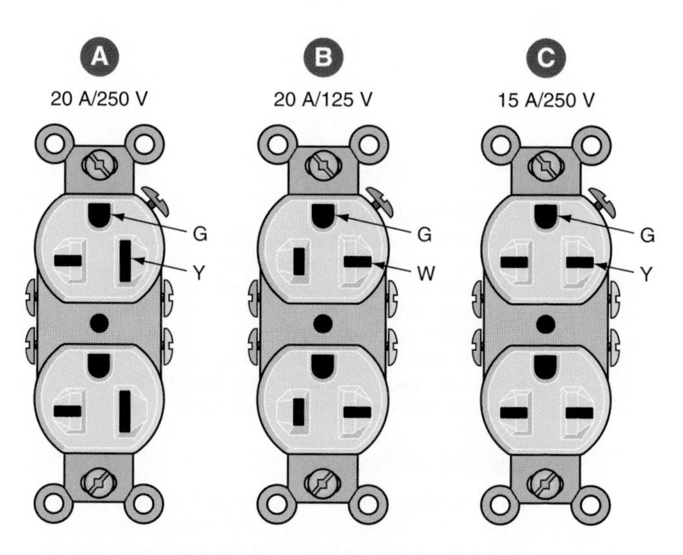

FIGURE 17-12 Types of receptacles used for air conditioners, *Diagram 1.*

CENTRAL HEATING AND AIR CONDITIONING

The residence used as an example in this textbook has central electric heating and air conditioning, consisting of an electric furnace and a central air-conditioning unit.

The wiring for central heating and cooling systems is shown in Figure 17-13. One branch circuit runs to the electric furnace, and another runs to the air conditioner or heat pump outside the dwelling. Extra-low-voltage wiring is used between the inside and outside units to provide control of the systems. The extra-low-voltage Class 2 circuit wiring must not be run in the same raceway as the power conductors, *Rule 16–212(1).*

SPECIAL TERMINOLOGY

These terms pertain to air conditioners, heat pumps, and other hermetically sealed motor-compressors.

Rated-load current is the current drawn when the unit is operating at rated load, at rated voltage, and at rated frequency.

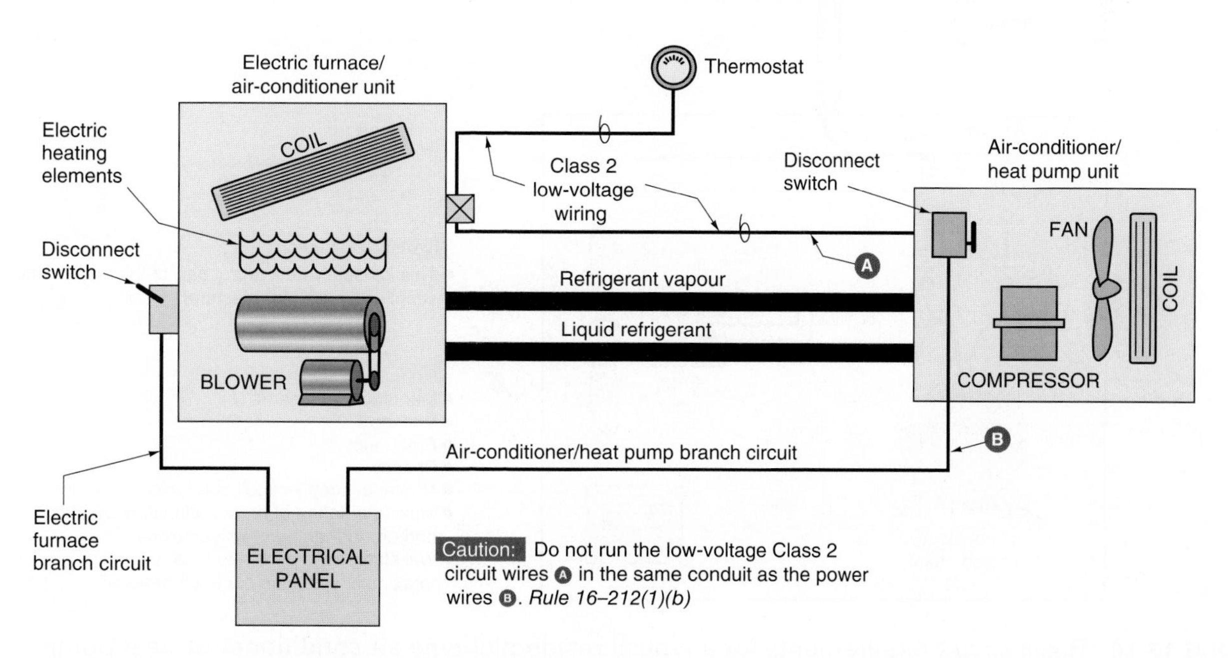

FIGURE 17-13 Connection diagram showing typical electrical furnace and air-conditioner/heat pump installation.

Branch-circuit conductor ampacity is based on the rated-load current, which must be marked on the label, *Rule 2–100(1)(d)*. This will determine conductor size, disconnect switch size, controller size, and short-circuit and ground-fault protective device ratings; see Figure 17-14.

Rule 2–100(4) requires that approved equipment be used only in accordance with instructions included in the original approval of the equipment. For example, the nameplate of an air-conditioning unit reads "Maximum size fuse 40 amperes," so only 40-ampere fuses can be used. Forty-ampere circuit breakers are not permitted. If the nameplate called for "Maximum overcurrent protection," either fuses or circuit breakers could be used. See Figure 17-15.

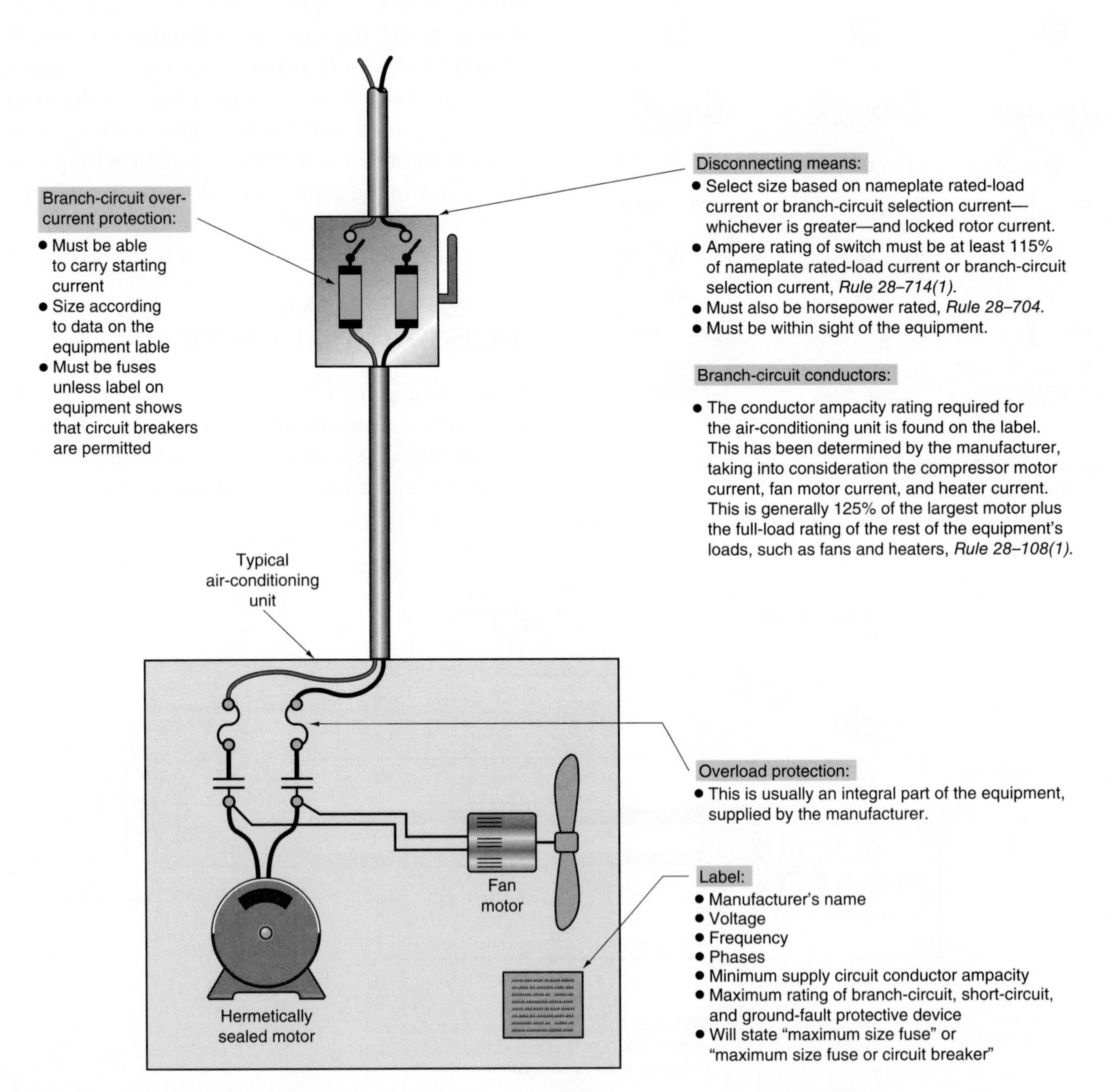

FIGURE 17-14 Basic circuit requirements for a typical residential-type air conditioner or heat pump. Reading the label is important because the manufacturer has determined the minimum conductor size and branch-circuit fuse size.

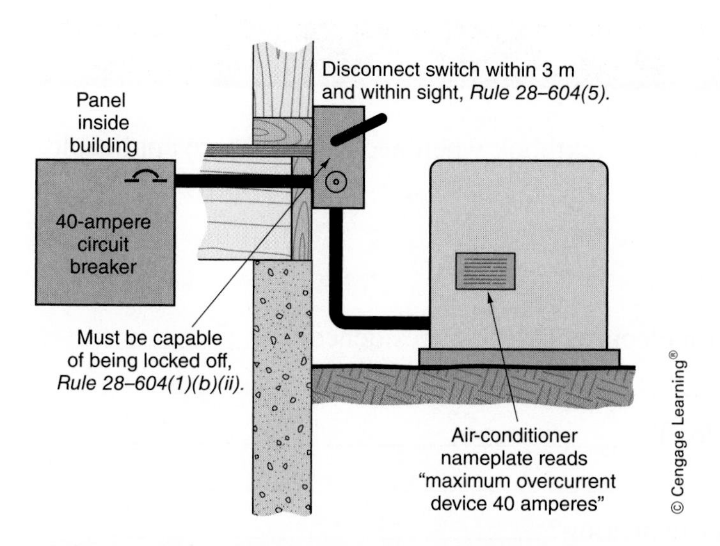

FIGURE 17-15 This installation conforms to *Rule 28–604.* The panel contains the 40-ampere circuit breaker called for on the air-conditioner nameplate as the branch-circuit protection. The disconnect must be capable of being locked off, *Rule 28–604(1)(b)(ii).* Neither the air conditioner nor its disconnect switch shall be located any closer than 1 m to a combustible gas vent, *Rule 2–324.*

The air conditioner requires a motor-rated disconnect to be installed within three metres and within sight of the unit, *Rule 28–604(5)*; see Figure 17-16.

LOADS NOT USED SIMULTANEOUSLY

Heating and air-conditioning loads are not likely to operate at the same time. Therefore, when calculating feeder sizes to an electric furnace and an air conditioner, you need to consider only the larger load. In *Rule 8–106(4),* the words "where it is known" refer to the fact that common sense would dictate that heating and air conditioning would not be in use at the same time. The larger connected load is used to calculate the demand on the service feeders.

If the air conditioning has the larger demand, this demand is used at 100%. If the heating demand is greater, this is used to calculate the demand watts according to *Rule 62–114.*

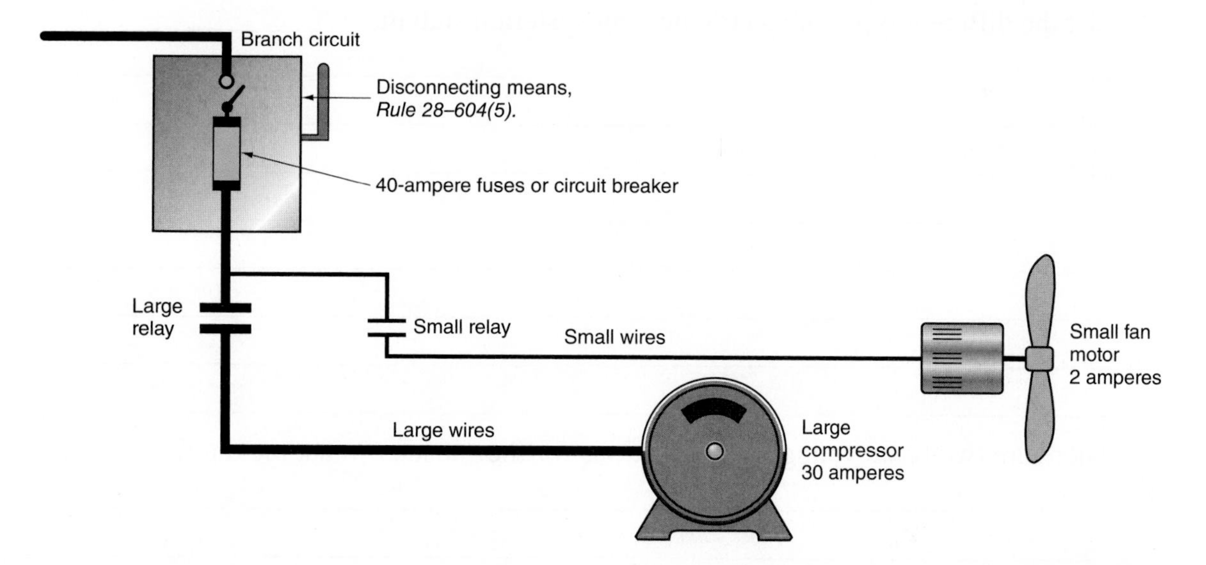

FIGURE 17-16 In a typical air-conditioner unit, the branch-circuit protective device is called on to protect the large components (wire, relay, compressor) and the small components (wire, relay, fan motor).

REVIEW

Note: Refer to the *CEC* or the blueprints provided with this textbook when necessary. Where applicable, responses should be written in complete sentences.

ELECTRIC HEAT

1. a. What is the allowance in watts made for electric heat in this residence? __________

   b. What is the value in amperes of this load? __________

2. What are some of the advantages of electric heating? __________

3. List the different types of electric heating system installations. __________

4. There are two basic voltage classifications for thermostats. What are they? __________

5. Which device is required when the total connected load exceeds the maximum rating of a thermostat? __________

6. The electric heat in this residence is provided by what type of equipment? __________

7. At what voltage does the electric furnace operate? ______________________________

____

8. A certain type of control connects electric heating units to a 120-volt supply or a 240-volt supply, depending on the amount of the temperature drop in a room. These controls are supplied from a 120/240-volt, three-wire, single-phase source. Assuming that this type of device controls a 240-volt, 2000-watt heating unit, what is the wattage produced when the control supplies 120 volts to the heating unit? Show all calculations. ______________________________

____

____

____

____

____

____

____

9. What advantages does a 240-volt heating unit have over a 120-volt heating unit?

____

____

____

____

____

____

____

10. a. The white wire of a cable may be used to connect to a hot circuit conductor only if

____

____

b. Quote the rule number. ______________________________

11. Electric baseboard heaters shall not be installed beneath wall receptacle outlets. Explain and quote the *CEC* rule. ______________________________

____

____

____

____

____

12. [Circle the correct answer.] The branch circuit supplying a heater shall have an ampacity of

a. not less than the connected load supplied

b. 125% of the current rating of the heater

13. Compute the current draw of the following electric furnaces. The furnaces are all rated 240 volts.
    a. 7.5 kW

    b. 15 kW

    c. 20 kW

14. The wattage output of a 240-volt electric furnace connected to a 208-volt supply will be about 75% of the wattage output if the furnace is connected to a 240-volt supply. Referring to Question 13, calculate the wattage output of (a), (b), and (c) if the furnace is connected to a 208-volt supply. ______

15. A central electric furnace heating system is installed in a home. Should a separate branch circuit supply this furnace? Explain. ______

16. What *CEC* rule provides the answer to Question 15?

AIR CONDITIONING

1. When calculating air-conditioner load requirements and electric heating load requirements, is it necessary to add the two loads together to determine the combined load on the service? Explain your answer. ______________________________

2. The total load of an air conditioner shall not exceed what percentage of a separate branch circuit? [Circle one of the following.]
   a. 75%
   b. 80%
   c. 125%

3. Do window air conditioners count as a connected load when calculating the demand watts on a service? ______________________________

4. Must an air conditioner installed in a window opening be bonded to ground?
   ____

5. A 120-volt air conditioner draws 12 amperes. What size is the circuit to which the air conditioner will be connected? ______________________________

6. When a central air-conditioning unit is installed and the label states "Maximum size fuse 50 amperes," is it permissible to connect the unit to a 50-ampere circuit breaker?
   ____

7. Which *CEC* rule prohibits installing Class 2 control circuit conductors in the same raceway as the power conductors? ______________________________

UNIT 18

Oil and Gas Heating Systems

OBJECTIVES

After studying this unit, you should be able to

- understand the basics of gas and oil heating systems
- identify some of the more important components of gas and oil heating systems
- interpret basic schematic wiring diagrams
- explain the principles of the thermocouple and the thermopile in a self-generating system
- understand and apply the requirements for control-circuit wiring, *CEC Section 16*

Unit 17 discussed the circuit requirements for electric heating (furnace, baseboard, ceiling cable). Unit 18 discusses typical residential oil and gas heating systems.

Heating a home with gas or oil is possible with

- warm air—fan-forced warm air or gravity (hot air rises naturally)
- hot water—hot water circulated through the piping system by a circulating pump; the system uses tubing embedded in the floor or radiators that look very much like electric baseboard heating units.

PRINCIPLES OF OPERATION

Most residential gas or oil heating systems operate as follows. A room thermostat is connected to the proper electrical terminals on the furnace or boiler, where the controls and valves are interconnected to provide safe and adequate heat to the home. Some typical wiring diagrams are shown in Figures 18-1, 18-2, and 18-3A and B.

Most residential gas or oil heating systems are "packaged units" in which the safety controls, valves, fan controls, limit switches, and so on are integral, preassembled prewired parts of the furnace or boiler.

For the electrician, the "field wiring" generally consists of installing and connecting the wire to the thermostat, and installing and connecting the power supply to the furnace or boiler. Refer to Figure 18-4.

MAJOR COMPONENTS

Gas-fired and oil-fired warm air systems and hot water boilers contain many components.

Aquastat. This direct immersion water temperature thermostat regulates boiler or tank temperature in hydronic heating systems, ensuring that the circulating water maintains a satisfactory temperature; see Figure 18-5.

Cad Cell. A photoconductive flame detector. Detector must be installed so that it can see flame. Connects to primary control system.

Circulating Pump. Circulates hot water in a hydronic central heating system.

Combustion Chamber. Surrounds the flame and radiates heat back into the flame to aid in combustion.

Combustion Head. Creates a specific pattern of air at the end of the air tube. The air is directed to force oxygen into the oil spray so the oil can burn.

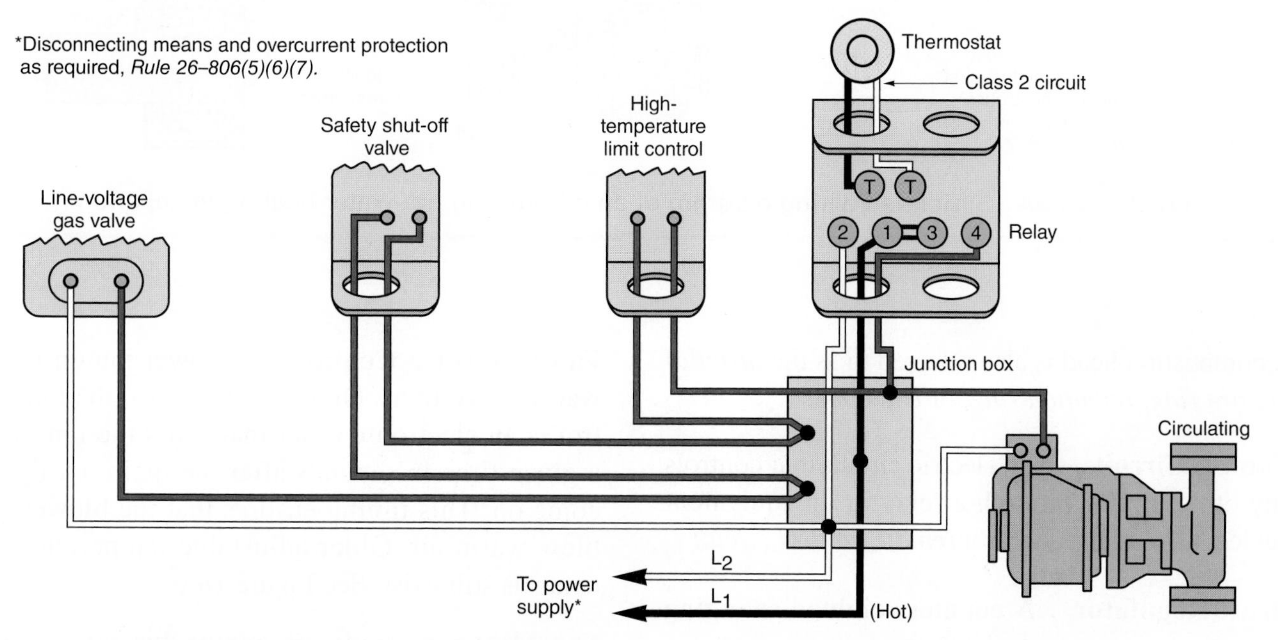

FIGURE 18-1 Simplified wiring diagram for a gas burner, forced hot water system.

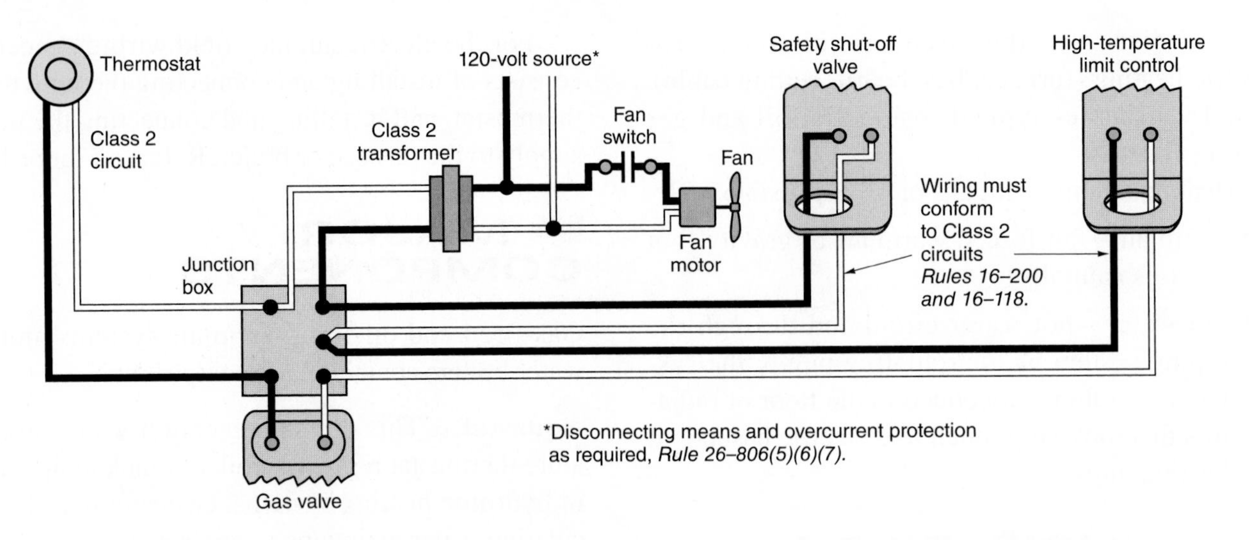

FIGURE 18-2 Simplified wiring diagram for a gas burner, forced warm air system.

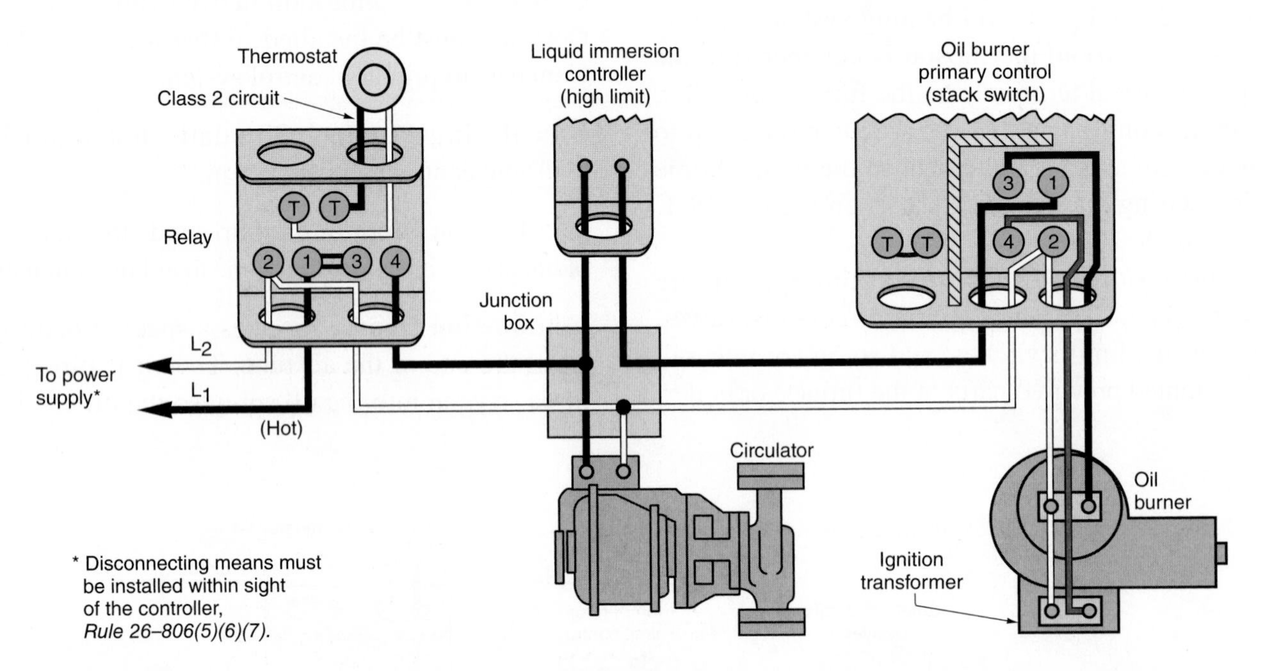

FIGURE 18-3A Simplified wiring diagram of an oil-burning hot water heating installation.

A combustion head is also referred to as the *turbulator, fire ring, retention ring,* or *end cone.*

Control Circuit. Any electric circuit that controls any other circuit through a relay or an equivalent device; also referred to as a *remote-control circuit.*

Draft Regulator. A counterweighted swinging door that opens and closes to help maintain a constant level of draft over the fire.

Fan Control. Controls the blower fan on forced warm air systems. On newer furnaces, the fan control is an electronic timer that starts the fan motor a given time in seconds after the main burner has come on. This timing ensures that the blower will blow warm air. Older adjustable temperature fan controls still exist. See Figure 18-6.

Fan Motor. An electric motor that forces warm air through the heating ducts. In typical residential

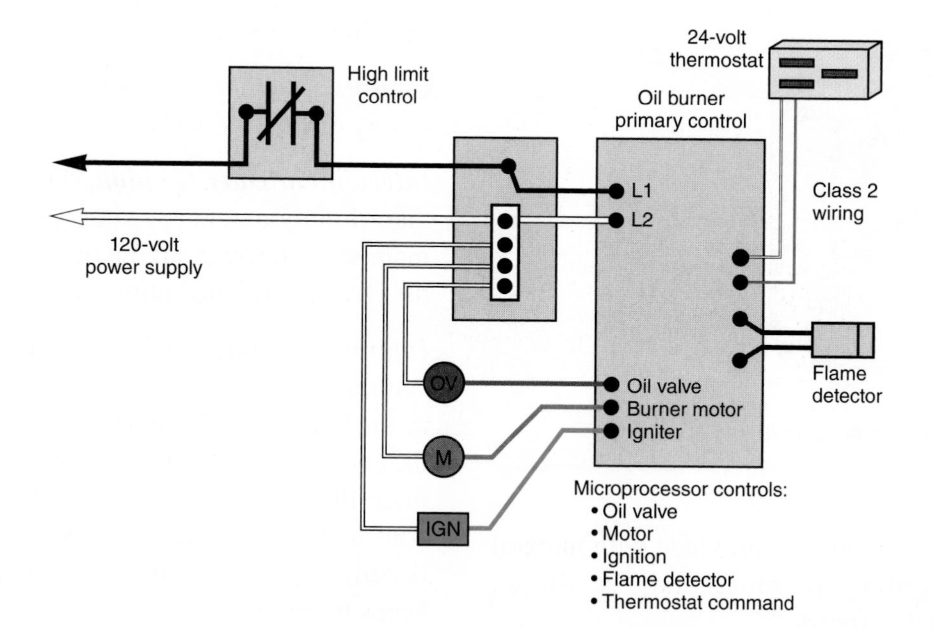

FIGURE 18-3B An oil burner featuring an electronic microprocessor that controls all facets of the burner operation. Most components are integral to the unit, so field wiring is kept to a minimum. All manufacturers provide detailed wiring diagrams for their products. Always consult the manufacturer's wiring diagrams for a particular model heating unit.

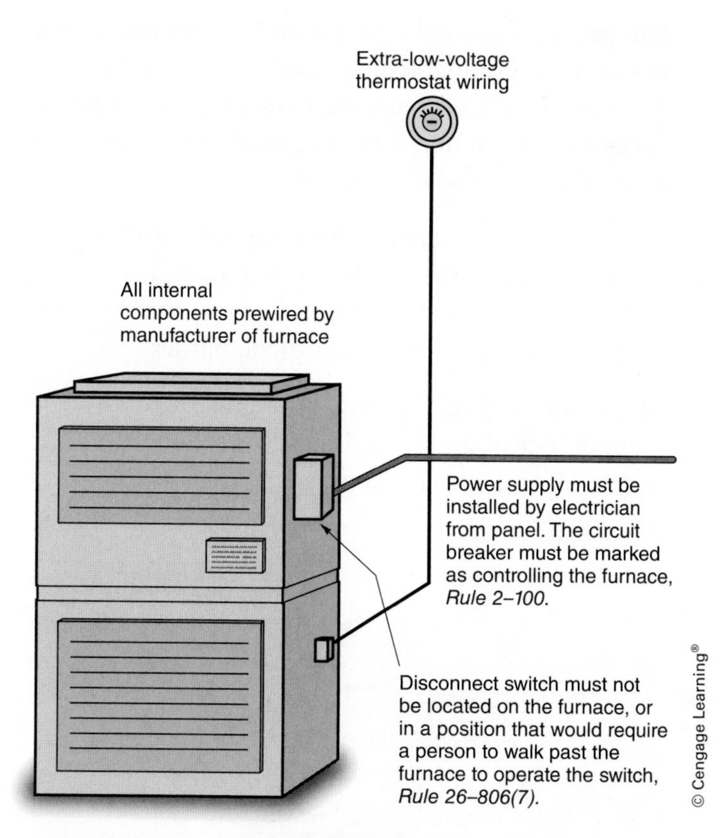

FIGURE 18-4 Diagram showing thermostat, power supply wiring, and disconnect switch.

FIGURE 18-5 Aquastat (liquid immersion controller).

FIGURE 18-6 Fan/limit control.

heating systems, the motor is provided with integral overload protection. The motor may be single, multiple, or variable speed.

Flow Valve. Prevents gravity circulation of water through the piping system.

Flue. A channel in a chimney or a pipe for conveying flame and smoke to the outer air.

Heat Exchanger. Transfers the heat energy from the combustion gases to the air in the furnace or to the water in a boiler.

High-Temperature Limit Control. A safety device that limits the temperature in the plenum of the furnace to a safe value predetermined by the manufacturer. When this temperature is reached, the control shuts off the power to the burner. In newer furnaces, the control is part of an integrated circuit board and is communicated to by a sensor strategically located in the plenum. Some older furnaces have individual high-temperature limit controls and a separate fan control; others have combination high-temperature limit/fan controls.

Hot Surface Igniter. Heats to a cherry red/orange colour in furnaces without a standing constant pilot light. When it reaches the proper temperature, the main gas valve is allowed to open. If the hot surface igniter does not come on, the main gas valve will not open. Most energy-efficient furnaces use this kind of ignition.

Hydronic System. A system of heating or cooling that involves transfer of heat by circulating fluid such as water in a closed system of pipes.

Ignition.

Continuous Pilot Ignition. The igniter is designed to stay on continuously.

Intermittent Duty Ignition. Defined by UL 296 as "ignition by an energy source that is continuously maintained throughout the time the burner is firing"; the igniter is on the entire time the burner is firing.

Interrupted Duty Ignition. Defined by UL 296 as "an ignition system that is energized each time the main burner is to be fired and de-energized at the end of a timed trial for ignition period or after the main flame is proven to be established"; the igniter comes on to light the flame. After the flame is established, the igniter is turned off and the main flame keeps burning.

Induced Draft Blower. When the thermostat calls for heat, this blower first expels any gases remaining in the combustion chamber from a previous burning cycle. It continues to run, pulling hot combustion gases through the heat exchangers, and then vents the gases to the outdoors.

Integrated Control. A printed circuit board that contains many electronic components. When the thermostat calls for heat, the integrated circuit board takes over to manage the sequence of events that allow the burner to operate safely.

Low-Water Control. Also called a *low-water cutoff,* this device senses low-water levels in a boiler. When a low-water situation occurs, this control shuts off the electrical power to the burner.

Main Gas Valve. Allows the gas to flow to the main burner of a gas-fired furnace; see Figure 18-7. It may also function as a pressure regulator, a safety

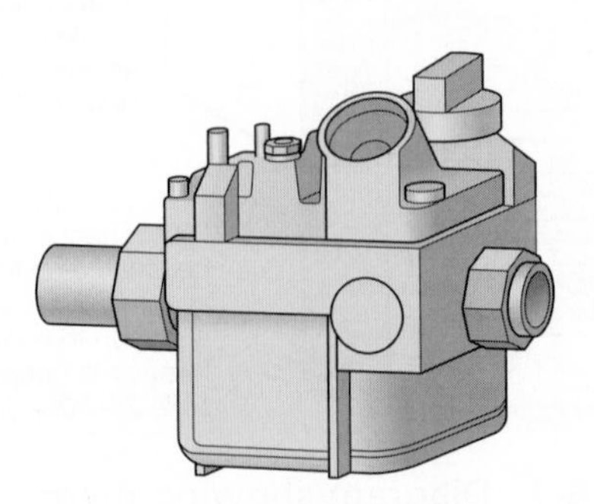

FIGURE 18-7 Main gas valve.

shut-off, and the pilot and main gas valve. Should there be a "flame-out," if the pilot light fails to operate, or in the event of a power failure, this valve will shut off the flow of gas to the main burner.

Nozzle. Produces the desired spray pattern for the particular appliance in which the burner is used.

Oil Burner. Breaks fuel oil into small droplets, mixes the droplets with air, and ignites the resulting spray to form a flame.

Primary Control. Controls the oil burner motor, ignition, and oil valve in response to commands from a thermostat; see Figure 18-3B. It monitors safety devices such as the flame detector and also monitors the status of the high-temperature limit. If the oil fails to ignite, the controller shuts down the oil burner.

Pump and Zone Controls. Regulate the flow of water or steam in boiler systems to specific "zones" in the building.

Safety Controls. Safety controls such as pressure relief valves, high-temperature limit controls, low-water cutoffs, and burner primary controls protect against appliance malfunction.

Solenoid. An electrically operated device; when the valve coil is energized, a magnetic field is developed, causing a spring-loaded steel valve piston to overcome the resistance of the spring and immediately pull the piston into the stem. The valve is now in the open position. When de-energized, the solenoid coil magnetic field instantly dissipates, and the spring-loaded valve piston snaps closed, stopping oil flow or gas flow to the nozzle. The flame is extinguished, allowing fuel to flow. Usually a spring returns the valve back to the closed position.

Spark Ignition. The spark comes on while pilot gas flows to the pilot orifice. Once the pilot flame is "proven" through an electronic sensing circuit, the main gas valve opens. Once ignition takes place, the sensor will monitor and prove existence of a main flame. If the main flame goes off, the furnace shuts down. Energy-efficient furnaces may use this for ignition.

Switching Relay. Used between the low-voltage wiring and line-voltage wiring; see Figure 18-8. When a low-voltage thermostat calls for heat, the relay

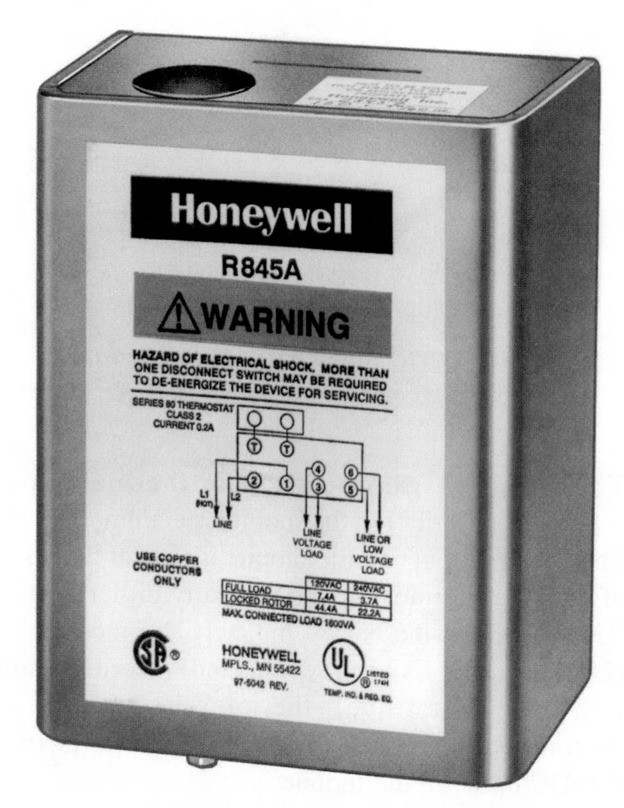

FIGURE 18-8 Switching relay.

"pulls in," closing the line-voltage contacts on the relay, which turns on the main power to the furnace or boiler. Usually they have a built-in transformer that provides the low-voltage Class 2 control circuit.

Thermocouple. In systems that have a standing pilot, a thermocouple holds open a gas valve pilot solenoid magnet. A thermocouple consists of two dissimilar metals connected together to form a circuit. The metals might be iron and copper, copper and iron constantan, copper-nickel alloy and chrome-iron alloy, platinum and platinum-rhodium alloy, or chromel and alumel. The types of metals used depend on the temperatures involved. When one of the junctions is heated, an electrical current flows; see Figure 18-9. To operate, there must be a temperature difference between the metal junctions. In a gas burner, the source of heat for the thermocouple is the pilot light. The cold junction of the thermocouple remains open and is connected to the pilot safety shut-off gas valve circuit. A single thermocouple develops a voltage of about 25 to 30 direct current (DC) millivolts. Because of this extremely low voltage, circuit resistance is kept to a very low value.

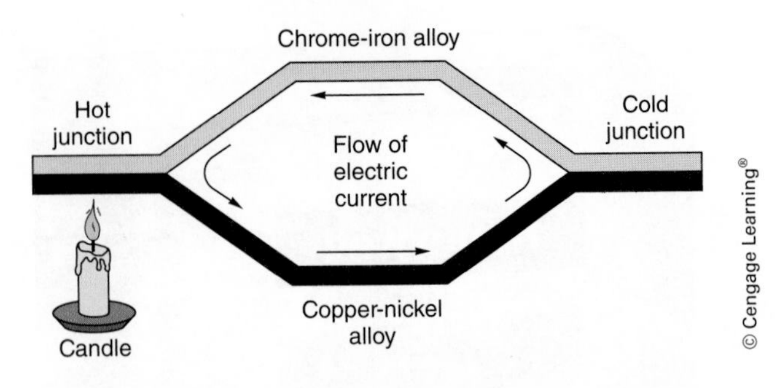

FIGURE 18-9 Principle of a thermocouple.

Thermopile. More than one thermocouple connected in series is a thermopile; see Figure 18-10. The power output of a thermopile is greater than that of a single thermocouple—typically either 250 or 750 DC millivolts. For example, 10 thermocouples (25 DC millivolts each) connected in series result in a 250 DC millivolt thermopile. Twenty-six thermocouples connected in series results in a 750 DC millivolt thermopile.

Thermostat. A thermostat turns the heating/cooling system on and off. Programmable thermostats might have day/night setback settings, temperature, humidity, and time displayed. Thermostats are usually mounted about 1.3 m above the finished floor. Do not mount thermostats in areas influenced by drafts or air currents from hot or cold air registers, near fireplaces, near concealed hot or cold water pipes or ducts, in the direct rays of the sun, or on outside walls; see Figure 18-11. Some thermostats contain a small vial of mercury that "tips," so the mercury closes the circuit by bridging the gap between the electrical contacts inside the vial. These thermostats must be kept perfectly level to ensure accuracy.

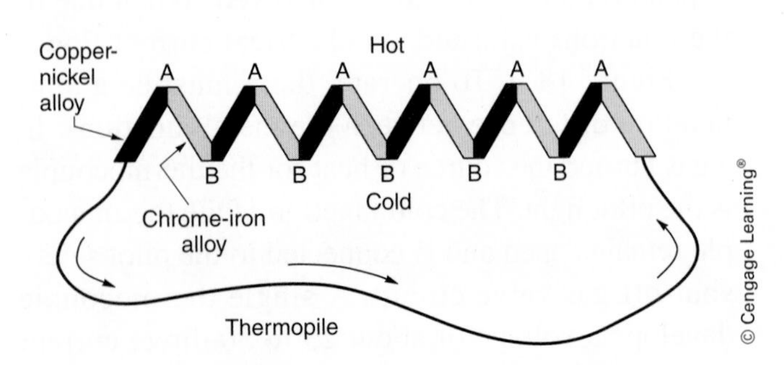

FIGURE 18-10 Principle of a thermopile.

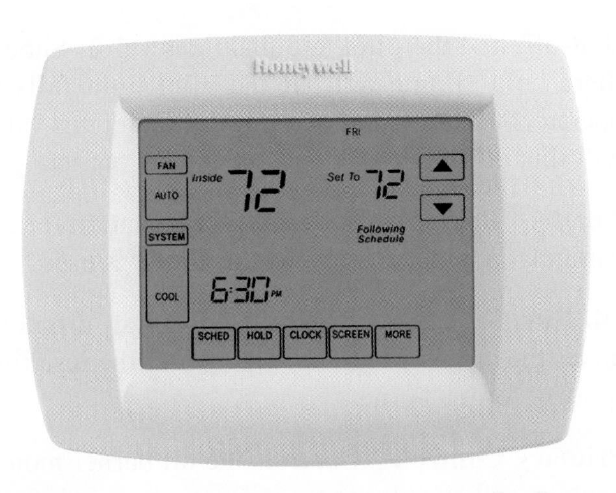

FIGURE 18-11 A digital electronic programmable thermostat provides proper cycling of the heating system, resulting in comfort and energy savings. This type of thermostat could be programmed to control temperature, time of operation, humidity, ventilation, filtration, circulation of air, and zone control. Installation and operational instructions are furnished with the thermostat.

WARNING: Do not throw mercury thermostats into the trash. Mercury is a hazardous waste material, and must be given to a waste-management organization for proper disposal.

Transformer. Converts (transforms) the branch-circuit voltage to low voltage; see Figure 18-12. For residential applications, these are Class 2 transformers, 120 volts to 24 volts. The low-voltage

FIGURE 18-12 Transformer.

Courtesy of Wilo SE

FIGURE 18-13 Water circulator.

wiring on the secondary of this transformer is Class 2 wiring.

Water Circulating Pump. In hydronic (wet) systems, a circulating pump circulates hot water through the piping system. Multizone systems have more than one pump, each controlled by a thermostat for a particular zone. In multizone systems, a high-temperature control (aquastat) might be set to maintain the boiler water temperature to a specific temperature. Some multizone systems have one circulating pump and use electrically operated valves that open and close on command from the thermostat(s) located in a particular zone(s). See Figure 18-13.

SELF-GENERATING (MILLIVOLT) SYSTEM

Several manufacturers of gas burners provide self-generating systems on decorative appliances such as gas fireplaces. These systems do not require an outside power supply. Figure 18-14 shows the wiring diagram of a typical self-generating system. The components are connected in series. The small amount of energy required by the gas valve is supplied by a thermopile. Many of these burners are prewired at the factory, so the only wiring at the site is done by the electrician, who runs a low-voltage cable to the thermostat. This system is not affected by a power failure because there is no outside power source.

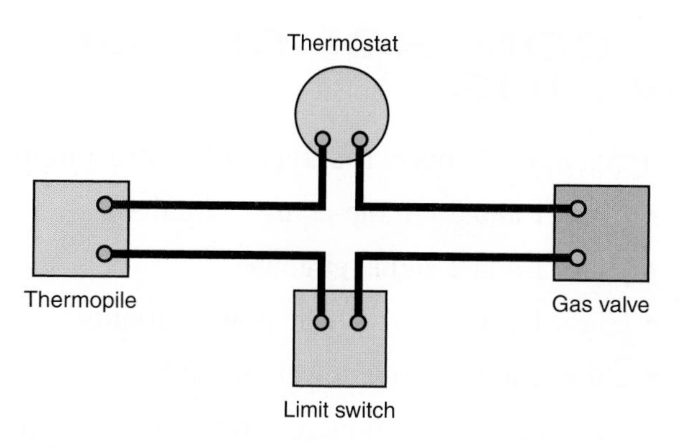

FIGURE 18-14 Wiring diagram of a self-generating system.

Self-generating systems operate in the millivolt range, typically 250 or 750 mV. A special thermostat is required to function at this low-voltage level. All other components of the system are matched to operate at the low voltage.

SUPPLY CIRCUIT WIRING

The wiring for the branch circuit for heating equipment that uses solid, liquid, or gaseous fuel is covered in *CEC Rules 26–800* through *26–808*.

The branch-circuit wiring must be adequate for the power requirements of the motors, fans, relays, pumps, ignition, transformers, and other power-consuming components of the furnace or boiler. The sizing of conductors, overcurrent protection, conduit fill, and so on has been covered in other units of this textbook.

The basic requirements for the branch circuit are

- The power for the heating unit and associated equipment must be supplied from a single circuit. For exceptions, see *Rule 26–806(2)(3)*.
- A suitable disconnecting means is required. The branch-circuit breaker in the panel may be used as the disconnecting means, if no one has to walk past the furnace to reach the breaker.
- If a separate disconnecting means is required, it cannot be mounted on the furnace or in a location that would require a person to walk past the furnace to operate it.
- The switch must be marked to indicate that it controls the furnace.

CONTROL-CIRCUIT WIRING

CEC Section 16 covers four types of control circuits:

- Class 1 and 2 remote-control circuits
- Class 1 and 2 signal circuits
- Class 1 extra-low-voltage power circuits
- Class 2 low-energy power circuits

As soon as the conductors for remote-control, signalling, or power-limited circuits leave the equipment, they are classified as Class 1 or Class 2 circuits; see Figure 18-15, *Rule 16–002*. There are different requirements for Class 1 and Class 2 circuits regarding wire sizes, de-rating factors, overcurrent protection, insulation, wiring methods, and materials that "differentiate" control-circuit wiring from regular (normal) light and power circuits, *Sections 4* and *12*.

In Part A in Figure 18-15, the control circuitry is contained completely within the equipment, so none of the control wiring is governed by *Section 16*. In Part B in Figure 18-15, the control wires leave the equipment. These conductors are classified as Class 1 or Class 2, depending on voltage, current, and other issues discussed in *Section 16*. Motor control circuits may be protected by control fuses in the controller.

When #16 and #18 conductors of a Class 1 circuit are run between pieces of equipment, they must be protected by overcurrent devices rated or set at not more than 10 and 5 amperes, respectively.

Figure 18-16 shows a Class 1 circuit. Where the failure of equipment or wiring, such as high-limit and low-water controls, would "introduce a direct fire or life hazard," *Rule 16–010*, the circuit is deemed to be a Class 1 circuit. Because of this potential hazard, *Rule 16–116* requires that the wiring be installed in rigid or flexible metal conduit, EMT, or armoured cable, or be otherwise suitably protected from physical damage. See *Rules 16–100* through *16–118* for requirements applicable to Class 1 circuits.

Class 1 remote-control and signalling circuits may be supplied from a source having a maximum voltage of 600 volts. Class 1 extra-low-voltage power circuits are limited to 30 volts and 1000 volt-amperes, *Rule 16–100*.

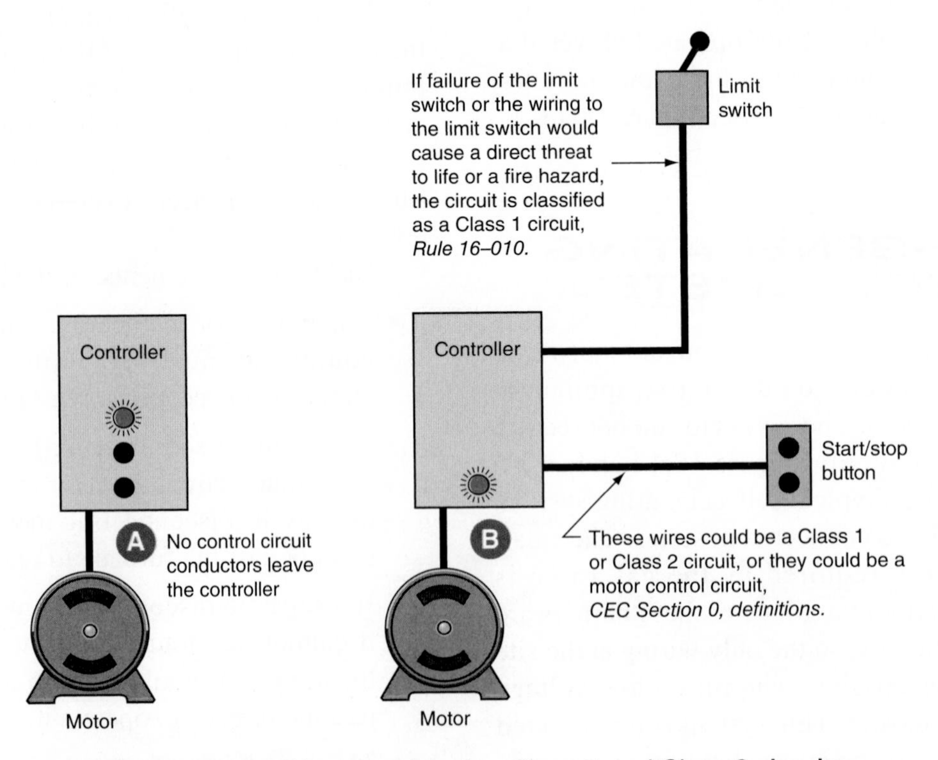

FIGURE 18-15 Diagram showing Class 1 and Class 2 circuits.

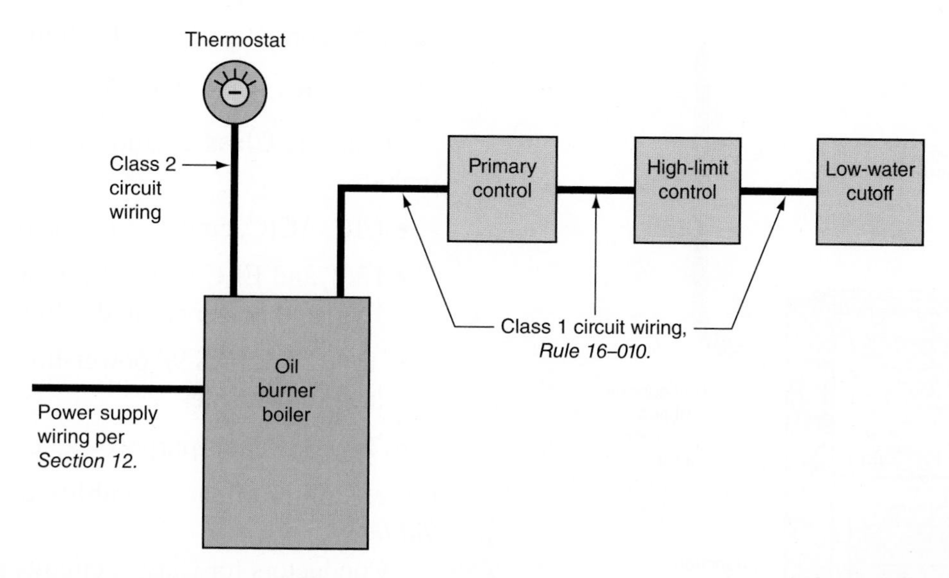

FIGURE 18-16 Any failure in Class 1 circuit wiring would introduce a direct fire or life hazard. A failure in the Class 2 circuit wiring would *not* introduce a direct fire or life hazard. See *Rule 16–010.*

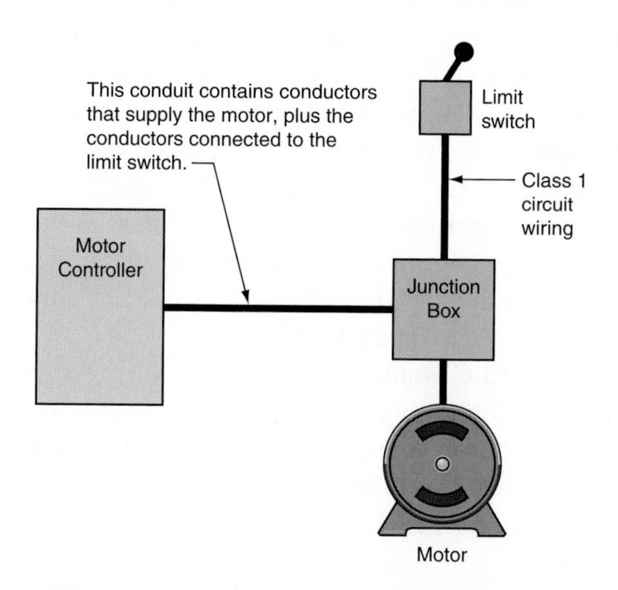

FIGURE 18-17A The power conductors supplying the motor and the Class 1 circuit remote-control conductors to the limit switch and controller may be run in the same conduit when the Class 1 conductors are insulated for the maximum voltage of any other conductors in the raceway.

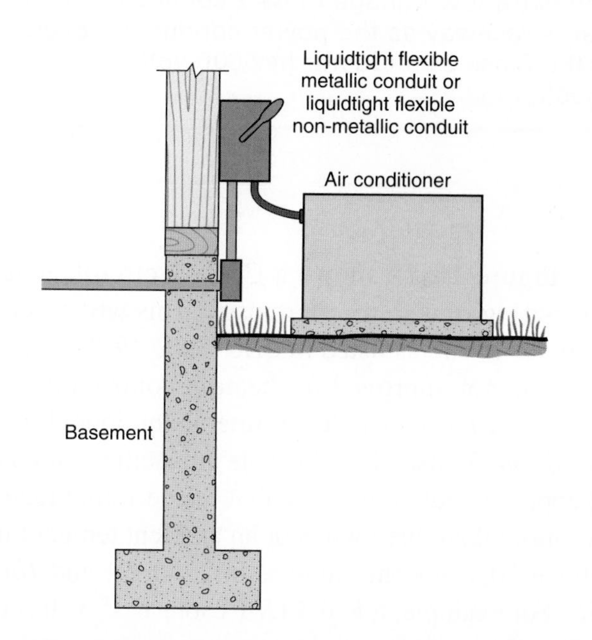

FIGURE 18-17B Class 2 wiring (LVT) cannot be run in the same raceway as power conductors. Typically, LVT is attached to the outside of the flexible conduit.

The conductors of Class 1 circuits and power conductors may be run in the same raceway only if they are connected to the same equipment and the Class 1 conductors have the same voltage rating as the highest rated wire in the raceway, *Rule 16–114* (Figure 18-17A). Most wiring diagrams for residential air-conditioning units indicate that their extra-low-voltage circuitry is Class 2, in which case the field wiring of the power and extra-low-voltage circuits must be run separately. See Figure 18-17B.

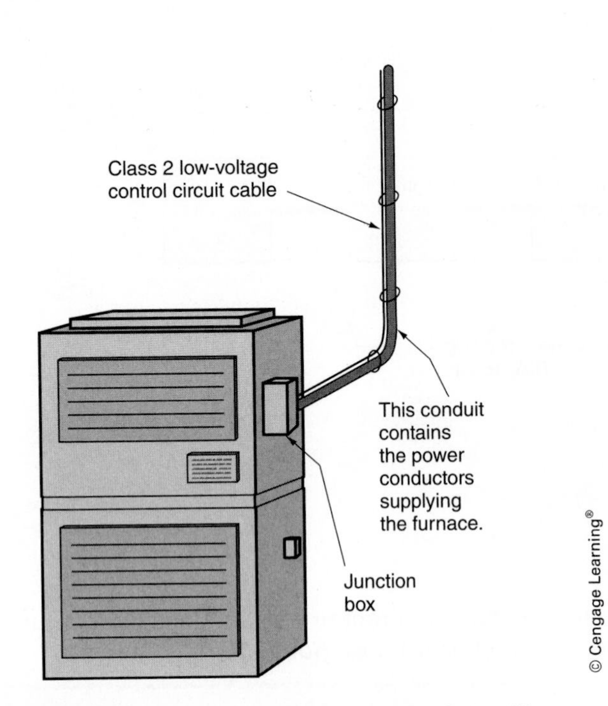

FIGURE 18-18 *Rule 16–212* prohibits installing the extra-low-voltage Class 2 conductor in the same raceway as the power conductors even if the Class 2 conductor has 600-volt insulation.

Figure 18-18 shows a Class 2 circuit, where the wiring runs to the thermostat. This will be done with LVT wire as listed in *CEC Table 19*. Type ELC cable is not approved for heating control circuits, *Rule 16–210*, but would be suitable for door chimes.

Table 57 lists the allowable ampacities for Class 2 copper conductors. Note that the de-rating factors for more than three wires or an ambient temperature above 30°C are the same as in *Table 5A* and *Table 5C*. For example, a #16/4 LVT cable at 35°C has the following ampacity:

Temperature de-rating from *Table 5A,* Column 2:

10 amperes × 0.82 = 8.2 amperes per conductor

Table 57 correction factor for four to six conductors:

8.2 amperes × 0.8 = 6.56 amperes per conductor

Class 1, Class 2, and communication cables include

- CIC, ACIC, control, and instrumentation cable
- LVT and ELC Class 2 remote-control, signalling, and power-limited cables
- Type FAS, FAS 90 power-limited fire protective signalling cables
- Type MP multipurpose cables.

Other communication cables are listed in *CEC Table 19*.

Conductors for Class 2 circuits are **not** permitted to be run in the same raceway, enclosure, box compartment, or fitting with the regular lighting circuits, power circuits, or Class 1 circuit wiring, *Rule 16–212(1,3)*.

Sometimes it is tempting to cut corners by installing Class 2 conductors in the same raceway as the power conductors. Even if the Class 2 conductors were insulated with 600-volt insulation, the same as the power conductors, this is a violation of *Rule 16–212(3)*; see Figure 18-18.

It is permitted to fasten a Class 2 extra-low-voltage control cable to a conduit containing the power conductors that supply a furnace or similar equipment; see Figure 18-18.

Type ELC conductors in Class 2 wiring are not permitted for heating control circuits, *Rule 16–210(3)*.

Conductor sizing and overcurrent protection requirements are found in *Subsection 16–100* for Class 1 circuits and *Subsection 16–200* for Class 2 circuits.

Transformers that are intended to supply Class 2 circuits are listed by UL and approved by the Canadian Standards Association (CSA) and are marked "Class 2 Transformer." Figure 18-19 illustrates the definitions of remote control, signalling, and power-limited terminology.

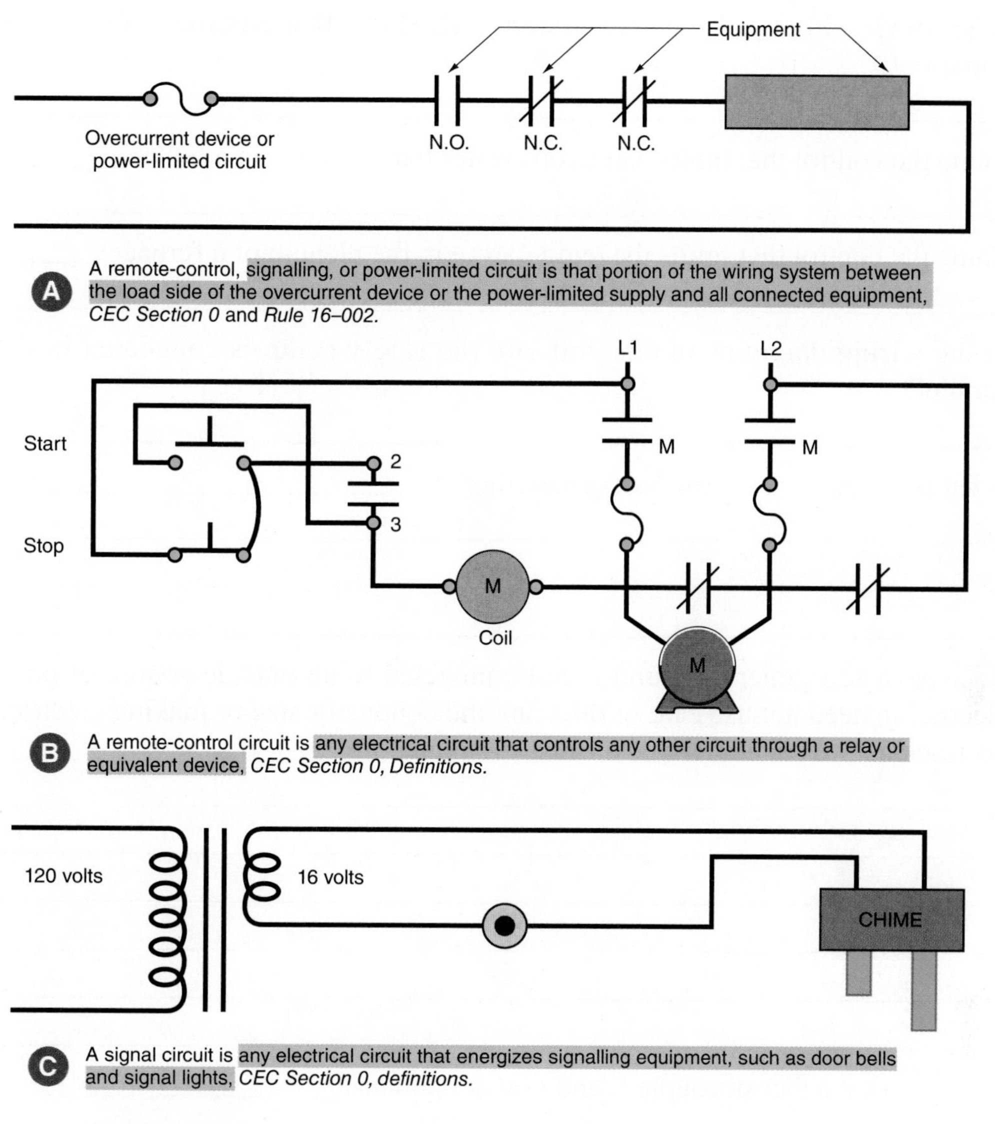

FIGURE 18-19 A remote-control circuit and a signalling circuit.

REVIEW

Note: Refer to the *CEC* or the blueprints provided with this textbook when necessary. Where applicable, responses should be written in complete sentences.

1. How is the residence in the text heated? ______________________
2. Where does the *CEC* require that a disconnecting means be installed for a furnace or boiler? ______________________

____

3. What device in an oil-burning system will shut off the burner in the event of a flame-out?

   ___

4. Name the control that limits dangerous water temperatures in a boiler. ___

   ___

5. Name the control that limits the temperature in the plenum of a furnace. ___

   ___

6. In the wiring diagrams in this unit, are the safety controls connected in series or parallel?

   ___

7. What is meant by the term "self-generating"? ___

   ___

   ___

   ___

8. Because a self-generating unit is not connected to an outside source of power, the electrician need not use care in selecting the conductor size or making electrical connections. Is this statement true or false? Explain. ___

   ___

   ___

   ___

   ___

   ___

   ___

9. Explain what a thermocouple is and how it operates. ___

   ___

   ___

   ___

   ___

   ___

   ___

10. Explain what a thermopile is. ___

    ___

    ___

    ___

11. If Class 2 extra-low-voltage conductors are insulated for 32 volts, and the power conductors are insulated for 600 volts, does the *CEC* permit pulling these conductors through the same raceway? ___

12. If Class 2 extra-low-voltage conductors are used on a 24-volt system but are actually insulated for 600 volts (T90 nylon #14, for example), does the *CEC* permit running these conductors together in the same raceway as the 600-volt power conductors?

____

____

13. Under what conditions may Class 1 conductors and power conductors be installed in the same raceway? ____

____

____

14. What is the *CEC* section and rule for Question 13? ____

____

____

____

UNIT 19

Recreation Room

OBJECTIVES

After studying this unit, you should be able to

- understand three-wire (multiwire) circuits—the advantages and the cautions
- understand how to install lay-in fixtures
- calculate watts loss and voltage drop in two-wire and three-wire circuits
- understand the term *fixture drops*
- understand the advantages of installing multiwire branch circuits
- understand problems that can be encountered on multiwire branch circuits as a result of open neutrals

RECREATION ROOM LIGHTING

The recreation room ceiling is finished with 13-mm gypsum board. Six lay-in 600-mm × 1200-mm fluorescent fixtures are installed into 600-mm by 1200-mm plaster frames that are fastened to the ceiling's framing. The fixtures rest inside of the plaster frames.

A suspended ceiling is another type of ceiling finish often used in recreation rooms. A T bar ceiling grid is installed below ceiling joists. Fluorescent fixtures can be installed directly on top of ceiling T bars. A typical suspended ceiling installation is shown in Figure 19-1. Figure 19-2 shows a device and wiring layout. When conduit is used, junction boxes are mounted above the dropped ceiling, usually within half a metre of the intended fixture location. One to two metres of AC90 armoured cable are installed between these junction boxes and the fixtures. These short cables are commonly called *fixture drops;* Figure 19-3. They must contain the correct type and size of conductor for the *Canadian Electrical Code (CEC)* temperature ratings and load requirements for recessed fixtures. In houses wired with non-metallic-sheathed cable, the luminaire may be wired directly using NMD 90.

Fluorescent fixtures of the type shown in Figure 19-3 might bear the label "Recessed fluorescent fixture," which might also state, "Suspended ceiling fluorescent fixture." Although they might look the same, there is a difference.

Recessed fluorescent fixtures are intended for installation in cavities in ceilings and walls and are to be wired according to *Rules 30–900* through *30–912*. These fixtures may also be installed in suspended ceilings if they have the necessary mounting hardware.

Suspended fluorescent fixtures are intended for installation only in suspended ceilings where the acoustical tiles, lay-in panels, and suspended grid are not part of the actual building structure.

The *National Building Code* and most electrical inspection departments require fixtures to be chained. This provides an independent means of support so, if the ceiling collapses, the fixtures remain supported.

Canadian Standards Association (CSA) Standard C22.2 E598 covers recessed and suspended ceiling fixtures in detail.

Important: To prevent the luminaire from inadvertently falling, the *Building Code* requires that (1) suspended ceiling framing members that support recessed luminaires must be securely fastened to each other, and must be securely attached to the building structure at appropriate intervals, and (2) recessed luminaires must be securely fastened to the suspended ceiling framing members by bolts, screws, rivets, or special listed clips provided by the manufacturer of the luminaire for the purpose of attaching the luminaire to the framing member.

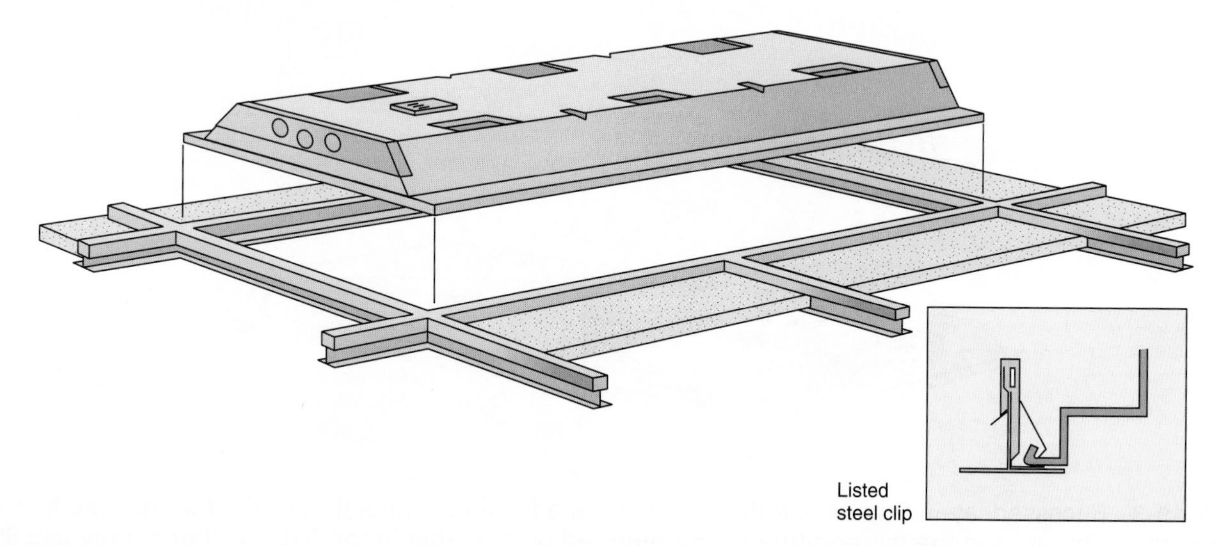

FIGURE 19-1 Lay-in fluorescent fixture commonly used in conjunction with dropped acoustical ceilings.

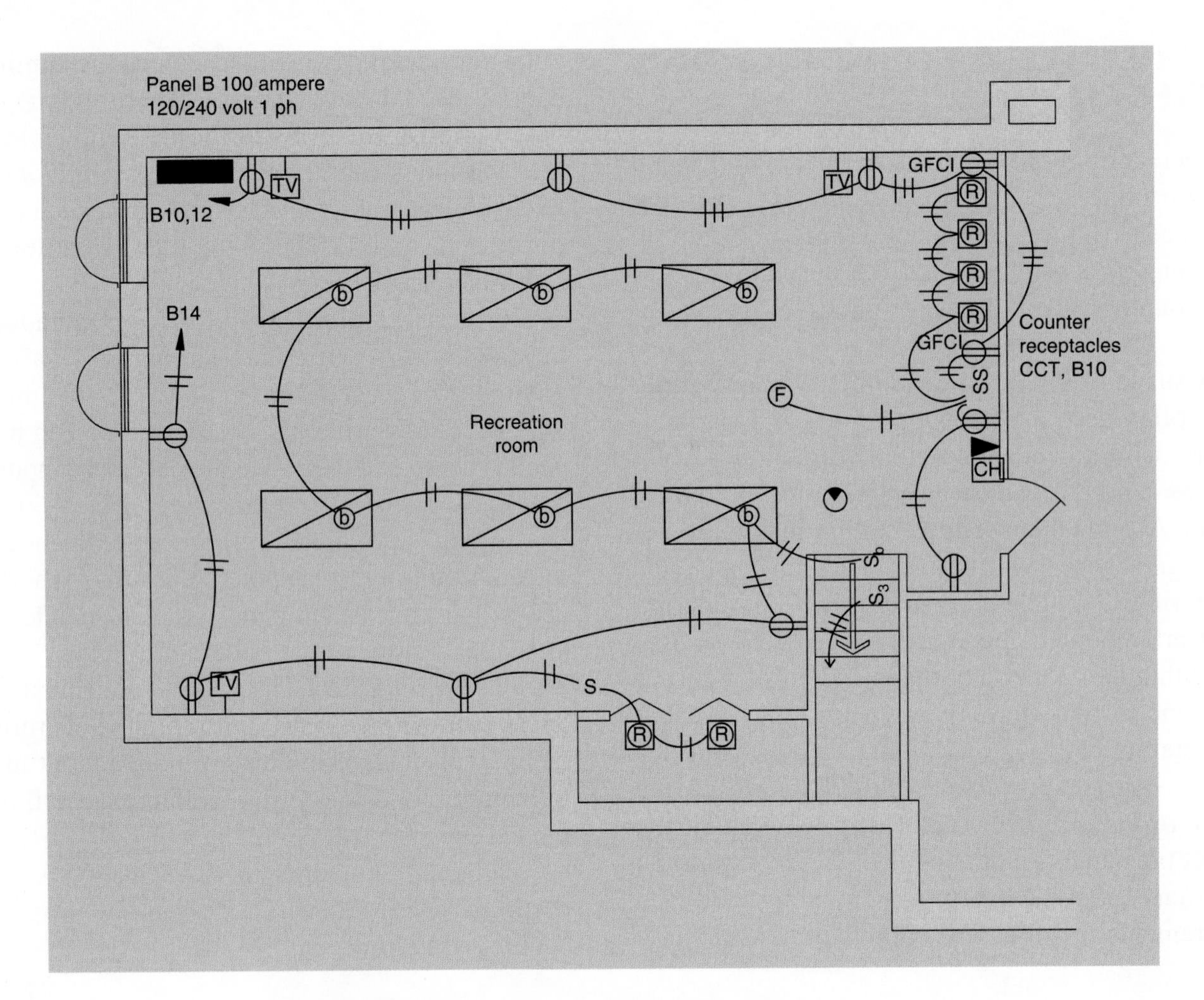

FIGURE 19-2 Cable layout for recreation room.

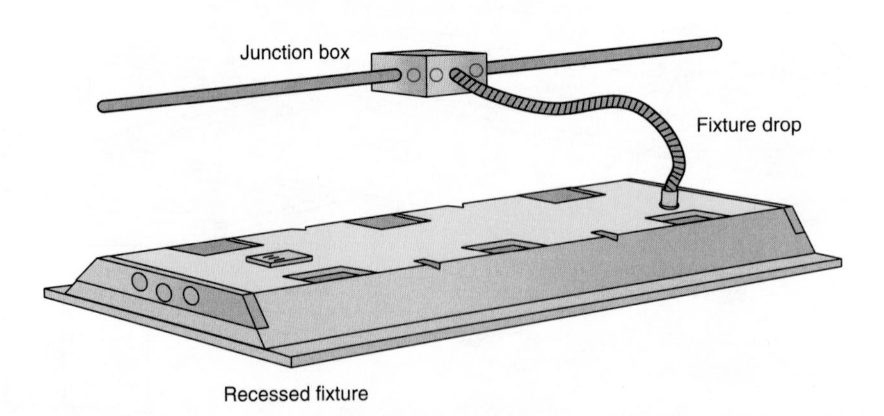

FIGURE 19-3 Recessed lay-in fluorescent fixture showing a flexible connection to the fixture. Conductors in fixture drops must have the temperature rating required on the label in the fixture: "For supply use 90°C conductors."

TABLE 19-1

Recreation room: Outlets and estimated load (three circuits).

DESCRIPTION	QUANTITY	WATTS	VOLT–AMPERES
Circuit B10			
Wet bar receptacles (GFCI-protected)	2	240	240
Total	2	240	240
Circuit B12			
Receptacles	5	600	600
Recessed lights	4	240	240
Fan	1	160	300
Total	10	1000	1140
Circuit B14			
Receptacles	4	480	480
Recessed lights	2	120	120
Lay-in fluorescent lights	6	336	366
Total	12	936	966

The fluorescent fixtures in the recreation room are four-lamp fixtures having warm white deluxe (WWX) or cool white deluxe (CWX) lamps installed in them. They are controlled by a single-pole switch located at the bottom of the stairs.

Circuit B14 feeds the fluorescent fixtures in the recreation room; see Figure 19-2. Table 19-1 summarizes the outlets and estimated load for the recreation room.

How to Connect Recessed Lay-In Fixtures

Figure 19-3 shows how to connect these fixtures. A junction box is located above the ceiling near the fixture. Fixture drops are usually made up of #14/2 AC90 cable if noncombustible construction is used. This cable must be supported a maximum of 300 mm from the box, *Rule 12–618*. NMD 90 cable can also be installed directly into fixtures and looped between fixtures if combustible construction is used.

Rule 12–1004(a) permits 12-mm flexible metal conduit in lengths not to exceed 1.5 m to make the connection between the junction box and the fixture. Bonding of the fixture is done according to *Rule 10–400*. When flexible metal conduit is used, a bonding conductor must be run as per *Rule 10–618(3)*.

A fixture drop of flexible metal conduit is not required to be supported if the drop is no more than 900 mm, *Rule 12–1010(3)*. Yet, many electricians secure the flex within 300 mm of the outlet box to avoid pulling the flex out of the connector while they are making up the electrical connections and installing the fixture in the lay-in ceiling.

RECEPTACLES AND WET BAR

The circuitry for the recreation room wall receptacles and the wet bar lighting area introduces a new type of circuit: a *multiwire*, or *three-wire, circuit*.

Using a multiwire circuit can save money since one three-wire circuit will do the job of two two-wire circuits. Also, if the loads are nearly balanced in a three-wire circuit, the neutral conductor carries only the unbalanced current. This results in less voltage drop and watts loss for a three-wire circuit compared to similar loads connected to two separate two-wire circuits.

Figures 19-4 and 19-5 illustrate the benefits of a multiwire circuit for watts loss and voltage drop. The example is a purely resistive circuit with equal loads.

EXAMPLE 1

TWO TWO-WIRE CIRCUITS (Figure 19-4)

Watts loss in each current-carrying conductor is

$$\text{Watts} = I^2R = 10 \times 10 \times 0.154 \text{ or } 15.4 \text{ watts}$$

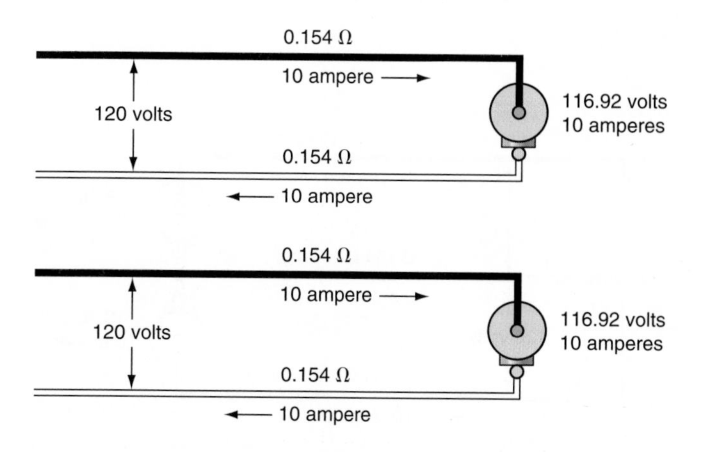

FIGURE 19-4 Diagram of Example 1.

Watts loss in all four current-carrying conductors is

$$15.4 \text{ watts} \times 4 = 61.6 \text{ watts}$$

Voltage drop in each current-carrying conductor is

$$\begin{aligned} E_d &= IR \\ &= 10 \times 0.154 \text{ volts} \\ &= 1.54 \text{ volts} \end{aligned}$$

E_d for both current-carrying conductors in each circuit is

$$2 \times 1.54 \text{ volts} = 3.08 \text{ volts}$$

Voltage available at load is

$$(120 - 3.08) \text{ volts} = 116.92 \text{ volts}$$

EXAMPLE 2

ONE THREE-WIRE CIRCUIT (Figure 19-5)

Watts loss in each current-carrying conductor is

$$\text{Watts} = I^2R = 10 \times 10 \times 0.154 \text{ watts} = 15.4 \text{ watts}$$

Watts loss in both current-carrying conductors is

$$15.4 \text{ watts} \times 2 = 30.8 \text{ watts}$$

Voltage drop in each current-carrying conductor is

$$\begin{aligned} E_d &= IR \\ &= 10 \times 0.154 \text{ volts} \\ &= 1.54 \text{ volts} \end{aligned}$$

E_d for both current-carrying conductors is

$$2 \times 1.54 \text{ volts} = 3.08 \text{ volts}$$

Voltage available at loads is

$$(240 - 3.08) \text{ volts} = 236.92 \text{ volts}$$

Voltage available at each load is

$$\frac{236.92}{2} \text{ volts} = 118.46 \text{ volts}$$

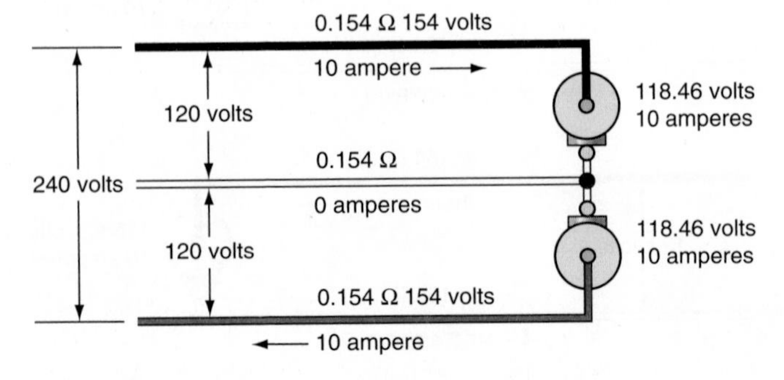

FIGURE 19-5 Diagram of Example 2.

The distance from the load to the source is 15 m. The conductor size is #14 AWG copper. The resistance of each 15-m length of #14 AWG is about 0.154 ohm.

The maximum distance that a #14/2 NMD 90 copper wire, based on a 3% voltage drop, may be run is 18.3 m, using *Table D3* at a 10-ampere load. Therefore, for runs longer than 20 m in residential applications, *Table D3* should be used to ensure that the wire is not too small, *Rule 8–102* and *Appendix D*.

It is apparent that a multiwire circuit can greatly reduce watts loss and voltage drop. This is the same reasoning that is used for the three-wire service entrance conductors and the three-wire feeder to Panel B in this residence.

In the recreation room, a three-wire cable carrying Circuits B10 and B12 feeds from Panel B to the closest receptacle. This three-wire cable continues to each receptacle wall box along the same wall. Circuit B10 feeds the two counter receptacles at the wet bar, and circuit B12 feeds the wall receptacles, fan, and pot lights over the wet bar counter.

The white conductor of the three-wire cable is common to both the receptacle circuit and the wet bar circuits. The black conductor feeds the receptacles. The red conductor is spliced straight through the boxes and feeds the wet bar area.

It is important to connect the neutral according to *Rule 4–036(4)*, which states that the continuity of the identified (white) conductor must be independent of the device connections. This means that a pigtail connection must be made at receptacles and lampholders; see Unit 11, Figure 11-8.

Take care when connecting a three-wire circuit to the panel. The black and red conductors *must* be connected to the opposite phases in the panel to prevent heavy overloading of the neutral (white) conductor. A two-pole breaker may be used for this purpose, but is not mandatory for multiwire circuits serving fixed lighting and not split receptacles, *Rule 14–010(b)*.

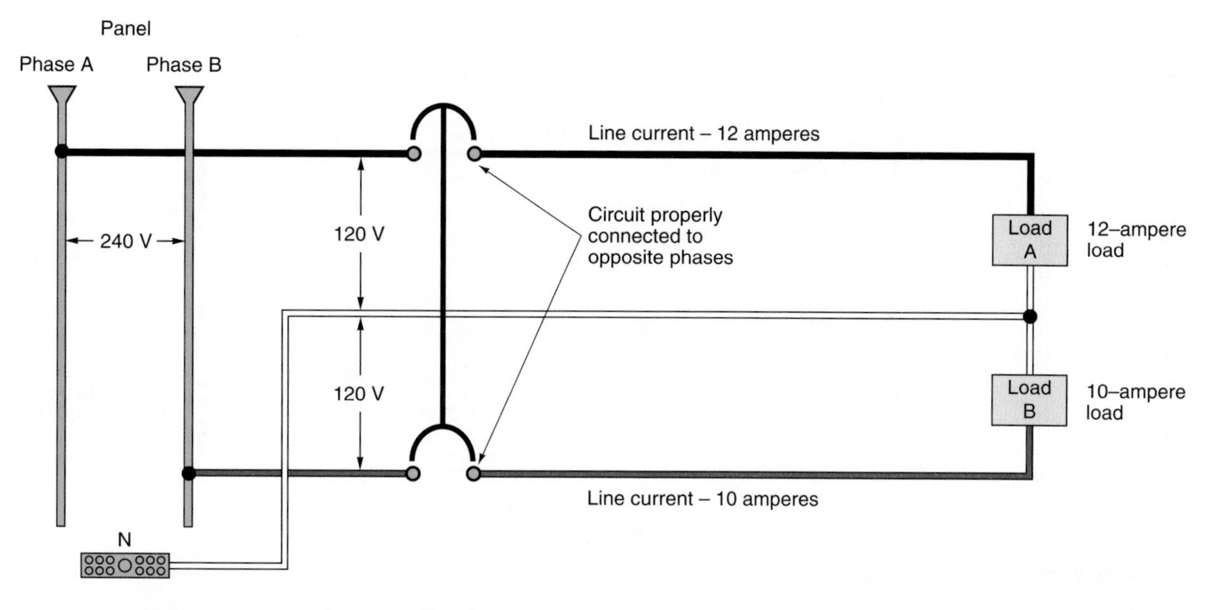

FIGURE 19-6 Correct wiring connections for a three-wire (multiwire) circuit.

The neutral conductor of the three-wire cable carries the unbalanced current. This current is the difference between the current in the black wire and the current in the red wire. For example, if one load is 12 amperes and the other is 10 amperes, the neutral current is the difference between these loads: 2 amperes; see Figure 19-6.

If the black and red conductors of the three-wire cable are connected to the same phase in Panel A (Figure 19-7), the neutral conductor must carry the total current of both the red and black conductors rather than the unbalanced current. As a result, the neutral conductor will be overloaded. All single-phase, 120/240-volt panels are clearly marked to help prevent an error in phase wiring. The electrician must check all panels for the proper wiring diagrams before beginning an installation.

Figure 19-7 shows how an improperly connected three-wire circuit results in an overloaded neutral conductor. If an open neutral occurs on a three-wire circuit, some electrical appliances in operation may experience voltages higher than normal for the circuit.

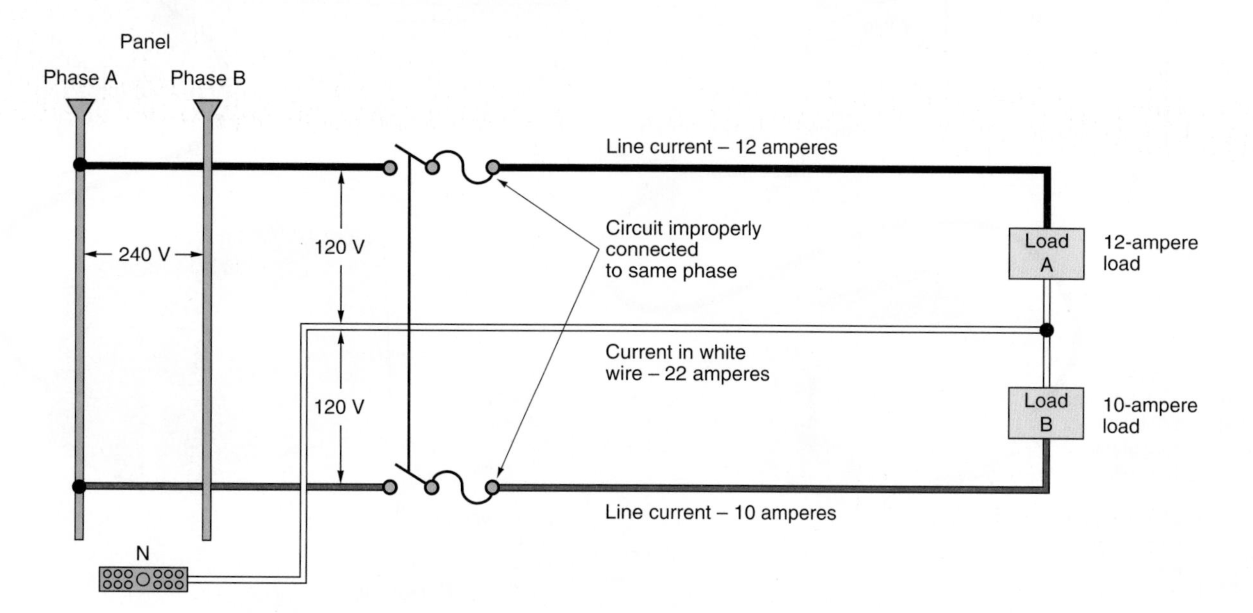

FIGURE 19-7 Improperly connected three-wire (multiwire) circuit.

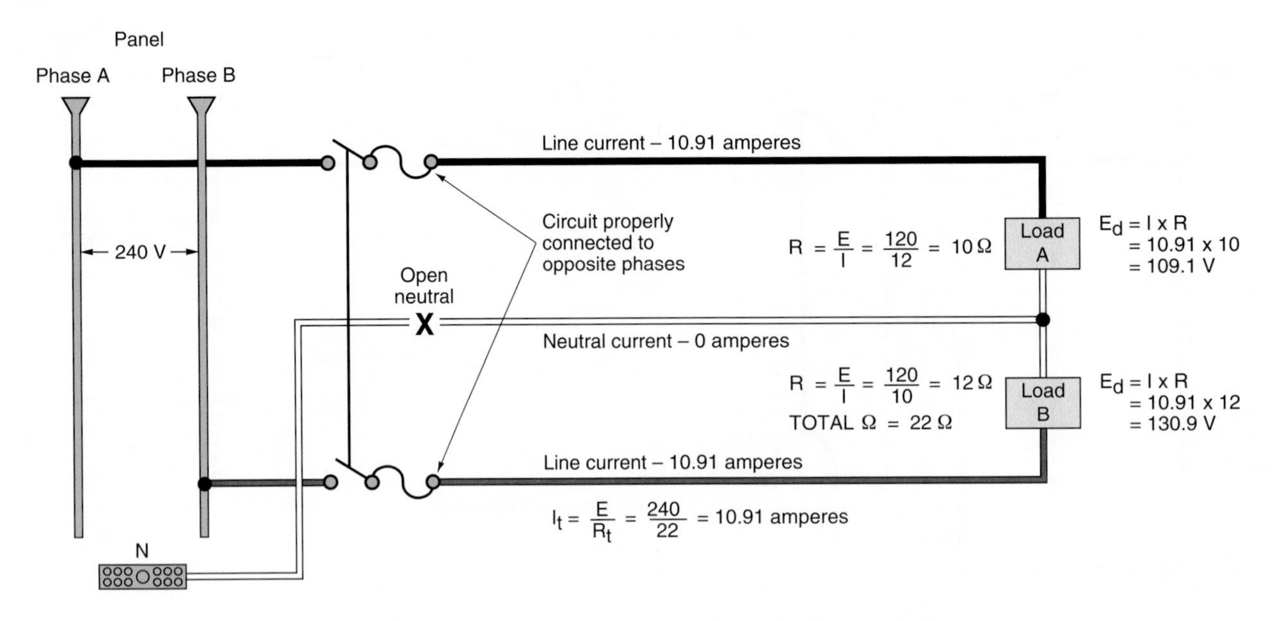

FIGURE 19-8 Example of an open neutral conductor.

Figure 19-8 shows that for an open neutral condition, the voltage across Load A decreases and the voltage across Load B increases. If the load on each circuit changes, the voltage on each circuit also changes. According to Ohm's law, the voltage drop across any device in a series circuit is directly proportional to the resistance of that device. Therefore, if Load B has twice the resistance of Load A, then Load B will be subjected to twice the voltage of Load A for an open neutral condition. To ensure the proper connection, take care when splicing the conductors.

Figure 19-9 shows what can occur if the neutral of a three-wire multiwire circuit opens. Trace the

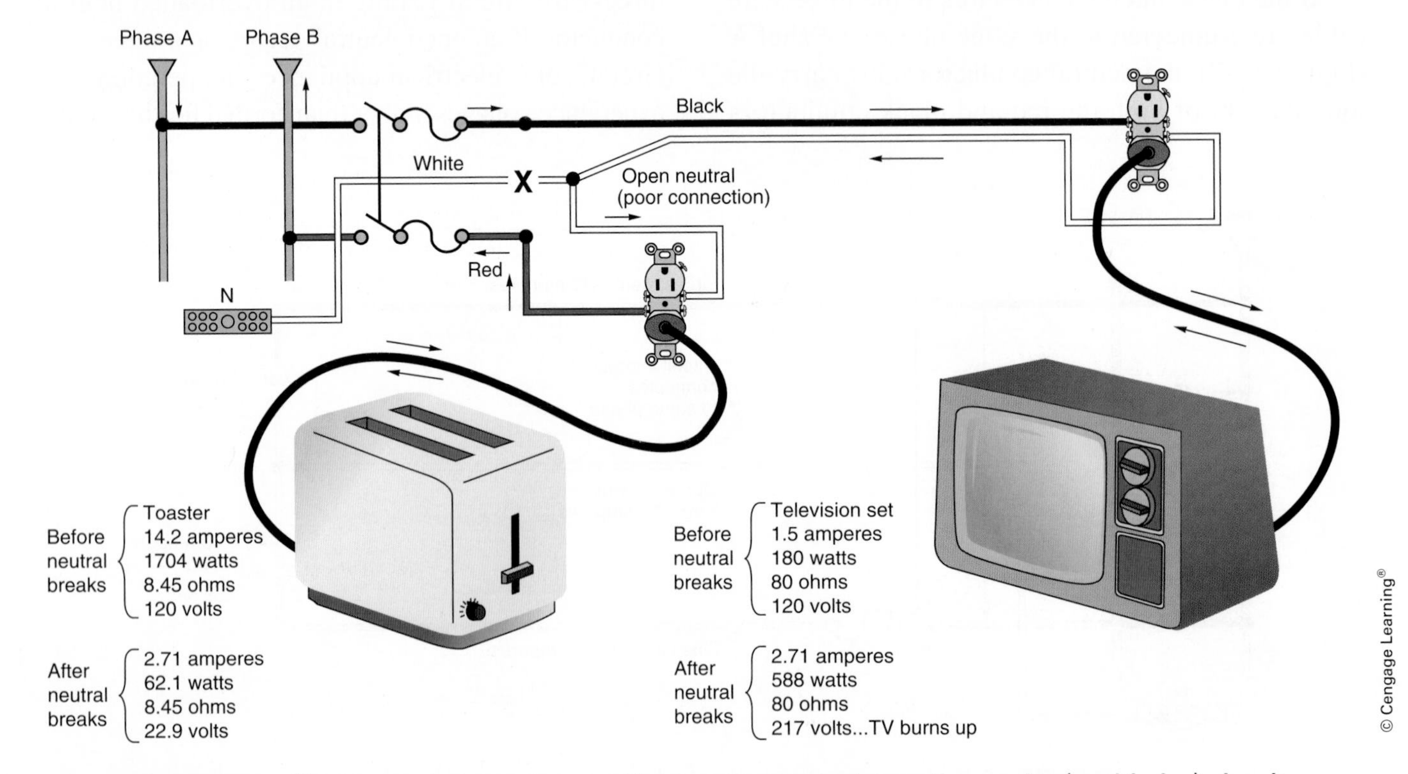

FIGURE 19-9 Problems that can occur with an open neutral on a three-wire (multiwire) circuit.

flow of current from Phase A through the television set, then through the toaster, then back to Phase B, thus completing the circuit. The following simple calculations show why the television set (or stereo or home computer) can be expected to burn up.

With the neutral broken, the TV and the toaster are connected in series across 240 volts. The total resistance (R_t) is now the sum of these two loads:

$$R_t = (8.45 + 80) \text{ ohms} = 88.45 \text{ ohms}$$

The total current of the circuit is the available voltage (240 volts) divided by the total resistance (88.45 ohms), therefore,

$$I = \frac{E}{R} = \frac{240}{88.45} \text{ amperes} = 2.71 \text{ amperes}$$

Voltage appearing across the toaster is

$$E = I \times R = 2.71 \times 8.45 \text{ volts} = 22.9 \text{ volts}$$

Voltage appearing across the television is

$$E = I \times R = 2.71 \times 80 \text{ volts} = 216.8 \text{ volts}$$

The toaster that is rated at 1700 watts is producing only 62 watts, whereas the TV that is rated at 180 watts is producing 587 watts. Therefore, the toaster will produce little heat, and the TV components will be damaged, with a possibility of fire. This example illustrates the problems that can arise with an open neutral on a three-wire, 120/240-volt multiwire branch circuit.

The same problem can arise when the neutral of the utility company's incoming service entrance conductors (underground or overhead) opens. The problem is minimized because the neutral of the service is solidly connected (grounded) to the building's water piping system or other approved grounding electrode, *Rule 10–106*. However, there are cases where poor service grounding has resulted in serious and expensive damage to appliances within the home because of an open neutral in the incoming service entrance conductors.

Three-wire circuits are used occasionally in residential installations. The two hot conductors of three-wire cable are connected to opposite phases. One white neutral (grounded) conductor is common to both hot conductors. All three-wire cables supplying multiwire circuits must be connected to a common-trip, two-pole breaker unless they feed fixed lighting loads or non-split receptacles, *Rule 14–010(b)*.

The two receptacle outlets above the wet bar in the recreation room are Ground-Fault Circuit Interrupter (GFCI)-protected as indicated on the plans. All receptacles within 1.5 m of a sink require GFCI protection, *Rule 26–700(11)*.

A single-pole switch to the right of the wet bar controls the four recessed fixtures above it. See Unit 10 for information on recessed fixtures.

An exhaust fan similar to the one in the laundry room is installed in the ceiling of the recreation room to exhaust stale, stagnant, and smoky air from the room.

The basic switch, receptacle, and fixture connections are similar to those discussed in previous units.

The selection of wall boxes for the receptacles is based on the measurements of the furring strips on the walls. Two-by-two furring strips will require the use of shallow device boxes. If the walls are furred with two-by-fours, regular sectional device boxes could be used. Select a box that can contain the number of conductors, devices, and connectors to meet *CEC* requirements, as discussed in this text *(CEC Table 23)*. After all, the installation must be safe.

REVIEW

Note: Refer to the *CEC* or the blueprints provided with this textbook when necessary. Where applicable, responses should be written in complete sentences.

1. What is the total current draw when all six fluorescent fixtures are turned on?

____

____

2. How many #14 AWG conductors will enter the junction box that will be installed above the dropped ceiling near the fluorescent fixtures closest to the stairway? ______

3. Why is it important that the hot conductors in a three-wire circuit be properly connected to opposite phases in a panel? ______

4. In the diagram below, Load A is rated at 10 amperes, 120 volts and Load B is rated at 5 amperes, 120 volts.

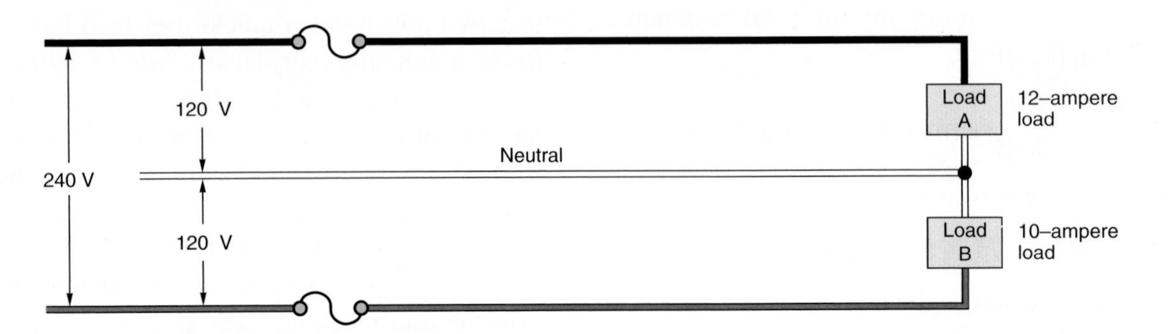

a. When connected to the three-wire circuit as indicated, how much current will flow in the neutral? ______

b. If the neutral should open, to what voltage would each load be subjected, assuming both loads were operating at the time the neutral opened? Show all calculations. ______

5. Calculate the watts loss and voltage drop in each conductor in the following circuit:

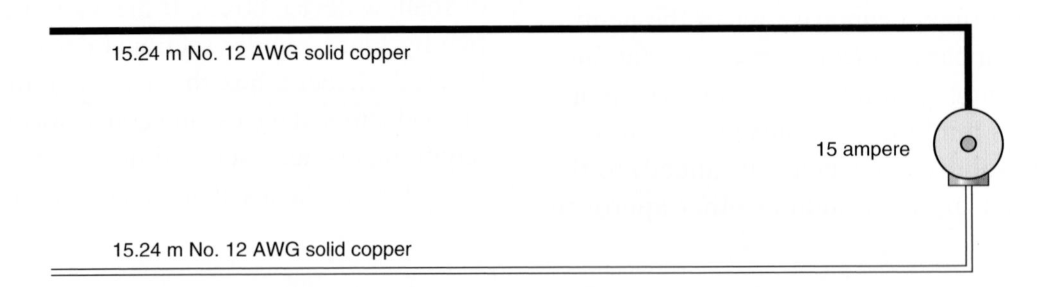

6. Unless specially designed, all recessed incandescent fixtures must be provided with factory-installed ______.

7. If the fluorescent fixtures in the recreation room were to be mounted on the ceiling, what sort of marking would you look for on the label of the fixture? The ceiling is low-density cellulose fibreboard. ______

8. What is the current draw of the recessed fixtures above the bar? ____________________

____

____

____

9. Calculate the total current draw for Circuits B10, B12, and B14.

10. Complete the wiring diagram below for the recreation room. Follow the suggested cable layout in Figure 19-2. Use coloured pencils or pens to identify the various colours of the conductor insulation.

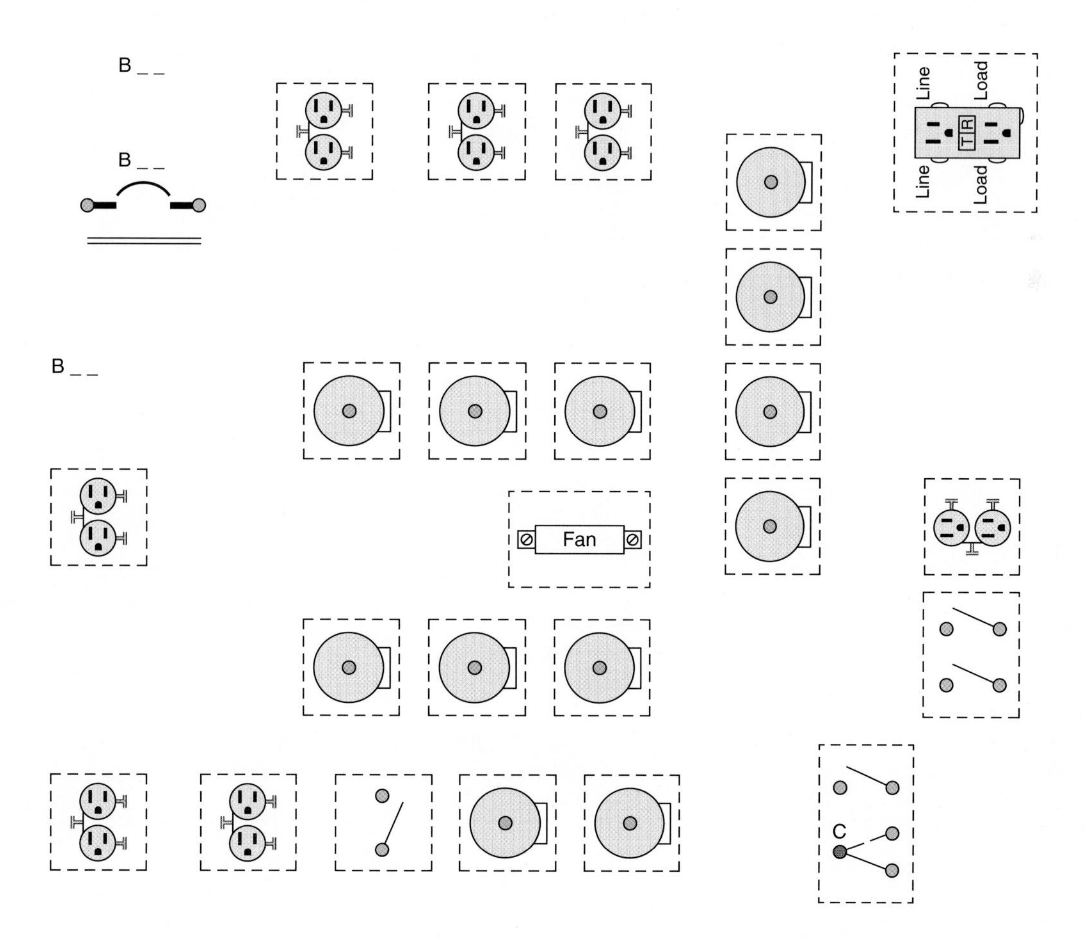

11. a. May a fluorescent fixture that is marked "Recessed fluorescent fixture" be installed in a suspended ceiling? ______________________

    b. May a fluorescent fixture that is marked "Suspended ceiling fluorescent fixture" be installed in a recessed cavity of a ceiling? ______________________

    c. What *CEC, Part 1*, section covers the installation of recessed fixtures?

    ____

    ____

12. Using *Table D3* in *Appendix D*, calculate the maximum distance for a #12/2 NMD 90 wire carrying 10 amperes at 240 volts if a 3% voltage drop is on the cable. Answer in metres. Show calculations. ______________________

    ____

    ____

13. Using the plans provided with this textbok, complete a device and wiring layout for the recreation room area to *minimum CEC* standards.

UNIT 20

Branch Circuits for Workshop and Utility Area

OBJECTIVES

After studying this unit, you should be able to

- understand the meaning, use, and installation of multioutlet assemblies
- list the requirements for a deep-well jet pump installation
- calculate (given the rating of the motor) the conductor size, conduit size, and overcurrent protection required for the pump circuit
- list the electrical circuits used for connecting electric water heaters
- describe the basic operation of the water heater
- make the proper grounding connections for the water heater

WORKSHOP

The workshop area is supplied by four circuits:

A12	Separate circuit for freezer
A17	Lighting
A19	Plug-in strip (multioutlet assembly)
A21	Two receptacles on window wall

A single-pole switch at the entry controls all four ceiling porcelain lampholders. However, note on the plans (Figure 20-1) that two of these lampholders have pull-chains to allow the homeowner to turn these lampholders on and off as needed, to save energy; see Figure 20-2.

In addition to supplying the lighting, Circuit A17 feeds the chime transformers and ceiling exhaust fan. The current draw of the chime transformer is extremely small. Table 20-1 shows the estimated load for lighting and Table 20-2 shows the load for the other three 15-ampere circuits. See also *Rules 30–104(1)* and *14–600.*

WORKBENCH LIGHTING

Two two-lamp, 34-watt fluorescent fixtures are mounted above the workbench to reduce shadows over the work area. The electrical plans show that these fixtures are controlled by a single-pole wall switch. Either a junction box is mounted immediately above or adjacent to the fluorescent fixture so that the connections can be made readily, or the fixture may be directly wired with NMD 90.

NMD 90 can be used to connect the fixture to the junction box. Flexible cord (Type S, SJ, or equivalent) cannot be used to connect the fixture to the junction box since it cannot be used as a substitute for the fixed wiring of structures, *Rule 4–012(3).* The fluorescent fixture must be grounded.

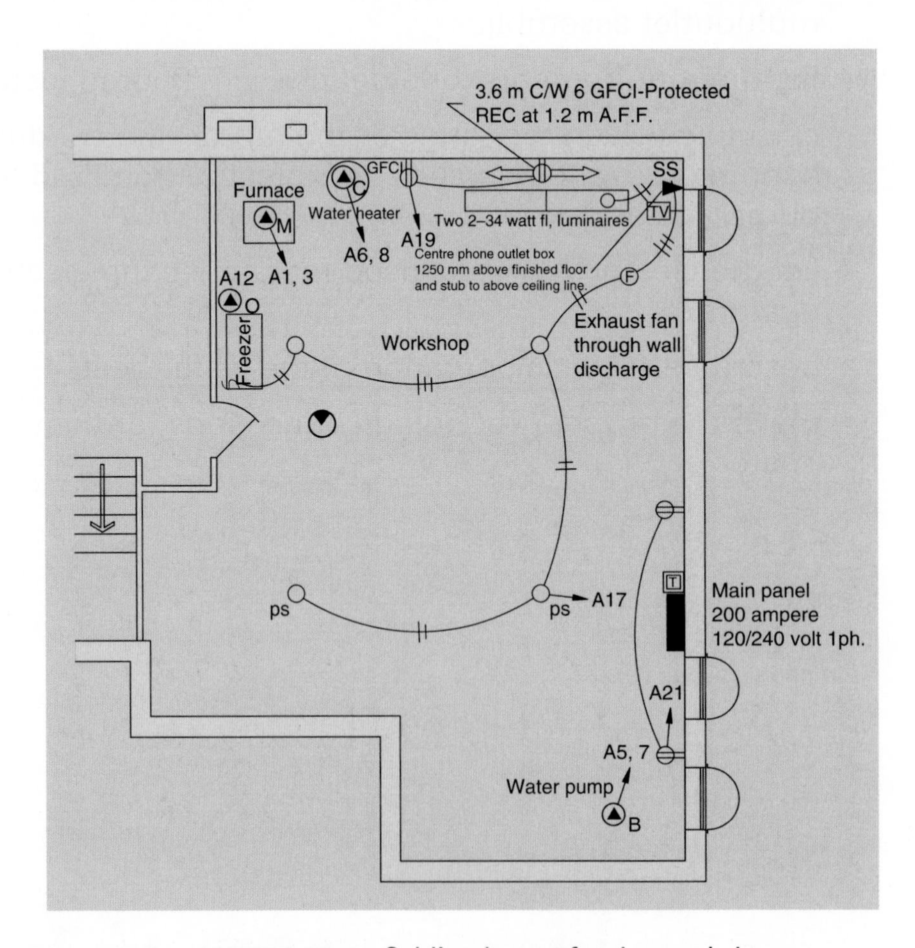

FIGURE 20-1 Cabling layout for the workshop.

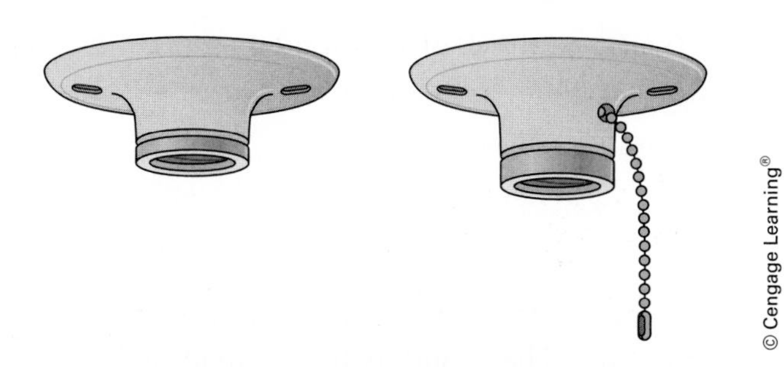

FIGURE 20-2 Porcelain lampholders.

TABLE 20-1

Workshop: Outlet count and estimated lighting load.

DESCRIPTION	QUANTITY	WATTS	VOLT–AMPERES
Ceiling lights @ 100 W each	4	400	400
Fluorescent fixtures: Two 34-W lamps each	2	136	156
Chime transformer	1	8	10
Exhaust fan	1	80	90
Totals	8	624	656

RECEPTACLE OUTLETS

Rule 26–710(e) stipulates that at least one duplex receptacle shall be provided in any unfinished basement area. *Rule 26–710(e)* requires that at least one duplex receptacle be provided for a central vacuum system, where the ductwork is already installed. In this residence, the outlet for the central vacuum is in the garage. In some jurisdictions, the installation of a sump pump in the basement is mandatory. A separate circuit is recommended for the receptacle feeding the sump pump and does not have to be Ground-Fault Circuit Interrupter (GFCI)-protected.

CABLE INSTALLATION IN BASEMENTS

Unit 7 covers the installation of non-metallic-sheathed cable (Romex) and armoured cable (BX).

TABLE 20-2

Workshop: Outlet count and estimated load for three 15-ampere receptacle circuits (wire with conduit). *Rule 8–304(4)* counts each 300 mm of multioutlet assembly as one outlet for areas where a number of appliances will be used simultaneously.

DESCRIPTION	QUANTITY	WATTS	VOLT–AMPERES
15-A circuit, A19			
Receptacle next to multioutlet assembly @ 1 A	1	120	120
A six-receptacle plug-in multioutlet assembly at 1 A per outlet (120 VA)	1	720	720
Totals	2	840	840
15-A circuit, A21			
Receptacles on window wall @ 120 W each	2	240	240
Totals	2	240	240
15-A circuit, A12			
Single receptacle for freezer (not GFCI)	1	360	696
Total of the three circuits	5	1440	1776

Running non-metallic-sheathed cable is covered by the *12–500* series of rules, and armoured cable (BX) installations by the *12–600* series of rules. When laying out the basement, consider the following:

- Cables cannot be run across the lower face of basement joists unless they are protected from mechanical injury.
- When run through holes in the joists, cables are considered to be supported.
- Cables for concealed installations must be kept at least 32 mm from the edge of the joist.
- Check with your local inspection authority to see how many cables may be run through holes in the joists before *Rule 4–004(11)* is applied.
- When cables are run near heating ducts, they must be 25 mm from the duct to prevent heat transfer from the duct to the cable. If it is not

possible to obtain a 25-mm separation, a thermal barrier must be installed. Running the cable below the ducts whenever possible will also help prevent heat transfer from the ducts.

- If the cable is run near a concrete or masonry chimney, a distance of 50 mm must be maintained between the conductor and a chimney, and 150 mm must be maintained between the conductor and a chimney cleanout.
- Some installations will require drilling of the basement joists. To make the cable easier to pull, start at one end of the basement and drill every second joist. At the end of the run, turn around and drill the remaining joists from the opposite direction. The cable will be much easier to pull through holes drilled this way.

MULTIOUTLET ASSEMBLY

A multioutlet assembly has been installed above the workbench. For safety and adequate wiring, any workshop receptacles should be connected to a separate circuit; see Figure 20-3. In the event of a malfunction in any power tool commonly used in the home workshop (saw, planer, lathe, or drill), only receptacle Circuit A19 is affected. The lighting in the workshop is not affected by a power outage on the receptacle circuit.

The installation of multioutlet assemblies must conform to the requirements of *Rule 12–3028.*

Load Considerations for Multioutlet Assemblies

A maximum of 12 outlets are permitted on any two-wire circuit. These outlets have a load of 1 ampere each. The maximum load, therefore, is 12 amperes, which is 80% of the 15-ampere circuit breaker that feeds Circuit A19. Where a number of appliances may be used simultaneously, each 300 mm of multioutlet assembly is considered to be one outlet, *Rule 8–304(4).*

Because the multioutlet assembly has been provided in the workshop for plugging in portable tools, a separate Circuit A19 is provided. This circuit has been included in the service entrance calculations as part of the general lighting and power load requirements for the residence.

Many types of convenience receptacles are available for various sizes of multioutlet assemblies, including duplex, single-circuit grounding, split-circuit grounding, and duplex split-circuit receptacles.

Wiring the Multioutlet Assembly

The electrician first attaches the base of the multioutlet assembly to the wall or baseboard. The receptacles are snapped into the base; see Figures 20-4 and 20-5.

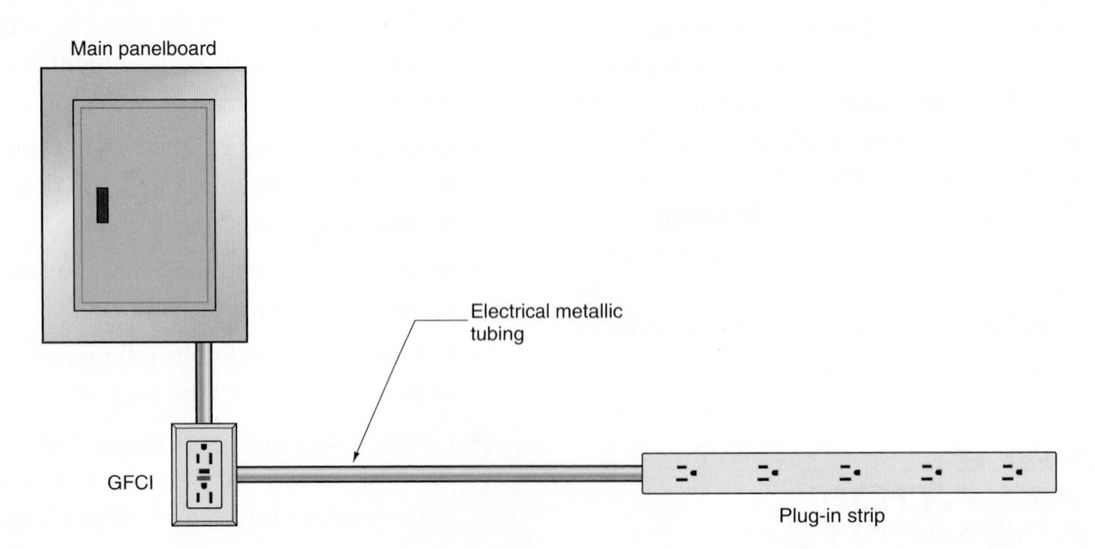

FIGURE 20-3 Detail of how the workbench receptacle plug-in strip is connected through the receptacle located adjacent to the workbench. This is Circuit A19.

Courtesy of Legrand/Wiremold

FIGURE 20-4 One type of multioutlet assembly showing track, receptacle, and track cover.

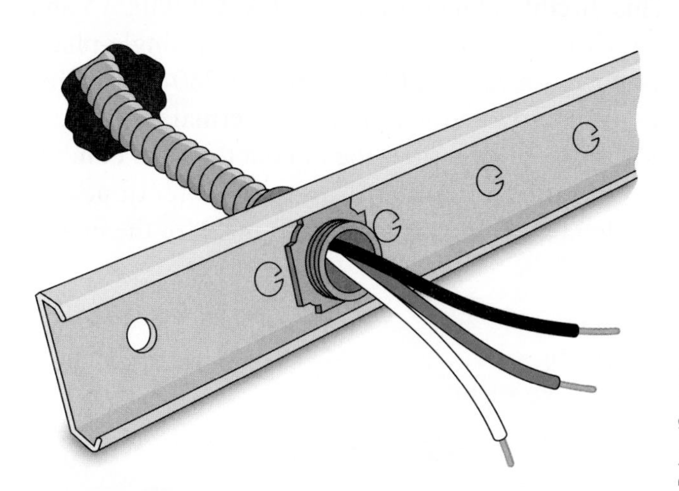

Delmar/Cengage

FIGURE 20-5 Multioutlet assembly starter is mounted on an outlet box and is used to supply power to the assembly track.

Covers are then cut to the lengths required to fill the spaces between the receptacles. Manufacturers of these assemblies can supply prefabricated covers to be used when evenly spaced receptacles are installed. These receptacles are usually installed 305 mm or 450 mm apart.

In general, multioutlet installations are simple because of the large number of fittings and accessories available. Connectors, couplings, ground clamps, blank end fittings, elbows for turning corners, and end entrance fittings can all be used to simplify the installation.

EMPTY CONDUITS

Although not truly part of the workshop wiring, Note 8 on the first-floor electrical plans indicates that two empty 16-mm EMT raceways are to be installed, running from the workshop to the attic. This is unusual but certainly handy when additional wiring might be required at some later date. The empty raceways will provide the necessary route from the basement to the attic, thus eliminating the need to fish through the wall partitions.

UTILITY ROOM

In the utility room, the water pump and water heater require special-purpose outlets.

WATER PUMP CIRCUIT ▲$_B$

All dwellings need a good water supply. City dwellings generally are connected to the city water system. In rural areas, where there is no public water supply, each dwelling has its own system in the form of a well. The residence plans show that a deep-well jet pump is used to pump water from the well to the various plumbing outlets. The outlet for this jet pump is shown by the symbol. ▲$_B$ An electric motor drives the jet pump. A circuit must be installed for this system.

JET PUMP OPERATION

Figure 20-6 shows the major parts of a typical jet pump. The pump impeller wheel ① forces water down a drive pipe ② at a high velocity and pressure to a point just above the water level in the well casing. (Some well casings may be driven to a depth of more than 30.5 m before striking water.) Just above the water level, the drive pipe curves sharply upward and enters a larger vertical suction pipe ③. The drive pipe terminates in a small nozzle or jet ④ in the suction pipe. The water emerges from the jet with great force and flows upward through the suction pipe.

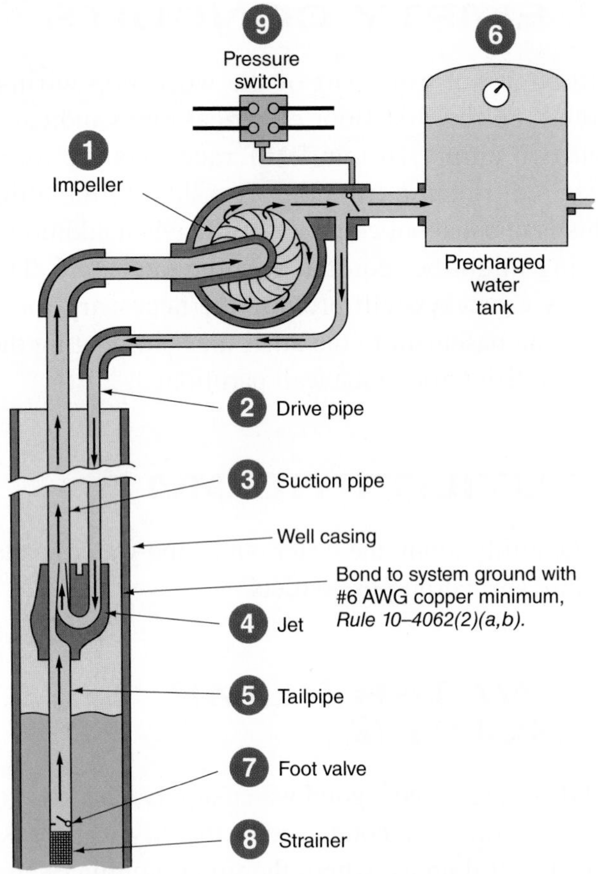

9 Pressure switch
6
1 Impeller
Precharged water tank
2 Drive pipe
3 Suction pipe
Well casing
Bond to system ground with #6 AWG copper minimum, *Rule 10–4062(2)(a,b).*
4 Jet
5 Tailpipe
7 Foot valve
8 Strainer

FIGURE 20-6 Components of a jet pump.

Water rises in the suction pipe, drawn up through the tailpipe ⑤ by the action of the jet. The water rises to the pump inlet and passes through the impeller wheel of the pump. Part of the water is forced down through the drive pipe again. The remaining water passes through a check valve and enters the storage tank ⑥.

The tailpipe is submerged in the well water to a depth of about 3 m. A foot valve ⑦ and strainer ⑧ are provided at the end of the tailpipe. The foot valve prevents water in the pumping equipment from draining back into the well when the pump is not operating.

When the pump is operating, the lower part of the storage tank fills with water, and air is trapped in the upper part. As the water rises in the tank, the air is compressed. When the air pressure is 276 kilopascals (kPa), a pressure switch ⑨ disconnects the pump motor. The pressure switch is adjusted so that as the water is used and the air pressure falls to 138 kPa, the pump restarts and fills the tank again.

The Pump Motor

A jet pump for residential use may be driven by a one-horsepower (hp) motor at 3400 rpm. A dual-voltage motor is used so that it can be connected to 120 or 240 volts. For a 1-hp motor, the higher voltage is preferred since the current at this voltage is half that at the lower voltage. It is recommended that a single-phase, capacitor-start motor be used for a residential jet pump. Such a motor is designed to produce a high starting torque, which helps to overcome the back pressure within the tank and the weight of the water being lifted from the well.

The Pump Motor Circuit

Figure 20-7 is a diagram of the electrical circuit that operates the jet pump motor and pressure switch. This circuit is taken from Panel A, Circuits A5 and A7. *CEC Table 45* indicates that a 1-hp single-phase motor has a rating of 8 amperes at 240 volts. This motor is provided with internal thermal protection. With this information, the conductor size, conduit size, branch-circuit overcurrent protection, and overload protection can be determined for the motor.

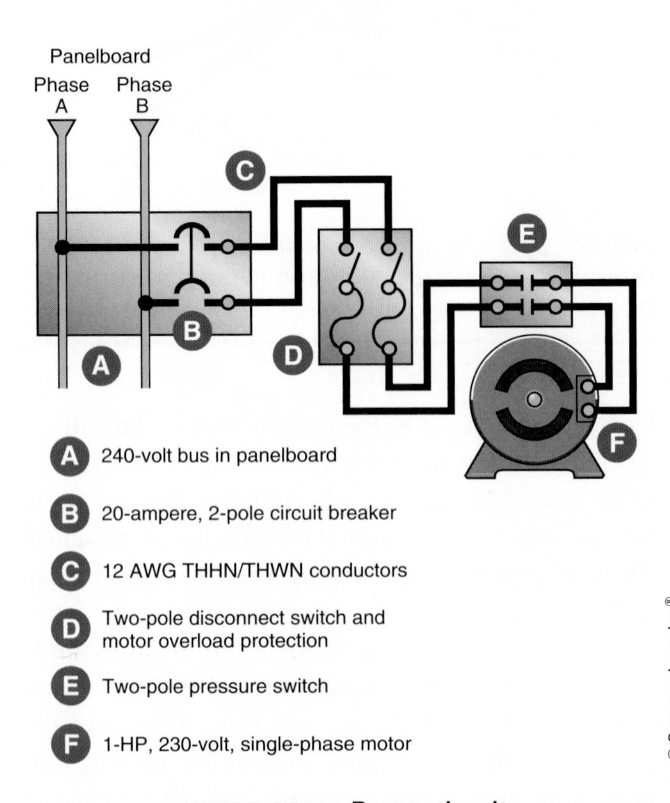

FIGURE 20-7 Pump circuit.

Conductor Size (*Rule 28–106(1)*)

Rule 28–106(1) requires that conductors supplying motors have an ampacity of not less than 125% of the motor full-load current rating. Therefore, the 8-ampere pump motor would require a minimum of a 10-ampere conductor, which would be a minimum #14 AWG. However, the schedule of special-purpose outlets in the specifications states that #12 AWG conductors are to be used for the pump circuit.

Conduit Size

If rigid PVC conduit is run from the house to the wellhead, two #12 and one #14 AWG T90/TWN75 conductors require 16-mm conduit, *Rules 12–1014* and *12–1120.*

Branch-Circuit Overcurrent Protection (*Rule 28–200, Tables 29* and *D16*)

Rule 28–200 requires the jet pump motor to be protected with an overcurrent device sized according to *Table 29.* Therefore, *Table 29,* line 1, can be used.

Motor circuit overcurrent sizing may be determined using *Table D16,* which is based on *Table 29.* However, since not all motor full-load current ratings are listed in this table, the overcurrent devices should be sized according to *Table 29* and *Rule 28–200* as follows:

- Nontime-delay fuses
  Maximum of 300% of the motor FLA (full-load amperes)
  $8 \times 300\% = 24$ amperes
  Therefore, use 20-ampere fuse.
- Time-delay fuses
  Maximum of 175% of motor FLA
  $8 \times 175\% = 14$ amperes.
  Therefore, use 15-ampere fuse, *Rule 28–200(b).*
- Circuit breakers—inverse time
  Maximum of 250% of motor FLA
  $8 \times 250\% = 20$ amperes
  Therefore, use 20-ampere breaker.

When overcurrent devices, as selected above, will not allow the motor to start, it is permissible to increase the overcurrent device according to *Rule 28–200(d)* as follows:

- Nontime-delay fuses
  Maximum of 400% of motor FLA
- Time-delay fuses
  Maximum of 225% of motor FLA
- Circuit breakers—inverse time
  Maximum of 400% of motor FLA for motors rated 100 hp or less

The previously mentioned percentage ratings, as indicated in *Table 29* and *Rule 28–200(d),* are maximum values. If the value calculated by this method does not correspond to a standard size or rating of overcurrent device, the next smaller size must be used. However, overcurrent devices smaller than 15 amperes are not required, *Rule 28–200(b).* Overcurrent devices may be sized smaller than the maximum allowable size if desired.

Running Overload Protection for the Motor (*Rule 28–300*)

All motors are required to have overload protection (*Rule 28–300*) unless they meet the requirements of *Rule 28–308.*

Overload devices can be time-delay fuses or thermal devices of the fixed or adjustable type. When sizing overload devices, it is important to note the motor service factor, which is indicated on the motor nameplate. If the motor has a marked service factor of 1.15 (115%) or greater, the overload devices are sized at 125% of the motor full-load current, *Rule 28–306(1).* If the service factor is less than 1.15 (115%) or is unknown, the overload devices will be sized at a maximum of 115% of the motor FLA.

Assuming that the pump has, as most modern motors do, a service factor of 1.15 or greater, the overloads will be sized as

$$8 \text{ amperes} \times 125\% = 10 \text{ amperes}$$

Wiring for the Pump Motor Circuit

The water pump circuit is run from Panel A Circuits A5 and A7, to a disconnect switch and controller mounted on the wall next to the pump. Thermal overload devices (also called *heaters*) are installed in the controller to provide running overload protection. Two No. 12 conductors are connected to the double-pole, 20-ampere branch-circuit overcurrent protective device in Panel A to form Circuit A5/A7. Dual-element, time-delay fuses may be installed in the disconnect switch. These fuses will serve the dual function of providing both branch-circuit overcurrent protection for the pump motor and backup protection for the thermal overload protecting the motor. These fuses may be rated at not more than 125% of the full-load running current of the motor.

A short length of flexible metal conduit is connected between the load side of the thermal overload switch and the pressure switch on the pump. This pressure switch is usually connected to the motor at the factory. Figure 20-7 shows a double-pole pressure switch at (E). When the contacts of the pressure switch are open, all conductors feeding the motor are disconnected.

SUBMERSIBLE PUMP

Figure 20-8 shows the major components of a typical submersible pump. A submersible pump consists of a centrifugal pump, ①, driven by an electric motor, ②. The pump and the motor are contained in one housing, ③, submersed below the permanent water level, ④, within the well casing, ⑤. The pump housing is a cylinder 3–5 in. in diameter and 2–4 ft. long. When running, the pump raises the water upward through the piping, ⑥, to the water tank, ⑦. Proper pressure is maintained in the system by a pressure switch, ⑧. The disconnect switch, ⑨, high-pressure safety cutoff switch ⑩, and controller, ⑪, are installed in logical and convenient locations near the water tank.

Some water tanks have a precharged air chamber containing a vinyl bag that separates the air from the water. This assures that air is not absorbed by the water. The absorption of air causes a slow reduction of the water pressure, and, ultimately, the tank will fill completely with water with no room for the air. The use of a vinyl bag to hold the air ensures that the initial air charge is always maintained.

Power to the motor is supplied by a cable especially designed for use with submersible pumps. Marked "submersible pump cable," this cable is

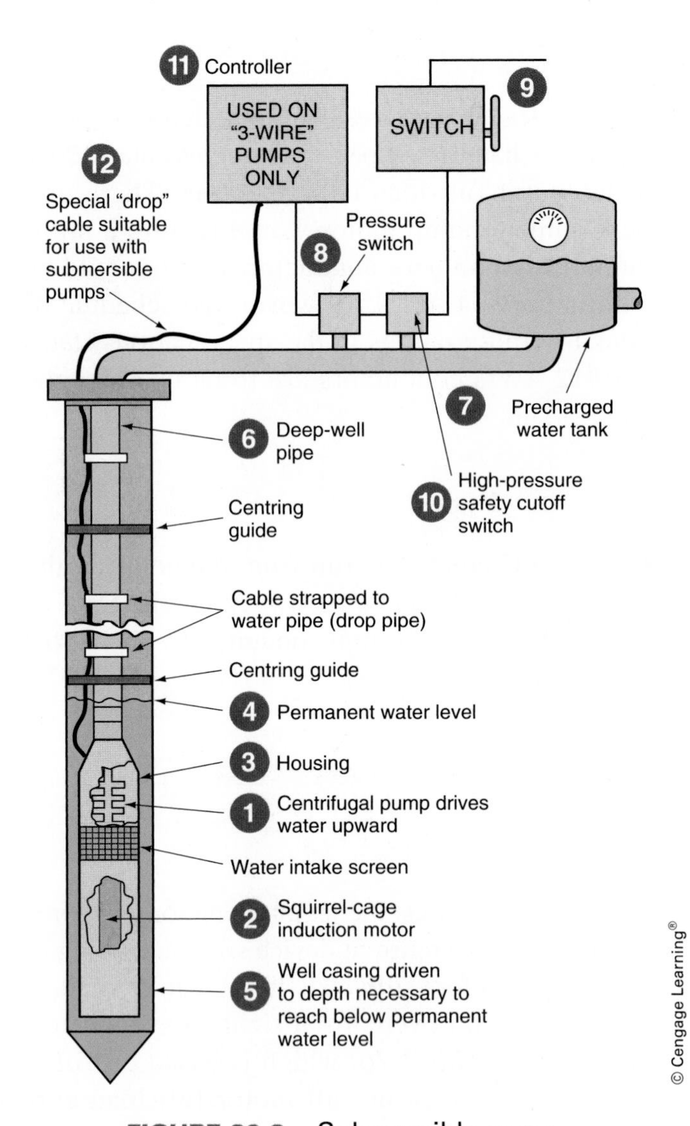

FIGURE 20-8 Submersible pump.

generally supplied with the pump and its other components. The cable is cut to the proper length to reach between the pump and its controller. When needed, the cable may be spliced according to the manufacturer's specifications.

Submersible water pump cable is "tag-marked" for use within the well casing for wiring deep-well water pumps where the cable is not subject to repetitive handling caused by frequent servicing of the pump unit.

The wiring from the wellhead to the main distribution panel must be in accordance with *CEC Section 12.* Should it be required to run the pump's circuit underground for any distance, it is necessary to install underground feeder cable type NMWU, or a raceway with suitable conductors, and then make up the necessary splices in an approved weatherproof junction box.

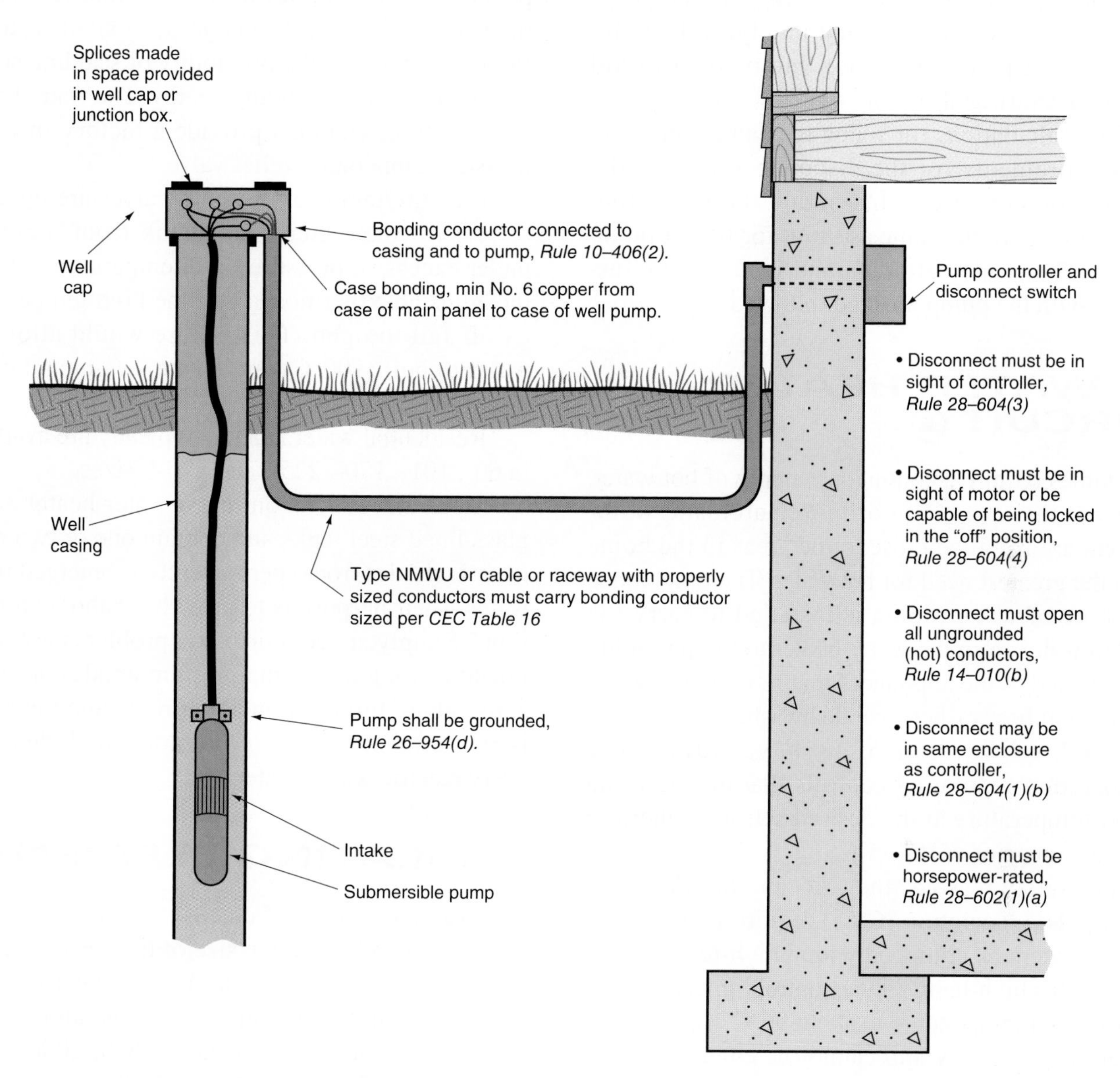

FIGURE 20-9 Bonding requirements for submersible pumps. *Rules 26–954(d)* and *10–406(2)* require that submersible pumps be bonded to ground.

Figure 20-9 illustrates how to provide proper bonding of a submersible water pump and the well casing. Proper grounding of the well casing and the submersible pump motor will minimize or eliminate stray voltage problems that could occur if the pump is not grounded.

Key *Canadian Electrical Code (CEC)* rules that relate to grounding water pumping equipment are

- *26–954(d):* Pumps shall be bonded to ground according to *Section 10.*
- *10–402(1a):* Motors operating at more than 30 volts must be grounded.
- *10–808:* Bonding conductor size.
- *10–406(2):* Bond water-piping system to ground with #6 copper.

See Unit 24 for information about underground wiring.

The controller contains the motor's starting relay, overload protection, starting and running capacitors, lightning arrester, and terminals for making the necessary electrical connections. Thus, there are no moving electrical parts with the pump itself, such as would be found in a typical single-phase, split-phase induction motor that would require a centrifugal starting switch.

The calculations for sizing the conductors, and the requirements for the disconnect switch, the motor's branch-circuit fuses, and the grounding connections are the same as those for the jet pump motor. The nameplate data and instructions furnished with the pump must be followed.

WATER HEATER CIRCUIT Ⓐ C

All homes require a continuous supply of hot water. To meet this need, one or more automatic water heaters are installed close to the areas in the home with the greatest need for hot water. The proper size and type of water piping is installed to carry the heated water from the water heater to the plumbing fixtures and to the appliances that require hot water, such as clothes washers and dishwashers.

Modern water heaters have adjustable temperature-regulating controls that maintain the water temperature at the desired setting, generally within a range of 45°C–75°C.

For safety reasons, the *CEC* and the Canadian Standards Association (CSA) require that electric water heaters be equipped with a high-temperature cutoff. This high-temperature control limits the maximum water temperature to about 96°C. This control is set at the factory and cannot be changed in the field. Electric water heaters shall be equipped with a temperature limiting means in addition to the control thermostat. This temperature limiting means must disconnect all ungrounded conductors.

In most residential water heaters, the upper thermostat is set so that it will not exceed 90°C, and the high-temperature cutoff is set so that the water temperature does not exceed 96°C. They are usually combined in one device.

Most customers will, for safety reasons, ask for a setting less than the maximum. See the section "Scalding from Hot Water" later in this unit.

The plumber usually installs pressure/temperature relief valves into an opening provided and marked for the purpose on the water heater. The plumber also installs tubing from the pressure/temperature relief valve downward to within 150 mm of the floor so that any discharge will exit close to the floor. This reduces the possibility of scalding someone who might be standing nearby. The water heater manufacturer can also provide a factory-installed pressure/temperature relief valve.

Pressure/temperature relief valves are installed to prevent the water heater tank from rupturing under excessive pressures and temperatures should the adjustable thermostat and the high-temperature cutoff fail to open. This failure would allow the water to boil (100°C) and create steam, and could lead to the tank rupturing.

Residential water hea ters typically are available in 60-, 101-, 170-, 225-, and 270-L sizes.

To reduce corrosion, most water heaters have glass-lined steel tanks and contain one or two magnesium anodes (rods) permanently submerged in the water. Their purpose is to provide "cathodic protection." Simply stated, corrosion problems are minimized as long as the magnesium anodes are in an active state. Information about replacing the anode is in the manufacturer's literature included with every electric water heater.

HEATING ELEMENTS

The wattage ratings of electric water heaters vary greatly, depending on the size of the heater in litres, the speed of recovery desired, local electric utility regulations, and local plumbing and building codes.

Wattage ratings are usually 1650, 2000, 2500, 3000, 3800, 4500, or 5500 watts; other wattage ratings are available.

The heating elements are generally rated 240 volts. Elements are also available with 120-volt ratings and 208-volt ratings. These elements will produce rated wattage output at rated voltage. When operated at less than rated voltage, the wattage output will be reduced. The effect of voltage variation is discussed later in this unit. Most heating elements will burn out if operated at voltages 5% higher than that for which they are rated.

To reduce the element burnout situation, some manufacturers will supply water heaters with heating elements that are actually rated for 250 volts, but will mark the water heater 240 volts with its

FIGURE 20-10 Heating elements.

corresponding wattage at 240 volts. This allows a cushion should higher than normal voltages be experienced. CSA Standard CAN/CSA-C22.2 No. 110-M90, entitled "Construction and Test of Electric Storage-Tank Water Heaters," details the specifications and test procedures for water heater storage tanks and elements.

Heating elements consist of a high-resistance nickel–chrome (nichrome) alloy wire that is coiled and embedded (compacted) in magnesium oxide in a copper-clad, or stainless steel, tubular housing; see Figure 20-10. The magnesium oxide is an excellent insulator of electricity, yet it effectively conducts the heat from the nichrome wire to the housing. The construction of heating elements is the same as that of the surface units of an electric range. Terminals are provided on the heating element for the connection of the conductors. A threaded hub allows the element to be screwed securely into the side of the water tank. Water heater resistance heating elements are in direct contact with the water for efficient transfer of heat from the element to the water.

Electric water heaters are available with one or two heating elements; see Figure 20-11. Single-element water heaters have the heating element located near the bottom of the tank. Two-element water heaters have one heating element near the bottom of the tank and the other about halfway up.

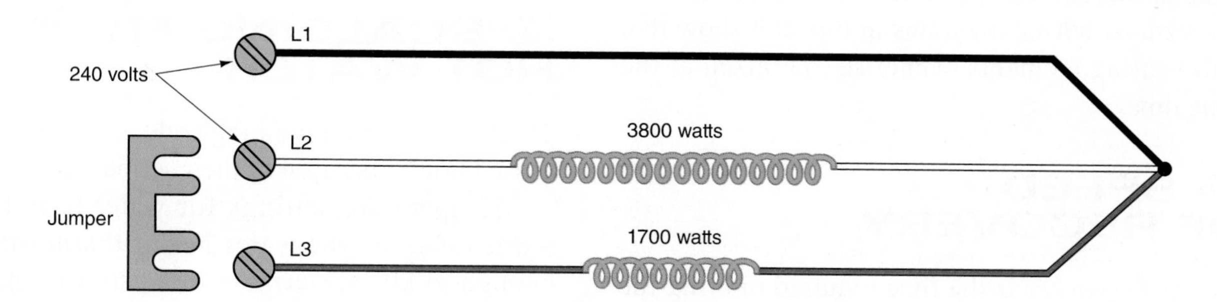

FIGURE 20-11 A dual-wattage heating element. The two wires of the incoming 240-volt circuit are connected to L_1 and L_2. These leads connect to the 3800-watt element. To connect the 1700-watt element, the jumper is connected between L_2 and L_3. With both heating elements connected, the total wattage is 5500 watts.

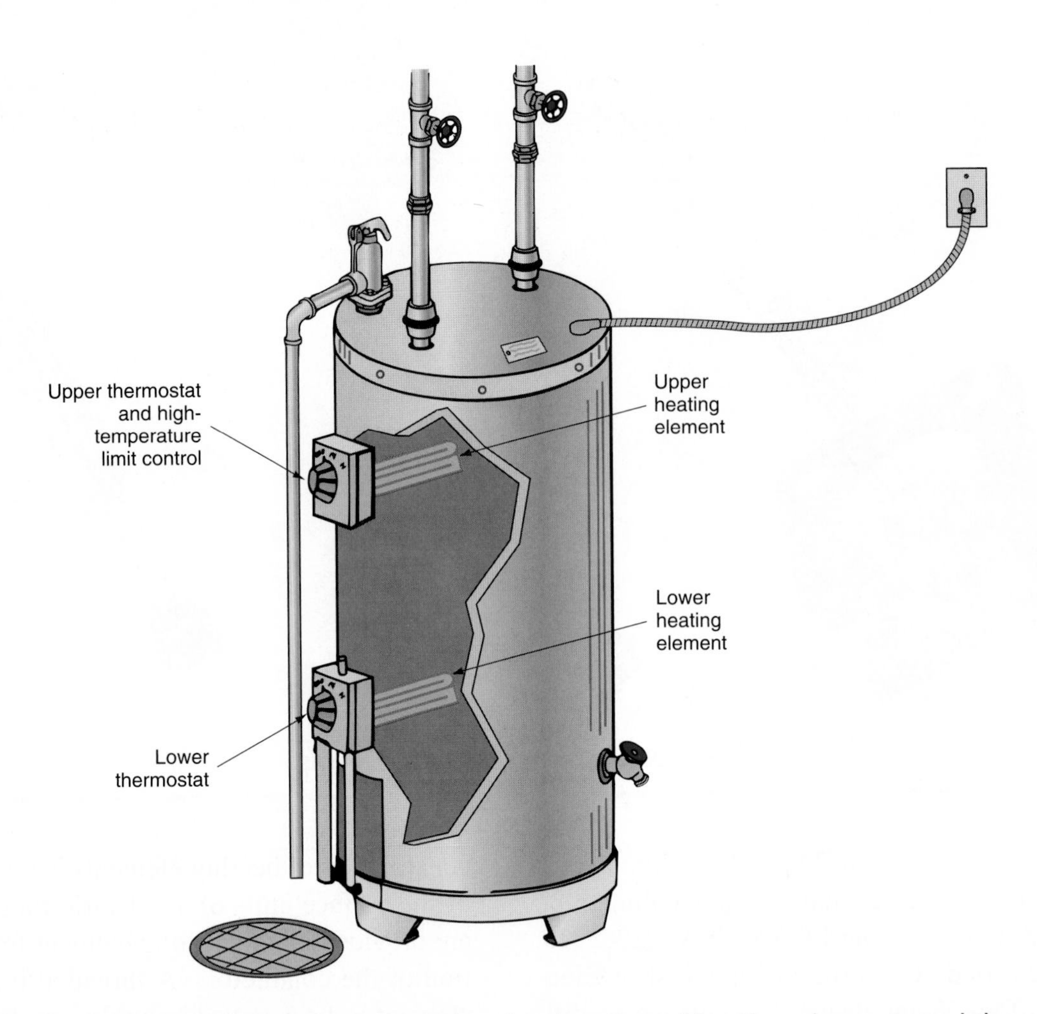

FIGURE 20-12 Typical electric water heater showing location of heating elements and thermostats.

Thermostats, as illustrated in Figure 20-12, turn the heating elements on and off, thereby maintaining the desired water temperature. In two-element water heaters, the top thermostat is an interlocking type, sometimes referred to as a *snap-over type.* The bottom thermostat is a single-pole, on/off type. Both thermostats have silver contacts and have quick make and break characteristics to prevent arcing. The various wiring diagrams in this unit show that both heating elements cannot be energized at the same time.

SPEED OF RECOVERY

Speed of recovery is the time required to bring the water temperature to satisfactory levels in a given time. The current accepted industry standard is based on a 32.22°C rise. For example, Table 20-3 shows that a 3500-watt element can raise the water temperature 90°F (32.22°C) of a little more than 16 gallons (about 60 litres) an hour. A 5500-watt element can raise water temperature 90°F (50°C) of about 25 gallons (about 95 litres) an hour. Speed of recovery is affected by the type and amount of insulation surrounding the tank, the supply voltage, and the temperature of the incoming cold water.

SCALDING FROM HOT WATER

The following are plumbing code issues but are discussed briefly because of their seriousness.

Temperature settings for water heaters pose a dilemma. Set the water heater thermostat high enough to kill bacteria on the dishes in the dishwasher and possibly get scalded in the shower. Or set the thermostat to a lower setting and not get scalded in the shower, but live with the possibility that bacteria will still be present on washed dishes.

TABLE 20-3

Electric water heater recovery rate per litre of water for various wattages.

Heating element wattage	LITRES PER HOUR RECOVERY FOR INDICATED TEMPERATURE RISE 30°C	35°C	40°C	45°C	50°C	55°C
750	21	18	16	14	13	12
1000	29	25	21	19	17	16
1250	36	31	26	24	22	20
1500	43	37	32	29	26	24
2000	57	49	43	38	34	31
2250	65	55	48	43	39	35
2500	72	61	54	48	43	39
3000	86	74	65	57	52	47
3500	100	86	75	67	60	55
4000	115	98	86	77	69	63
4500	129	111	97	86	77	70
5000	143	123	108	96	86	78
5500	158	135	118	105	95	86
6000	172	147	129	115	103	94

Notes: If the incoming water temperature is 5°C, use the 45°C column to raise the water temperature to 50°C.

For example, about how long would it initially take to raise the water temperature in a 160-litre electric water heater to 50°C when the incoming water temperature is 5°C? The water heater has a 2500-watt heating element. Answer: 160 + 48 = 3.33 hours (about 3 hours and 20 minutes)

In the United States, the Consumer Product Safety Commission reports that 3800 injuries and 34 deaths occur each year due to scalding from excessively hot tap water. Unfortunately, the victims are mainly the elderly and children under the age of 5. They recommend that the setting should not be higher than 49°C. Some state laws require that the thermostat be preset at no higher than 49°C. Residential water heaters are all preset by the manufacturer at 49°C.

Manufacturers post caution labels on their water heaters that read something like this:

SCALD HAZARD: WATER TEMPERATURE OVER 49°C CAN CAUSE SEVERE BURNS INSTANTLY OR DEATH FROM SCALDS. SEE INSTRUCTION MANUAL BEFORE CHANGING TEMPERATURE SETTINGS.

TABLE 20-4

Time/temperature relationships related to scalding an adult subjected to moving water. For children, the time to produce a serious burn is less than for adults.

TEMPERATURE (°C)	TIME TO PRODUCE A SERIOUS BURN
37	Safe temperature for bathing
48	5 minutes
51	3 minutes
52	1 minute
56	15 seconds
60	5 seconds
64	2 seconds
68	1 second
71	0.5 seconds

To reduce the possibility of scalding, water heaters can be equipped with a thermostatic non-scald mixing valve or pressure-balancing valve that holds a selected water temperature to within one degree, regardless of incoming water pressure changes. This prevents sudden unanticipated changes in water temperatures.

Table 20-4 shows time/temperature relationships relating to scalding.

WHAT ABOUT WASHING DISHES?

Older automatic dishwashers required incoming water temperature of 60°C or greater to get the dishes clean. A water temperature of 49°C is not hot enough to dissolve grease, activate power detergents, and kill bacteria. That is why commercial and restaurant dishwashers boost temperatures to as high as 82°C.

Newer residential dishwashers have their own built-in water heaters to boost the incoming water temperature in the dishwasher from 49°C to 60°C to 63°C. This heating element also provides the heat for the drying cycle. Today, a thermostat setting of 49°C on the water heater should be satisfactory.

SEQUENCE OF OPERATION

Water heaters can be connected in many ways, usually depending on the local power company's regulations. This book illustrates only three types of connections. Consult your power utility and/or your supplier of electric water heaters to find out what connections are available in your area.

Figures 20-13 through 20-15 illustrate wiring for water heaters that are two-element, limited-demand operation type.

Figures 20-13 and 20-14 show the wiring for flat rate connection, where electricity to the water heater is "off" during peak load periods of the day. This is controlled by the hydro utility. Generally, the flat rate charged depends on the size of the hot water tank in the residence. Figure 20-15 shows the wiring for connection to the main panel, so that the kilowatt-hour power consumption of the water heater is metered, but has the advantage of having 24-hour operation.

When all of the water in the tank is cold, the upper element heats only the water in the upper half of the tank. The upper element is energized through the lower contacts of the upper thermostat. The lower thermostat also indicates that heat is needed, but heat cannot be supplied because of the open condition of the upper contacts of the upper thermostat.

When the water in the upper half of the tank reaches a temperature of about 65°C, the upper thermostat snaps over; the upper contacts are closed

FIGURE 20-13 Electric water heater controls. This seven-terminal control combines water temperature control plus high-temperature limit control. This is the type commonly used to control the upper heating element and provide a high-temperature limiting feature. The two-terminal control is the type used to control water temperature for the lower heating element.

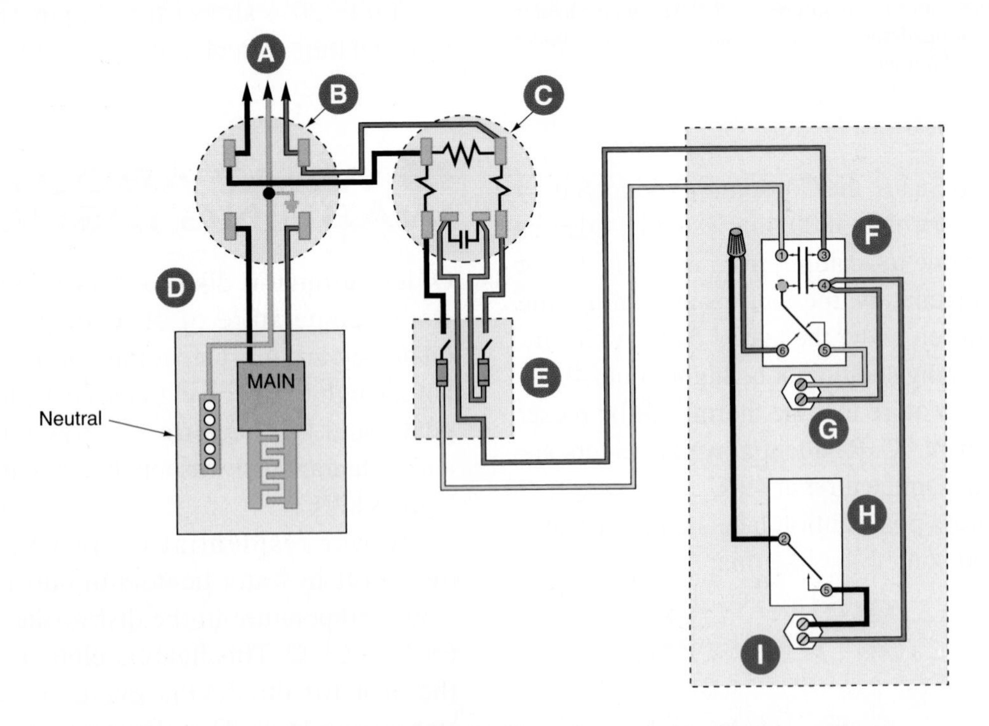

FIGURE 20-14 Wiring for typical off-peak water heating installation.

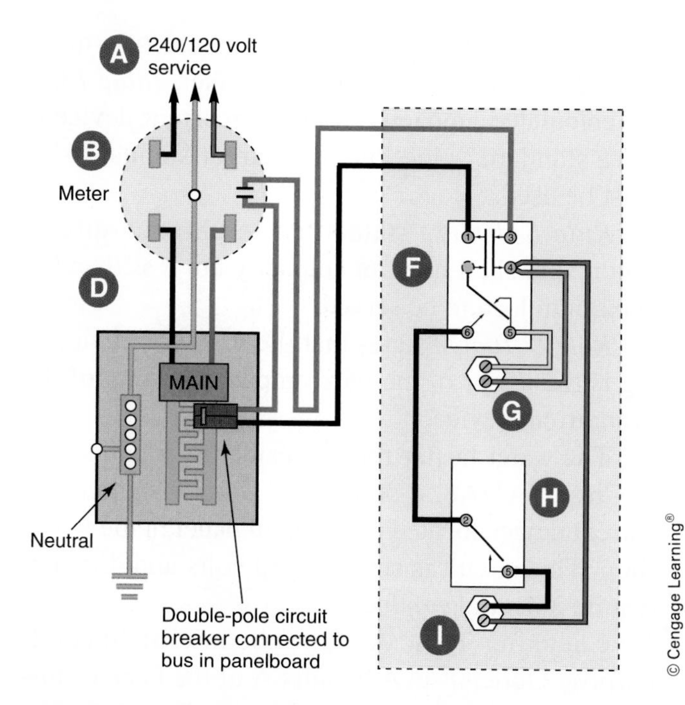

FIGURE 20-15 Feedthrough wiring for off-peak control.

and the lower contacts of the upper thermostat are opened. The lower element now begins to heat the cold water in the bottom of the tank. When the entire tank is heated to 65°C, the lower thermostat shuts off. Electrical energy is not being used by either element at this point. The heat-insulating jacket around the water heater is designed to keep the heat loss to a minimum. Thus, it takes many hours for the water to cool only a few degrees.

Most of the time, the bottom heating element will keep the water at the desired temperature. If a large amount of hot water is used, the top element comes on, resulting in fast recovery of the water in the upper portion of the tank, which is where the hot water is drawn from.

METERING AND SEQUENCE OF OPERATION

Electric water heaters are available in a number of styles:

- Single element
- Double element, simultaneous operation
- Double element, nonsimultaneous operation

There are several ways that a utility company can provide metering and for the electrician to make the connections:

1. Connect the water heating to one of the two-pole, 240-volt circuits in the main distribution panel; see Figure 20-15. This is probably the most common method because the wiring is simple and additional switches, breakers, time clocks, or meters are not required. This is the connection for the residence used as an example in this textbook. This method provides electricity to the water heater 24 hours of every day.
2. Connect the water heater in a flat rate arrangement; see Figure 20-13. The utility supplies power to the consumer via a four-wire cable. This contains the three feeder lines and a fourth wire to control the water heater during peak demand periods. This wire is connected to a separate terminal block located at the 3 o'clock or 9 o'clock position in the meter base. The electrician then connects two red #10 AWG solid wires to the terminal block at the 3 o'clock position and to the water heater lug on the line side of the meter base. These wires pass through the service conduit and through the main disconnect into a separate 30-ampere, fusible disconnect. The #10/2 AC90 or NMD 90 then connects to the water heater.
3. Connect the water heater circuit, as in Figure 20-15, but install a time clock in the circuit leading to the water heater. The time clock would be set to be "off" during certain times of the day deemed peak hours by the utility. The utility gives special incentives to the homeowner for agreeing to this arrangement. This is a good conservation method that many supply authorities are promoting.
4. Some utilities use radio-controlled switching to control water heater loads. They send a carrier signal through the power line, which is received by the electronic controlling device, thereby turning the power to the water heater "on" and "off" as required by the utility; see Figure 20-14. This is commonly called a *flat rate water heater.* The utility will charge a fixed rate regardless of the energy used. This is still very common in many centres.

WATER HEATER LOAD DEMAND

The following *CEC* rules apply to load demands placed on an electric water heater branch circuit.

Rule 8–104(2) requires that a branch circuit shall have a rating not less than the calculated load of the heater.

Rule 8–104(6)(a) requires that a continuously loaded appliance must have a branch-circuit rating not less than 125% of the marked rating of the appliance.

Disconnecting Means

Rule 26–746 requires that the appliance have a means of disconnect. This would normally be located in the panel (load centre), or it could be provided by a separate disconnect switch. *Rule 14–416* allows the unit to have a switch that is an integral part of the appliance having an "off" position and capable of disconnecting all ungrounded conductors from the source, to be considered the required disconnecting means. In many local codes, however, this is not acceptable.

Rules 14–300 and *14–408* require that switches and circuit breakers used as the disconnecting means be of the indicating type, meaning they must clearly show that they are in the "on" or "off" position.

Branch-Circuit Overcurrent Protection

The rating or setting of overcurrent devices shall not exceed the allowable ampacity of the conductors they protect, *Rule 14–104*.

EXAMPLE

A water heater nameplate indicates 4500 watts, 240 volts. What is the maximum size of overcurrent protection permitted by the *CEC*?

SOLUTION

$$\frac{4500}{240} \text{ amperes} = 18.75 \text{ amperes}$$

$$18.75 \times 1.25 \text{ amperes} = 23.44 \text{ amperes}$$

Therefore, the *CEC* permits the use of a minimum 25-ampere fuse or circuit breaker, *Table 13*. If the calculated ampacity of an overcurrent device is not a standard value, the next larger standard size must be used.

Rule 8–104(2) states that the branch-circuit conductors shall have an ampacity not less than the maximum load to be served.

Rule 8–104(1) states that the branch-circuit rating is the lesser of the wire ampacity or that of the overcurrent device.

The water heater for the residence is connected to Circuit A6/A8, a 20-ampere, double-pole overcurrent device located in the main panel in the workshop. This circuit is straight 240 volts and does not require a neutral conductor.

Checking Table A–1, "Schedule of Special-Purpose Outlets," in Appendix A at the back of this textbook, we find that the electric water heater has two heating elements: a 2000-watt element and a 3000-watt element. Because of the thermostatic devices on the water heater, both elements cannot be energized at the same time. Therefore, the maximum load demand is

$$\text{Amperes} = \frac{\text{Watts}}{\text{Volts}} = \frac{3000 \text{ watts}}{240 \text{ volts}} = 12.5 \text{ amperes}$$

For water heaters with two heating elements connected for nonsimultaneous (limited demand) operation, the nameplate on the water heater will be marked with the largest element's wattage at rated voltage. For water heaters with two heating elements connected for simultaneous operation, the nameplate on the water heater will be marked with the total wattage of both elements at rated voltage.

Conductor Size

CEC Table 2 indicates that the conductors will be #12 AWG. The specifications for the residence call for conductors to be Type T90/TWN75 or NMD 90.

Bonding

If flexible conduit is run from a junction box to the water heater, the water heater is bonded through a separate bonding conductor sized according to *Table 16A,* which is run in the flex to the box.

EFFECT OF VOLTAGE VARIATION

The heating elements in the heater in this residence are rated 240 volts. They will operate at a lower voltage with a reduction in wattage. If connected to voltages above their rating, they will have a very short life.

Ohm's law and the wattage formula show how the wattage and current depend on the applied voltage:

$$R = \frac{E^2}{W} = \frac{240 \times 240}{3000} \text{ ohms} = 19.2 \text{ ohms}$$

$$I = \frac{E}{R} = \frac{240}{19.2} \text{ amperes} = 12.5 \text{ amperes}$$

If a different voltage is substituted, the wattage and current values change accordingly. At 220 volts,

$$W = \frac{E^2}{R} = \frac{220 \times 220}{19.2} \text{ watts} = 2521 \text{ watts}$$

$$I = \frac{W}{E} = \frac{2521}{220} \text{ amperes} = 11.46 \text{ amperes}$$

At 230 volts,

$$W = \frac{E^2}{R} = \frac{230 \times 230}{19.2} \text{ watts} = 2755 \text{ watts}$$

$$I = \frac{W}{E} = \frac{2755}{230} \text{ amperes} = 11.98 \text{ amperes}$$

In resistive circuits, current is directly proportional to voltage and can be simply calculated using a ratio and proportion formula. For example, if a 240-volt heating element draws 12.7 amperes at rated voltage, the current draw can be determined at any other applied voltage, say 208 volts

$$\frac{208}{240} = \frac{x}{12.7}$$

$$\frac{208 \times 12.7}{240} \text{ amperes} = 11.01 \text{ amperes}$$

To calculate this example if the applied voltage is 120 volts

$$\frac{120}{240} = \frac{x}{12.7}$$

$$\frac{120 \times 12.7}{240} \text{ amperes} = 6.35 \text{ amperes}$$

Also, in a resistive circuit, wattage varies as the square of the current. Therefore, when the voltage on a heating element is doubled, the current also doubles and the wattage increases four times. When the voltage is reduced to one-half, the current is halved, and the wattage is reduced to one-fourth.

According to these formulas, a 240-volt heating element connected to 208 volts will have about three-quarters of the wattage rating it has at the rated voltage.

REVIEW

Note: Refer to the *CEC* or the blueprints provided with this textbook when necessary. Where applicable, responses should be written in complete sentences.

WORKSHOP

1. a. What circuit supplies the workshop lighting? ______________________________
   b. What circuit supplies the plug-in strip over the workbench? ____________________
   c. What circuit supplies the freezer receptacle? ______________________________

2. What type of lighting fixtures are installed on the workshop ceiling? ____________

3. Does a receptacle for a freezer have to be GFCI-protected? Explain. ____________

4. Does a receptacle for a sump pump have to be GFCI-protected? Explain. ____________

5. When a 120-volt receptacle is provided for the laundry equipment in a basement, what is the *minimum* number of additional receptacle outlets required? What is the *CEC* reference? ____________

6. When a single receptacle is installed on an individual branch circuit, the receptacle must have a rating ____________ the rating of the branch circuit.

7. Calculate the total current draw of Circuit A17 if all lighting fixtures and the exhaust fan were turned on. ____________

8. If a #10 NMD 90 cable is located in an ambient temperature of 42°C, what would be the temperature correction factor? ____________

9. What size of box would you use where Circuit A17 enters the junction box on the ceiling?

10. The following is a layout of the lighting circuits for the workshop. Using the conduit layout in Figure 20-1, make a complete wiring diagram. Use coloured pencils or pens to indicate conductors.

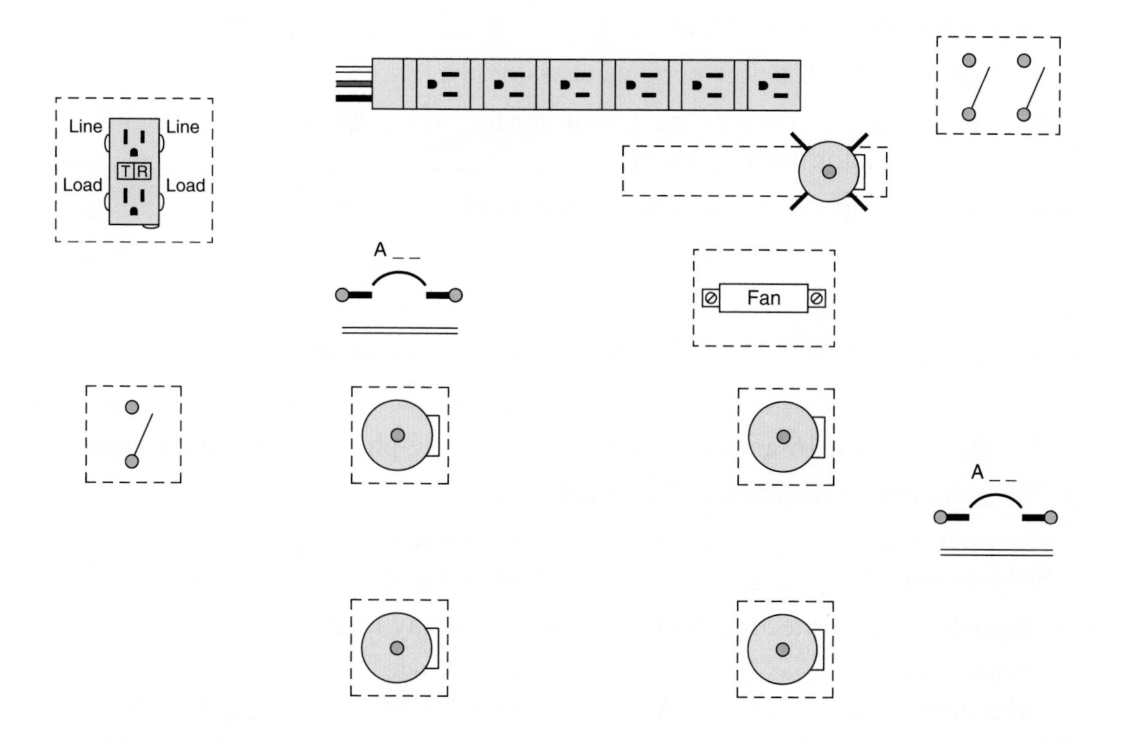

WATER PUMP CIRCUIT ⓐB

1. How is the motor disconnected if pumping is no longer required? ________________

____

____

____

2. Why is a 240-volt motor preferable to a 120-volt motor for use in this residence?

____

____

____

3. How many amperes does a 1-hp, 240-volt, single-phase motor draw? (See *Table 45.)*

____

4. What size are the conductors used for this circuit? ________________

5. a. What is the branch-circuit protective device? ________________

   b. What furnishes the running protection for the pump motor? ________________

   ____

   c. What is the maximum ampere setting required for running protection of the 1-hp, 240-volt pump motor? ________________

6. [Circle correct answer.] Submersible water pumps operate with the electrical motor and actual pump located
   a. above permanent water level.
   b. below permanent water level.
   c. half above and half below permanent water level.
7. Since the controller contains the motor starting relay and the running and starting capacitors, the motor itself contains ____________________.
8. What type of pump moves the water inside the deep-well pipe? ____________________
   ____
   ____
9. Proper pressure of the submersible pump system is maintained by a ____________________
   ____
10. Fill in the data for a 16-ampere electric motor, single-phase, 1.15 service factor.
    a. Branch-circuit protection nontime-delay fuses.
       Normal size __________ A Switch size __________ A
       Maximum size __________ A Switch size __________ A
    b. Branch-circuit protection dual-element, time-delay fuses.
       Normal size __________ A Switch size __________ A
       Maximum size __________ A Switch size __________ A
    c. Branch-circuit protection instant-trip breaker.
       Maximum setting __________ A
    d. Branch-circuit protection—inverse time breaker.
       Maximum setting __________ B
    e. Branch-circuit conductor size.____________________
       Ampacity __________ A
    f. Motor running protection using dual-element time-delay fuses.
       Maximum size __________ A
11. The *CEC* is very specific in its requirement that submersible electric water pump motors be grounded. Where is this specific requirement found? ____________________
    ____
12. Does the *CEC* allow submersible pump cable to be buried directly into the ground?
    ____
    ____

WATER HEATER CIRCUIT Ⓐ C

1. At what temperature is the water in a residential heater usually maintained? __________
2. Magnesium rods are installed inside the water tank to reduce ____________________.
3. Is a separate meter required to record the amount of energy used to heat the water?
   ____

4. What is meant by "flat rate" installation? ______________________

5. What is the most common method of metering water heater loads?______________________

6. Two thermostats are generally used in an electric water heater.
   a. What is the location of each thermostat? ______________________
   b. What type of thermostat is used at each location? ______________________

7. a. How many heating elements are provided in the heater in the residence discussed in this text? ______________________
   b. Are these heating elements allowed to operate at the same time? ______________________

8. When does the lower heater operate? ______________________

9. The *CEC* states that water heaters must have a temperature regulating device set at ________°C maximum. Secondary protection is provided by a temperature cutoff set at the factory at ________°C.

10. Why does the storage tank hold the heat so long? ______________________

11. The electric water heater is connected for "limited demand" so that only one heating element can be on at one time.
    a. What size of wire is used to connect the water heater? ______________________
    b. What size of overcurrent device is used? ______________________

12. a. If both elements of the water heater are energized at the same time, how much current will they draw? (Assume the elements are rated at 240 volts.)

Element No. 1 = 3000 watts

Element No. 2 = 2000 watts

b. What size of wire is required for the load of both elements? Show your calculations.

13. a. How much current do the elements in Question 12 draw if connected to 220 volts? Show your calculations.

b. What is the wattage value at 220 volts? Show your calculations.

14. A 240-volt heater is rated at 1500 watts.
    a. What is its current rating at rated voltage? ________________________
    b. What current draw would occur if connected to a 120-volt supply?

    c. What is the wattage output at 120 volts?

15. A customer asks the electrician to set the water heater thermostat 30°C below the maximum permitted by *CEC*. The thermostat is set at which of the following temperatures? [Circle one]
    a. 70°C
    b. 66°C
    c. 60°C
    d. 49°C

UNIT 21

Heat and Smoke Detectors, Carbon Monoxide Detectors, and Security Systems

OBJECTIVES

After studying this unit, you should be able to

- identify the Underwriters Laboratories of Canada (ULC) standard that refers to household fire warning equipment
- name the two basic types of smoke detectors
- discuss the location requirements for the installation of heat and smoke detectors
- list the locations where heat and smoke detectors may not be installed
- describe the wiring requirements for the installation of heat and smoke detectors
- list the major components of a typical residential security system
- identify the *CEC Rule 32* requirements for the installation of residential fire alarm systems
- identify the requirement for the installation of carbon monoxide detectors

THE IMPORTANCE OF HEAT AND SMOKE DETECTORS

Fire is the third leading cause of accidental death. Home fires account for the biggest share of these fatalities, most of which occur at night during sleeping hours.

In recent years, the standards of the ULC, combined with the *National Building Code of Canada* and associated provincial and municipal codes, have made mandatory the installation of smoke alarms, smoke detectors, and/or heat detectors in critical areas of a home. A sufficient number of detectors must be installed to provide full coverage.

Smoke alarms are installed in a residence to give the occupants an early warning of the presence of a fire. Fires produce smoke and toxic gases, which can overcome the occupants while they sleep. Most fatalities result from the inhalation of smoke and toxic gases rather than from burns. Heavy smoke also reduces visibility.

Fire-sensing devices commonly used in a residence are heat detectors (Figure 21-1) and smoke detectors (Figure 21-2). These must be connected to a central fire alarm control panel that will process the information and send an output signal to alarm bells or buzzers.

The Underwriters Laboratories of Canada

The Underwriters Laboratories of Canada Standard CAN/ULC-S524-M91 discusses the proper selection, installation, operation, and maintenance of fire warning equipment commonly used in residential occupancies. Minimum requirements are stated in the standard.

Courtesy of Sandy F Gerolimon

FIGURE 21-1 Heat detector.

© Cengage Learning®

FIGURE 21-2 Combination smoke detector/alarm.

In some provincial jurisdictions, it is mandatory that qualified technicians install, verify, and maintain fire alarm systems. The certification to become a fire alarm technician is available through a variety of government-approved courses, through colleges, and through local electrical unions, such as the International Brotherhood of Electrical Workers (IBEW) and the Canadian Fire Alarm Association (CFAA).

Smoke alarms must be installed outside each separate sleeping area in the immediate vicinity of the bedrooms, and on each additional storey of the dwelling, including basements but excluding crawl spaces and unfinished attics.

Residential-type smoke alarms incorporate both the initiating circuit (detector) and the alarm circuit (buzzer) in the same device. If more than one of these are installed, they must be interconnected so that all will go into alarm mode at the same time.

It is recommended that smoke and heat detectors be placed strategically to provide protection for living rooms, dining rooms, bedrooms, hallways, attics, furnace and utility rooms, basements, and heated attached garages. Detectors in kitchens must not be directly above the stove.

TYPES OF SMOKE DETECTORS

Two types of smoke detectors commonly used are the photoelectric type and the ionization type.

Photoelectric Smoke Detectors

In the photoelectric type of smoke detector, a light sensor measures the amount of light in a chamber. When smoke is present, an alarm sounds, indicating that light is being reflected off the smoke in the chamber. This type of sensor detects smoke from burning materials that produce great quantities of white or grey smoke, such as furniture, mattresses, and rags. It is less effective for gasoline and alcohol fires, which do not produce heavy smoke. It is also not as effective with black smoke, which absorbs light rather than reflecting it as grey or white smoke does.

Ionization Smoke Detectors

The ionization type of smoke detector contains a low-level radioactive source that supplies particles that ionize the air in the smoke chamber of the detector. Plates in this chamber are oppositely charged. Because the air is ionized, an extremely small amount of current (millionths of an ampere) flows between the plates. Smoke entering the chamber impedes the movement of the ions, reducing the current flow, which causes an alarm to go off. This type of detector is able to sense products of combustion in the earliest stages of a fire. It can detect airborne particles (aerosols) as small as 5 microns in diameter.

The ionization type of detector is effective for detecting small amounts of smoke, as in gasoline and alcohol fires, which are fast-flaming. It is also more effective for detecting dark or black smoke than the photoelectric type of detector.

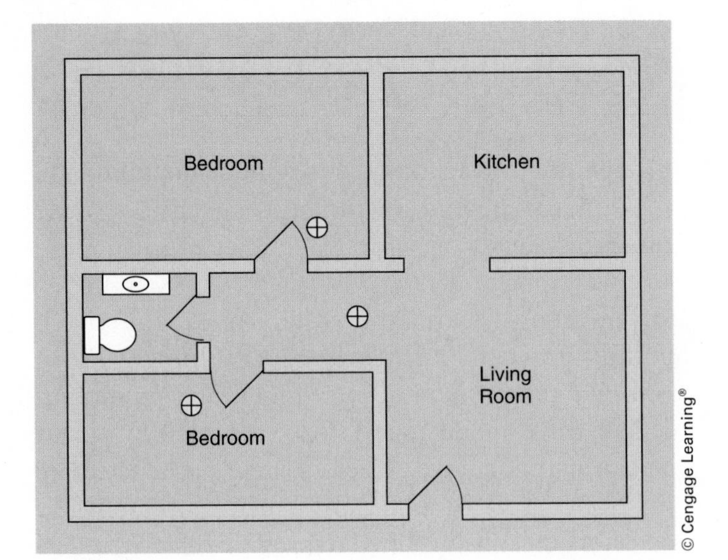

FIGURE 21-3 Recommended location of heat or smoke detector between the sleeping areas and the rest of the house. Locate outside the bedrooms, but near the bedrooms.

INSTALLATION REQUIREMENTS

The following information concerns specific recommendations for installing smoke or heat detectors; see Figures 21-3, 21-4, and 21-5. Complete data are found in the ULC standards.

It is **mandatory to install smoke detectors and alarms** in all new homes (*National Building Code of Canada* 3.2.4.21.1/9.10.18.1).

- **Do** install smoke alarms on each floor of a residence for *minimum* protection. Make sure that one of these alarms is located between the sleeping area and the rest of the house. These alarms must be interconnected so that if one alarm detects smoke, they all sound.
- **Do** install smoke alarms in all rooms, hallways, attics, basements, and storage areas of a residence to provide protection that *exceeds* the minimum recommendation above.
- **Do** install the detector on the ceiling as close as possible to the centre of the room or hallway.
- **Do** install a smoke detector on a basement ceiling close to the stairway to the first floor.
- **Do** place a detector within 900 mm of the peak on a sloped ceiling that has greater than 300 mm rise per 2400 mm horizontally (1:8).
- **Do** install smoke detectors and heat detectors at the *top* of an *open* stairway, because heat and smoke travel up.
- **Do** install hard-wired detectors in a *new* construction. These detectors are directly connected to a 120-volt alternating current (AC) source. For existing homes, either battery-powered units or hardwired 120-volt AC units may be used.
- **Do not** install detectors in the *dead* airspace at the top of a stairway that can be closed off by a door.

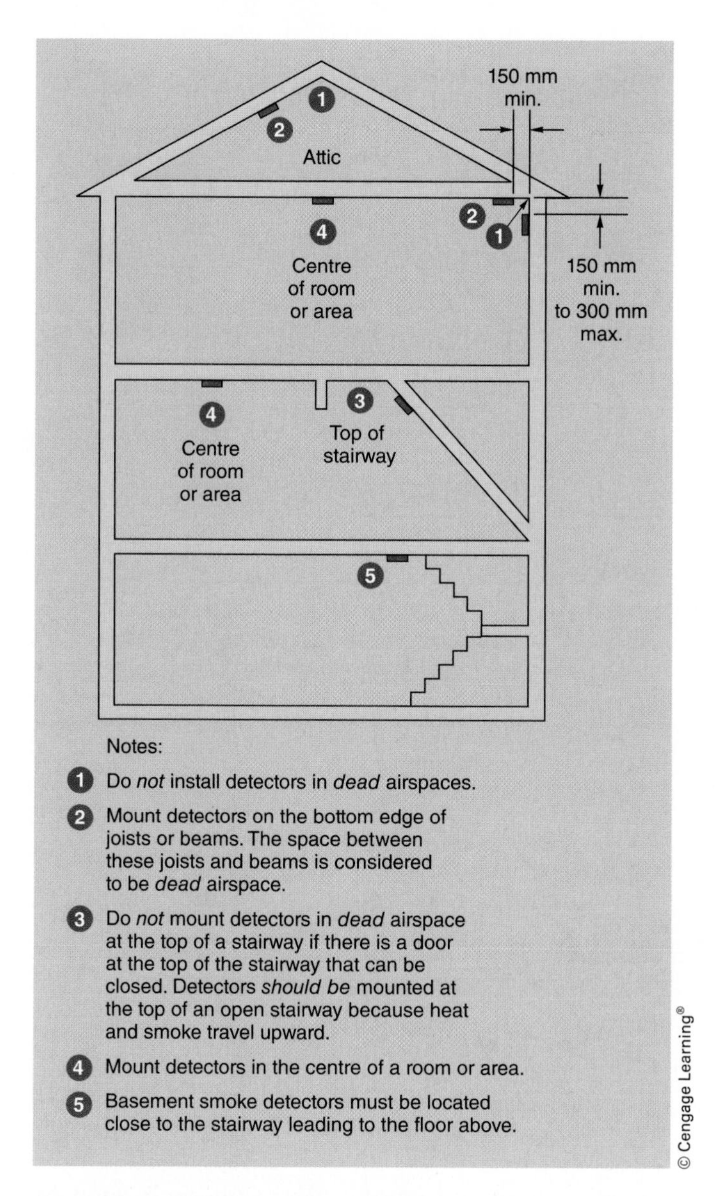

FIGURE 21-4 Recommendations for the installation of heat and smoke detectors.

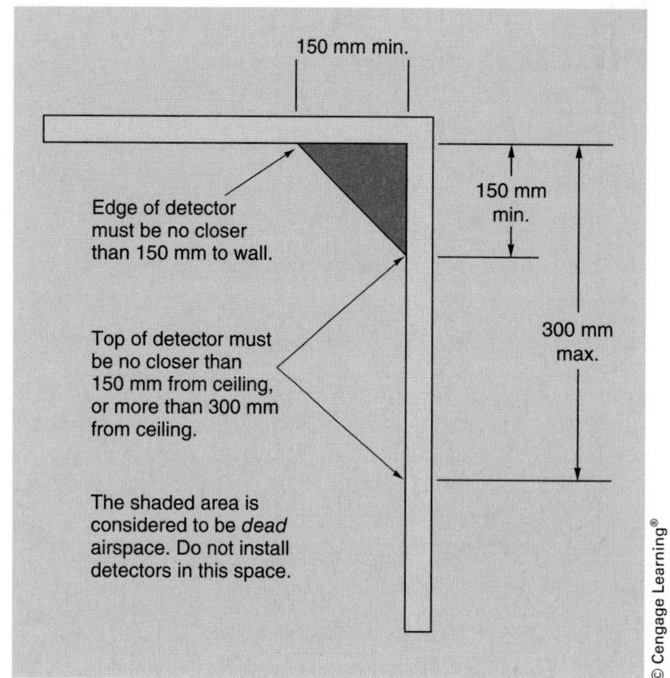

FIGURE 21-5 Do not mount detectors in the *dead* airspace where the ceiling meets the wall.

- **Do not** place the edge of a detector mounted on the ceiling closer than 150 mm from the wall.
- **Do not** place a wall-mounted detector closer than 150 mm from the ceiling and not farther than 300 mm from the ceiling. The area where the ceiling meets the wall is considered dead airspace where smoke and heat may not reach the detector.
- **Do not** connect detectors to wiring that is controlled by a wall switch.
- **Do not** connect detectors to a circuit that is protected by a ground-fault circuit interrupter (GFCI) or an arc-fault circuit interrupter (AFCI), *Rule 32–110(a)(ii)*.
- **Do not** install detectors where relative humidity exceeds 85%, for example, adjacent to showers, laundry areas, or other areas where large amounts of visible water vapour exist.
- **Do not** install detectors in front of air ducts, air conditioners, or any high-draft areas where the moving air will keep the smoke or heat from entering the detector.
- **Do not** install detectors in kitchens where the accumulation of household smoke can result in false alarms with certain types of detectors. The photoelectric type *may* be installed in kitchens, but must *not* be installed directly over the range or cooking appliance.
- **Do not** install where the temperature can fall below 0°C or rise above 50°C.
- **Do not** install in garages (vehicle exhaust can set off the detector).
- **Do not** install detectors in airstreams that will pass air originating at the kitchen cooking appliances across the detector. False alarms will result.

MANUFACTURERS' REQUIREMENTS

Manufacturers of smoke and heat detectors have certain responsibilities:

- The power supply must be capable of operating the signal for at least 4 minutes continuously.
- Battery-powered units must be capable of a low-battery warning "beep" of at least one beep per minute for seven consecutive days.
- Direct-connected 120-volt AC detectors must have a visible indicator that shows "Power On."
- Detectors must not signal when a power loss occurs or when power is restored.

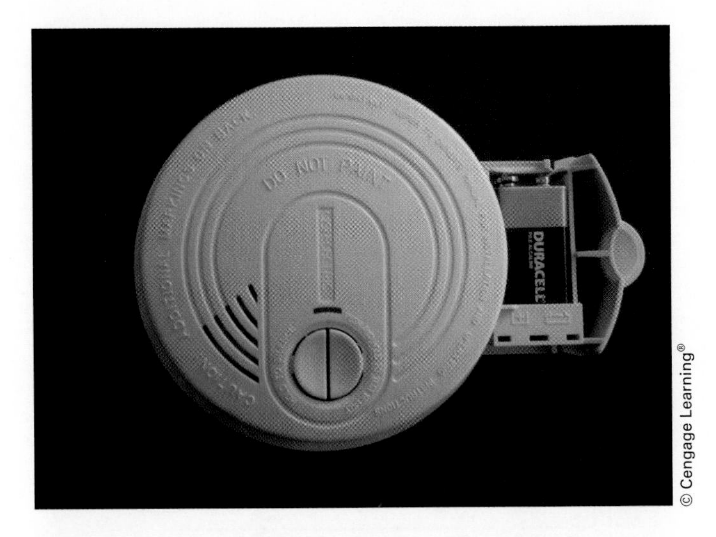

FIGURE 21-6 An AC–DC smoke alarm that operates on 120-volts AC as the primary source of power and a 9-volt battery as the secondary source of power. In a power outage, the alarm continues to operate on the battery.

Features of Smoke Alarms and Heat Detectors

Smoke alarms may contain an indicating light to show that the unit is functioning properly. They may also have a test button to test the circuitry and alarm capability: When the button is pushed, it is the circuitry and alarm that are tested, not the smoke-detecting ability.

Heat detectors are available that sense a specific *fixed* temperature, such as 57°C or 90°C. Available also are *rate-of-rise* heat detectors that sense rapid changes in temperature (6°C per minute), such as those caused by flash fires.

Fixed and rate-of-rise temperature detectors are available as a combination unit. Combination smoke, fixed, and rate-of-rise temperature detectors are also available in one unit.

The spacing of heat detectors shall be as recommended by the manufacturer, because detectors are capable of sensing heat only within a limited space in a given time.

Wiring Requirements

Direct-Connected Units. All new construction requires smoke alarms to be connected to a 120-volt circuit that feeds a lighting circuit, *Rule 32–110(a)*. Smoke alarms must never be connected to a dedicated circuit. This type of smoke alarm will usually be of the interconnected type and may include other features such as battery backup or a built-in carbon monoxide detector. A direct connect unit with battery backup is shown in Figure 21-6.

When using interconnected smoke alarms, it is important to refer to the manufacturer's literature for the maximum number of devices allowed. Each manufacturer and model may allow different maximum numbers. When devices are interconnected, all alarms will sound at the same time, regardless of which device actually detected smoke. All new residential construction requires an interconnected smoke alarm on each level of the home and near the bedrooms. If bedrooms are located at opposite ends of a home, it may be necessary to install more than one alarm on a level.

Interconnected smoke alarms have one white wire, one black wire, and one red or yellow wire, which is used to interconnect them. Refer to Figure 21-7 for wiring these devices.

Battery-Operated Units. These units require no wiring; they are simply mounted in the desired locations. Batteries last for about a year. Battery-operated units should be tested periodically, as recommended by the manufacturer.

Always follow the installation requirements and recommendations of the unit manufacturer. Refer to the ULC standards for additional technical data relating to the installation of smoke and heat detectors.

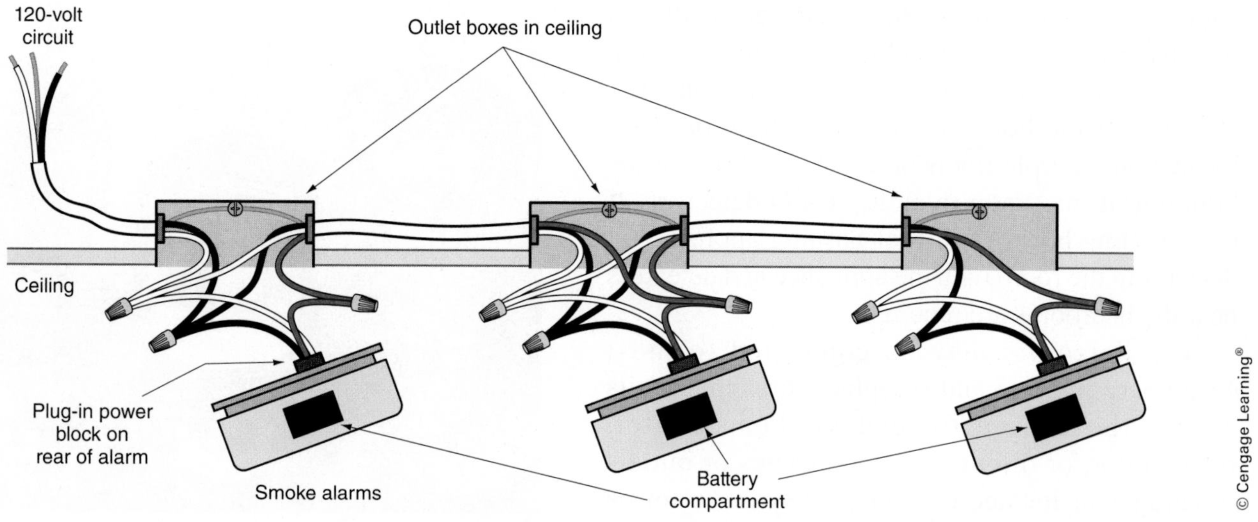

FIGURE 21-7 Interconnected smoke alarms. The two-wire branch circuit is run to the first alarm, then three-wire non-metallic-sheathed cable interconnects the other alarms. If one alarm in the series is triggered, all of the other alarms will sound off.

COMBINATION DIRECT/BATTERY/ FEEDTHROUGH DETECTORS

Combination units are connected directly to a 120-volt AC circuit (not GFCI- or AFCI-protected) and are equipped with a 9-volt direct current (DC) battery; see Figure 21-6. In a power outage, these detectors will continue to operate on the 9-volt DC battery. The ability to interconnect a number of detectors offers "whole house" protection, because when one unit senses smoke, it triggers all of the other interconnected units into alarm mode as well.

Because of the vital importance of smoke and fire detectors to prevent loss of life, you should very carefully read the application and installation manuals that are prepared by the manufacturer of the smoke and heat detectors.

The residence featured in this textbook has four smoke detectors, hard-wired directly to a 120-volt AC branch circuit. They are interconnected feedthrough units that will set off signalling in all units should one unit trigger. These interconnected smoke detectors are also powered by a 9-volt DC battery that allows the unit to operate in the event of a failure in the 120-volt AC power supply.

CARBON MONOXIDE DETECTORS

The *National Building Code of Canada* may require the installation of carbon monoxide (CO) detectors in dwelling units. These may be free-standing units, permanently wired to a source of supply, or part of a smoke detector. If the building code requires a CO detector, a battery-operated or plug-in device is not acceptable. It must be permanently wired.

A CO detector is required when

- The dwelling has a solid-fuel-burning appliance such as a wood stove, wood-burning fireplace, or wood or corn furnaces.
- The dwelling contains a fuel-burning appliance such as a furnace or fireplace operating from a non-solid fuel such as natural gas, propane, or oil.
- The dwelling shares a wall with a garage.

Location and Connection of Carbon Monoxide Detectors

If a solid-fuel appliance is used, a detector must be located on or near the ceiling in the room or area where the appliance is located. When the dwelling uses another source of fuel or shares a wall with a garage, a detector must be located in each bedroom or outside each bedroom within 5 m of the bedroom doors, this distance being measured along the corridor. If the house is large and bedrooms are located on multiple floors or at opposite ends of the building, it may require multiple CO detectors. If the dwelling has a solid-fuel-burning appliance, a detector in the room with the appliance and detectors near the bedrooms are required.

CO detectors may be connected to most 15-ampere lighting and receptacle branch circuits except those used to feed kitchen counter, refrigerator, or dining area receptacles or outlets in garages or located outdoors. They must not be connected to GFCI- or AFCI-protected circuits or circuits that supply only receptacles. A switching or disconnecting means between the device and the circuit breaker is not allowed.

FIGURE 21-8 The wireless key allows the end user to disarm and arm the security system with the press of a button. There is no need to remember a security code, and it can also be used to operate lights and appliances.

SECURITY SYSTEMS

It is beyond the scope of this text to cover every type of residential security system; we focus on some of the features available from the manufacturers of these systems.

Installed security systems range from simple to very complex. Features and options include intruder detectors for doors and windows, motion detectors, infrared detectors, and under-the-carpet floor mat detectors for "space" protection. Inside and outdoor horns, bells, electronic buzzers, and strobe lights can provide audio as well as visual alarm. Telephone interconnection to preselected telephone numbers, such as the police or fire departments, is also available. (See Figures 21-8 to 21-15.)

FIGURE 21-9 Professionally monitored smoke/heat detectors are on 24/7, even if the burglary protection is turned off.

The homeowner and electrician should discuss specific features and benefits of security systems. Most electricians are familiar with a particular manufacturer's selection of systems. Figure 21-16 shows the range of devices available for such a system.

These systems are generally on display at lighting fixture stores that offer the homeowner an opportunity to see a security system "in action."

Wiring for a security system consists of small, easy-to-install, extra-low-voltage, multiconductor cables made up of #18 AWG conductors. The actual installation of these conductors should be done after the regular house wiring is completed to prevent damage to these smaller cables. Usually the wiring can be done at the same time as the chime wiring is being installed. Security system wiring comes under the scope of *Section 16*.

When such detectors as door-entry, glass-break, floor mat, and window foil are used, circuits are electrically connected in series so that if any part

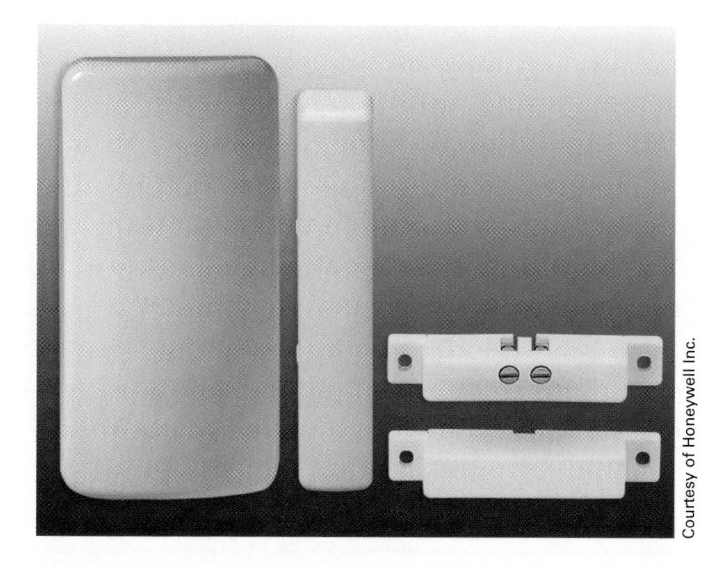
Courtesy of Honeywell Inc.

FIGURE 21-10 Magnetic contacts are often located at vulnerable entry points such as doors.

Courtesy of Honeywell Inc.

FIGURE 21-11 Sirens, such as these placed inside the home, alert families to emergencies.

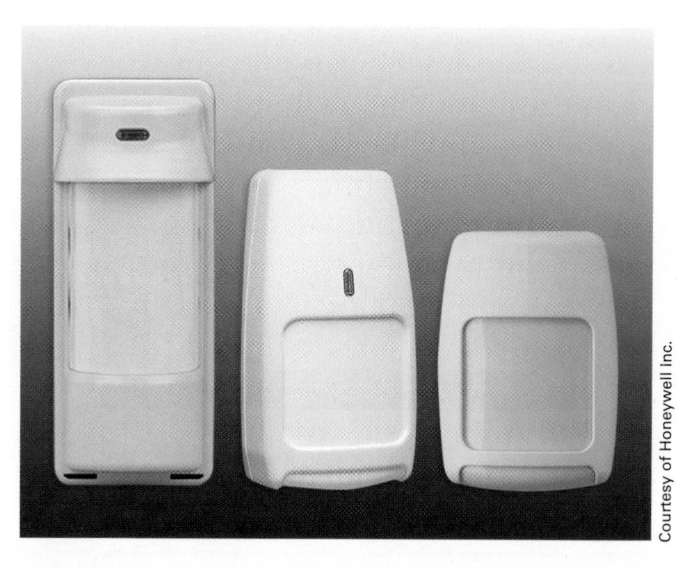
Courtesy of Honeywell Inc.

FIGURE 21-12 Motion detectors are interior security devices that sense motion inside the home.

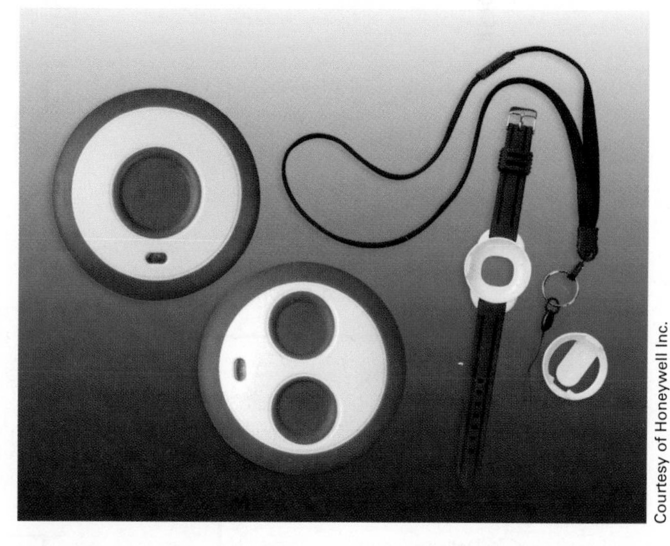
Courtesy of Honeywell Inc.

FIGURE 21-13 A wireless personal panic transmitter allows the user to press a button to signal for help.

of the circuit is opened, the security system will detect the open circuit. These circuits are generally referred to as "closed" or "closed loop." Alarms, horns, and other signalling devices are connected in parallel, since they will all signal at the same time when the security system is set off. Heat detectors and smoke detectors are generally connected in parallel because all of these devices will "close" the circuit to the security master control unit, setting off the alarms. Fire detectors are available for series connection for use in security systems that require normally closed contacts for circuit operation.

The instructions furnished with all security systems cover the installation requirements in detail, alerting the installer to *Canadian Electrical Code (CEC)* regulations, clearances, suggested locations, and mounting heights of the systems components.

Always check with the local electrical inspection authority to determine if there are any special requirements in your locality for the installation of security systems.

FIGURE 21-14 The wireless keypad is portable so you can arm or disarm the system from the yard or driveway.

FIGURE 21-15 Glass-break detectors, available in a variety of sizes, will sound the alarm when they detect the sound of breaking glass.

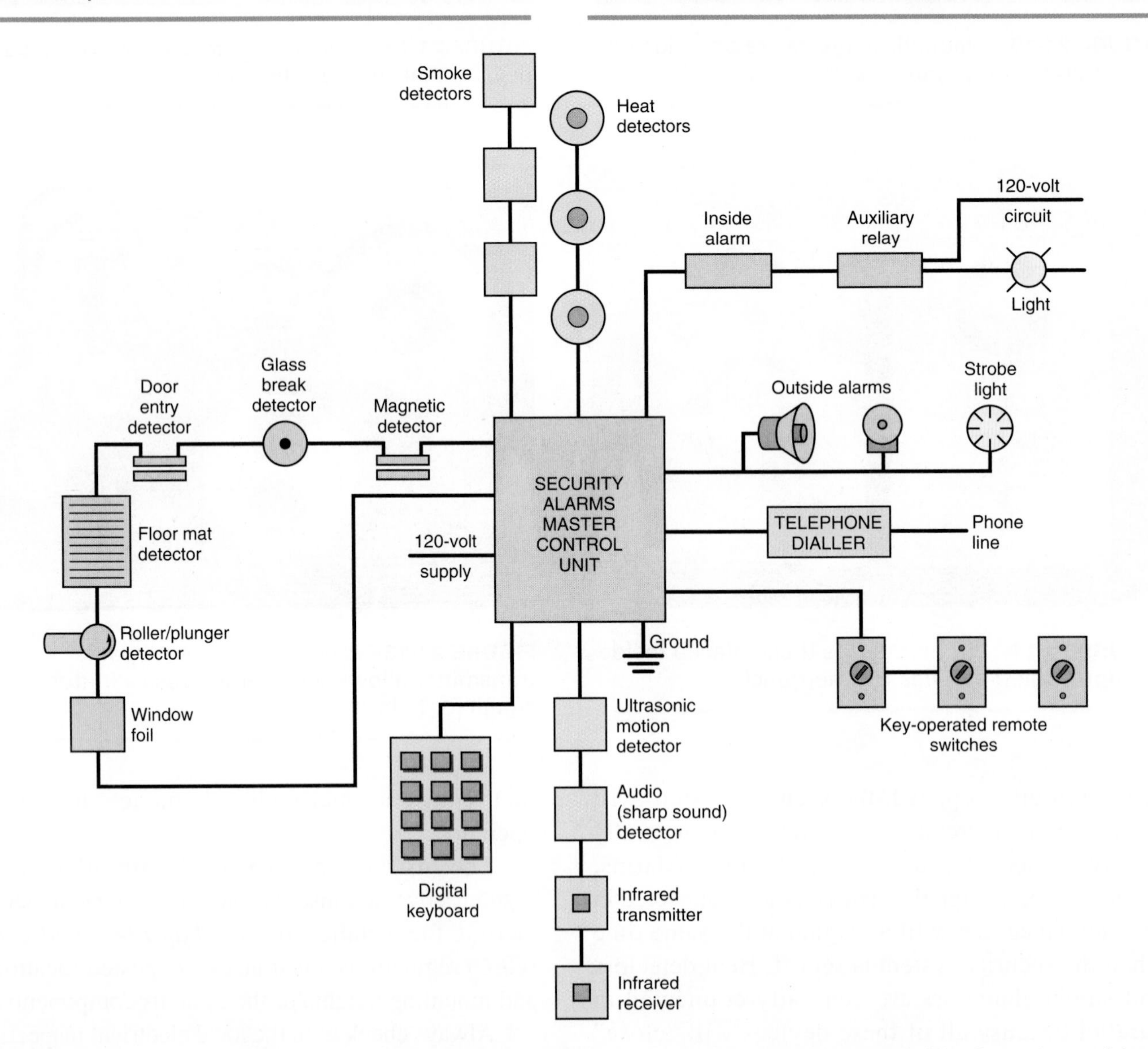

FIGURE 21-16 Typical residential security system showing some of the devices available. Complete wiring and installation instructions are included with these systems. Check local code requirements and follow the detailed instructions furnished with the system. Most of the interconnecting conductors are #18 AWG.

REVIEW

Note: Refer to the *CEC* or the blueprints provided with this textbook when necessary. Where applicable, responses should be written in complete sentences.

1. What is the name and number of the ULC standard that gives information about smoke and heat detectors? ______________________________

   ____

2. Name the two basic types of smoke detectors.______________________________

   ____

   ____

3. Why is it important to mount a smoke alarm on the ceiling not closer than 150 mm to a wall? ______________________________

   ____

   ____

   ____

   ____

4. [Circle the correct answer.] A basic rule is to install smoke alarms
   a. between the sleeping area and the rest of the house.
   b. at the top of a basement stairway that has a closed door at the top of the stairs.
   c. in a garage that is subject to subzero temperatures.

5. List two types of thermal detectors that are used in residential fire alarm systems.

   ____

   ____

6. Although the ULC gives many rules for the installation of smoke and heat detectors, always follow the installation recommendations of ______________________________

7. Security systems are usually installed with wire that is much smaller than normal house wires. These wires are generally # _____ AWG, which is similar in size to the wiring for chimes.

8. [Circle the correct answer.] Because the wire used in wiring security systems is rather small and cannot stand rough service and abuse, security wiring should be installed during the rough-in stages of a new house (before) (after) the other power-circuit wiring is completed.

9. [Circle the correct answer.] When several smoke alarms are installed in a new home, these units must be
   a. connected to separate 120-volt circuits.
   b. connected to one 120-volt circuit.
   c. interconnected so that if one smoke alarm is set off, the others will also sound a warning.

10. [Circle the correct answer.] It is advisable to connect smoke alarms to circuits that are protected by GFCI devices. True False
11. [Circle the correct answer.] Because a smoke alarm might need servicing or cleaning, be sure to connect it so that it is controlled by a wall switch. True False
12. [Circle the correct answer.] Smoke alarms of the type installed in homes must be able to sound a continuous alarm when set off for at least
    a. 4 minutes.
    b. 30 minutes.
    c. 60 minutes.
13. If a carbon monoxide detector is located in the hall outside the bedrooms, what is the maximum distance that the detector may be from the bedroom doors? ____________

    ____
14. If a home has an attached garage, is a CO detector required? ____________

    ____
15. In a home with a wood-burning fireplace, where must the CO detector(s) be located?

    ____

    ____

UNIT 22

Swimming Pools, Spas, and Hot Tubs

OBJECTIVES

After studying this unit, you should be able to

- recognize the importance of proper swimming pool wiring for human safety
- discuss the hazards of electric shock associated with faulty wiring in, on, or near pools
- describe the differences between permanently installed pools and pools that are portable or storable
- understand and apply the basic *Canadian Electrical Code (CEC)* requirements for the wiring of swimming pools, spas, hot tubs, and hydromassage bathtubs

POOL WIRING (*SECTION 68*)

For easy reference, a detailed drawing of the *CEC* requirements for swimming pool wiring is included with this textbook (Sheet 8 of the residence plans provided with this textbook). Refer to this drawing as you study this unit.

Swimming pools, wading pools, hydromassage bathtubs, therapeutic pools, decorative pools, hot tubs, and spas must be wired according to *CEC Section 68*. To protect people using such pools, rules specific to pool wiring have been developed over the years. Extreme care is required when wiring all equipment associated with pools.

ELECTRICAL HAZARDS

A person can suffer an immobilizing or lethal shock in a residential-type pool in two ways:

1. An electrical shock can be transmitted to someone in a pool who touches a live wire or the live casing or enclosure of an appliance, such as a hair dryer, radio, or extension cord; see Figure 22-1.
2. If an appliance falls into the pool, an electrical shock can be transmitted to a person in the water by means of voltage gradients in the water. Refer to Figure 22-2 for an illustration of this life-threatening hazard.

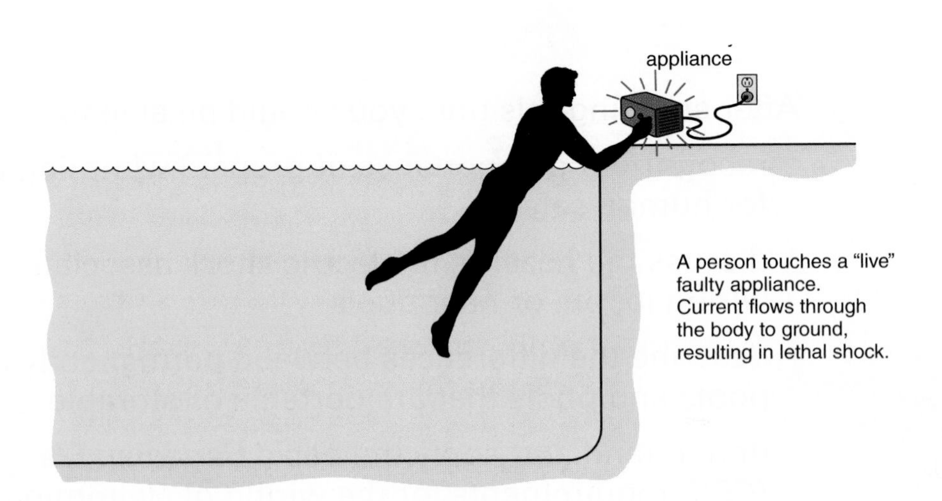

FIGURE 22-1 Touching a "live" faulty appliance can cause a lethal shock.

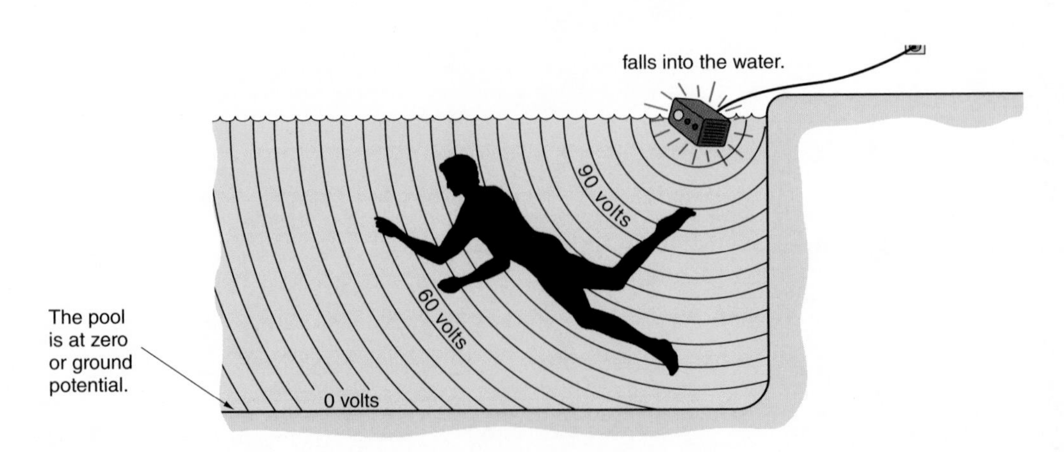

FIGURE 22-2 Voltage gradients surrounding a person in the pool can cause severe shock, drowning, or electrocution.

As Figure 22-2 shows, "rings" of voltage radiate outwardly from the radio to the pool walls. These rings can be likened to the rings that form when a rock is thrown into the water. The voltage gradients range from 120 volts at the radio to zero volts at the pool walls. The pool walls are assumed to be at ground or zero potential. The gradients, in varying degrees, are found in the entire body of water.

Figure 22-2 shows voltage gradients in the pool of 90 volts and 60 volts. (This figure is a simplification of the actual situation, in which there are many voltage gradients.) In this case, the voltage differential, 30 volts, is an extremely hazardous value. The person in the pool, who is surrounded by these voltage gradients, is subject to severe shock, immobilization (which can result in drowning), or actual electrocution. Tests conducted over the years have shown that a voltage gradient of 1.5 volts per 300 mm can cause paralysis.

Study Sheet 8 of the house plans supplied with this text. Note that, in general, underwater swimming pool lighting fixtures must be positioned not more than 600 mm below the normal water level, so that when a person is in the water close to an underwater lighting fixture, the relative position of the fixture is well below the person's heart. Fixtures that are permitted to be installed no less than 10 mm below the normal water level have undergone tougher, abnormal impact tests, such as those tests given to cleaning tools or other mechanical objects. The lenses on these fixtures can withstand these abnormal impact tests, whereas the standard underwater fixtures that are required to be 600 mm below the water level are subjected to normal impact tests that duplicate the impact of a person kicking the lens.

The shock hazard to a person in a pool is quite different from that of the normal "touch" shock hazard to a person not submersed in water. The water makes contact with the entire skin surface of the body rather than just at one "touch" point. Also, skin wounds, such as cuts and scratches, reduce the body's resistance to shock to a much lower value than that of the skin alone. Body openings such as ears, nose, and mouth further reduce the body resistance. As Ohm's law states, for a given voltage, the lower the resistance, the higher the current.

Another hazard associated with spas and hot tubs is prolonged immersion in hot water. If the water is too hot and/or the immersion too long, lethargy (drowsiness, hyperthermia) can set in. This can increase the risk of drowning. The Canadian Standards Association (CSA) standard establishes maximum water temperature at 40°C. The suggested maximum time of immersion is generally 15 minutes.

Instructions furnished with spas and hot tubs specify the maximum temperature and time permitted for using the spa or hot tub. Hot water, in combination with drugs and alcohol, presents a hazard to life. Caution must be observed at all times.

The figures in this unit show the *CEC* rules that relate to safety procedures for wiring pools.

Wiring Methods

Throughout *Section 68*, you will find that the wiring must be installed in rigid metal conduit, flexible metal conduit, rigid non-metallic conduit, flexible cord for wet locations, or, in some cases, EMT. For wet-niche fixtures, brass, copper, or other corrosion-resistant metal must be used, *Rule 68–066(5)*.

Rule 68–100(3) allows any type of wiring method recognized in *Section 12* for that portion of the interior wiring of one-family dwelling installations that supplies pool-associated pump motors, provided the wiring method contains an equipment bonding conductor no smaller than that required by *Table 16A*.

Be sure to read the requirements of *Rule 68–100* closely to determine the proper wiring method to be used for a particular situation.

CEC-DEFINED POOLS

The *CEC* describes a *permanently installed swimming pool* as one that cannot be disassembled readily for storage, *Rule 68–050*.

The *CEC* describes a *storable swimming pool* as one that can be disassembled for storage and reassembled to its original form, *Rule 68–050*.

Pools that have non-metallic inflatable walls are considered to be storable pools regardless of their dimensions.

GROUNDING AND BONDING OF SWIMMING POOLS

Bonding to Ground (*Rule 68–058*)

Rule 68–058 requires that all of the following items *must* be bonded to ground, as illustrated in Figure 22-3:

- wet- and dry-niche lighting fixtures
- any electrical equipment within 1.5 m of the inside wall of the pool
- pool-reinforcing steel
- any metal parts of the pool located within 1.5 m of the inside wall of the pool
- all electrical equipment associated with the recirculating system of the pool
- junction boxes
- transformer enclosures
- ground-fault circuit interrupters
- panelboards that are not part of the service equipment and supply the electrical equipment associated with the pool

Proper grounding and bonding ensures that all metal parts in and around the pool area are at the

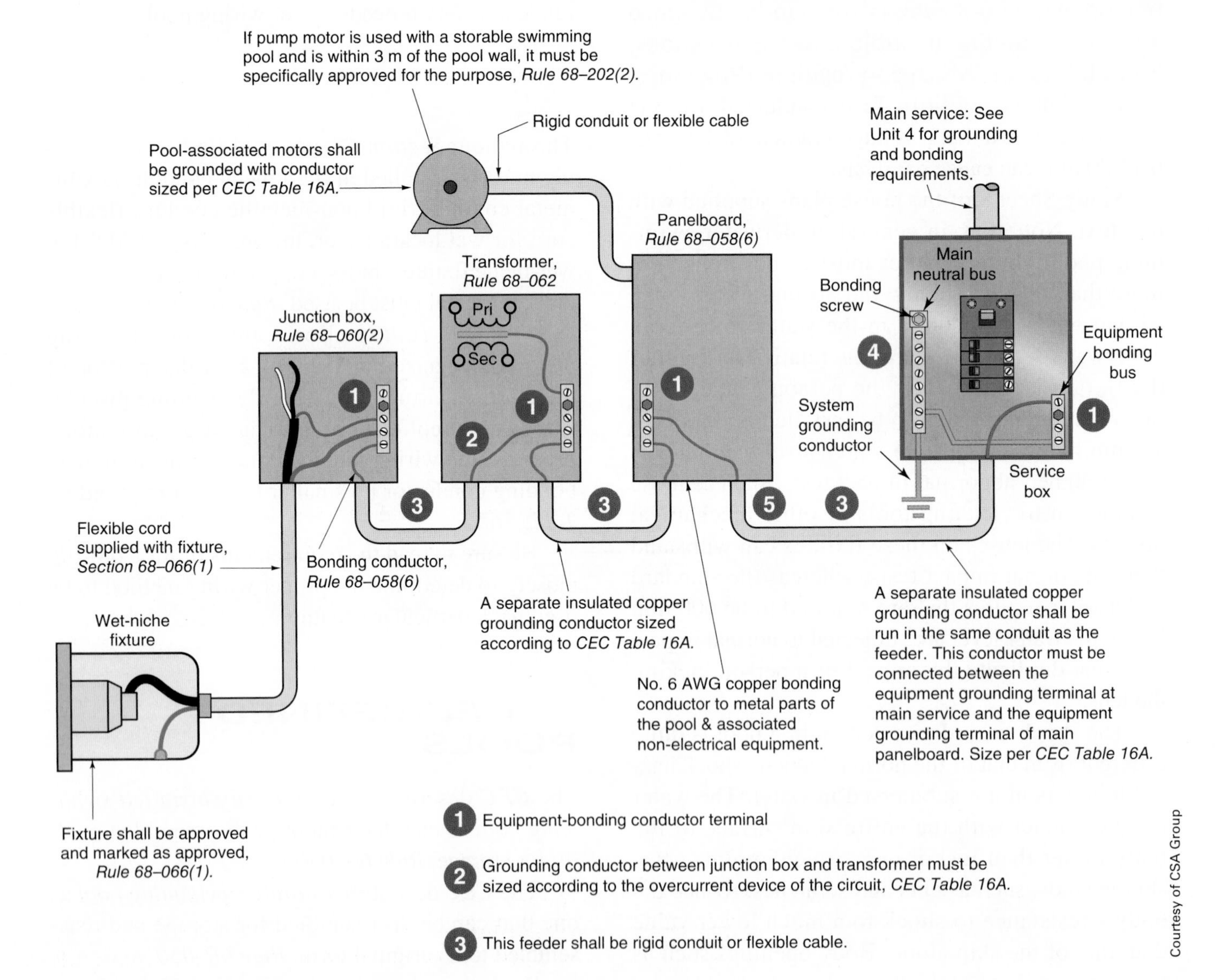

FIGURE 22-3 Grounding and bonding of important metal parts of a swimming pool, as covered in *Rule 68–058*. Refer to Sheet 8 of the plans supplied with this text.

same ground potential, thus reducing the shock hazard. Proper grounding and bonding practices also facilitate the opening of the overcurrent protective device (fuse or circuit breaker) in the event of a fault in the circuit. Grounding is covered in other units of this text.

Bonding Conductors. Bonding conductors *must*

- when smaller than #6 AWG, be run in the same conduit with the circuit conductors, or be part of an approved flexible cord assembly, as used for the connection of wet-niche underwater lighting fixtures;
- terminate on equipment-bonding terminals provided in the junction box, transformer, ground-fault circuit interrupter, subpanel, or other specific equipment.

Metal conduit by itself is *not* considered to be an adequate grounding means for bonding in and around pools.

Bonding (*Rule 68–058*)

Rule 68–058 requires that all metal parts of a pool installation be bonded together. For in-ground pools, a #6 AWG copper conductor is required. For above-ground pools a copper conductor sized in accordance with *Table 16A* is acceptable; see Figure 22-4. The #6 or larger copper bonding wire must be run to the main distribution panel board and "tie everything together." This helps keep all metal parts in and around the pool at the same voltage potential, reducing the shock hazard brought about by stray voltages and voltage gradients.

Bonding Conductors. The bonding conductor must be connected to

- the steel reinforcing bars in the concrete using a minimum of four equally spaced connections; when epoxy-coated reinforcing steel is used in the construction of a swimming pool, placing a loop of #6 AWG copper around the pool below the normal water level could be an alternative to bonding the reinforcing steel at four locations, *Rule 68–058(3)*
- the wall of a bolted or welded metal pool
- any deck boxes
- ladders
- other metal parts of the pool or non-electrical equipment

Bonding conductors

- need *not* be installed in conduit
- may be connected directly to the equipment that requires bonding, by means of brass, copper, or copper-alloy clamps.

LIGHTING FIXTURES UNDER WATER

Rule 68–066 covers underwater lighting fixtures for permanently installed pools. There are three types of underwater lighting fixtures:

1. *Dry-niche* lighting fixtures are mounted in the side walls of the pool and are designed to be relamped from the rear. These fixtures are waterproof.
2. *Wet-niche* lighting fixtures are mounted in the side wall of the pool and are designed to be relamped from the front. The supply cord is stored inside the fixture. The cord should be long enough to reach the top of the pool deck for lamp replacement. The forming shell to which the fixture attaches is intended to fill up with pool water.
3. *No-niche* lighting fixtures are mounted on a bracket on the inside wall of the pool. The supply conduit terminates on this bracket. The supply cord runs through this conduit to a deck box for connection to the circuit wiring. The extra supply cord is stored in the space behind the no-niche fixture. The cord should be long enough to reach the top of the pool deck for lamp replacement. This type of fixture is mainly used in retrofit installations and for above-ground pools.

Extra-low-voltage lighting has recently become more popular for pool installations. This type of lighting receives its power from an extra-low-voltage transformer fed from a Ground-Fault Circuit Interrupter (GFCI)-protected circuit. The transformer and the GFCI are generally located over 3 m from the pool.

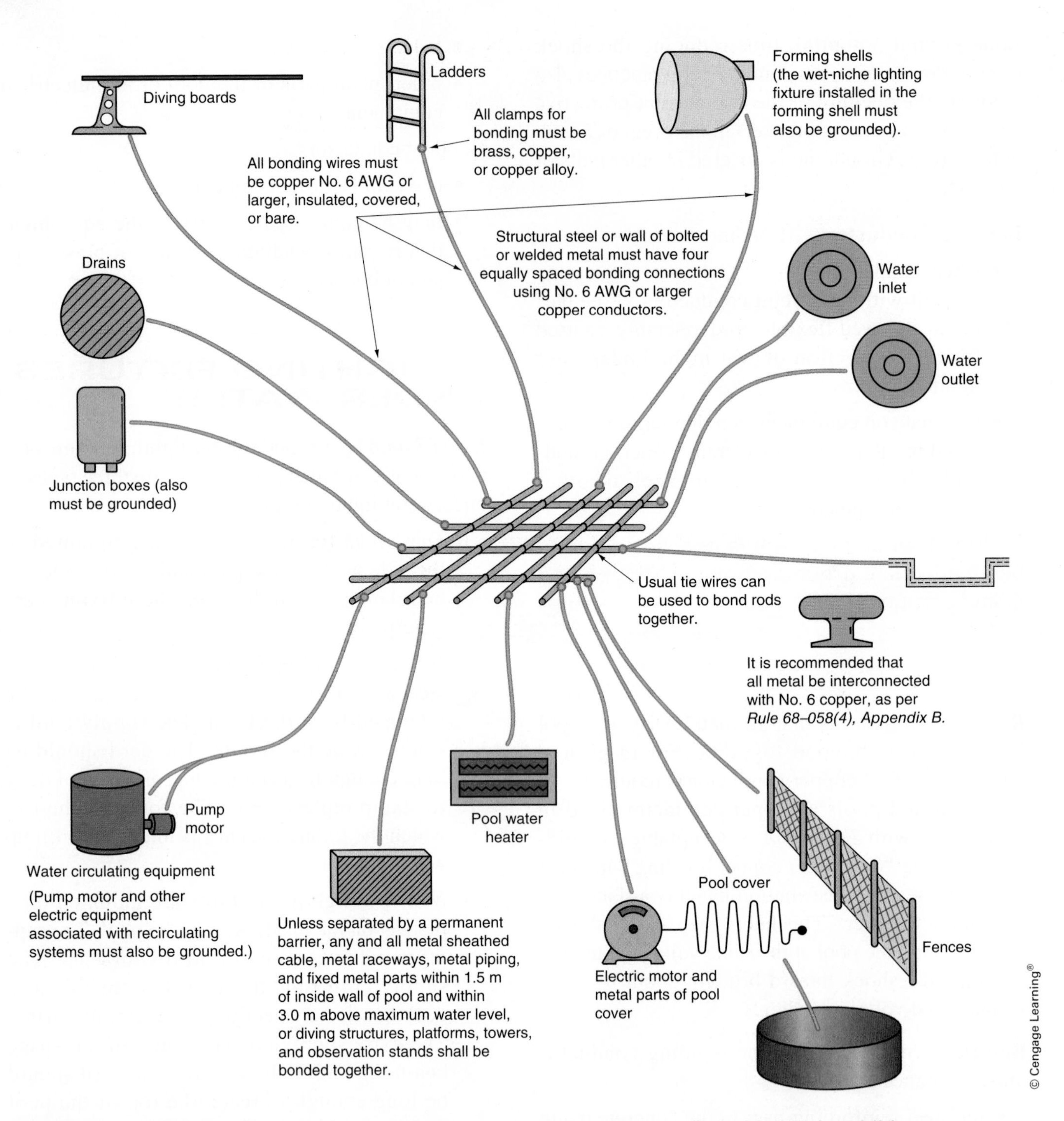

FIGURE 22-4 Bonding the metal parts of swimming pool installations, *Rule 68–058.* In addition to bonding, some of the above items must also be grounded. Refer to *Rules 68–058* and *68–060* for specifics. See Sheet 8 of the plans supplied with this text.

ELECTRIC HEATING OF SWIMMING POOL DECKS

Recommendations for deck-area electric heating units are illustrated in Figure 22-5.

Circuit sizing for electric heaters: The branch-circuit overcurrent protective devices shall be rated not less than 125% of the heater's nameplate rating, according to *Rule 62–114(8).* The branch-circuit conductors must have an ampacity not less than the connected load supplied, and not less than 80% of

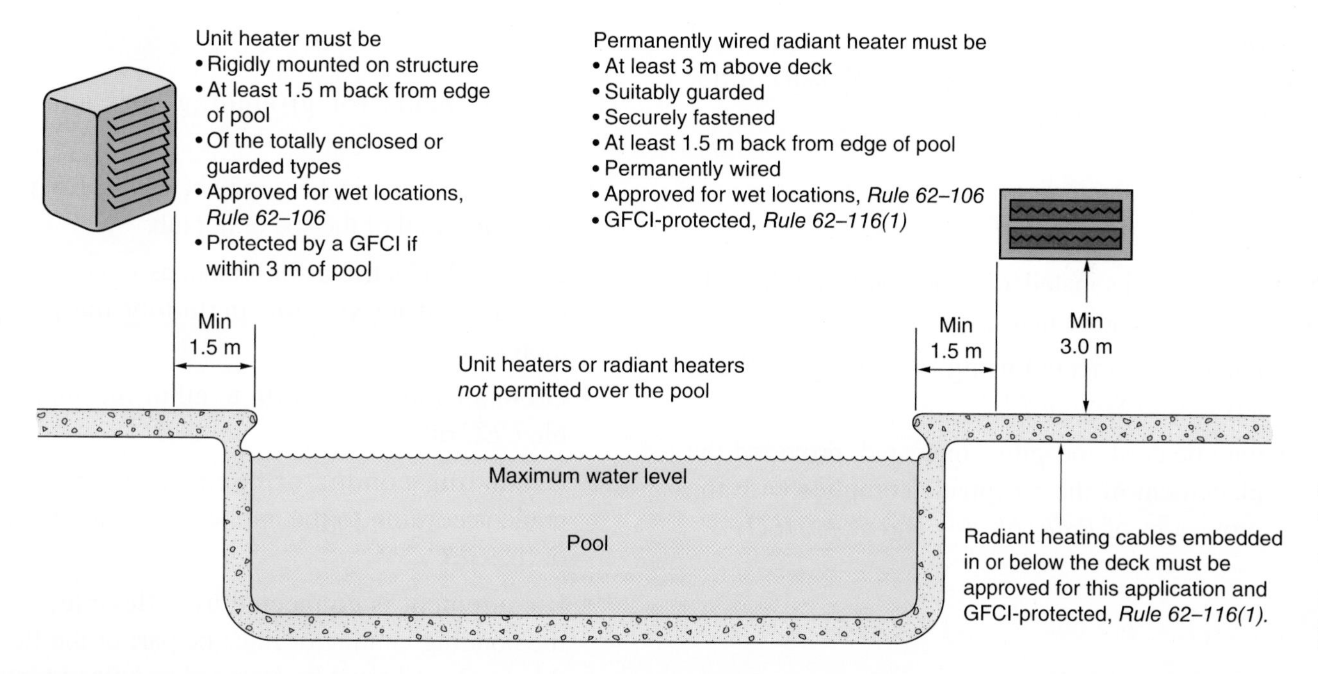

FIGURE 22-5 Recommendations for deck-area electric heating 3 m to 6 m from the inside edge of swimming pools.

the rating of the overcurrent devices protecting the circuit, *Rules 62–110(1)(b)* and *62–114(6)(b)*.

All heating cable sets used for surface heating sidewalks, driveways, and pool decks must be fed from a GFCI-protected circuit, *Rule 62–116(1)*.

SPAS AND HOT TUBS (*RULES 68–400* THROUGH *68–408*)

The basic difference between a spa and a hot tub is that a hot tub is constructed of wood, such as redwood, teak, cypress, or oak, whereas a spa is made of plastic, fibreglass, concrete, tile, or other man-made products. See Figure 22-6.

Spas and hot tubs contain electrical equipment for heating and recirculating water. They have no provisions for direct connection to the building plumbing system.

Spas and hot tubs are intended to be filled and used. They are not drained after each use, as in the case of a regular bathtub.

Some spas and hot tubs are furnished with a cord-and-plug set when shipped by the manufacturer. If these can be converted from cord-and-plug connection to "hard wiring," they are listed and identified as "convertible" spas and hot tubs. The manufacturer must provide clear instructions on how to convert from cord-and-plug set to hard wiring.

FIGURE 22-6 A typical spa.

Outdoor Installation

Spas and hot tubs installed outdoors *must* conform to the installation requirements for regular swimming pools, *Rule 68–404*.

The metal parts of spas and hot tubs must be bonded together and to ground in accordance with *Rule 68–058*.

Rule 68–402(2) states that the metal bands and hoops that secure the wooden staves of a spa or hot tub do not have to be bonded.

Indoor Installation

Spas and hot tubs installed indoors *must* conform to the following requirements:

- must be connected using the wiring methods covered in *Section 12*.
- may be cord-and-plug connected, provided the placement of the receptacle complies with the provisions of *Rules 68–064* and *68–404(2)*.

Receptacles (*Rule 68–064*)

- Any receptacles located between 1.5 and 3 m of the inside walls of a pool, hot tub, or spa shall be protected by a GFCI of the Class A-type, *Rule 68–064(2)*.
- No receptacle shall be located less than 1.5 m from the edge of the pool, *Rule 68–064(1)*.
- Any receptacles that supply power to a spa or hot tub must be GFCI-protected no matter how far they are from the spa or hot tub.

Lighting Fixtures (*Rule 68–066*)

1. All lighting outlets and fixtures located within 3 m of the inside wall or the water surface of the spa or hot tub must be GFCI-protected. GFCI protection is not required for fixtures more than 3 m above maximum water level, *Rule 68–066(6)*.
2. Recessed lighting fixtures may be installed less than 3 m above the hot tub or spa if the fixtures are GFCI-protected and are suitable for use in damp locations.
3. Surface-mounted fixtures may be installed less than 3 m above the hot tub or spa if the fixtures are GFCI-protected and are suitable for use in damp locations.
4. If any underwater lighting is to be installed, the rules for regular swimming pools apply.

Bonding to Ground (*Rule 68–402*)

The requirements for grounding spas and hot tubs are

- Ground all electrical equipment within 1.5 m of the inside wall of the spa or hot tub.
- Ground all electrical equipment associated with the circulating system, including the pump motor.
- Grounding *must* conform to all of the applicable *CEC* rules.
- Grounding conductor connections *must* be made according to the applicable requirements of the *CEC*.
- If equipment is connected by a flexible cord, the bonding conductor must be part of the flexible cord, and must be fastened to a fixed metal part of the equipment.

Bonding (*Rule 68–402*)

The bonding requirements for spas and hot tubs are similar to those for regular swimming pools. Bond together

- all metal fittings within or attached to the spa or hot tub
- all metal parts of electrical equipment associated with the circulating system, including pump motors
- all metal pipes, conduits, and metal surfaces within 1.5 m of the inside edge of the spa or hot tub (this bonding is not required if the materials are separated from the spa or hot tub by a permanent barrier, such as a wall or building)
- all electrical devices and controls *not* associated with the spa or hot tub located less than 1.5 m from the inside edge of the spa or hot tub. If located 1.5 m or more from the hot tub or spa, bonding is not required

Bonding is to be accomplished by means of threaded metal piping and fittings, metal-to-metal mounting on a common base or frame, or by means of a solid, insulated, covered, or bare bonding conductor sized according to *Table 16A*.

Electric Water Heaters (*Rule 68–408[2]*)

Electric water heaters must be approved for the purpose.

They must have overcurrent protection rated not less than 125% of the total load, as indicated on the nameplate of the heater, due to the panel being marked for 80% continuous operation.

Wiring and Circuit Protection

Most spas and hot tubs require a 120/240-volt circuit with an overcurrent device ranging from 40 to 60 amperes. All spas and hot tubs must be fed from a GFCI-protected circuit. As with pools, this GFCI device must be located more than 3 m from the pool water. Since the spa or hot tub has a motor, a motor-disconnecting means must also be supplied. This is typically accomplished by installing the GFCI circuit breaker in a weatherproof enclosure, located in a convenient location for testing purposes, and within sight of the hot tub or spa.

HYDROMASSAGE BATHTUBS (*SUBSECTION 68–300*)

Hydromassage bathtubs, also called *whirlpool tubs* or *whirlpool bathtubs,* together with their associated electrical components, must be GFCI-protected.

Because a hydromassage bathtub does not present any more of a shock hazard than a regular bathtub, the *CEC* permits all other wiring (fixtures, switches, receptacles, and other equipment) in the same room but not directly associated with the hydromassage bathtub to be installed according to all the normal *CEC* requirements covering installation of that equipment in bathrooms.

This residence has a hydromassage tub in the master bathroom. It is powered by a ½-hp, 115-volt, 3450-rpm, single-phase motor rated at 10 amperes.

A hydromassage tub and all of its associated equipment (generally provided by the manufacturer) must be connected to a circuit that has GFCI protection. Some manufacturers provide a GFCI device built into the hydromassage tub control system. Otherwise, the GFCI device must be supplied and installed by the electrician. This can be a feedthrough device installed in a device box in the bathroom, a GFCI receptacle located under the tub, or a GFCI circuit breaker. The device *must* be located where it is accessible for testing.

If the hydromassage tub is cord-cap connected, the receptacle located under the tub must be located at least 300 mm above the floor, *Rule 68–306(1).*

Figure 13-15 in Unit 13 illustrates one manufacturer's hydromassage bathtub.

By definition in *Rule 68–050,* a hydromassage bathtub is intended to be filled (used) and then drained after each use, whereas a spa or hot tub is filled with water and not drained after each use.

FOUNTAINS

The home discussed in this text does not have a fountain. When fountains share water with a regular swimming pool, the fountain wiring must conform to pool wiring requirements. If the house has a water pump for a waterfall or decorative pool in the garden, the pump must be protected by a GFCI, *Rule 26–952.*

A decorative pool is one in which any dimension is greater than 1.5 m, *Rule 68–050.* Decorative pools must be wired in the same manner as swimming pools.

Leakage Current Collectors (*Rule 68–406*)

Leakage current collectors must be installed in the water inlets or water outlets of spas or hot tubs. They must be electrically insulated from the tub and bonded to the control panel of the unit or to the main service ground with a minimum #6 copper bonding conductor, or the bonding conductor may be sized from *Table 16A*, provided it is mechanically protected in the same manner as the circuit conductors.

SUMMARY

The figures in this unit and the building plans included with this book present a detailed overview of the *CEC* requirements for pool wiring.

REVIEW

Note: Refer to the *CEC* or the blueprints provided with this textbook when necessary. Where applicable, responses should be written in complete sentences.

1. The *CEC* rule that covers most of the requirements for wiring swimming pools is *Rule* ____________________.
2. Name two ways in which a person can receive an electrical shock when in a pool.
____
____
____
____
3. *Rule 68–050* discusses types of pools. Name and describe two types. ____________
____
____
____
4. Use *Section 68* to determine if the following items must be bonded.

	YES	NO
a. Wet- and dry-niche lighting fixtures	____	____
b. Electrical equipment located beyond 1.5 m of the inside edge of the pool	____	____
c. Electrical equipment located within 3 m of the inside edge of the pool	____	____
d. Recirculating equipment and pumps	____	____
e. Lighting fixtures installed more than 7.5 m from the inside edge of pool	____	____
f. Junction boxes, transformers, and GFCI enclosures	____	____
g. Panelboards that supply the electrical equipment for the pool	____	____
h. Panelboards 6 m from the pool that do not supply the electrical equipment for the pool	____	____

5. [Circle the correct answer.] Bonding conductors (must) (may) be run in the same conduit as the circuit conductors.
6. [Circle the correct answer.] Bonding conductors (may) (may not) be spliced with the wire-nut types of wire connectors.
7. The purpose of grounding and bonding is to ________________________
____
____
____

8. What parts of a pool must be bonded together? ______________________________

9. May electrical wires be run above the pool? Explain. ______________________________

10. What is the closest distance that a receptacle may be installed to the inside edge of a pool?

11. Receptacles located between 1.5 m and 3 m from the inside edge of a pool must be protected by a ______________________________.

12. [Circle the correct answer.] Lighting fixtures installed over a pool must be mounted at least (3 m) (4 m) (4.5 m) above the maximum water level, *Rule 68–054*.

13. Grounding conductor terminations in wet-niche metal-forming shells, as well as the conduits entering junction boxes or transformer enclosures where the conduit runs directly to the wet-niche lighting fixture, must be ____________________ with a (an) ____________________ to prevent corrosion to the terminal and to prevent the passage of water through the conduit, which could result in corrosion, *Rule 68–060(7)*.

14. [Circle the correct answer.] Wet-niche lighting fixtures are accessible (from the inside of the pool) (from a tunnel) (on top of a pole).

15. [Circle the correct answer.] According to *Rule 68–066*, wet-niche lighting fixtures shall not be submersed to a depth of more than
    a. 600 mm
    b. 300 mm
    c. 450 mm
    d. 1 m

16. [Circle the correct answer.] Dry-niche lighting fixtures are accessible (from a tunnel, or passageway, or deck) (from the inside of the pool) (on top of the pole).

17. [Circle: True or False] In general, it is *not* permitted to install conduits under the pool or within 1.5 m measured horizontally from the inside edge of the pool, *Rule 68–056, Table 61*. Explain. ____

____

____

____

____

____

____

____

18. [Circle the correct answer.] Junction boxes, transformers, and GFCI enclosures have one thing in common. They all (are made of bronze) (have threaded hubs) (must have a bonding wire or terminal screw).
19. [Circle the correct answer.] Lighting fixtures are permitted to be mounted less than 3 m measured horizontally from the inside edge of the pool only if they are (approved for wet locations) (rigidly fastened to an existing structure) (GFCI-protected).
20. The following statements about indoor spas and hot tubs are either true or false. Check one.

	TRUE	FALSE
a. Receptacles may be installed within 1.5 m of the edge of the spa or hot tub.	_____	_____
b. All receptacles within 3 m of the spa or hot tub must be GFCI-protected.	_____	_____
c. Any receptacles that supply power to pool equipment must be GFCI-protected if within 3 m of the pool.	_____	_____
d. Wall switches must be located at least 1 m from the spa or hot tub.	_____	_____
e. Lighting fixtures above the pool or within 1.5 m from the inside edge of the pool must be GFCI-protected.	_____	_____

21. [Circle the correct answer.] Bonding and grounding of electrical equipment in and around spas and hot tubs (are required by the *CEC*) (are not required by the *CEC*) (are decided by the electrician).
22. When installing a hot tub, a control shall be located behind a barrier not less than ____ m from the tub.
23. [Circle the correct answer.] For the circuit supplying a hydromassage or whirlpool bathtub, be sure that the circuit (does not have GFCI protection that could result in nuisance tripping) (is GFCI-protected).
24. [Circle: True or False] Leakage current collectors shall be bonded to the control panel with a minimum #6 AWG copper bonding conductor.

25. The maximum voltage that wet-niche fixtures are intended to be connected to is
    a. 120 volts.
    b. 15 volts.
    c. 240 volts.
26. Swimming pool pump motors shall be specifically approved, and, if within 3 m of the pool wall, shall be ____

    ____

    ____

    Quote rule numbers.

UNIT 23

Television, Telephone, Data, and Home Automation Systems

OBJECTIVES

After studying this unit, you should be able to

- install television outlets, antennas, cables, and lead-in wires
- list cable television installation requirements
- describe the basic installation of satellite antennas
- install telephone conductors, outlet boxes, and outlets
- install and terminate computer network cable
- understand the basics of "wireless"
- understand the basic terminology of home automation
- identify types of devices that can be controlled by a home automation system

TELEVISION TV

Television signals can be received by means of antenna, cable, or satellite dish. Because some systems may not be available in the region of the country or the area in which you will be working, the text discusses all three types of installations.

General Wiring

Television outlets may be connected in several ways, depending on their proposed locations. In general, for installation in dwellings, the electrician uses standard sectional switch boxes or 102-mm square, 38-mm deep outlet boxes with single-gang raised plaster covers. A box is installed at each point on the system where an outlet is located. Use non-metallic boxes if the system is to be wired with unshielded cable; see Figure 23-1.

Television Installation Methods

Figure 23-2 shows a master amplifier distribution system. The lead-in wire from the antenna is connected to an amplifier placed in an accessible area, such as the basement. Twin-lead, 300-ohm cable runs from the amplifier to each of the outlet boxes. The cable is connected to tap-off units, which have a 300-ohm input and a 300-ohm output. The tap-off units have terminals or plug-in arrangements so that the 300-ohm, twin-lead cable can be connected between the television outlet and the television.

Amplified systems are generally not necessary in typical residential installations, where cable television is available.

With a *multiset coupler* the lead-in wire from the antenna is connected to a pair of terminals on the coupler. The cables to each television are connected to other pairs of terminals. Two to four

FIGURE 23-1 Non-metallic boxes and a non-metallic raised cover.

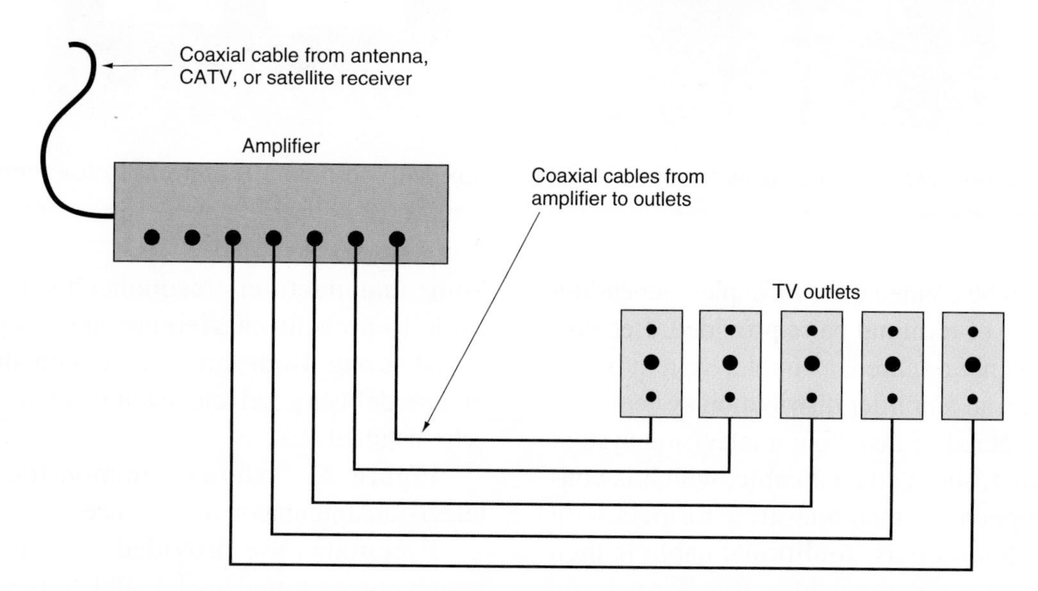

FIGURE 23-2 Television master amplifier distribution system. A television master amplifier distribution system might be needed when many TV outlets are to be installed. This will minimize the signal loss.

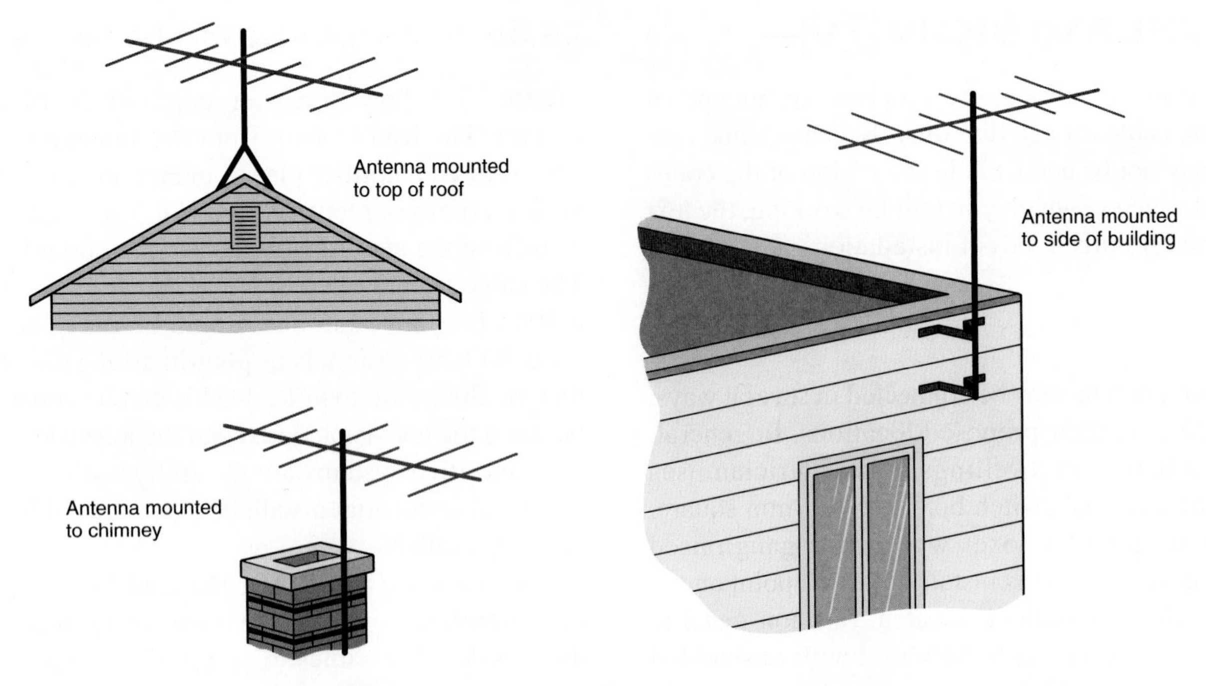

FIGURE 23-3 Typical mounting of television antennas.

FIGURE 23-4 Mounting of power/communication box with coaxial outlet and duplex receptacle.

televisions can be connected to a coupler, depending on the number of terminal pairs provided. Because an amplifier is not required, a coupler system is less expensive than an amplifier distribution system.

A third method of installing a television system uses shielded 75-ohm coaxial cable, which is connected to impedance-matching transformers with 75- or 300-ohm outputs. Additional cable is then connected between the matching transformer and the television. Many older televisions have 300-ohm inputs; newer models generally have 75-ohm inputs. Some manufacturers recommend shielded coaxial cable to prevent interference and keep the colour signal strong. Both shielded and non-shielded lead-in wire deliver good television reception when properly installed.

Figure 23-3 shows common locations for a television antenna on a residence.

Faceplates are provided for tap-off units to match conventional switch and convenience outlet faceplates. Combination two-gang faceplates also are available for tap-off (Figures 23-4 and 23-5).

FIGURE 23-5 Coaxial (F connector) and telephone combination wall jacks and cover plate.

Both a 120-volt convenience receptacle and a television outlet can be mounted in one outlet box using this combination faceplate.

However, a combination installation requires that a metal barrier be installed between the 120-volt section and the television section of the outlet box, *Rule 54–400(3)(a)*. This barrier prevents line voltage interference with the television signal and keeps the 120-volt wiring away from the television wiring. In addition, line voltage cables and lead-in wire must be separated to prevent interference when they are run in the same space within the walls and ceilings. The line voltage cables should be fastened to one side of the space and the lead-in cable to the other side.

CEC Rules for Cable Television (*Section 54*)

Cable systems, also called *community antenna television* (CATV) or just *cable TV*, are installed both overhead and underground in a community. To supply an individual customer, the cable company runs coaxial cables through the wall of a residence at a convenient point. Up to this point of entry, inside the building, and underground, the cable company must conform to the requirements of *Section 54*, plus local codes if applicable. These are some of the key rules:

1. The outer conductive shield of the coaxial cables must be grounded as close as possible to the point of entry, *Rule 54–200(1)*.
2. Coaxial cables shall not be run in the same conduits or box as electric light and power conductors, *Rule 54–400(3)*.
3. The grounding conductor *(Rule 54–300)*:
   a. must be insulated
   b. must not be smaller than #14 AWG
   c. may be solid or stranded but must be of copper
   d. must be run in a line as straight as practicable to the grounding electrode
   e. shall be guarded from mechanical damage
   f. shall be connected to the nearest accessible location on the building or structure grounding electrode (*Rules 10–700* and *54–304*); the grounded interior water pipes (*Rule 10–700*); or the grounding electrode conductor
   g. If none of the options in (f) are available, then ground to any of the electrodes per *Rule 10–702,* such as the metal frame of the building, a concrete-encased electrode, or a plate electrode.
   h. If none of the options in (f) and (g) are available, then ground to any of the electrodes per *Rule 54–302.* The electrodes shall be driven to a depth of at least 2 m. The connection may be by ground clamp or by a wire lead with a pressure connector. Refer to Canadian Standards Association (CSA) Standard C22.1–09.

CAUTION: If any of the preceding results in one grounding electrode for the CATV shield and another grounding electrode for the electrical system, bond the two electrodes together with a bonding jumper no smaller than #6 copper, *Rule 10–702*. Grounding and bonding the coaxial cable shield to the same grounding electrode as the main service ground prevents voltage differences between the shield and other grounded objects, such as a water pipe.

SATELLITE ANTENNAS

A satellite antenna system consists of a parabolic dish, a low-noise block amplifier with integral feed (LNBF), mounting hardware, and a receiver. At one time, satellite systems were most popular in the rural areas where choices of stations on a conventional antenna were limited. The dish was 2 or 3 m in diameter and was usually mounted on a large steel pole cemented into the ground (Figure 23-6). Many of these systems had rotors that rotated and pointed the dish to different satellites. These systems were prone to lightning strikes and were very expensive to repair.

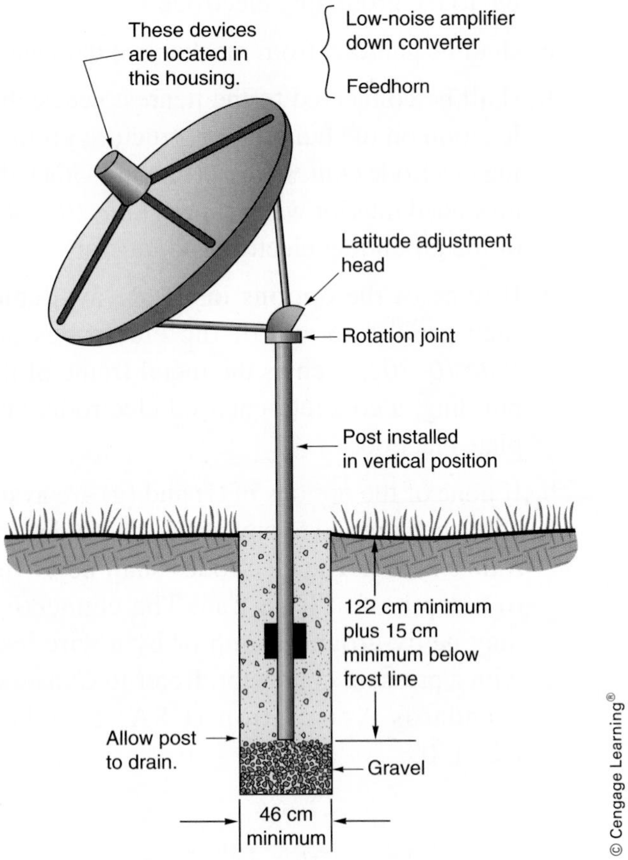

These devices are located in this housing.
Low-noise amplifier down converter
Feedhorn
Latitude adjustment head
Rotation joint
Post installed in vertical position
122 cm minimum plus 15 cm minimum below frost line
Allow post to drain.
Gravel
46 cm minimum

FIGURE 23-6 Satellite antenna.

Modern satellite systems have dishes ranging from 450 to 600 mm. They are light and easy to install, with many users choosing to perform their own installation. These satellite systems are "fixed" on one satellite, which is in a geosynchronous orbit above the earth (Figure 23-7). Because of their light weight, many dishes are installed on the side of the residence, where they are less likely to be struck by lightning.

Selecting a Mounting Location for the Dish

The dish can be mounted on nearly any surface, including brick, concrete, masonry block, wood siding, roof, or a pole. When selecting the mounting location, check the documentation included with the system. The dish must have an unobstructed "view" of the satellite. Be sure to consider future growth of trees and possible construction of adjacent buildings. Be sure to protect the antenna from wind, as strong winds can bend mounting hardware or pull out mounting screws. The mounting surface must also be strong enough to prevent movement of the dish during high winds as even the smallest amount of movement may affect the signal reception. Because the system will be grounded to earth, it will provide a low resistance path to ground for lightning. Locating the dish high on the building or on a pole increases the likelihood of a lightning strike.

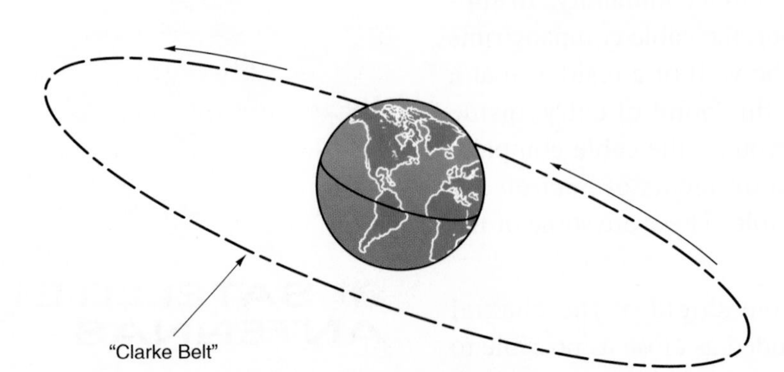

FIGURE 23-7 Television satellites appear to be stationary in space because they are rotating at the same speed as the earth. This is called *geosynchronous orbit.* The satellite receives the uplink signal from earth, amplifies it, and transmits it back to earth. The downlink signal is picked up by the satellite antenna.

Mounting the Satellite Dish

Each manufacturer of satellite dishes provides detailed installation instructions with the systems. The mounting base must be securely fastened to the mounting surface. The mounting pole is attached to the mounting base and adjusted as recommended by the manufacturer. The support arm is attached to the mounting pole with the dish and the LNBF attached to the arm. After installation of the system is complete, aim the dish as recommended by the manufacturer.

Grounding of a Satellite System

The system must be grounded according to the manufacturer's recommendations and all applicable codes. To provide proper grounding, install a grounding block in the coaxial cable that enters the building from the dish. The distance from the copper ground rod and the grounding block should be as short as possible, and the grounding block must be connected to the rod using a copper grounding conductor of not less than #14 AWG (Figure 23-8).

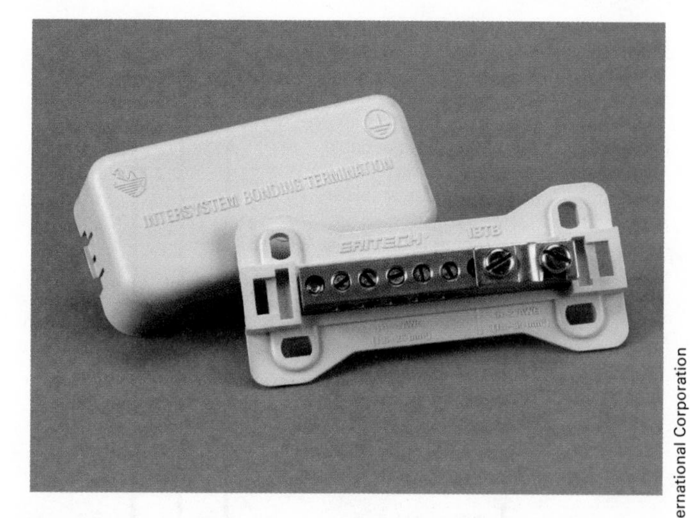

FIGURE 23-8 Intersystem bonding termination equipment.

Installation of the Satellite Receiver

The satellite receiver is a remote-controlled device the viewer uses to control the channel. It therefore must be installed close to the television. More than one TV may be connected to this system; however, the same program will be viewed on all. It is possible to view up to two different programs at the same time on different televisions, but this will require that the LNBF be a dual type and a second receiver must be purchased. Also, a second coaxial cable must be supplied from the dish to the second receiver. Satellite television companies offer pay-per-view service on their systems, which require the receiver to be connected to a telephone line. Manufacturers recommend that the phone line be left connected to the receiver at all times. Refer to Figure 23-9 for a diagram of the installation. The coaxial cable from the satellite dish to the receiver must be Type RG-6. After the receiver, Type RG-59 may be used to distribute the signal throughout the residence. However, the higher-quality video provided by digital and high-definition television (HDTV) require RG-6 throughout the home.

Many signals received by a satellite dish provide the audio portion of the program in stereo. Many satellite services even supply specialty audio (radio) stations. The satellite receiver has audio output jacks on the rear panel for connection to a conventional stereo system for improved audio. This connection usually requires two audio patch cables with RCA-type jacks on both ends.

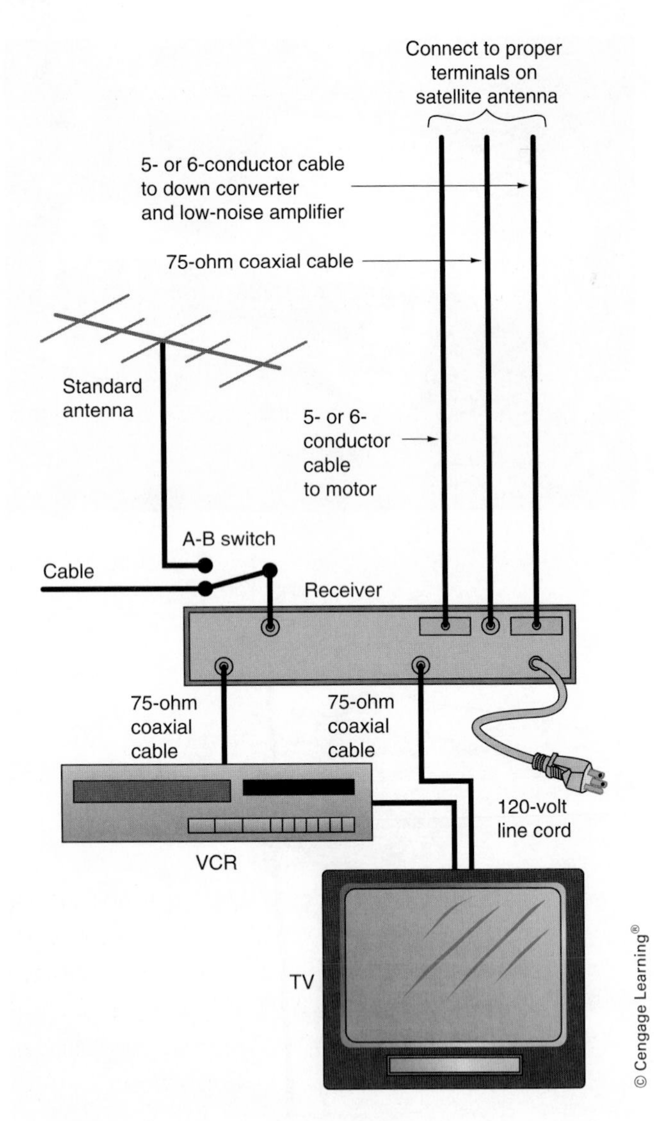

FIGURE 23-9 Typical wiring diagram for a dual-receiver satellite system.

CEC RULES FOR INSTALLING ANTENNAS AND LEAD-IN WIRES (*SECTION 54*)

Home television sets and AM and FM radios generally come complete with built-in antennas. For those locations in outlying areas, out of reach of strong signals, it is quite common to install a separate antenna system.

Either indoor or outdoor antennas may be used with televisions. The front of an outdoor antenna is aimed at the television transmitting station. When transmitting stations are located in different directions, a rotor on the mast turns the antenna to face the direction of each transmitter. The rotor is controlled from inside the building. The controller's cord is plugged into a regular 120-volt receptacle to obtain power. A four-wire cable is usually installed between the rotor motor and the control unit. The wiring for a rotor may be installed during the roughing-in stage of construction by running the rotor's four-wire cable into a device wall box, allowing 1.5–1.8 m of extra cable, and then installing a regular single-gang switch plate when finishing.

Section 54 covers this subject. Although instructions are supplied with antennas, be sure to follow these *Canadian Electrical Code (CEC)* rules:

1. Antennas and lead-in wires shall be securely supported, *Rule 54–108*.
2. Antennas and lead-in wires shall not be attached to the electric service mast.
3. Antennas and lead-in wires shall be kept away from all light and power conductors to avoid accidental contact with them.
4. Lead-in wires shall be securely attached to the antenna.
5. Outdoor antennas and lead-in conductors shall not cross over light and power wires.
6. Lead-in conductors shall be kept at least 600 mm away from open light and power conductors.
7. Where practicable, antenna conductors shall not be run under open light and power conductors.
8. On the outside of a building:
   a. position and fasten lead-in wires so they cannot swing closer than 300 mm to light and power wires having a maximum of 300 volts between conductors.
   b. keep lead-in conductors at least 2 m away from the lightning rod system, or bond together according to *Rule 10–702*.
9. On the inside of a building:
   a. keep antenna and lead-in wires at least 50 mm from other open wiring (as in old houses) unless the other wiring is in a metal raceway or in cable armour.

b. keep lead-in wires out of electric boxes unless there is an effective, permanently installed barrier to separate the light and power wires from the lead-in wire.

10. Grounding:

a. The grounding wire must be copper.

b. The grounding wire need not be insulated. It must be securely fastened in place; may be attached directly to a surface without the need for insulating supports; shall be protected from physical damage or be large enough to compensate for lack of protection; and shall be run in as straight a line as is practicable.

c. The grounding conductor shall be connected to the nearest accessible location on the building or structure grounding electrode, *Rule 54–200(1);* or the grounded interior water pipe, *Rule 54–302;* or the metallic power service raceway; or the service equipment enclosure; or the grounding electrode conductor or its metal enclosure.

d. If none of the options in (c) is available, ground to any one of the electrodes per *Rule 10–700,* such as a metal underground water pipe, an independent metal water-piping system, or a concrete-encased electrode.

e. The grounding conductor may be run inside or outside the building, but must be protected where exposed to mechanical injury.

f. The grounding conductor shall not be smaller than #14 copper.

CAUTION: If any of the preceding rules results in one grounding electrode for the antenna and another grounding electrode for the electrical system, bond the two electrodes together with a bonding jumper not smaller than #6 copper or its equivalent.

Grounding and bonding to the same grounding electrode as the main service ground prevents voltage differences between the two systems.

TELEPHONE WIRING ▲ (*SECTION 60*)

Since the deregulation of telephone companies, residential do-it-yourself telephone wiring has become quite common.

A telephone company installs the service line to a residence up to a protector device, which protects the system from hazardous voltages. The company decides if the protector is mounted outside or inside the home. Always check with the phone company before starting your installation.

The telephone company's wiring ends and the homeowner's interior wiring begins at the *demarcation point.* The telephone wiring inside the home may be done by the telephone company, the homeowner, or the electrician. Whatever the case, the rules in *Section 60* apply. Refer to Figures 23-10 through 23-12.

Section 60 covers the installation of communication systems. For the typical residence, the electrician or homeowner will rough in the boxes wherever a telephone outlet is wanted. The most common interior telephone wiring for a new home is with four-conductor #22 or #24 cable, known as *station wire*. The outer jacket is usually of a thermoplastic material, most often in a neutral colour that blends in with decorator colours in the home. This is particularly important in existing homes where the telephone cable may have to be exposed.

Voice communication does not require a high-quality cable within the residence. However, many consumers use their telephone lines for data communication through a modem connected to their computer. The phone lines themselves are sometimes used as a computer data network within the home, using special equipment. In these cases, the performance of station wire may not be satisfactory. In these situations, a three- or four-pair unshielded twisted pair (UTP) cable of the Category 3 type (or higher) is recommended. Twisting the conductors within the pairs and twisting the pairs themselves help to reduce electromagnetic interference with data communication.

Cables, mounting boxes, junction boxes, terminal blocks, jacks, adapters, faceplates, cords, hardware,

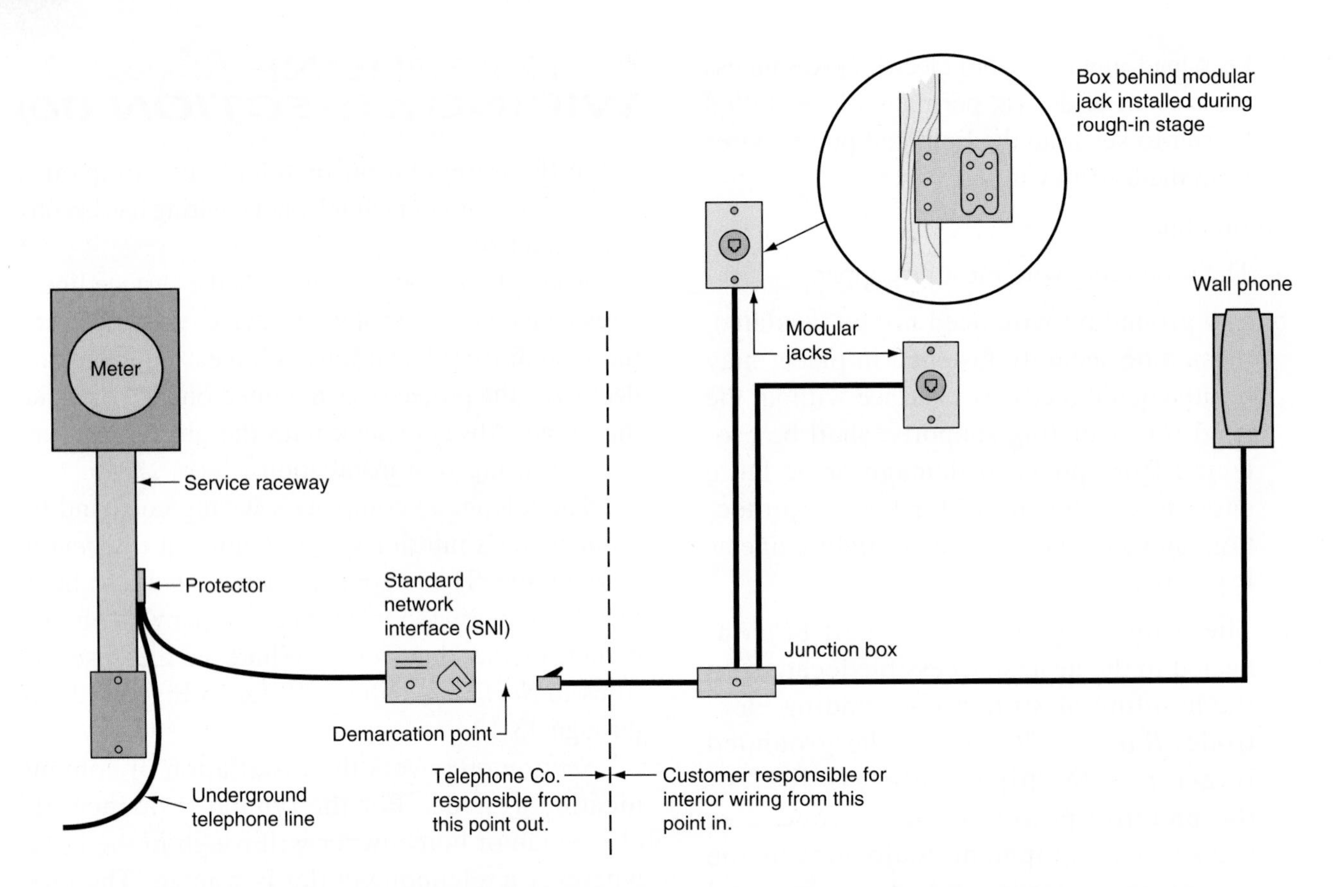

FIGURE 23-10 Typical telephone installation. The telephone company installs and connects its underground cable to the protector. In this illustration, the protector is mounted to the service raceway. This establishes the ground that offers protection against hazardous voltages such as a lightning surge on the incoming line. A cable is then run from the protector to a standard network interface (SNI). The electrician or homeowner plugs the interior wiring into the SNI by means of the modular plug. The term *modular* means that most of the phones are plugged-in rather than hard-wired connections.

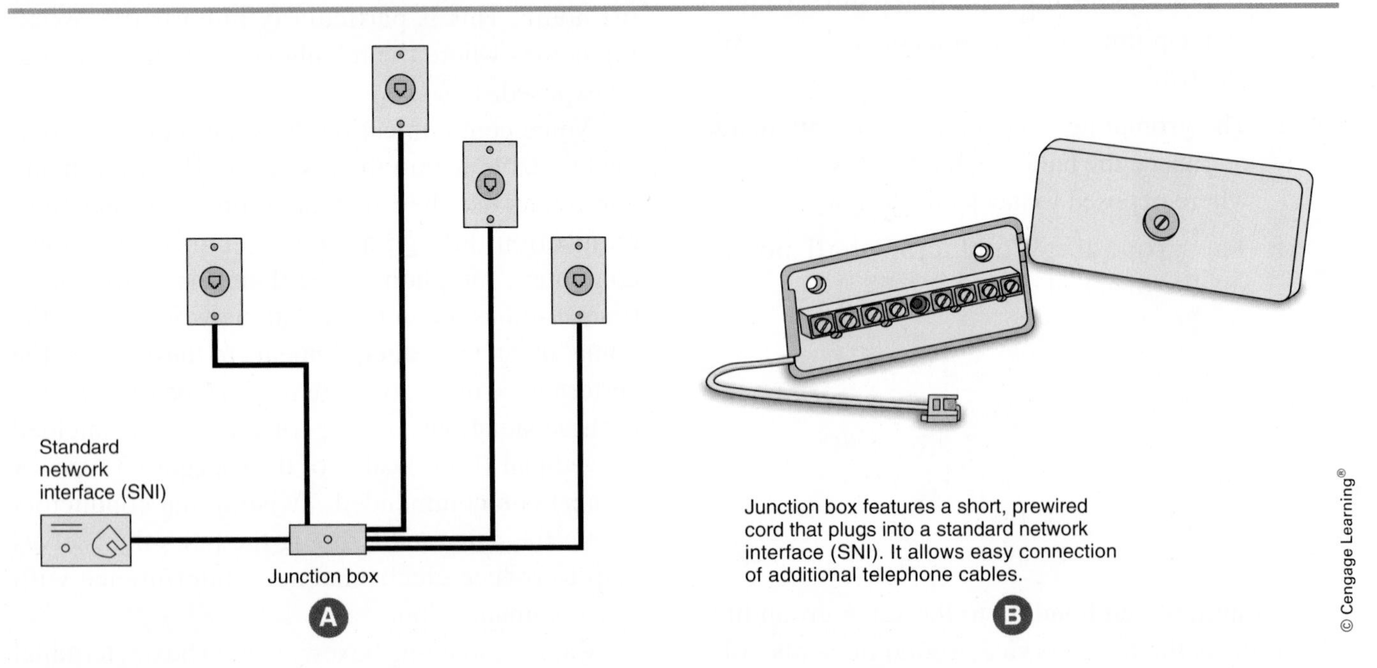

FIGURE 23-11 Individual telephone cables run to each telephone outlet from the telephone company's connection point.

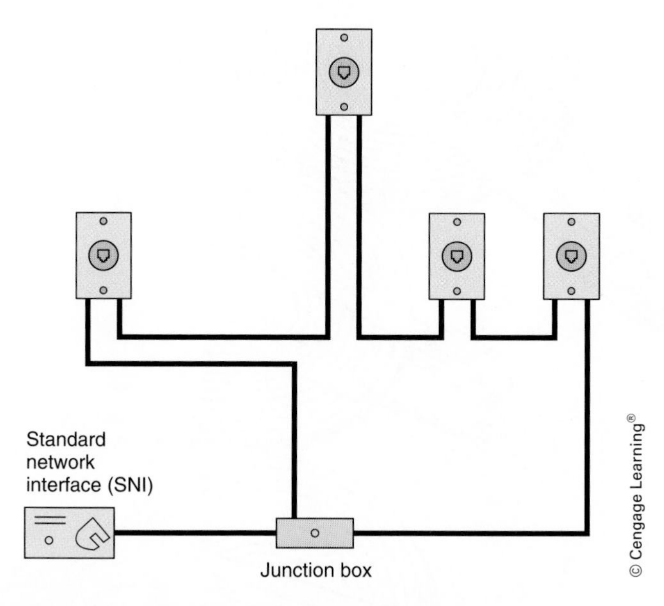

FIGURE 23-12 A complete "loop system" of the telephone cable. If something happens to one section of the cable, the circuit can be fed from the other direction.

plugs, and so on are all available through electrical distributors, builders' supply outlets, telephone stores, electronic stores, hardware stores, and various wholesale and retail outlets.

Telephone Conductors

The colour coding of telephone cables is shown in Figure 23-13. Figure 23-14 shows some types of telephone cable. The colour coding of telephone cables is also shown in Figure B–2 in Appendix B at the back of this book.

Cables are available in these sizes:

2 pair (4 conductors)
3 pair (6 conductors)
6 pair (12 conductors)
12 pair (24 conductors)
25 pair (50 conductors)
50 pair (100 conductors)

Telephone circuits require a separate pair of conductors from the telephone all the way back to the phone company's central switching centre. To minimize interference that could come from other electrical equipment, such as electric motors and fluorescent fixtures, each pair of telephone wires is twisted together.

The recommended circuit lengths are:

#24 gauge: Not over 61 m
#22 gauge: Not over 76 m

When telephones had electromechanical ringers, one line had enough power to ring up to five

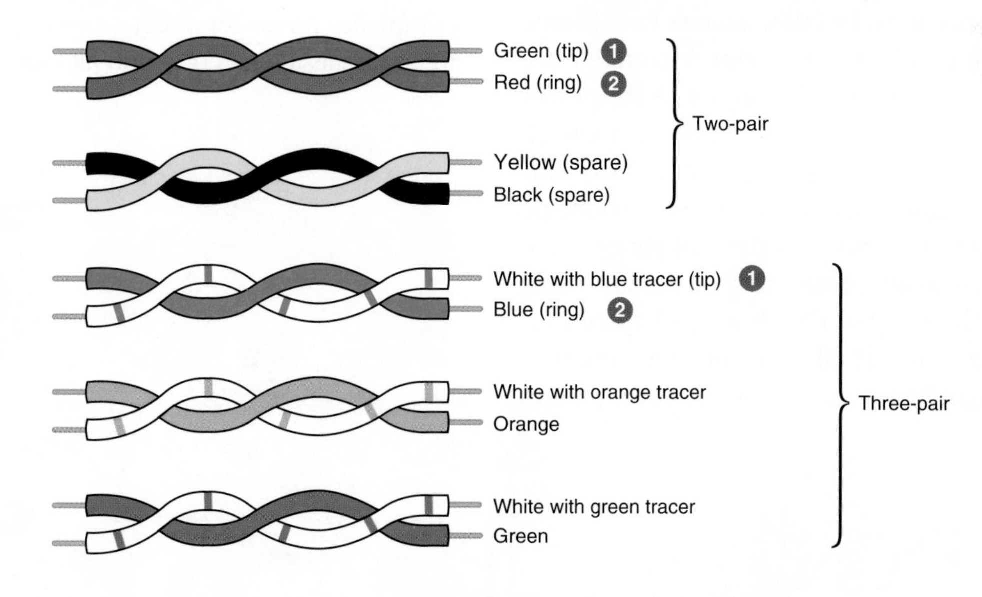

FIGURE 23-13 Colour coding of telephone cables.

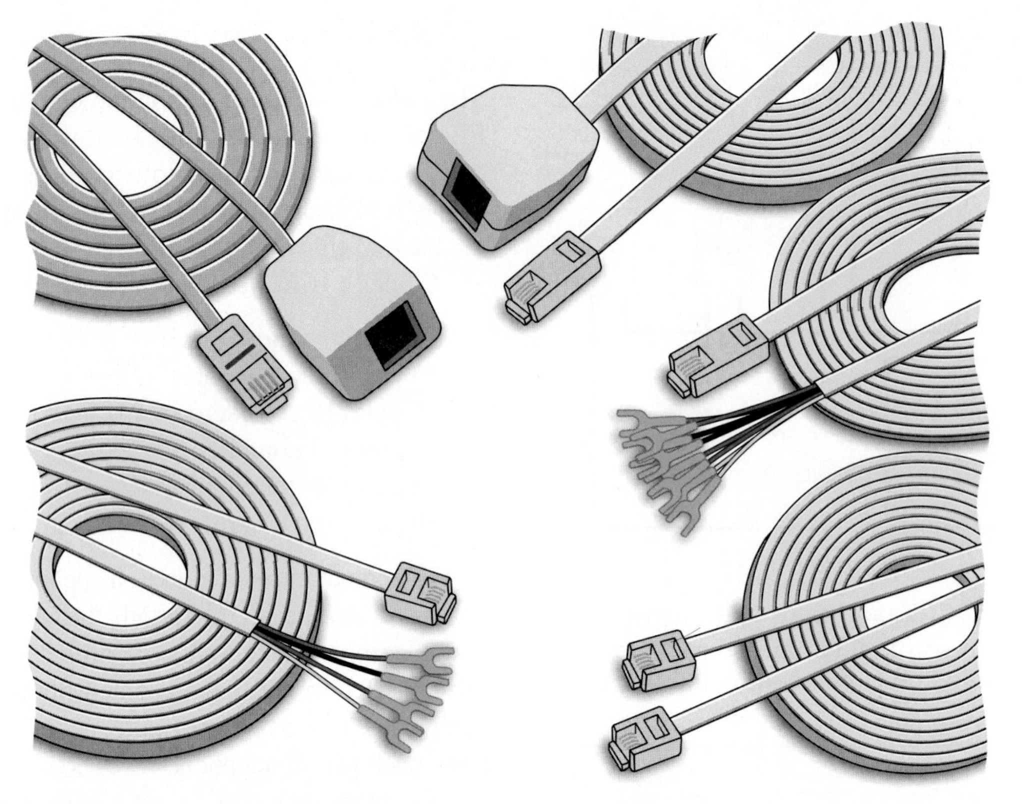

FIGURE 23-14 Telephone cables. These cords are available in round and flat configurations, depending on the number of conductors in the cable.

phones. Most telephones now have electronic ringers, which use very little power; therefore, more than five phones can be installed on one line. (Some telephones are labelled with a ringer equivalence number [REN], with one REN equal to the power required by one electromechanical ringer.) Check this with the telephone company before installation.

It is recommended that at least one telephone be permanently installed, probably the wall phone in the kitchen. The remaining phones may be portable, able to be plugged into any of the phone jacks; see Figures 23-15 and 23-16. If all of the phones were portable, there might be a time when none of the phones were connected to phone jacks. As a result there would be no audible signal. However, most telephone companies no longer direct-connect phones, resulting in all phones being cord- and jack-connected.

FIGURE 23-15 Telephone modular jacks.

FIGURE 23-16 Three styles of wall plates for modular telephone jacks: rectangular stainless steel, weatherproof for outdoor use, and circular.

The electrical plans show that nine telephone outlets are provided, one in each of these areas:

Front bedroom
Kitchen
Laundry
Living room
Master bedroom
Rear outdoor patio
Recreation room
Study/bedroom
Workshop

Installation of Telephone Conductors, *Rules 60–300* through *60–334*

Telephone conductors should be as listed below (the items below that are marked with an asterisk are recommendations from residential telephone manufacturers' installation data):

1. Shall be of a type listed in *Table 19* as suitable for communications and suitable for the installation environment
2. Shall be separated by at least 50 mm from light and power conductors operating at 300 volts or less, unless the light and power conductors are in a raceway, in non-metallic-sheathed cable, or Type AC90 cable, *Table 19*
3. Shall not be placed in any conduit or boxes with electric light and power conductors unless the conductors are separated by a partition
4. May be terminated in either metallic or non-metallic boxes—check with the local telephone company or local electrical inspector for specific requirements
5. Should not share the same stud space as the electrical branch-circuit wiring* *Rule 60–308(1)*
6. Should be kept at least 300 mm away from the electrical branch-circuit wiring where the telephone cable is run parallel to the branch-circuit wiring to guard against induction noise* *Rule 60–308(1)*
7. Should not share the same bored holes as electrical wiring, plumbing, or gas pipes
8. Should be kept away from hot-water pipes, hot-air ducts, and other heat sources that might harm the insulation* *Rule 60–302*
9. Should be carefully attached to the sides of joists and studs with insulated staples* *Rule 60–302*

*These are recommendations from residential telephone manufacturers' installation data.

Conduit

In some communities, electricians prefer to install regular device boxes at a telephone outlet location, then "stub" a 27-mm conduit to the basement or attic from this box. At a later date, the telephone cable can be "fished" through the conduit.

Grounding (*Rule 60–202*)

The telephone company will provide the proper grounding of its incoming cable sheath and protector, generally using wire not smaller than #14 AWG copper or equal. Or the company might clamp the protector to the grounded metal service raceway conduit to establish ground. The grounding requirements are practically the same as for CATV and antennas, which were covered earlier in this unit.

Safety

Open-circuit voltage between telephone conductors of an idle pair can range from 50 to 60 volts direct current (DC). The ringing voltage can reach 90 volts alternating current (AC) at 20 cycles per second. Therefore, always work carefully with insulated tools and stay clear of bare terminals and grounded surfaces. Disconnect the interior telephone wiring if you must work on the circuit, or take the phone off the hook, in which case the DC voltage level will drop and there should be no AC ringing voltage delivered.

DATA SYSTEMS

As more families enter the world of computers and the Internet, they quickly discover the usefulness of this technology. It provides the opportunity to find information on nearly any subject to engage in multiplayer computer games, and to buy or sell just about anything. Many families have more than one computer and opt to install a home data network similar to those installed in small businesses, which allows all of the computers to share peripheral devices such as drives, printers, and scanners, as well as Internet access and files.

Courtesy of Honeywell Inc.

FIGURE 23-17 Typical bundled cables showing both coaxial and Category 5e cables.

Data Rate

Many forms of networks have been used over the years. The most popular of these is Ethernet, using UTP cable (Figure 23-17). Ethernet began with a data rate of 10 megabits per second (Mbps), which is quite adequate for most residential applications.

Ethernet expanded to a data rate of 100 Mbps and is now up to 1000 Mbps, or 1 Gbps (gigabits per second). Equipment for gigabit networks is still quite expensive, while 100-Mbps equipment is only slightly more expensive than 10-Mbps equipment.

We recommend 100-Mbps equipment as a good compromise between communication speed and affordability. The equipment is not permanently installed and can be easily, although not necessarily inexpensively, replaced. The wiring, which is inside the walls, is considerably more difficult to replace when it becomes inadequate.

Data Cable Type

Category 5 cable has been the standard for Ethernet applications for many years, providing data communication rates of 100 Mbps. Category 5e (Cat 5e), or Category 5 enhanced, cable can support data rates up to 1 Gbps. Cat 6, which supports 1 Gbps but at a higher frequency, allows for greater efficiency in the network. Cat 6 is recommended for all new business applications but is considered excessive for most residential networks.

Category 5e cabling is recommended for these reasons:

1. Cat 5e cable and terminations are only slightly more expensive than Category 5.
2. Cat 5e cabling and terminations are relatively simple to install. Cat 6 terminations require a higher level of expertise to install successfully.
3. Affordable network interface cards (NIC) for the computers and network switches (Figure 23-18) or hubs can be used. These devices can be either 10-Mbps or 100-Mbps devices.

Eugene Shapovalov/Shutterstock.com

FIGURE 23-18 A popular choice for the home network. This router includes a four-port 10/100-Mbps switch, a print server for printer sharing, and a WAN port for connection to the Internet by cable or ADSL.

4. If the consumer later needs a 1000-Mbps network, the installed cable and terminations can support the higher speed.

Installation of Data Cable

There are only a few basic rules to remember for a successful installation of a computer data network:

1. All cable and terminations must be of the same type: Cat 5e cable requires Cat 5e terminations. If cable and terminations of different categories are used, the system will perform to only the lowest category standard.
2. The maximum pulling force applied to a data cable is 11 kg.
3. When installing UTP cable, maintain a separation of 300 mm from the cable and any electrical equipment whenever possible.
4. A separation of 300 mm must be maintained between the data cable and ballasts of any type of fixtures, including fluorescent and transformers of low-voltage spotlights and track lights.
5. Data cable may be supported in many ways, including straps and plastic ties. Take care to not deform the cable when installing supports. If the supports are too tight, the twisting of the conductors inside the cable will be altered and the performance of the system will decrease.
6. The minimum radius of the bend of data cable is to be not less than 25 mm. A smaller bend radius can damage the conductors or insulation and alter the twist rates of the conductors or pairs.

Installation of Termination Jacks

Jacks for data network installations are available in many forms from many manufacturers (Figure 23-19). Some require the use of a punchdown tool, shown in Figure 23-20, while others snap together. Jacks are available as both surface and flush-mount. All jacks are clearly identified as to the colour coding for termination. Two configurations are used for termination: 568A is the most common in Canada and 568B is the most common in the United States. Either configuration is acceptable if you use the same one throughout the network. Some manufacturers make jacks that are specifically 568A or 568B: Be sure to order carefully and check that they are all the same when the order is received. Other manufacturers make jacks that can be used for either configuration and provide both colour coding layouts.

Courtesy of Sandy F. Gerolimon

FIGURE 23-19 Category 5e data jacks.

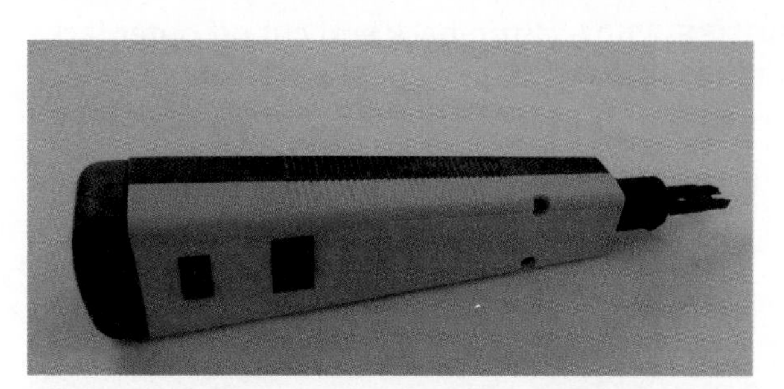

Courtesy of Sandy F. Gerolimon

FIGURE 23-20 Punchdown tools.

When terminating the conductor to the jack, ensure that no more than 12 mm of the conductor is exposed from the jacket. Each pair of conductors in UTP is twisted together at different numbers of twists per unit length. The four twisted pairs are then twisted together. No more twisting shall be removed than is necessary to complete the termination. Do not alter the twist rates of the pairs in any way. Figure 23-24 demonstrates the use of a punchdown tool to terminate a jack, while Figures 23-21 through 23-25 show a detailed installation of a modular type of snap-together jack.

Wireless Data Networks

Wireless data networks are now affordable for the average consumer. The equipment used in

Courtesy of Sandy F. Gerolimon

FIGURE 23-21 Strip back and cut off outer jacket and nylon cord.

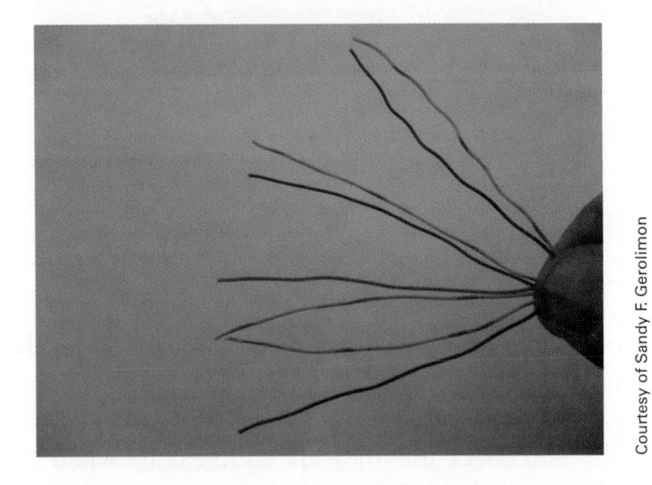

Courtesy of Sandy F. Gerolimon

FIGURE 23-22 Fan out individual conductors.

Courtesy of Sandy F. Gerolimon

FIGURE 23-23 Lay conductors in modular wiring jig and cut off excess.

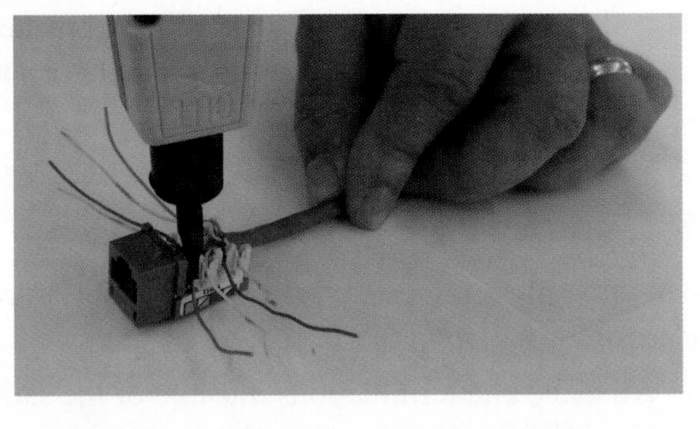

Courtesy of Sandy F. Gerolimon

FIGURE 23-24 Termination of punchdown-type data jack.

Courtesy of Sandy F. Gerolimon

FIGURE 23-25 Assemble wiring jig and jack body.

wireless systems, such as routers and wireless network adapters, are only slightly more expensive than their wired counterparts. However, the up-front cost of the installation of the network cabling can be eliminated.

With a wired network system, the consumer is restricted to a location with the computer. Relocating the computer may require installing additional cabling and network jacks or using a long network cable to connect to an existing jack. Most consumers do not want the long cables lying on their floors because of the appearance and the trip hazard. Wireless networks allow the consumers to locate their computers anywhere they like.

The advantages of wireless networks are

- Increased flexibility when locating computers and peripherals such as printers.

- Portability when using laptop or portable computing devices, even into the yard surrounding the building.
- No wiring costs.
- Ease of adding devices to the network.

Its disadvantages are

- Equipment is slightly more expensive than comparable wired equipment.
- The network can be tapped into by outside computers. It is **essential** that the security features supplied with the equipment be used.
- Wireless systems generally have a slower data transfer rate than wired systems; however, new technologies allow connection speeds as high as 108 Mbps.
- Connection speed is affected by the signal strength of the transmitter: It drops as the distance from the transmitter increases.
- Network connectivity is affected by building construction. When heating system ductwork, refrigerators, or other metal objects are located between the user and the wireless transmitter, the metal will cast a "shadow" that makes wireless connection not possible. Wood frame construction is generally better than concrete or steel for the use of wireless networking.
- Connectivity is affected by other systems operating in the same wireless frequency range, such as microwave ovens and wireless telephones.

Wireless routers are small in size. A typical wireless router is shown on Figure 23-26.

FIGURE 23-26 Wireless router.

EIA/TIA 568A & 568B STANDARD

There are two accepted standards for eight-wire data network jacks (commonly and incorrectly called *RJ-45*). See Appendix C for wire colour pin layout.

ANSI/TIA/EIA-568-B "Commercial Building Telecommunications Cabling Standard" lists both wiring configurations. T568B is the most prevalent for commercial installations, and was used by AT&T for the original Merlin phone systems. To help you remember, associate "B" with "Bell."

ANSI/TIA/EIA-570-B "Residential Telecommunications Cabling Standards" recommends T568A:

- If the installation is residential, choose T568A unless other conditions apply. The two inner pairs of 568A are wired the same as a two-line phone jack.
- If there is pre-existing voice/data wiring (remodel, moves, adds, changes), duplicate this wiring scheme on any new connection.
- If project specifications are available, use the specified wiring configuration.
- If components used within the project are internally wired for either T568A or T568B, use that wiring scheme.
- Make sure both ends of a cable are wired the same way.

Cable terminal pin layout connections for EIA/TIA 568A and 568B are shown in Figure 23-27. A 10Base-T crossover cable is represented in Figure 23-28.

Crossover Cable	
RJ-45 Pin	**RJ-45 Pin**
1 Rx+	3 Tx+
2 Rc−	6 Tx−
3 Tx+	1 Rc+
6 Tx−	2 Rc−
568A	568B

A

Straight-Through Cable	
RJ-45 Pin	**RJ-45 Pin**
1 Tx+	1 Rc+
2 Tc−	2 Rc−
3 Rx+	3 Tx+
6 Rc−	6 Tx−

B

A Shows the crossover cable pin assignments.

B Shows the straight-through cable pin assignments.

FIGURE 23-27 Pin Layout for EIA/TIA 568A and 568B.

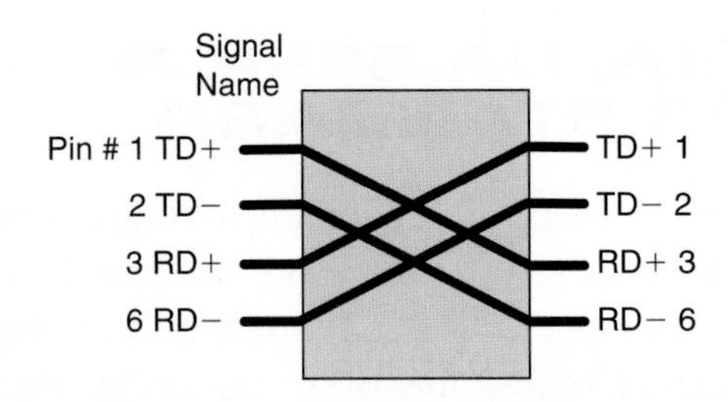

FIGURE 23-28 A 10Base-T crossover cable.

HOME AUTOMATION

The X10 System

In the home automation field, X10 has been around the longest. X10 technology was invented in Scotland in 1975. The original X10 patent expired in 1997, so it is now an open standard for any manufacturers to apply to their products. X10 technology is simple and is the least costly home automation system, especially for retrofit wiring in existing homes. For the most part, special wiring is not required for basic functions. For many applications, all that is necessary is to plug in the various components as needed.

This system transmits carrier wave (120 kilohertz [kHz]) signals that are superimposed on the regular 120/240-volt, 60-cycle branch-circuit wiring in the home. Components communicate with each other over the existing electrical wiring without a central controller.

Transmitting devices take a small amount of power from the 120-volt power line, modulate or change it to a higher frequency, and then superimpose this high-frequency signal back onto the AC circuit. At the other end, a receiving device responds to the high-frequency burst. X10 is basically a one-way communications system. Some devices "talk" and others "listen."

The control of appliances, audio/video equipment, outdoor and indoor lighting, and receptacles, as well as the arming of a security system, is possible from just about any place in the home. You might call it a "plug-and-play" system. Components are easily added as needed.

An X10 system might include programmers; receiving modules (Figure 23-29); transmitters; tabletop controller (Figure 23-30); dimmers; single-pole, three-way, four-way, and double-pole switches; on/off/dim lamp modules; on/off appliance modules;

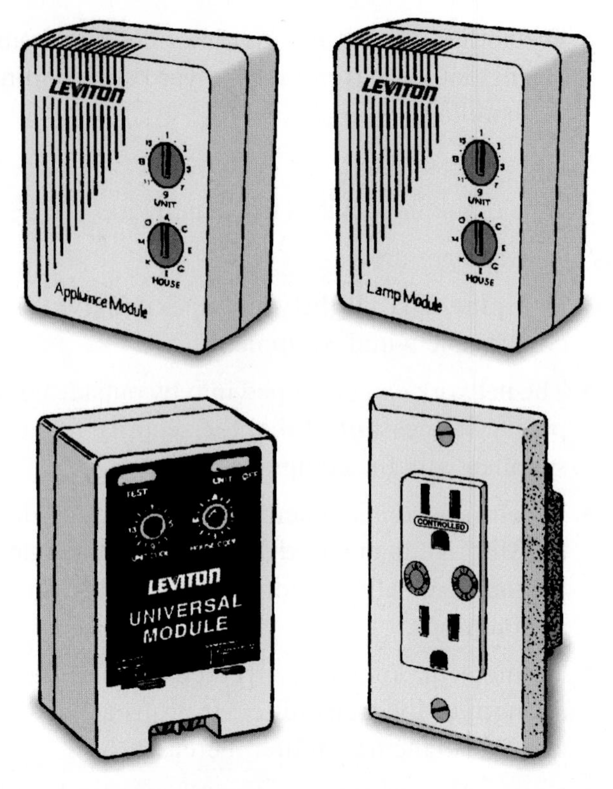

FIGURE 23-29 Various types of typical receiver modules used with X10 systems.

FIGURE 23-30 A typical X10 tabletop keypad controller makes it easy to transmit preprogrammed signals to receiver modules.

receptacle modules; thermostats; timers; surge protection devices; burglar alarm devices; motion sensors (turn on lights, sound an alarm); wireless controllers; handheld remotes; photocells for light/darkness sensing; drapery controllers; telephone responders; tie-in with a personal computer; voice activation; and so on.

A typical system of this type will have the capability of 256 easily adjusted different "addresses,"

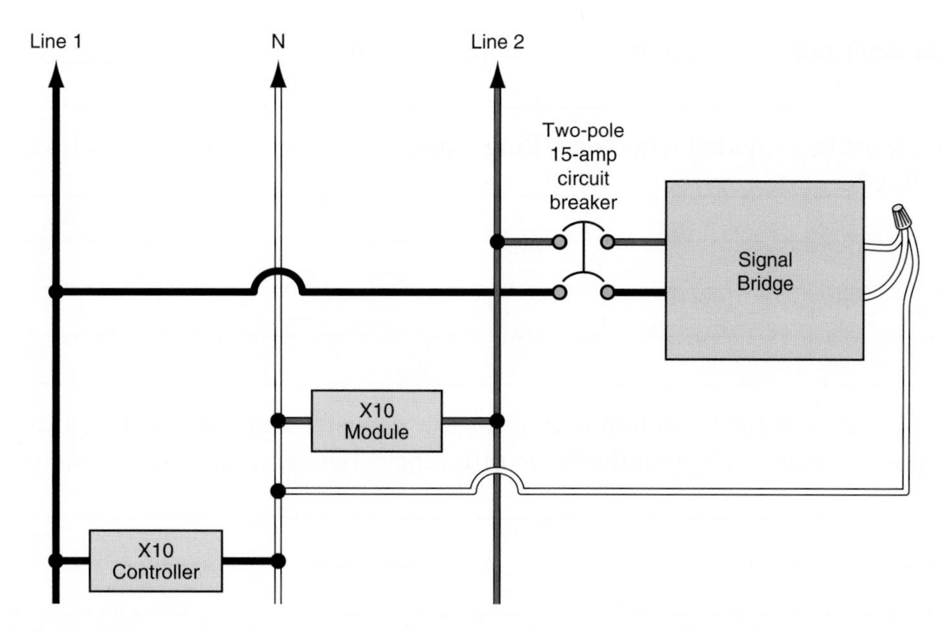

FIGURE 23-31 A signal bridge phase coupler connected to the main panel to ensure proper signals on both sides of an AC line.

more than adequate for just about any size of home. Receivers and switches generally mount in standard wall boxes. X10 wall switches and receptacles can replace conventional switches and receptacles. Make sure that the wall box has ample space for the larger X10 devices so the wires in the wall box will not be jammed.

To ensure trouble-free operation, a phase coupler (Figure 23-31) connected at the main panel is used to provide the proper strength of command signals to both sides of the AC wiring.

Limitations of X10 Technology. X10 modules are susceptible to interference from an AC power line. Many things can generate "noise" on 120-volt power lines, including fluorescent luminaire ballasts, appliances, and wireless intercoms. And because standard home wiring (most often Type NM cable) is unshielded, it can pick up unintended induced signals in the same way that an antenna does, even from devices that are not attached to the same circuit.

In fact, when two or more houses are connected to the secondary of the same utility transformer, signals could be transmitted from one house to another. X10 filters are available that install at the service panel to block interference from outside the house. X10 signals are relatively slow. The maximum data rate is about 60 bits per second (bps), compared to systems that handle 10 000 bps. This means there can be a noticeable delay between the time you activate a control and the time that a controlled action takes place.

REVIEW

Note: Refer to the *CEC* or the blueprints provided with this textbook when necessary. Where applicable, responses should be written in complete sentences.

TELEVISION CIRCUIT

1. How many television outlets are installed in this residence? ______________________
2. What types of boxes are recommended when unshielded lead-in wire is used?

____

3. What determines the design of the faceplates used? ______________________________

4. What must be provided when installing a television outlet and receptacle outlet in one wall box? ______________________________

5. Which system is more economical to install: a master amplifier distribution system or a multiset coupler? Explain the basic differences between these two systems.

6. What precautions should you take when installing a television antenna and its lead-in wires? ______________________________

7. List the requirements for cable television inside the house. ______________________________

8. What section of the *CEC* explains the requirements for cable television antenna installation? ______________________________

9. Grounding and bonding together all metal parts of an electrical system and the metal shield of the cable television cable to the same grounding reference point in a residence will keep both systems at the same voltage level should a surge, such as lightning, occur. If the CATV company installer grounds the metal shield of the incoming cable (to a house) to a driven ground rod, does this installation conform to the *CEC?*

10. The basics of cable television are that a transmitter on earth sends a(n) ______________ signal to a(n) ______________ where the signal is reamplified and sent back to earth via the ______________ signal. This signal is picked up by the ______________ on the satellite antenna, which is sometimes referred to as a(n) ______________. The feed-horn funnels the signal into a(n) ______________, then onto a(n) ______________, where the high-frequency signal is converted into a(n) ______________-frequency signal suitable for the television receiver.
11. [Circle the correct answer.] All television satellites revolve above the earth in (the same orbit) (different orbits).
12. [Circle the correct answer.] Television satellites are set into orbit (16 093) (28 970) (36 532) km above the earth, which results in their revolving around the earth at (precisely the same) (different) rotational speed as/from the earth. This is done so that the satellite dish can be focused on a specific satellite (once) (one time each month) (whenever the television set is used).

TELEPHONE SYSTEM

1. How many locations are provided for telephones in the residence? ______________
____
2. At what height are the telephone outlets mounted? ______________
3. Sketch the symbol for a telephone outlet.

4. Is the telephone system regulated by the *CEC?* ______________
5. a. Who is to furnish the outlet boxes required at each telephone outlet? ______________
____
   b. Who is to furnish the faceplates? ______________
6. Who is to furnish the telephones? ______________
7. Who does the actual installation of the telephone equipment? ______________
____
8. How are the telephone cables concealed in this residence? ______________
____
____
____

9. The point where the telephone company's cable enters the residence and the interior telephone cable wiring meet is called the ______________________ point.
10. What are the colours contained in a four-conductor telephone cable assembly and what are they used for? ______________________
11. List the *CEC* rules for the installation of telephone cables in a residence.
12. If finger contact was made between the red conductor and green conductor at the instant a "ring" occurs, what voltage of shock would be felt? ______________________

DATA SYSTEMS

1. What is the data rate for Ethernet networks as discussed in this chapter?
2. What precautions must be taken when installing data cable?
3. Which termination configuration is most commonly used in Canada?
4. What is the maximum data rate of a Category 5e network?
5. What is the maximum pulling force that may be exerted on a data cable?
6. When supporting data cable, why is it important not to strap or tie the cable too tightly?
7. What is the minimum radius of the bend of data cable and why?

8. List the factors that affect the connectivity and data transfer rate of wireless network systems.

9. What other equipment can affect a wireless data network?

HOME AUTOMATION

1. What systems can be controlled using an X10-based home automation system?

2. Is additional wiring between the controller and controlled device required? Explain why or why not.

3. If the home is very large, what additional device may need to be installed?

4. With the alphanumeric addressing system, how many unique devices can be controlled?

UNIT 24

Lighting Branch Circuit for the Garage and Outdoor Lighting

OBJECTIVES

After studying this unit, you should be able to

- understand the fundamentals of providing proper lighting in residential garages
- understand the *Canadian Electrical Code (CEC)* requirements for underground wiring, with either conduit or underground cable
- complete the garage circuit wiring diagram
- discuss typical outdoor lighting, including *CEC* requirements
- describe how conduits and cables are brought through cement foundations to serve loads outside the building structure
- understand the application of Ground-Fault Circuit Interrupter (GFCI) protection for loads fed by underground wiring
- make a proper installation for a residential overhead garage door opener
- make a proper installation for a residential central vacuum system

LIGHTING BRANCH CIRCUIT

Circuit B14 originates at Panel B in the recreation room and is brought into the wall receptacle box on the rear wall of the garage. Figure 24-1 shows the cable layout for Circuit B14. From this point, the circuit is carried to the switch box adjacent to the side door of the garage. From there, the circuit conductors plus the switch leg are carried upward to the ceiling fixtures and then to the wall receptacle outlets on the right-hand wall of the garage.

The garage ceiling's porcelain lampholders are controlled from switches located at all three entrances to the garage.

The post light feed from Circuit A15 is controlled by a switch in the front entrance, while the post light turns on and off by a photocell, an integral part of the light.

The fixtures on the outside next to the overhead garage door are controlled by two three-way switches, one just inside the overhead door and the other at the door entering the service entrance. These fixtures are not connected to the garage lighting circuit.

All of the wiring in the garage will be concealed because the walls and ceiling are to be covered with ¾-hour fire-rated drywall.

Table 24-1 summarizes the outlets and the estimated load for the garage circuit.

TABLE 24-1

Garage outlet count and estimated load (Circuit B14).

DESCRIPTION	QUANTITY	WATTS	VOLT-AMPERES
Receptacles @ 180 W each	3	540	540
Ceiling lampholders @ 100 W each	3	300	300
Outdoor garage bracket luminaire (rear)	1	100	100
Post light	1	100	100
Totals	8	1040	1040

LIGHTING A RESIDENTIAL GARAGE

To provide adequate lighting in a residential garage, you should install a light fixture on the ceiling above each side of an automobile (Figure 24-2).

Figure 24-3 shows typical lampholders commonly mounted in garages.

- For a one-car garage, a minimum of two fixtures is recommended.
- For a two-car garage, a minimum of three fixtures is recommended.
- For a three-car garage, a minimum of four fixtures is recommended.

Lighting fixtures arranged this way eliminate shadows between automobiles, which can hide objects that might cause a person to trip or fall.

These ceiling fixtures should be mounted toward the front end of the automobile as it is usually parked in the garage. This provides better lighting where it is most needed. Do not place lights where they will be covered by the open overhead door.

RECEPTACLE OUTLETS

At least one receptacle must be installed for each car space in an attached residential garage or carport, *Rule 26–714(b)*. The garage in this residence has three receptacles, and if standard receptacles are installed, they must be of a tamper-resistant nature to comply with *Rule 26–712(g)*. These must be on a separate circuit, but the garage door opener and the lighting fixtures for the garage may be on the same circuit, *Rule 26–726(b)*. When electricity is provided in a detached garage, all pertinent *CEC* rules become effective and must be followed. These include rules relating to grounding, lighting outlets, receptacle outlets, GFCI protection, and so on. If the home owner requests electric vehicle-charging equipment, a 5-20R receptacle will be needed, as specified in *Rule 86–306*.

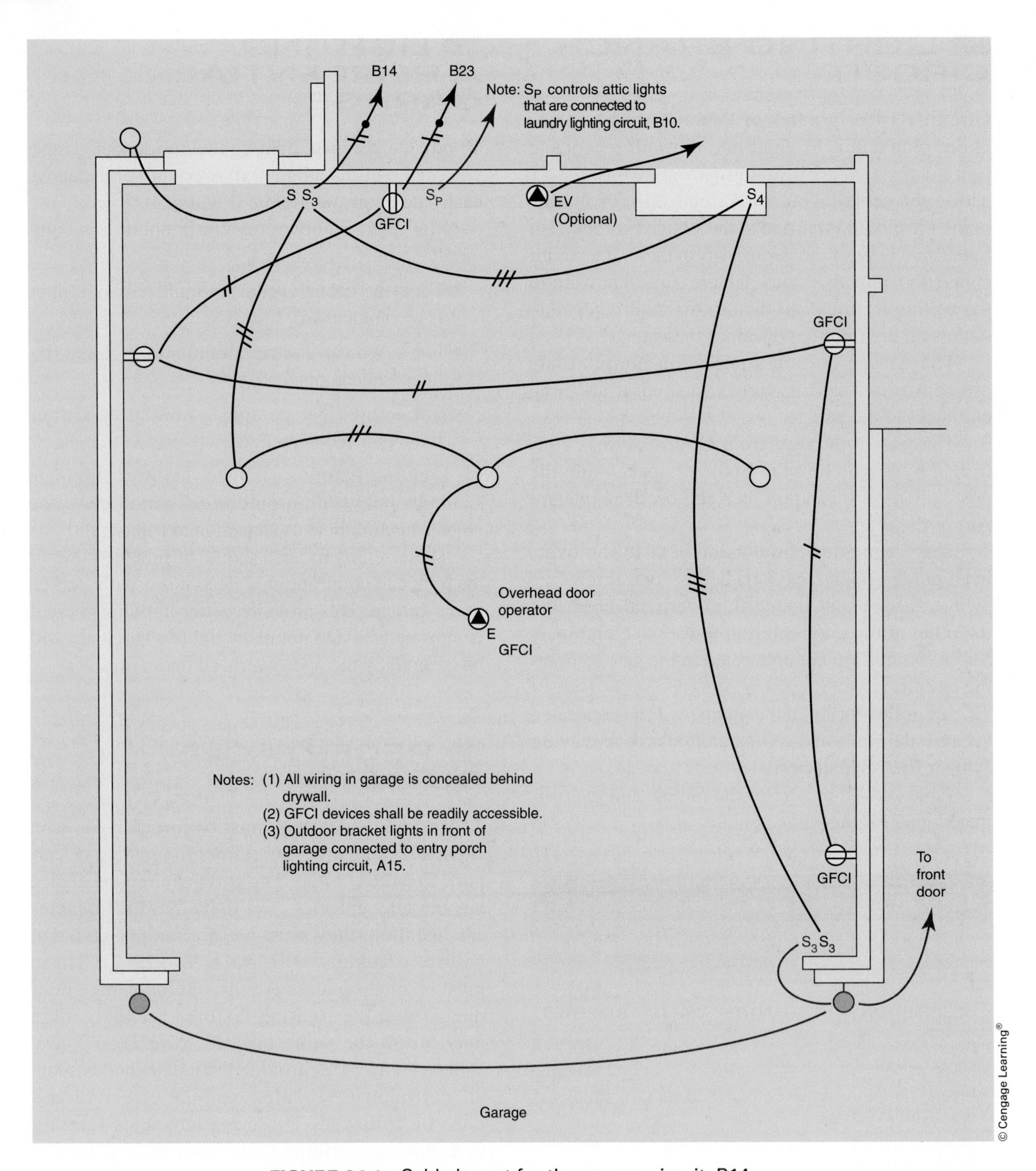

FIGURE 24-1 Cable layout for the garage circuit, B14.

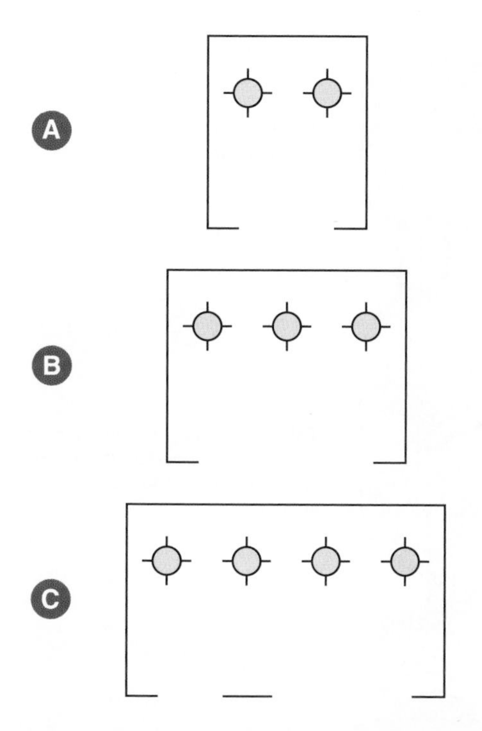

FIGURE 24-2 Positioning lights in (A) a one-car garage, (B) a two-car garage, and (C) a three-car garage.

OUTDOOR WIRING

If the homeowner requests that 120-volt receptacles be installed away from the building structure, the electrician can provide weatherproof receptacle outlets; see Figure 24-4. These must be GFCI-protected if within 2.5 m of ground level, *Rule 26–710(n)*.

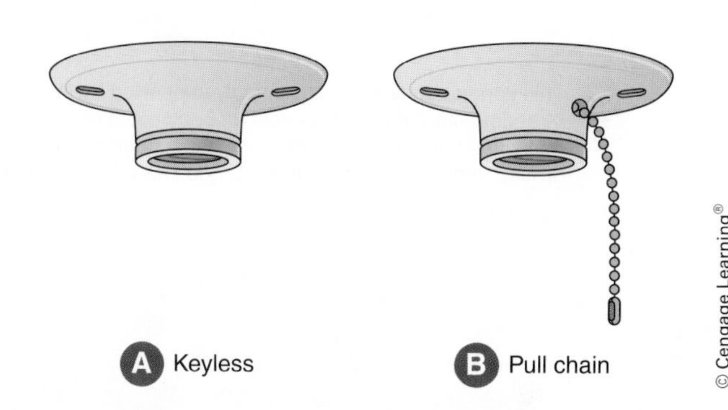

FIGURE 24-3 Typical plastic lampholders of the type installed in garages, attics, basements, crawl spaces, and similar locations.

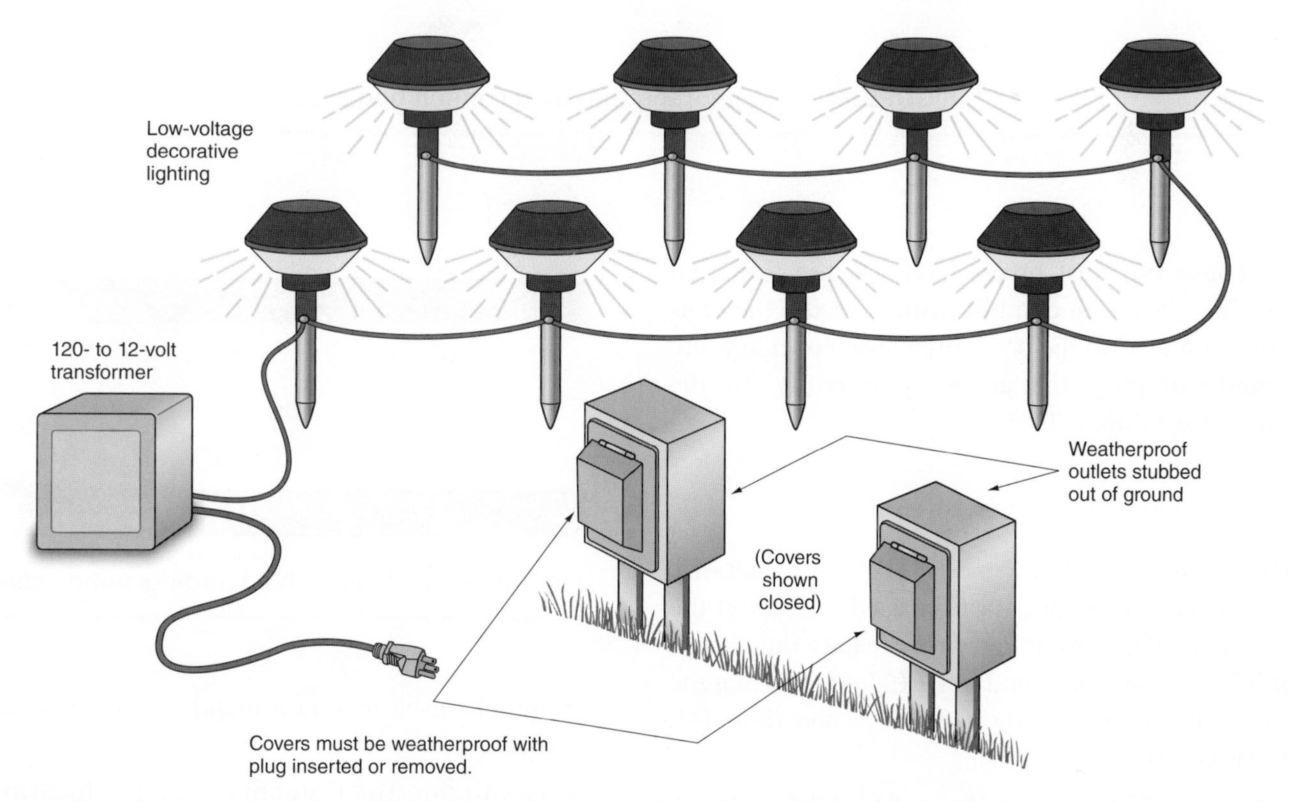

FIGURE 24-4 Weatherproof receptacle outlets stubbed out of the ground: Either low-voltage decorative lighting or 120-volt PAR lighting fixtures may be plugged into these outlets.

FIGURE 24-5 Supporting threaded conduit bodies.

These types of boxes have 16-mm threaded openings in which conduit fittings secure the conduit "stub-ups" to the box. Any unused openings are closed with plugs that are screwed tightly into the threads; see Figure 24-5.

Wiring with Type NMWU Cable

The plans show that Type NMWU non-metallic sheathed cable (Figure 24-6) is used to connect the post light. The current-carrying capacity (ampacity) of NMWU cable is found in *CEC Table 2*, using the 60°C column. According to *Table 19* and *Table D1*, NMWU cable

- is available in sizes from #14 AWG through #2 (for copper conductors) (see *Table 2* for the ampacity ratings)
- may be used with non-metallic-sheathed cable fittings
- is suitable for Category 1 and 2 locations, *Table 19, Section 22*
- is moisture- and fungus-resistant

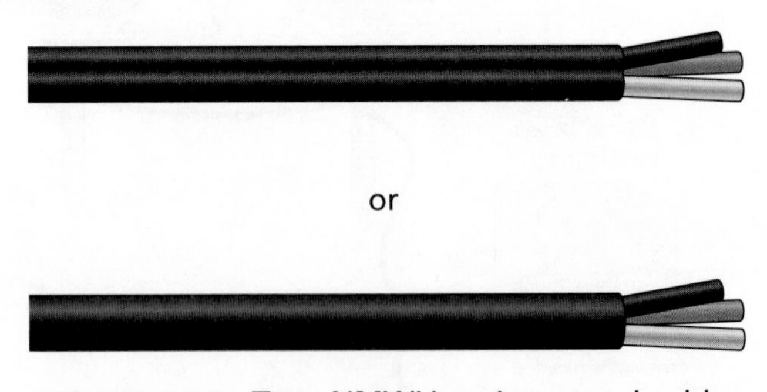

FIGURE 24-6 Type NMWU underground cable.

- may be buried directly in the earth
- may be used for interior wiring in wet, dry, or corrosive areas
- is installed using the methods for non-metallic-sheathed cable, *Rule 12–500*
- must not be used as service entrance cable
- must not be embedded in concrete, cement, or aggregate

The grounding of equipment fed by Type NMWU cable is accomplished by properly connecting the bare bonding conductor in the cable to the equipment to be grounded; see Figure 24-7.

UNDERGROUND WIRING

Underground wiring is common in such residential applications as decorative landscape lighting and wiring to post lamps and detached buildings such as garages or tool sheds.

Table 53 shows the minimum depths required for the various types of wiring methods; see Figures 24-8A and 24-8B.

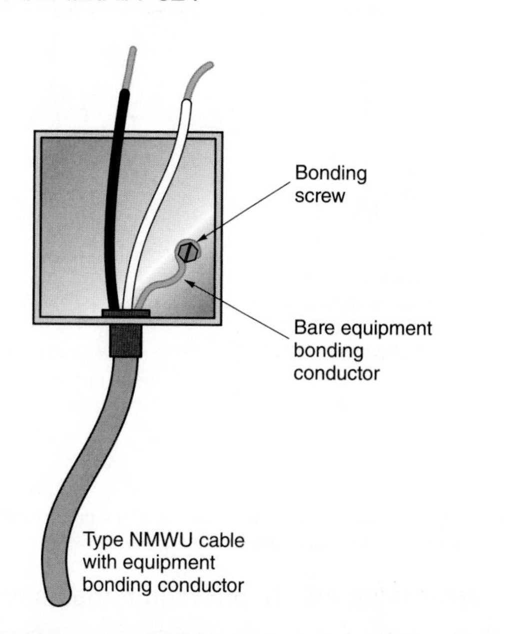

FIGURE 24-7 How the bare equipment bonding conductor of a Type NMWU cable is used to bond the metal box.

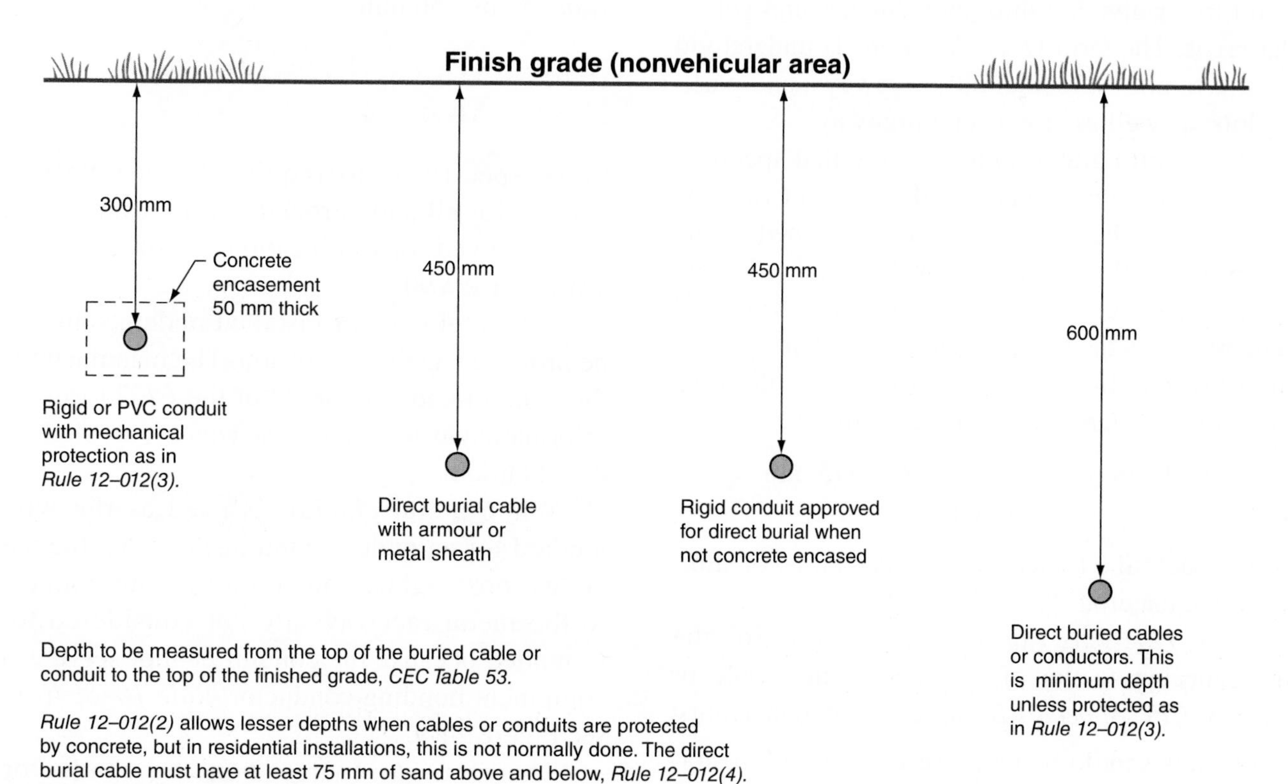

FIGURE 24-8A Minimum depths for cables and conduits installed underground. See *CEC Table 53* for depths for other conditions.

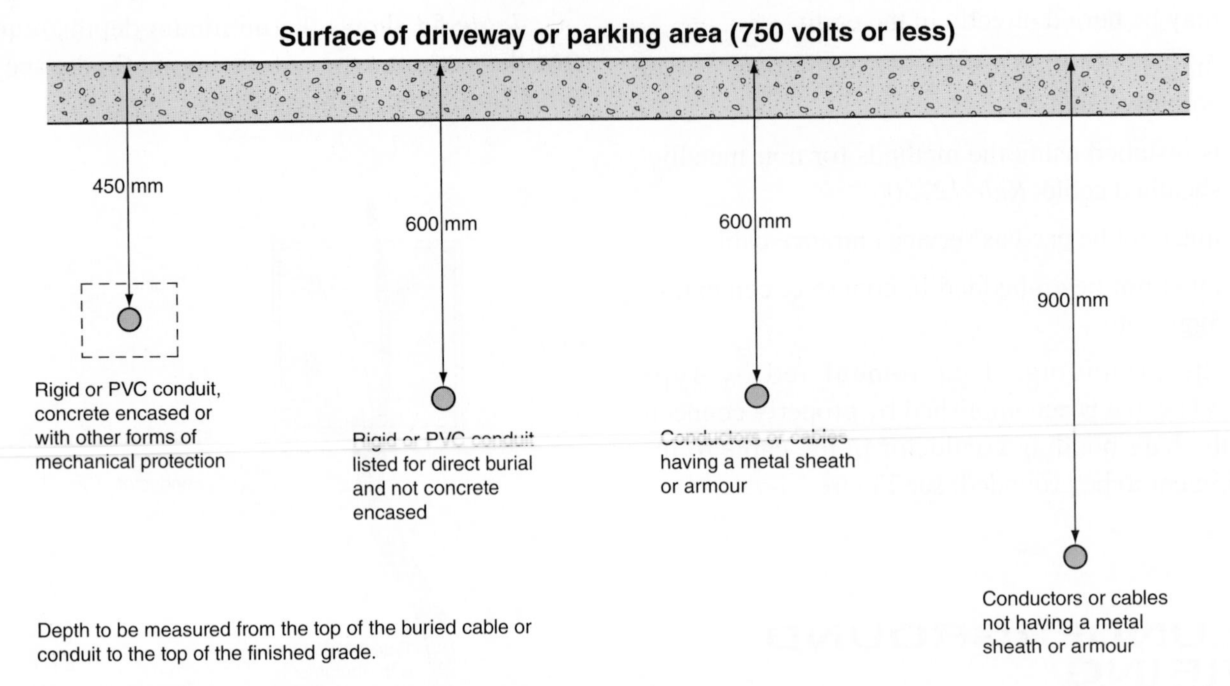

FIGURE 24-8B Depth requirements under residential driveways and parking areas, *Rule 12–012.*

The *CEC* table specifies the minimum cover in millimetres for the three approved wiring methods, with further division into nonvehicular and vehicular areas. The term *vehicular areas* is understood to refer to alleys, residential driveways, and parking lots, as well as streets and highways.

Pay careful attention to the note that specifies that *minimum cover* means the distance between the top surface of the conductor and the finished grade. For example, the basic rule for direct buried cable is that the cover must be a minimum of 600 mm in depth. Thus the trench must have a depth of 600 mm (cover), plus the diameter of the cable, plus 75 mm for screened sand or earth, *Table 53.*

$$\begin{aligned}&600\text{ mm} + \text{Cable diameter} + 75\text{ mm}\\&= \text{Required depth}\end{aligned}$$

Do not backfill a trench with rocks, debris, or similar coarse material, *Rule 12–012(10).*

Rule 12–012(3) describes methods for the mechanical protection of buried conductor, cable, or raceway. *CEC Appendix B* adds two important points:

1. Planks should be pressure-treated with pentachlorophenol if used to provide protection.
2. Black poly water pipe is acceptable for mechanical protection, *Rule 12–012(3)(e).*

Rule 12–012(2) indicates that approved mechanical protection can reduce the cover requirements of *Table 53* by 150 mm.

Installation of Conduit Underground

Some specifications require the installation of conduit for all underground wiring, using conductors approved for wet locations, such as Type TW, TW75, or RW90.

All metal conduit installed underground must be protected against corrosion. The manufacturer of the conduit and *Section 12* of the *CEC* furnish the information as to whether the conduit is suitable for direct burial.

When metal conduit is used as the wiring method for an underground installation, the metal boxes, post lights, and so on that are connected to the metal raceways are not considered to be grounded because the conduit cannot serve as the equipment bonding conductor, *Rule 10–804(c)(ii);* see Figure 24-9.

Therefore, a separate bonding conductor is required when either metallic or non-metallic raceways are installed. The bonding conductor may be bare if the length of the run does not exceed

15 m and has no more than two quarter bends, or is insulated, *Rule 10–808(3)*. When an insulated conductor is used, it must be green or green with a yellow stripe. The size of a bonding conductor is determined from the ampacity of the circuit conductors and *Table 16A;* Figure 24-10 and *Rule 10–814*.

Figure 24-11 shows two methods of bringing the conduit into the basement: (1) It can be run

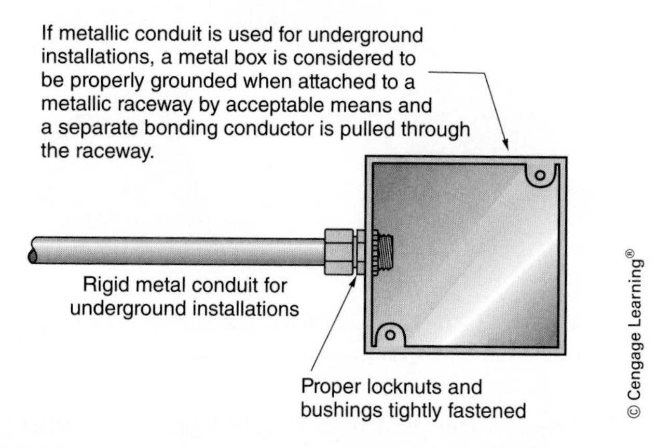

FIGURE 24-9 For direct earth burial, the *CEC* requires that a separate bonding conductor be installed in all rigid metal conduits, *Rule 10–804(c)*.

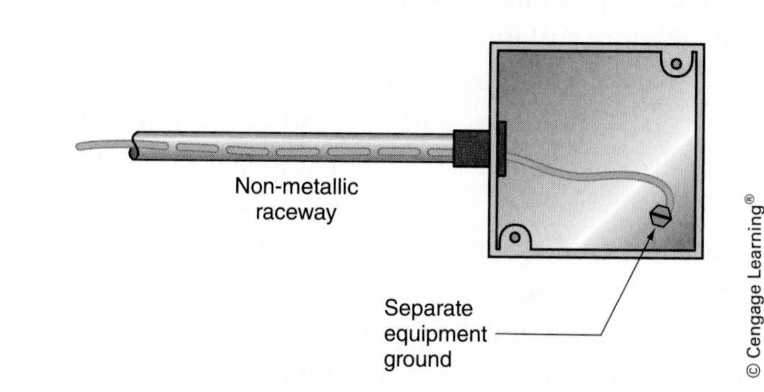

FIGURE 24-10 Bonding a metal box with a separate equipment bonding conductor when PVC conduit is used, *Rules 10–404(1)* and *10–814(1)*.

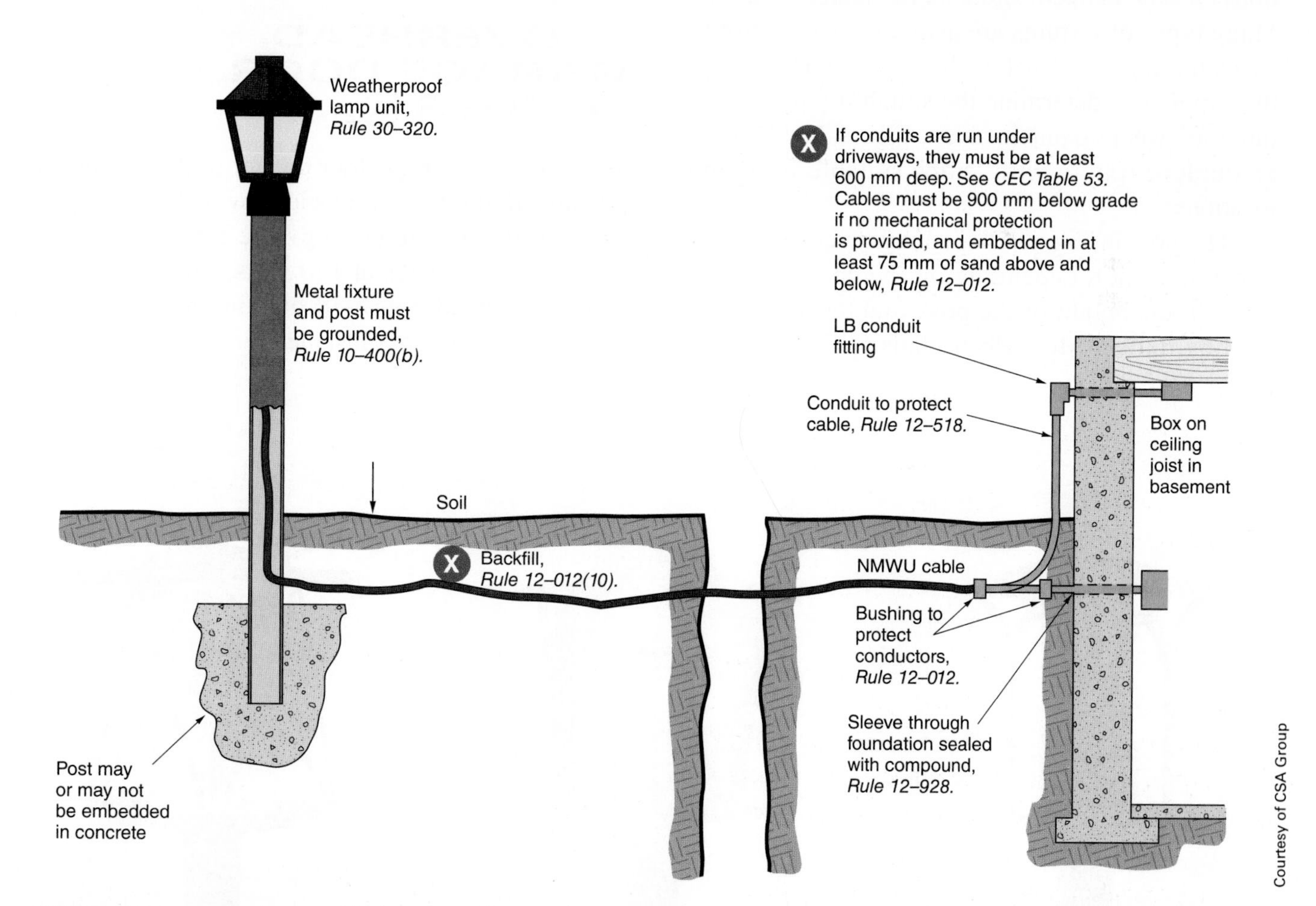

FIGURE 24-11 Methods of bringing cable and/or conduit through a concrete wall and/or upward from a concrete wall into the hollow space within a framed wall.

below ground level and then brought through the basement wall, or (2) it can be run up the side of the building and through the basement wall at the ceiling joist level. When the conduit is run through the basement wall, the opening must be sealed to prevent moisture from seeping into the basement. The electrician must decide which of the two methods is more suitable for the installation.

Section 0 defines a wet location as an area where electrical equipment will be exposed to liquids that may drip, splash, or flow against the equipment.

Outdoor fixtures must be constructed so that water cannot enter or accumulate in lampholders, wiring compartments, or other electrical parts. These fixtures must be marked "Suitable for Wet Locations" or approved for outdoor use.

The *CEC* defines partially protected areas, such as roofed open porches or areas under canopies, as damp locations. Fixtures to be used in these locations must be marked "Suitable for damp locations." Many types of fixtures are available, so be sure to check the approval label on the fixture or check with the supplier to determine the suitability of the fixture for a wet or damp location. Figure 24-12 shows an outdoor fixture with lamps suitable for damp locations.

The post in Figure 24-11 may be embedded in concrete or not, depending on the consistency of the soil, the height of the post, and the size of the lighting fixture. Most electricians prefer to embed the base of the post in concrete to prevent rotting of wood posts and rusting of metal posts. Figure 24-13 illustrates some typical post lights.

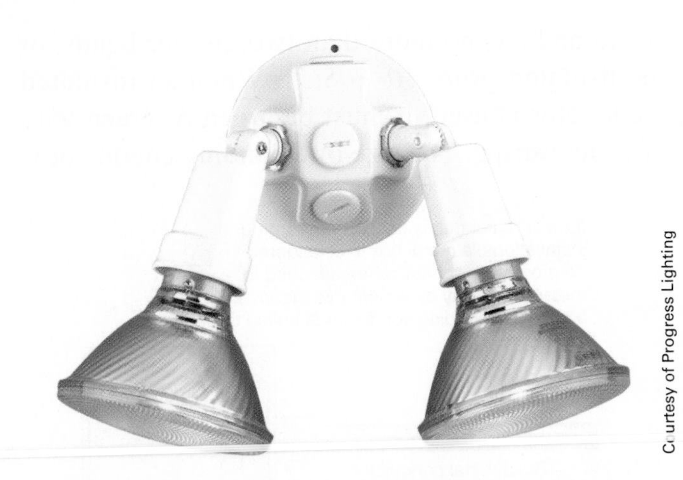

Courtesy of Progress Lighting

FIGURE 24-12 Outdoor fixture with lamps.

OVERHEAD GARAGE DOOR OPENER ▲$_E$

The overhead garage door opener in this residence is plugged into the receptacle provided for this purpose on the garage ceiling. The opener is ¼ HP, rated at 5.8 amperes at 120 volts. The installation of the overhead door will be done by a specially trained overhead door installer.

All photos: Courtesy of Progress Lighting

FIGURE 24-13 Post lights.

Principles of Operation

The overhead door opener contains a motor, gear reduction unit, motor reversing switch, and electrical clutch. This unit is preassembled and wired by the manufacturer.

Almost all residential overhead door openers use split-phase, capacitor-start motors in sizes from ¼ to ½ hp. The motor size depends on the size (weight) of the door to be raised and lowered. If springs are used to counterbalance the weight of the door, a small motor with a gear reduction unit can lift a fairly heavy door.

There are two ways to change the direction of a split-phase motor: (1) Reverse the starting winding with respect to the running winding or (2) reverse the running winding with respect to the starting winding. Although method (2) is correct, it is not commonly used. The motor can be run in either direction by properly connecting the starting and running winding leads of the motor to the reversing switch. When the motor shaft is connected to a chain drive or screw drive, the door can be raised or lowered according to the direction in which the motor is rotating.

The electrically operated reversing switch can be controlled by several means, including pushbutton stations (marked "open-close-stop," "up-down-stop," or "open-close"); indoor and outdoor weatherproof pushbuttons; and key-operated stations. Radio-operated controllers are popular because they permit overhead doors to be controlled from an automobile.

Wiring of Door Openers

Wiring overhead door openers for residential use is quite simple, since these units are completely prewired by the manufacturer. The electrician must provide a 120-volt circuit close to where the overhead door operators are to be installed. This 120-volt circuit can be wired directly into the overhead door unit, or a receptacle can be provided into which the cord on the overhead door operator is plugged. The outlet for the door opener is normally placed about 3.2 m in from the centre of the door opening.

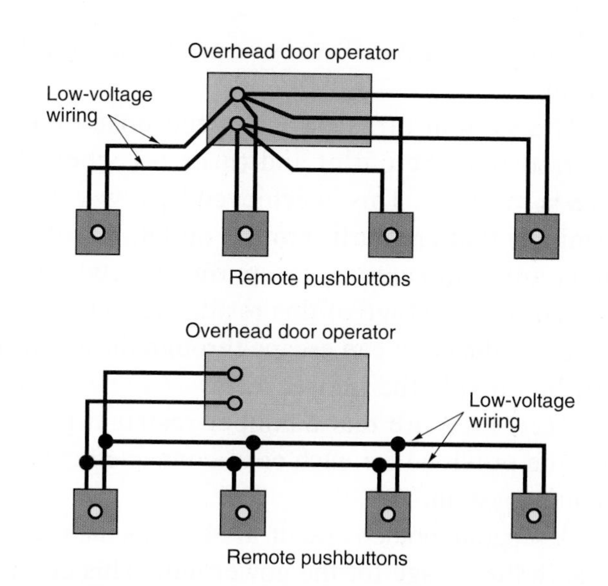

FIGURE 24-14 Pushbutton wiring for overhead door operator.

The pushbuttons used to operate a residential overhead door opener can be placed in any convenient location. They are wired with low-voltage wiring, usually 24 volts; see Figure 24-14. This is the same type of wire used for wiring bells and chimes. From each button, the electrician runs a two-wire, low-voltage cable back to the control unit. Since the residential openers use a simple control system, a two-wire cable is all that is necessary between the operator and the pushbuttons. Commercial overhead door openers can require three or more conductors.

It is desirable to install pushbuttons at each door leading into the garage. The actual connection of the low-voltage wiring between the controller and the pushbuttons is a parallel circuit connection; see Figure 24-14.

CENTRAL VACUUM SYSTEM (▲)Q

It is common practice to make provision for a central vacuum system in new homes, and to install it when renovating or upgrading existing homes. The ductwork is installed soon after the house is framed and provides a series of centrally located outlets for vacuuming with the 10-m hose provided with the system.

The system turns on when the hose-end is inserted into the outlet and turns off when the hose is removed. This is achieved by low-voltage terminals that are built into the outlet. All of the ducts are connected to the main unit, which is located in the garage of this residence so that any noise and dust that can escape through the exhaust will be outside the house. *Rules 26–710(1)* and *26–712(g)* require that a tamper-resistant receptacle be provided for each cord-connected central vacuum system.

A separate branch circuit, B24, feeds the receptacle in the garage for the power unit. This circuit is required by *Rule 26–722(e)* and provides power for the motor rated at 12 amperes, 120 volts; see Figure 24-15.

Courtesy of Jeff Wampler

FIGURE 24-15 The central vacuum system requires a separate circuit for the unit in the garage, which is a 2-hp motor and draws 12 amperes at 120 volt.

REVIEW

Note: Refer to the *CEC* or the blueprints provided with this textbook when necessary. Where applicable, responses should be written in complete sentences.

1. Which circuit supplies the garage? ____________
2. What is the circuit rating? ____________
3. How many receptacle outlets are connected to the garage circuit? ____________
4. a. How many cables enter the wall switch box located next to the side garage door? ____________
   b. How many circuit conductors enter this box? ____________
   c. How many bonding conductors enter this box? ____________
5. Show calculations on how to select a proper box for Question 4. What kind of box would you use? ____________
6. a. How many lights are recommended for a one-car garage? ____________ For a two-car garage? ____________ For a three-car garage? ____________

b. Where are these lights to be located? ______________________________

____

7. From how many points in the garage of this residence are the ceiling lights controlled?

____

8. List three methods by which the cover requirements of *CEC Table 53* can be reduced.

____

____

____

____

____

9. The total estimated load of the garage circuit draws how many amperes? Show your calculations. ______________________________

____

____

____

10. How high from the floor are the switches and receptacles to be mounted? __________

____

11. What type of cable feeds the post light? ______________________________

12. Describe the appropriate method of backfilling trenches containing cables, conduits, and/or direct burial conductors. Quote the *CEC* rule: ______________________________

____

____

____

13. In the spaces provided, fill in the cover (depth) for the following residential underground installations (120/240 volt)

a. Type NMWU with no other protection. ________ mm

b. Type NMWU below the driveway. ________ mm

c. Rigid conduit under the lawn. ________ mm

d. Rigid 240-volt circuit conduit under the driveway. ________ mm

e. Cable in poly pipe between the house and detached garage. ________ mm

f. TECK cable under the lawn; circuit is 120 volts, 20 amperes, GFCI protected. ________ mm

g. TECK cable with an approved plank covering under the driveway; the circuit is 240 volts AC. ________ mm

h. Rigid non-metallic conduit, concrete-encased (50 mm); the circuit is 120 volts, 20 amperes, GFCI protected. ________ mm

i. Rigid conduit embedded in rock and grouted with concrete to the level of the rock surface. ________ mm

14. What rule of the *CEC* prohibits embedding type NMWU cable in concrete? ________
____
____

15. The following is a layout of the garage circuit. Using the suggested cable layout, make a complete wiring diagram of this circuit, using coloured pens or pencils to indicate conductor insulation colours.

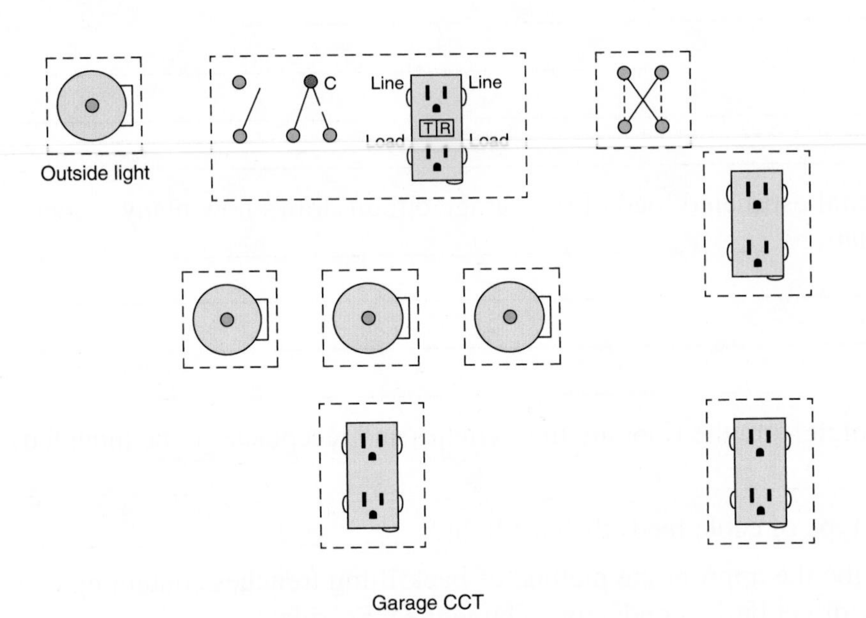

Garage CCT

16. What type of motor is generally used for garage door openers? ________
____

17. How is the direction of a split-phase motor reversed? ________
____
____
____

UNIT 25

Standby Power Systems

OBJECTIVES

After studying this unit, you should be able to

- discuss some of the safety issues concerning optional standby power systems
- understand the basics of standby power
- describe the types of standby power systems
- identify wiring diagrams for portable and standby power systems
- explain transfer switches, disconnecting means, and sizing recommendations

This unit provides only a general discussion and overview of home-type temporary generator systems. It is not possible in the available space to cover all of the variables and details involved with the many types of standby power systems available in the marketplace today.

CAUTION: Exhaust gases contain deadly carbon monoxide, the same as an automobile engine. One 5.5-kilowatt generator produces as much carbon monoxide as six idling automobiles. Carbon monoxide can cause severe nausea, fainting, or death. It is particularly dangerous because it is an odourless, colourless, tasteless, nonirritating gas. Typical symptoms of carbon monoxide poisoning are dizziness, headache, light-headedness, vomiting, stomach ache, blurred vision, and the inability to speak clearly. If you suspect carbon monoxide poisoning, remain active. Do not sit down, lie down, or fall asleep. Breathe fresh air—fast!

Precautions to Follow When Operating a Generator

- Always install equipment such as generators, power inlets, transfer switches, and panelboards that are listed by a nationally recognized testing laboratory.
- Always carefully read, understand, and follow the manufacturer's installation instructions.
- Always turn off the power when hooking up standby systems. Do not work on "live" equipment.
- Do not stand in water or have wet hands when working on electrical equipment.
- Do not touch bare wires.
- Store gasoline only in approved red containers clearly marked "GASOLINE."
- Do not operate a generator inside a building.

Operate a portable generator only outside in the open air, not close to windows, doors, or other vents. Exhaust fumes from the generator could infiltrate your house or a neighbour's house.

Death is just around the corner when the above warnings are not heeded!

WHY STANDBY (TEMPORARY) POWER?

Everyone has experienced a power outage, sometimes more often than one would like. You are left in the dark. Your furnace does not operate. You are concerned about water pipes freezing in the winter. Your refrigerator and freezer are not running. Your sump pump is not operating, and your basement is flooding. These are all important concerns. A standby generator system can address many of these concerns.

Before selecting a system, a few questions have to be answered. In your home, which loads are critical and must continue to operate in the event of a utility power outage? You need to know this so you can select a generator panelboard with adequate space for branch circuits. Only you can make this decision. Another question that must be answered is: How large must the generator be to serve the loads that are considered critical in your home? Still another question is: How simple or complicated a standby system do you want?

WHAT TYPES OF STANDBY POWER SYSTEMS ARE AVAILABLE?

The Simplest

Home supply centres usually carry the simplest and most economical portable generators, as shown in Figure 25-1. These consist of a gasoline-driven motor/generator set that must be started manually. Some models have the recoil "pull-to-start" feature; others have battery electric start. These are the types of portable generators that construction workers use for temporary power on job sites. Depending on the size of the fuel tank and the load being supplied, these smaller generators might be capable of running for 4 to 8 hours. This type of portable generator is commonly referred to as *backup power*.

These small portable generators consist of a *gasoline* engine driving a small electrical generator, just like a gasoline engine drives the blades of

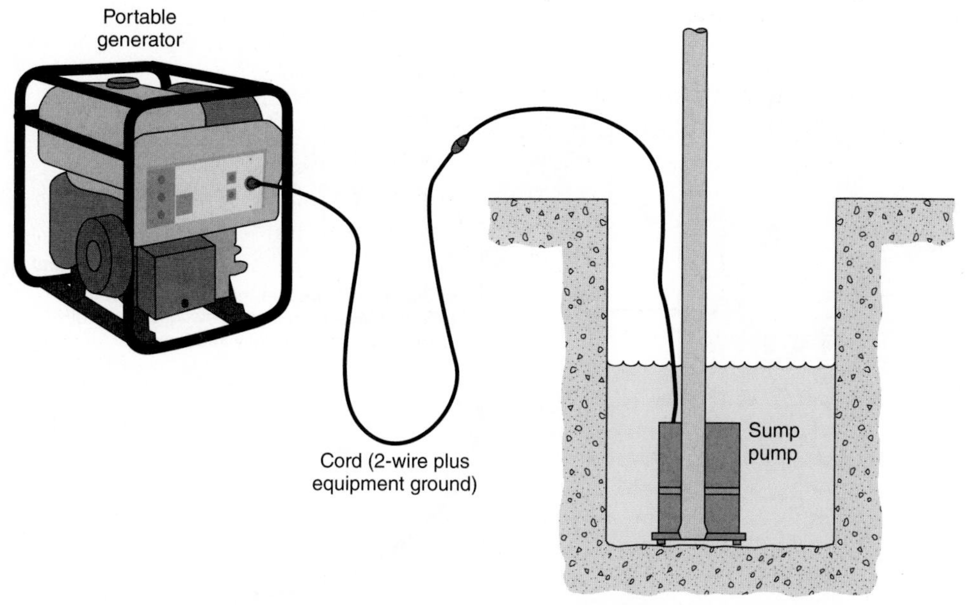

Portable generator

Cord (2-wire plus equipment ground)

Sump pump

FIGURE 25-1 A portable generator serving standby power to a 120-volt sump pump.

a lawn mower or snow thrower. Some generator sets have one or more standard 15- or 20-ampere single or duplex grounding-type receptacles, some have Ground-Fault Circuit Interrupter (GFCI) receptacles, and others have 20- and 30-ampere twist-lock-type receptacles. Some are rated 120 volts, whereas others are rated 120/240 volts. These generators are available in sizes up to about 7000 watts. Also available are models with more bells and whistles, such as oil alerts, adjustable output voltage, larger fuel tanks, and so on.

The procedure for this type is to start up the generator and then plug an extension cord into the receptacle and run it to whatever critical cord-and-plug-connected load needs to operate. If you are not home when a power outage occurs, you are out of luck!

The Next Step Up

Standby home generator systems are available in both permanent wiring and cord-and-plug-connected wiring. The cord-and-plug-connected system is by far the most popular when it comes to standby power for homes. In the event of a power outage, you must manually start the generator, flip the transfer switching device over from normal power to standby power, and plug in the power "patch" cord (Figure 25-2). The cord-and-plug-connected part of the installation consists of merely a flexible power patch cord that runs between the generator and the power inlet receptacle. This cord and the generator are set up as needed.

The permanent wiring part of the installation involves connecting specific branch circuits to a separate generator panelboard (usually located right next to the main panelboard), installing and properly connecting a transfer switch (discussed later), and installing a power inlet receptacle (also discussed later). The generator panelboard serves those circuits that you have selected as critical to operate in the event of a power outage. This panelboard is sometimes referred to as a *generator panelboard,* an *emergency panelboard,* and sometimes as a *critical load panelboard* (Figure 25-3).

These are gasoline-powered generators that use a flexible four-wire rubber cord (12 AWG for 5000-watt generators, 10 AWG for 7500-watt generators) that plugs into a female polarized twistlock receptacle on the generator. The other end of the cord plugs into a polarized twistlock receptacle mounted on the outside of the house in a weatherproof power inlet box. This polarized twistlock receptacle is permanently connected to transfer equipment or to a special electrical generator panelboard in which the critical branch

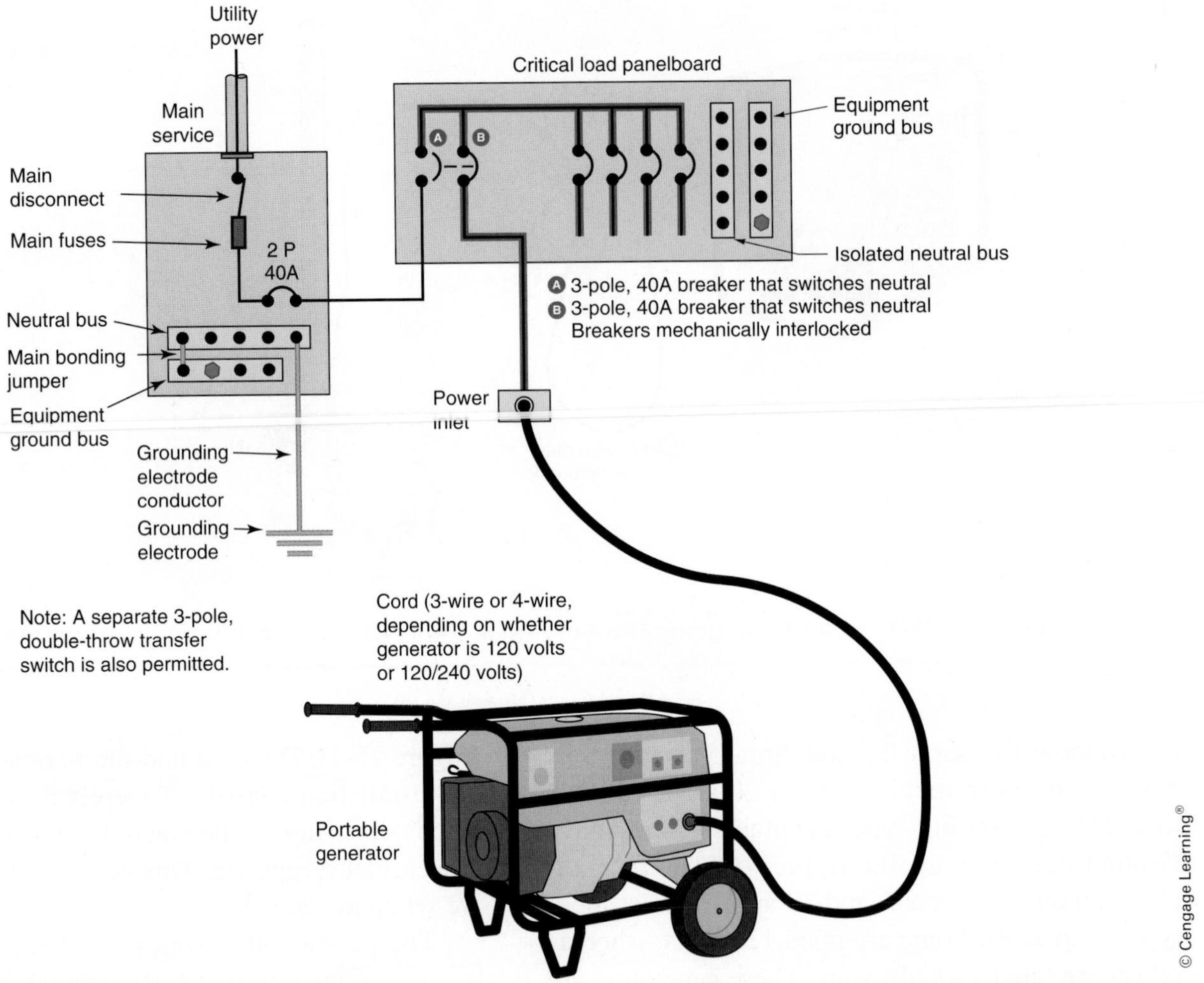

FIGURE 25-2 A portable generator supplies standby power to a critical load panelboard. The circuit breakers in the critical load panelboard must be manually turned to the temporary power position. These breakers are equipped with a mechanical interlock so there will be no feedback of electricity between the generator power and the utility normal power.

circuits are connected. The generator panelboard will have a transfer mechanism plus from 4 to 20 branch-circuit breakers; it all depends on the wattage rating of the generator. Some generator panelboards have the polarized twistlock receptacle mounted as part of and just below the generator panelboard.

During normal operation, this generator panelboard is fed from a two-pole circuit breaker in the main panelboard. This two-pole breaker might be rated 30, 40, 50, or 60 amperes, depending on the number of critical branch circuits to be supplied. When utility power is lost, this generator panelboard is fed from the generator. When the electric utility loses power, you must *manually* turn the transfer switch from its normal power position to its temporary power position.

One manufacturer of electrical equipment furnishes a panelboard that contains two circuit breakers that are mechanically interlocked. These are required to be three-pole breakers because the neutral from a portable generator must be switched. These breakers serve as the transfer switching means. Only one breaker can be in the ON position at a time. One of these breakers brings the normal power to the generator panelboard. The second breaker

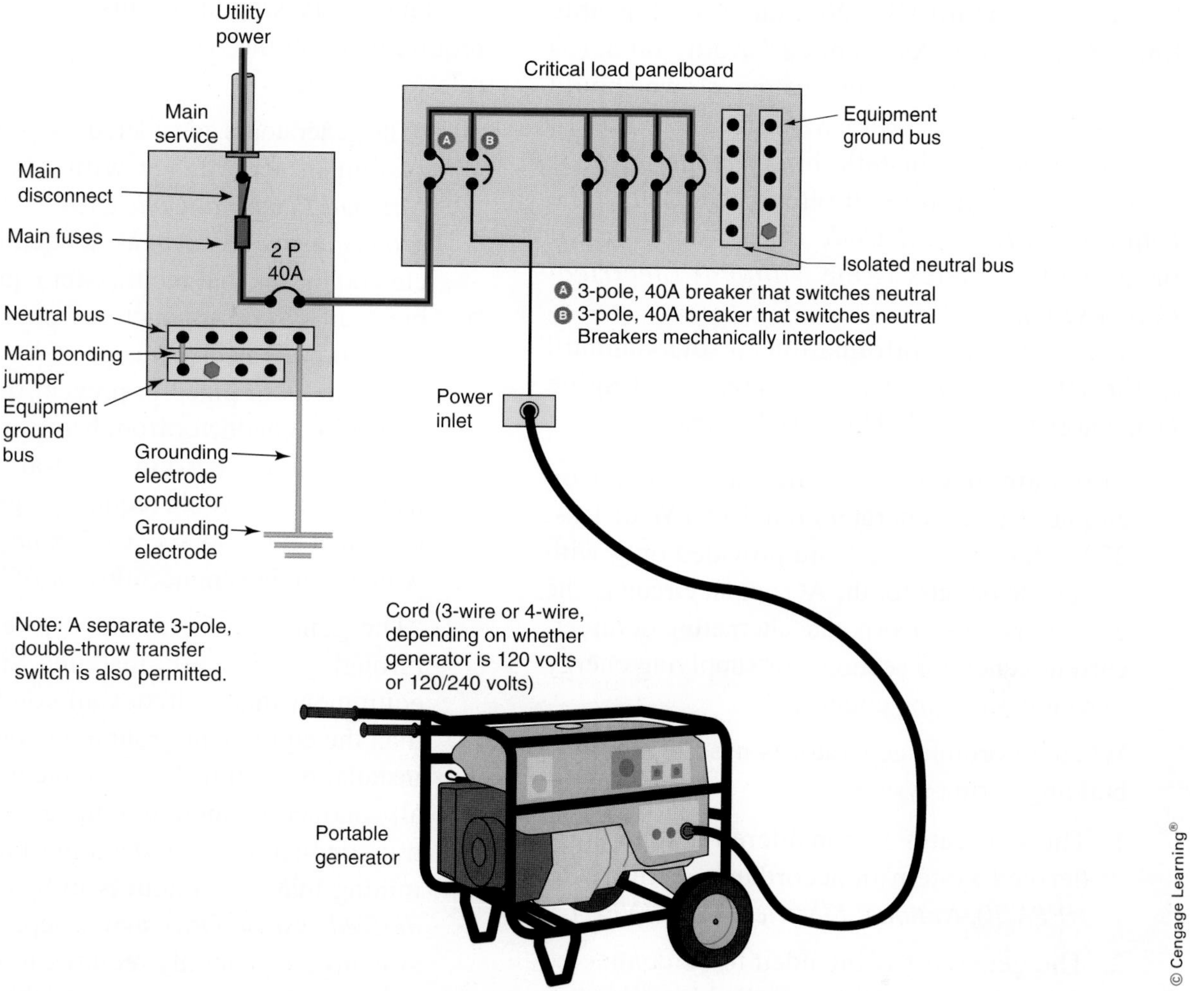

FIGURE 25-3 A critical load panelboard served by normal power from the main service panelboard. A portable generator is shown plugged into the power inlet. To switch to standby power, the circuit breakers in the critical load panelboard must be manually turned to the standby position. These breakers are equipped with a mechanical interlock so there will be no feedback of electricity between the generator power and the utility normal power.

brings the power from the generator to the generator panelboard. When one breaker is turned off, the other is turned on. This panelboard will have four or more branch-circuit breakers (Figures 25-2, 25-3).

With this system, the procedure generally is to first plug the four-wire cord into the polarized twist-lock receptacle on the generator; plug the other end into the power inlet receptacle; and then start the generator, which is always located outdoors. The transfer switch in the generator panelboard is then switched to the GEN position.

When normal power is restored, turn the transfer switch in the generator panelboard back to its normal position, turn off the generator according to the manufacturer's instructions, unplug the power cord from the generator, and finally, unplug the other end of the cord from the power inlet receptacle.

STANDARD REQUIREMENTS FOR STANDBY POWER SYSTEMS

Two standards are used in North America: Canadian Standards Association CSA C22.2 No. 100-14, entitled "Motor and Generators," and

UL safety standard UL 2201, entitled "Portable Engine-Generator Assemblies." Both standards are very similar. Standards change, so we should always refer to the latest CSA C22.2 No. 100-14 update before any installation. For the purpose of our discussion, we will refer to UL 2201. The *Canadian Electrical Code (CEC)* counterpart in the United States is the *National Electrical Code (NEC).*

The guide card information for the standard in the UL *White Book* in the category "Engine Generators for Portable Use (FTCN)" reads:

> This category covers internal-combustion engine-driven generators rated 15 kW or less, 250 volt or less, which are provided only with receptacle outlets for the AC output circuits. The generators may incorporate alternating or direct-current generator sections for supplying energy to battery-charging circuits.
>
> When a portable generator is used to supply a building wiring system:
>
> 1. The generator is considered a separately derived system in accordance with *ANSI/NFPA* 70, *National Electrical Code (NEC).*
> 2. The generator is intended to be connected through permanently installed Listed transfer equipment that switches all conductors other than the equipment grounding conductor.
> 3. The frame of a listed generator is connected to the equipment grounding conductor and the grounded (neutral) conductor of the generator. When properly connected to a premises or structure, the portable generator will be connected to the premises or structure grounding electrode for its ground reference.
> 4. Portable generators used other than to power building structures are intended to be connected to ground in accordance with the *NEC.*[*]

[*]Reprinted from the *White Book* with permission from Underwriters Laboratories Inc.® Copyright © 2010 Underwriters Laboratories Inc.®

The safety standard adds to or supplements requirements in the *NEC.*[†] Let's look at these UL rules one by one.

1. "The generator is considered a separately derived system in accordance with *ANSI/NFPA 70, National Electrical Code* (*NEC*)." The term *separately derived system* as defined in *NEC* Article 100 means that the transfer equipment must break all system conductors from the generator, including the neutral, as per *CEC Rule 4–028.* This rule is in place to prevent the equipment grounding conductor from being in parallel with the neutral and carrying neutral current. Normally, this rule would require the generator to be grounded at the source. *CEC Rule 10–204* deals with grounding connections for AC systems.
2. "The generator is intended to be connected through permanently installed Listed transfer equipment that switches all conductors other than the equipment grounding conductor." The installation of transfer equipment that switches all conductors, including the grounded or neutral conductor, is the deciding factor in determining that the system is in fact a *separately derived system.* Once again, separately derived systems are generally required to have the neutral grounded (connected to earth) at the source. This is not the case for portable generators.
3. "The frame of a Listed generator is connected to the equipment-grounding conductor and the grounded (neutral) conductor of the generator. When properly connected to a premises or structure, the portable generator will be connected to the premises or structure grounding electrode for its ground reference." These connections ensure that a ground-fault return path is established back to the source, which is the generator winding. Be very cautious when selecting a generator. Some manufacturers produce generators with a floating neutral.

[†]Reprinted with permission from NFPA 70® - 2011, *National Electrical Code®*, Copyright © 2010, National Fire Protection Association, Quincy, MA. This reprinted material is not the complete and official position of the NFPA on the referenced subject, which is represented only by the standard in its entirety. NFPA 70®, *National Electrical Code* and *NEC®* are registered trademarks of the National Fire Protection Association, Quincy, MA.

As a result, it is impossible for an overcurrent device to operate on a ground fault because no circuit for fault current can be established. The connection of the neutral and equipment grounding conductor to the frame of the generator is required in NEC 250.34(A)(2) and (C). See *CEC Rule 10–402(1)(a).*

4. "Portable generators used other than to power building structures are intended to be connected to ground in accordance with the *NEC*." This sentence might better be read as "Portable generators used other than to power building structures are intended to be connected to ground (*grounded*) if required by the *NEC*." The *NEC* contains special requirements for grounding of portable generators in 250.34. The *CEC* requires grounding for connections of an AC system that the portable generator delivers. *CEC Rule 10–204* covers this grounding issue. As long as the generator supplies loads only through receptacles mounted on the generator and the equipment grounding conductor is connected to the generator frame, the frame of the portable generator is not required to be connected to a grounding electrode.

When connected to the service-supplied electrical system by cord-and-plug connection, the equipment grounding conductor serves as a ground or earth reference for the electrical system because the conductor is extended to the windings of the generator.

CAUTION: If you intend to have a GFCI breaker-protected branch circuit transferred to the generator panelboard, be sure to connect this circuit to a GFCI breaker in the generator panelboard.

Because GFCIs and AFCIs require a neutral conductor, make sure the neutral conductor is properly connected between the main panelboard neutral bus and the neutral bus in the critical panelboard.

The neutral bus in the generator panelboard is insulated from the generator panelboard enclosure.

The ground bus in the generator panelboard is bonded to the generator panelboard enclosure.

Standby generators of this type can have run times as long as 16 hours. Again, it depends on the loading. The manufacturer's installation and operating instructions will provide this information.

Permanently Installed Generator

A permanently installed generator is truly standby power. These types of generator systems are covered under UL 2200 and CSA C22.2 No. 100-14.

Standby power is a system that allows you to safely provide electrical power to your home in the event of a power outage. Instead of running extension cords from a portable generator located outside the house to critical loads (lights, sump pumps, refrigerators, etc.), you select the branch circuits that you feel are critical or essential to satisfactorily operate your home electrical system. These branch circuits are placed in a panelboard that will be transferred to the generator in the event of a power outage from the electric utility. The diagrams in this chapter show this.

The generator for this type of system is permanently installed on a concrete pad at a suitable location outside the home that is convenient for making up all of the electrical connections to serve the selected critical loads. These systems generally run on natural gas (Figure 25-4) or propane (liquid petroleum gas—LPG), and in rare cases, on diesel fuel. All of the conductors between the generator, the transfer switch, and the generator panelboard installed for the selected critical loads are permanently wired.

The automatic transfer switch monitors the incoming electric utility's voltage. Should a power outage occur, the standby generator automatically starts, the transfer switch "transfers," and the selected loads continue to function. When normal power is restored, the transfer switch automatically "transfers" back to the normal power source, and the generator shuts down. Because the fuel supply is a permanently installed natural gas line, these systems can run indefinitely in conformance with the manufacturer's specifications.

To be totally automatic, an electronic controller that is part of the transfer switch immediately senses the loss of normal power and "tells" the standby generator

Courtesy of Kohler Power Systems

FIGURE 25-4 A natural gas-fueled generator.

to start and "tells" the transfer switch to transfer from normal power to standby power. This controller senses when normal power is restored, allowing the transfer switch to return to normal power and to shut off the standby generator. All of the circuitry is reset, and it is now ready for the next power interruption.

WIRING DIAGRAMS FOR A TYPICAL STANDBY GENERATOR

For simplicity, the diagrams in Figures 25-5 and 25-6 are one-line diagrams. Actual wiring consists of two ungrounded conductors and one grounded "neutral" conductor. The neutral bus in a panelboard that serves selected loads must *not* be connected to the metal panelboard enclosure. Connecting the neutral conductor, the grounding electrode conductor, and the equipment grounding conductors together is permitted only in the main service panelboard, *CEC Rule 10–204*. The generator neutral shall be solidly connected to the normal supply isolated neutral bus in the transfer switch. The neutral shall not be switched.

If you were to bond the neutral conductor and the metal enclosures of the equipment (panelboard, transfer switch, and generator) together beyond the main service panelboard, you would create a parallel path. A parallel path means that some of the normal return current and fault current will flow on the grounded neutral conductor and some will flow on the metal raceways and other enclosures. This is not a good situation!

Manufacturers of generators in most cases do not connect the generator neutral conductor lead to the metal frame of the generator. Instead, they connect it to an isolated terminal. Then it is up to you to determine how to connect the generator in compliance with local electrical codes.

Most inspectors will permit the internal neutral bond in a portable generator to remain in place. In fact, when you purchase a portable generator, you should insist that the neutral-to-case bond be in place. Without the bond, an overcurrent device cannot function on a ground fault because a return path does not exist. For permanently installed generators, inspectors will generally not permit a bond between the generator neutral and the metal frame of the generator.

A sign must be placed at the service entrance main panelboard indicating where the standby generator is located and what type of standby power it is, *Rule 84–030*. As required (*Rules 10–400 and 10–408*), an equipment bonding conductor shall be installed with

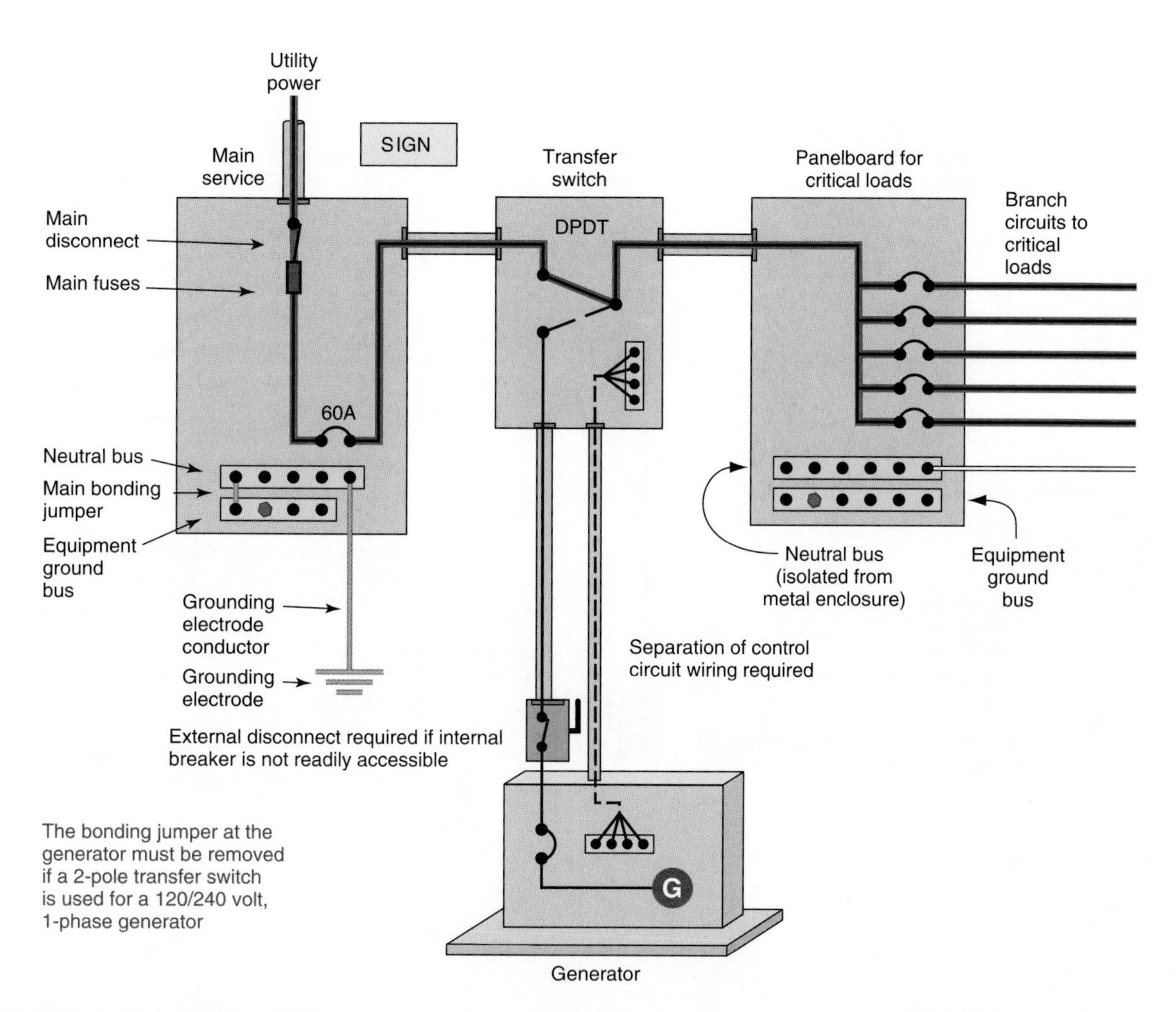

FIGURE 25-5 A natural gas-driven generator connected through a transfer switch. The transfer switch is in the normal power position. Power to the critical load panelboard is through the main service panelboard, then through the transfer switch.

the circuit conductors to bond all noncurrent-carrying metal parts of the installation, including generator frame, transfer switch, and main service panelboard.

The total time for complete transfer to standby power is approximately 45-60 seconds.

To further give the homeowner assurance that the standby generator will operate when called on, some systems provide automatic "exercising" of the system periodically, such as once every 7 or 14 days for a run time of 7 to 15 minutes.

CAUTION: When an automatic type of standby power system is in place and set in the automatic mode, the engine may crank and start at any time without warning. This would occur when the utility power supply is lost. To prevent possible injury, be alert, aware, and very careful when working on the standby generator equipment or on the transfer switch. Always turn the generator disconnect to the OFF position, then lock out and tag out the switch, warning others not to turn the switch back to ON. In the main panelboard, locate the circuit breaker that supplies the transfer equipment, turn it to OFF, then lock out and tag out the circuit breaker feeding the transfer switch.

TRANSFER SWITCHES OR EQUIPMENT

A transfer switch or equipment shifts electrical power from the normal utility power source to the standby power source. A transfer switch isolates the

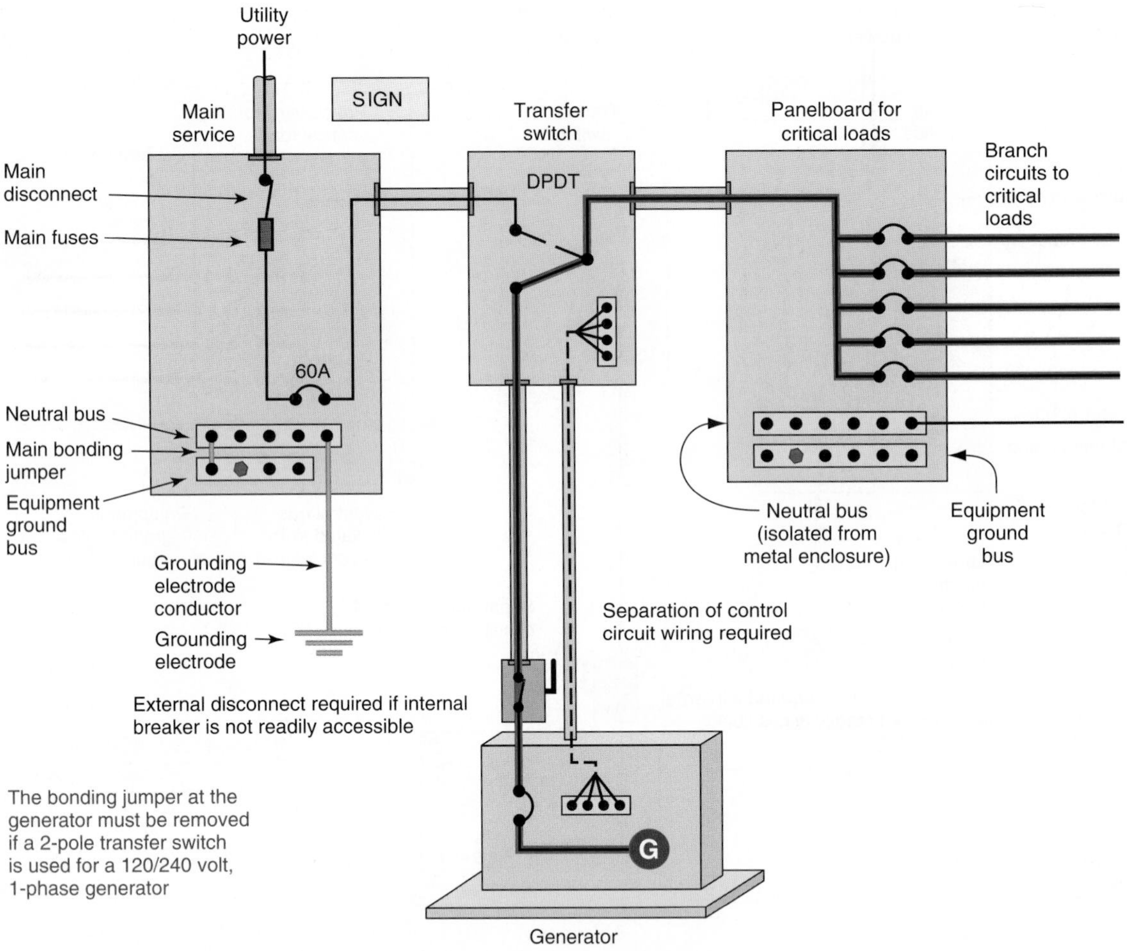

FIGURE 25-6 A natural gas-driven generator connected through a transfer switch. The transfer switch is in the standby position, feeding power to the critical load panelboard.

utility source from the standby source in such a way that there is no feedback from the generator to the utility's system, or vice versa. When normal power is restored, the automatic transfer switch resets itself and is ready for the next power outage.

Transfer switches or equipment for typical residential applications might have ratings of 40 to 200 amperes. Some have wattmeters for balancing generator loads.

For home installations, transfer switches may be three-pole double throw (TPDT) or double-pole, double throw (DPDT), depending on how the type of system is being installed. As stated above for listed portable generators, a three-pole, double-throw transfer switch or equipment is required.

For permanently installed generators, either a two-pole or three-pole transfer switch or equipment is permitted as long as the generator-produced electrical system is coordinated with the transfer equipment.

Two-pole transfer switches or equipment do not break the "neutral" conductor of a 120/240-volt single-phase system. Technically speaking, this type of system is a *nonseparately derived system.* A nonseparately derived system is properly grounded through the grounding electrode system of the normal premises wiring at the service equipment. A system where the neutral is disconnected by the transfer switch is referred to as a *separately derived system* and must be regrounded. Separately derived systems are common in commercial and industrial installations and are required for connection of portable generators.

If a transfer switch is connected on the line side of the main service disconnect, it must be listed as being suitable for use as service equipment. In Figures 25-5 and 25-6, the transfer switch is not on the line side of the main service disconnect and therefore does not have to be listed for use as service equipment.

A transfer switch must be a "break before make." Without the break-before-make feature, a dangerous situation is present and could lead to destruction of the generator, personal injury, or death. If the transfer switch does not separate the utility line from the standby power line, utility workers working on the line could be seriously injured. For low-cost, nonautomatic transfer systems, the transferring means might consist of 2 two-pole or three-pole circuit breakers that are mechanically interlocked so both cannot be on at the same time.

Capacity and Ratings of Transfer Equipment

The rating of the transfer switch must be capable of safely handling the load to be served. Otherwise, the transfer switch could get hot enough to cause a fire. This is nothing new; it is the same hazard as overloading a conductor. Specific rules on the capacity of the generator are dependent on the type of the transfer switch used, as follows:

1. If manual transfer equipment is used, an optional standby system is required to have adequate capacity and rating for the supply of all equipment intended to be operated at one time. The user of the optional standby system is permitted to select the load that is connected to the system.
2. If automatic transfer equipment is used, an optional standby system must comply with parts (a) or (b):
   a. **Full Load.** The standby source shall be capable of supplying the full load that is transferred by the automatic transfer equipment.
   b. **Load Management.** If a system is employed that will automatically manage the connected load, the standby source must have a capacity sufficient to supply the maximum load that will be connected by the load management system.

A transfer switch must also be capable of safely interrupting the available fault current that the generator or utility is capable of delivering. Listed transfer switches or equipment may be provided with or without integral overcurrent protection. The suitability of listed transfer equipment for interrupting or withstanding short-circuit current is marked on the transfer equipment.

A typical transfer switch might take 10 to 15 seconds to start the generator. This eliminates nuisance start-ups during momentary utility power outages. After start-up, another 10- to 15-second delay is provided to allow the voltage to stabilize. After start-up and warm-up, full transfer takes place. Similar time-delay features are used for the return to normal power. Some systems can do the full transfer in a few seconds.

DISCONNECTING MEANS

A disconnecting means is required for a generator. *Rule 84–020* states: "Disconnecting means shall be provided to disconnect simultaneously all ungrounded conductors of any electric power production source of an interconnected system from all circuits supplied by the electric power production source equipment." Most electrical inspectors will allow the cord-and-plug connection for a portable generator to serve as the disconnecting means.

When the normal power main service disconnecting means is located inside the home, *CEC Rules 84–024* and *28–900* require that a disconnect for the standby power from the generator be installed near the main service equipment.

Having one disconnecting means inside the home and another outside the home presents quite a challenge for firefighters or others wanting to totally shut off the power to the home. This is why *Rule 84–030* requires that a sign be placed at the service entrance equipment that indicates the type and location of on-site optional standby power sources.

GENERATOR SIZING RECOMMENDATIONS

No matter what type or brand of generator you purchase, you will have to size it properly. There is no better way to do this than to follow the manufacturer's recommendations. Generators for home use are generally rated in watts. The manufacturer of the generator has taken into consideration that:

Watts = Volts × Amperes. Some manufacturers suggest that after adding up all of the loads to be picked up by the generator, another 20% capacity should be added for future loads.

As stated above, the type of transfer switch installed determines the minimum capacity required for the generator. If a manual transfer switch or equipment is installed, a generator is required to have capacity for all of the equipment intended to be operated at one time. The user of the optional standby system is permitted to select the load connected to the system.

If the generator is supplied with an automatic transfer switch or equipment, it is required to be capable of supplying the full load that is transferred by the automatic transfer equipment. A load management system is permitted to be installed, in which case the generator must be sized not smaller than required to supply the maximum load that the load management system will allow for it to be connected to the equipment.

Resistive loads such as toasters, heaters, and lighting do not have an initial high inrush surge of current. Motors, on the other hand, do have a high inrush surge of starting current, which lasts for only a few seconds until the motor gets up to speed. This inrush must be taken into consideration when selecting a generator.

Table 25-1 shows some typical *approximate* wattage requirements for household loads. For heating-type appliances, the typical operating wattage values are used. For motor-operated appliances, the starting wattage values should be used. Verify actual wattage ratings by checking the nameplate on the appliance. Verify that the connected loads do not exceed the generator's marked capacity. Most generators are capable of handling a momentary inrush of an extra 50% of their rating. Here again, check the manufacturer's specifications.

Permits

Permanently installed generators require a considerable amount of electrical work and gas line piping. The installation will require applying for a permit so that proper inspection by the authorities can be made.

Sound Level

Because generators produce a certain level of noise (decibels) when running, you will want to check with your local building authority to make sure the generator you choose is in compliance with local codes relative to any sound ordinance that might be applicable. Sound-level information is provided by manufacturers in their descriptive literature.

TABLE 25-1

Approximate typical wattage requirements for appliances.

APPLIANCE	TYPICAL OPERATING WATTAGE REQUIREMENTS	STARTING WATTAGE REQUIREMENTS
Blender	600	700
Broiler	1600	1600
Central air conditioner		
10 000 BTU	1500	4500
20 000 BTU	2500	7500
24 000 BTU	3800	11 000
32 000 BTU	5000	15 000
40 000 BTU	6000	18 000
Coffeemaker	900–1750	900–1750
CD player	50–100	50–100
Clothes dryer		
Electric	5000 @ 240 volts	6500
Gas	700	2200
Computer	300–800	300–800

(*continues*)

TABLE 25-1 *(Continued)*

APPLIANCE	TYPICAL OPERATING WATTAGE REQUIREMENTS	STARTING WATTAGE REQUIREMENTS
Curlers	50	50
Dehumidifier	250	350
Dishwasher	1500	2500
Electric		
Drill	250–750	300–900
Frying pan	1300	1300
Blanket	50–200	50–200
Space heater	1650	1650
Water heater	4500–8000 @ 240 volts	4 500–8000 @ 240 volts
Electric range		
6-inch surface unit	1500	1500
8-inch surface unit	2100	2100
Oven	4500	4500
Lights: Add wattage of bulbs	Add wattage of bulbs	
Furnace fan		
1/8 horsepower motor	300	900
1/6 horsepower motor	500	1500
1/4 horsepower motor	600	1800
1/3 horsepower motor	700	2100
1/2 horsepower motor	875	2650
Garage door opener		
1/3 horsepower	725	1400
1/4 horsepower	550	1100
Hair dryer	300–1650	300–1650
Iron	1200	1200
Microwave oven	700–1500	1000–2300
Radio	15 to 500 (with components)	15 to 500
Refrigerator	700–1000	2200
Security, home	25–100	25–100
Sump pump		
1/3 horsepower	800	2400
1/2 horsepower	1050	3150
Television	300	300
Toaster		
Four-slice	1500	1500
Two-slice	950	950
Automatic washer	1150	3400
Well pump motor		
1/3 horsepower	800	2400
1/2 horsepower	1050	3150
1 horsepower	1920 @ 240 volts	5500
Vacuum cleaner	1100	1600
VCR	150–250	150–250
Window fan	200	300

REVIEW

Note: Refer to the *CEC* or the blueprints provided with this textbook when necessary. Where applicable, responses should be written in complete sentences.

1. The basic safety rule when working with electricity is to ____________________
2. What would be the logical location in which to run a portable generator? __________
3. [Circle the correct answer.] The best advice to follow is to always use (listed) (cheapest) (smallest) equipment.
4. How would you define the term *standby power*? ____________________
5. Describe in simple terms the three types of standby power systems.

 ____

 ____

 ____

6. Briefly explain the function of a transfer switch.

 ____

 ____

 ____

7. [Circle the correct answer.] When a transfer switch transfers to standby power, the electrical connection inside the switch
   a. maintains connection to the normal power supply as it makes connection to the standby power supply.
   b. breaks the connection to the normal power supply before it makes connection to the standby power supply.
8. [Circle the correct answer.] A typical transfer switch for residential application is
   a. three-pole, double-pole switch (TPDT).
   b. double-pole, double-throw switch (DPDT).
   c. single-pole, double-throw switch (SPDT).
   d. double-pole, single-throw switch (DPST).
9. The *CEC* requires that a ____________________ means be provided for standby power systems. In the case of portable gen sets, this might be as simple as pulling out the plug on the extension cord that is plugged into the receptacle on the gen set. For permanently installed standby power generators, this might be on the gen set and/or separately provided inside or outside the home.

UNIT 26

Residential Utility-Interactive Solar Photovoltaic Systems

OBJECTIVES

After studying this unit, you should be able to

- identify the components of a residential utility-interactive solar photovoltaic system
- recognize the electrical hazards unique to solar photovoltaic systems

Electricity from sunlight? Is this possible? This is not only possible, but it has become very practical through the installation of utility-interactive solar photovoltaic (PV) systems. A combination of factors recently has made photovoltaic systems very popular:

- increased use of electricity throughout Canada
- increased costs for electricity production
- environmental concerns about the use of fossil fuels
- concern over dependence on foreign sources of energy
- increased efficiency of photovoltaic systems

Installation of photovoltaic systems is being encouraged by government and electric utilities through tax incentives and rebates. Electric utilities can construct large central photovoltaic generating plants to encourage the needs set by government incentives. Generation of electricity by the consumer at the point of use (distributed generation) decreases the need for utility-generated power. Excess electricity will be supplied to the utility grid for use by other customers. Electric utility companies are required to purchase customer-generated electricity at predetermined rates.

Depending on size, a utility-interactive photovoltaic system can supply most or all of a home's electrical load. Photovoltaic systems are being retrofitted to existing homes and even installed in new homes as they are constructed. Battery storage of the generated electricity may also be included, but is not as common.

Electrical Hazards

Electrical work for the installation of photovoltaic systems is not complex, but there are significant differences when compared to typical residential wiring. First and foremost, a utility-interactive photovoltaic system is a supply of electricity for the home, not another load. Photovoltaic modules on the roof generate electricity when exposed to sunlight. Although the utility disconnect (main circuit breaker) for a house can be turned off, dangerous levels of electricity will still be present in the dwelling as long as the sun is shining.

Modules on the roof of a house are connected in series in strings that operate at up to 600 volts. Electricity generated by the photovoltaic modules is direct current (DC). All conductors and components must be listed for use with DC voltage. The sun can shine for more than three hours at a time, so all conductors/components must be sized for continuous currents (three hours or longer of operation). Because string conductors of the correct type are permitted to be exposed on the roof of a dwelling, good workmanship is critical. There are open conductors, exposed to the elements, operating at up to 600 volts of continuous current (VDC) that cannot be turned off!

If a short circuit does occur, it is not always obvious. Module short-circuit current is only slightly higher than normal operating current. Contrast this with the short-circuit current from sources such as a utility or even 12-VDC vehicle batteries. Shorting out these sources results in extremely high-fault currents. High-fault currents are easier to detect. High-fault currents will also cause an overcurrent device (fuse or circuit breaker) to open quickly. The connection of the utility-interactive PV system to the existing service panel is of concern. Service-disconnecting means, grounding, and proper labelling must be accomplished to maintain a safe electrical installation.

It is obvious that there are some special considerations for the installation of residential photovoltaic systems. Local jurisdictions may have amendments to *Section 64*, which was added to the *Canadian Electrical Code* in 2015 for renewable energy systems. *Rules 64–200* through *64–222* provide a guide for photovoltaic systems. Always check with the local authority having jurisdiction (AHJ) before starting an installation in a new area.

THE BASIC UTILITY-INTERACTIVE PHOTOVOLTAIC SYSTEM

Several components are required in order to convert sunlight into useful amounts of electricity. A basic utility-interactive system will consist of modules, mounting racks, combiner/transition boxes, inverter(s), and several disconnects. A grounding electrode system and connection to the existing service panel will be required. See Figures 26-1 and 26-2 for examples of basic PV system components and arrangements.

Solar Cells, Modules, and Arrays

A photovoltaic module is the basic unit of power production in the system. A module is a manufactured unit made up of many semiconductor PV cells encased in a protective covering and mounted to an aluminum frame. All modules are required by the *Canadian Electrical Code (CEC)* to be listed to nationally recognized standards. Individual photovoltaic modules are mounted to either a support rack that is connected to roof members or to ground-mounted structural supports.

Individual modules are wired together in a series circuit. Factory-installed leads are provided by the manufacturer for this purpose. Type RPV or RPVU conductors can be spliced to module leads to facilitate circuiting. A separate equipment-grounding conductor is used to ground each module.

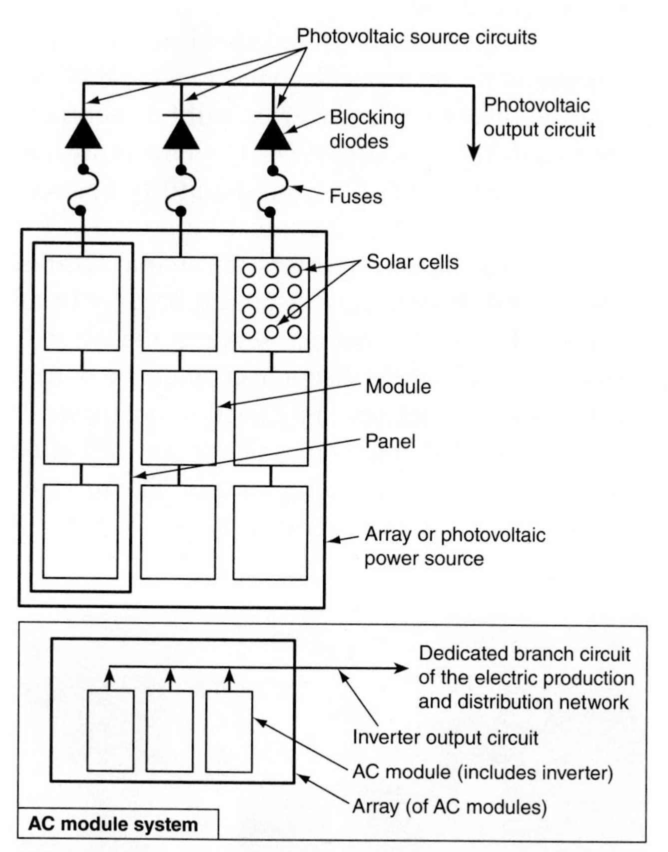

Notes:
1. These diagrams are intended to be a means of identification for photovoltaic system components, circuits, and connections.
2. Disconnecting means required by *Rule 64–060*.
3. System grounding and equipment grounding are not shown.

FIGURE 26-1 Identification of solar photovoltaic system components.

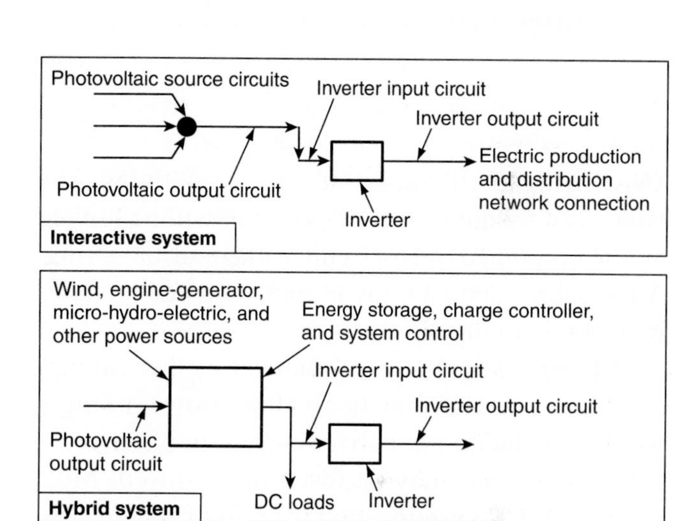

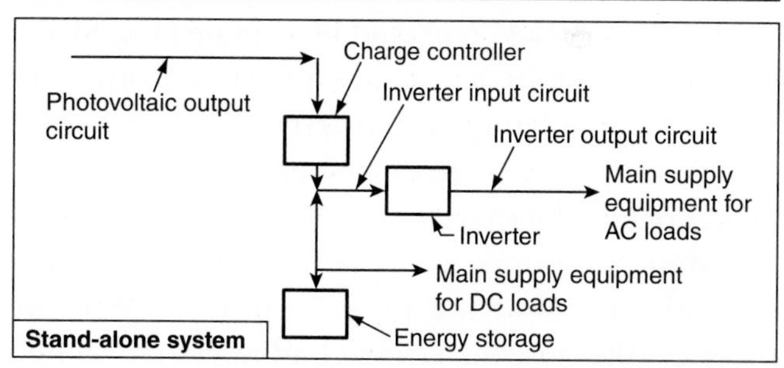

Notes:
1. These diagrams are intended to be a means of identification for photovoltaic system components, circuits, and connections.
2. Disconnecting means required by *Rule 64–060*.
3. System grounding and equipment grounding are not shown.
4. Custom designs occur in each configuration, and some components are optional.

FIGURE 26-2 Identification of solar photovoltaic system components in common system configurations.

FIGURE 26-3 Rooftop array of modules.

Most strings will consist of 8 to 15 modules. The number of modules in a series circuit will be limited by the combined open-circuit voltage of the string. A typical residential array of roof-mounted modules is shown in Figure 26-3.

Multiple strings are combined together (in parallel) at the combiner box. This allows a single pair of conductors to deliver current to an inverter. Combiner boxes may be fused or nonfused. Most residential PV systems are made up of three to six strings that can be combined in a single box. String fusing is generally not required for three strings or fewer. Module ratings will determine this, but some designers specify fused combiner boxes even when these are not required.

Many inverters have the ability to combine several strings at the input terminals. With an inverter like this, the string conductors are routed to the inverter without being combined first. The combiner box would not be required. A junction box, known as a *transition box*, is used to splice the open-string conductors to a wiring method for connection of the inverter. The entire assembly of racks, modules, and combiner/junction boxes is known as a *photovoltaic array*. Available space, sunlight, and the ultimate amount of electricity desired will determine the number of modules in the array. Obviously, the larger the system, the more expensive the installation, so the budget must be considered.

A disconnect for the ungrounded DC string conductor(s) is required before the conductors enter the dwelling unless a metallic raceway is used as the wiring method. Use of a metallic wiring method such as EMT or FMC is common so that string conductors can be routed through a dwelling unit attic. Metallic raceways provide more protection for the 400–600 VDC string conductors, which remain energized until the sun goes down. Firefighters responding to a house fire will open the service-disconnecting means on arrival. This does not de-energize the photovoltaic string conductors if the sun is still shining. Metallic raceways will provide a level of protection for the firefighters who may encounter the DC wiring method when responding to a fire.

Use of a metallic raceway eliminates the requirement for an array disconnect on the roof but not at the inverter. The inverter will be energized from two different sources: DC from the array and alternating current (AC) from the utility connection. Disconnecting means must be provided for both sources. Some inverters are manufactured with integral disconnects. A circuit breaker in an adjacent electrical panel can serve as the AC disconnect, but all safety disconnects must be within sight of the inverter and may need to be grouped. The DC grounded conductor is not permitted to be opened by any disconnecting means. The inverter in Figure 26-4 has an integrated DC disconnect.

FIGURE 26-4 A residential utility-interactive inverter.

The Inverter

The inverter will need to be installed in a location with sufficient working space. Larger systems may use two or more inverters with output combined in a separate AC panelboard before supplying the service. Grounding/bonding of the utility-interactive photovoltaic system is required and is usually accomplished at the inverter location.

Both the AC and the DC systems contain grounded current-carrying conductors. Grounding of the DC conductor is accomplished at the inverter. A new grounding electrode must be installed (and bonded to the existing premises' grounding electrode) or the existing grounding electrode for the dwelling must be accessed for connection. Correct termination of the grounding electrode conductor, equipment-grounding conductors, and grounded AC/DC conductors at the inverter is critical. Utility-interactive inverters are required to provide ground-fault protection for the array. Proper operation of the ground-fault protection system is dependent on the correct terminations of the grounding and grounded conductors.

Inverter design will dictate whether the positive or the negative DC conductor is the grounded current-carrying conductor for the array. Inverter size is based on the capacity of the array. Most residential inverters are in the 2-kW to 10-kW range. There are advantages to installing an inverter indoors. Cooler temperatures result in better operating efficiency, but outdoor installation is also common.

Safety Features of Inverters

Just as with modules and combiner boxes, inverters are required to be manufactured to nationally recognized standards. Inverters that are listed by a nationally recognized testing laboratory (NRTL) to Underwriters Laboratory of Canada/Other Recognized Document - Canadian (ULC/ORD-C) 1741 have met this standard, and CSA C22.2 No.107.1, Clause 15, applies to utlity-interconnected inverters. Both rules name this as *anti-islanding*, which means that an inverter must automatically turn off AC output when utility power is lost. *CEC Appendix B, Rule 84–024*, addresses this also. This feature prevents an inverter from supplying electricity into the utility grid when there is an outage. Without this feature, electric utility workers risk being electrocuted by inverter-supplied electricity.

Connection of Inverters

The two options for connection of the inverter to the electric service panel are known as a *supply-side* or a *load-side connection*. The supply-side connection is on the utility side of the main disconnect. A load-side connection will supply electricity on the customer side of the main disconnect. A dedicated circuit breaker (suitable for backfeed) or fusible disconnect is required. The total supply of current to a panelboard is limited to 120% of the busbar rating.

The service panel is supplied by both the utility (through the main circuit breaker) and the inverter (through a back-fed circuit breaker). The sum of the ampere ratings of the main circuit breaker and the inverter circuit breaker cannot exceed the rating of the panelboard bus multiplied by 1.2 (120%). A circuit breaker used for connection of the inverter will have to be located at the opposite end of the bus from the main circuit breaker to avoid overloading of the bus.

Building-Integrated Photovoltaic Modules

Photovoltaic modules that also serve as an outer protective finish for a building are known as *building-integrated photovoltaic (BIPV) modules*. This type of module is often installed in the form of roofing tiles intended to blend in with surrounding non-photovoltaic roof tiles. See Figure 26-5. BIPV modules are investigated through the listing process for conformance to appropriate fire-resistance and waterproofing standards that apply to roofing tiles, along with the electrical standards of ULC/ORD C-1703-01. A module the size of a roofing tile is obviously much smaller than the standard photovoltaic module. Many more of the smaller modules are required to create strings.

FIGURE 26-5 Building-integrated photovoltaic roof tiles.

FIGURE 26-6 A utility-interactive micro-inverter.

Micro-Inverters and AC Photovoltaic Modules

A single small inverter connected to each photovoltaic module is known as a *micro-inverter*. Instead of connecting multiple modules to a single inverter, each module will have its own attached inverter. The output of each module is connected directly to the micro-inverter with the existing module leads. Multiple micro-inverters are connected in parallel on a single circuit, which then supplies the service panel of the home. Micro-inverters are required to be listed, just as with the larger string inverters. Ground-fault protection, anti-islanding, and the other requirements of utility-interactive inverters are applicable. Output current for a single micro-inverter is approximately 0.8 amps. This would permit up to 16 inverters on a single 20-amp circuit. The inverter shown in Figure 26-6 is a micro-inverter.

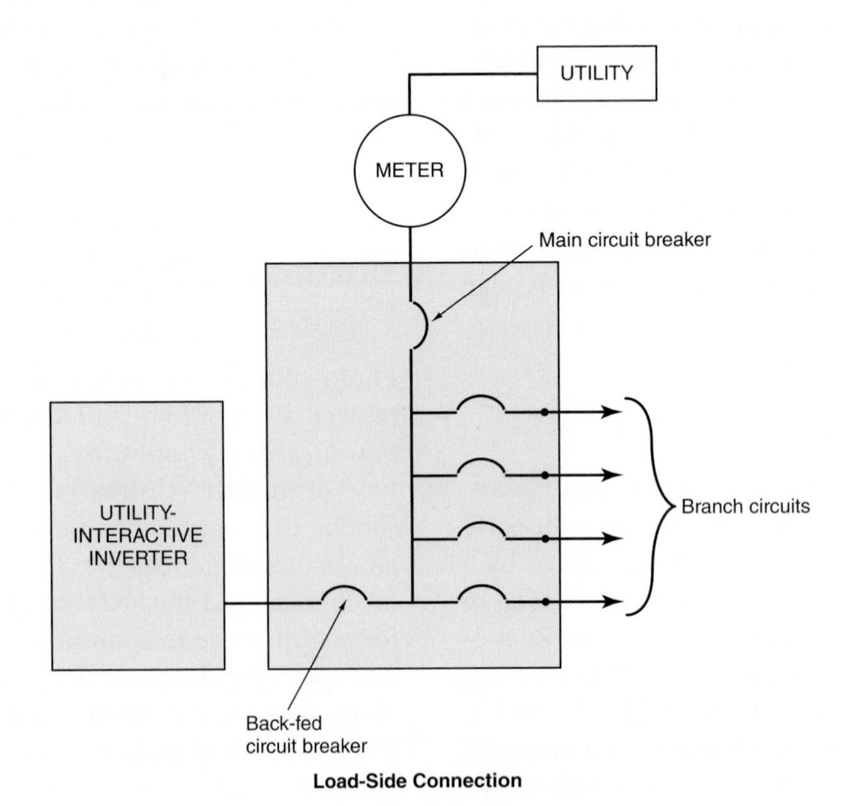

FIGURE 26-7 Load-side connection.

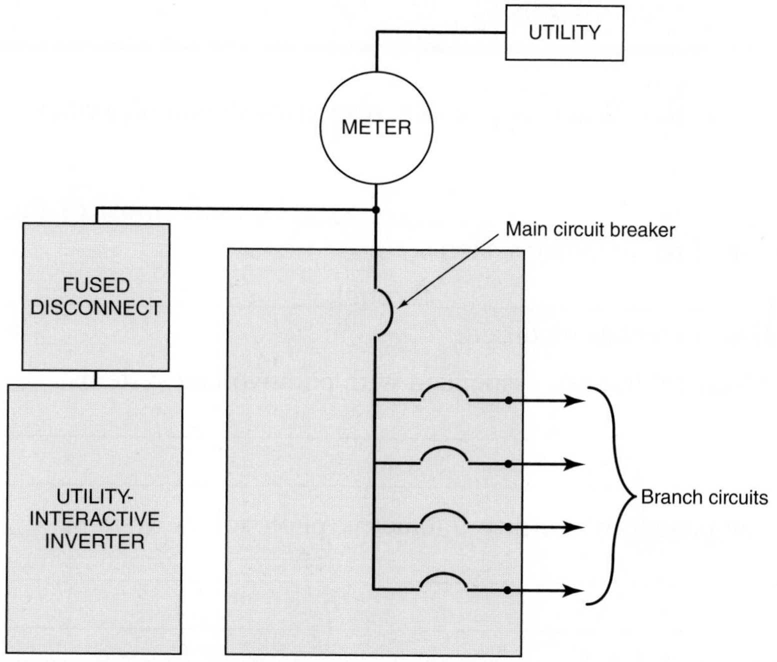

FIGURE 26-8 Supply-side connection.

The *CEC* permits different methods for connection of a utility-interactive photovoltaic system to the electrical service. Requirements for both methods are found in *Rule 64–112*, "Point of Connection." A connection on the line side of the service disconnecting means is known as a supply-side connection, *Rule 64–112(2)*. The size of the photovoltaic system is virtually unlimited, but the rules of *CEC Section 6* will apply to these service conductors. Requirements for load-side connections are found in *Rule 64–112(3)*. The photovoltaic system size is limited by the panel bus rating and main circuit-breaker size. A dedicated PV system circuit breaker, suitable for backfeed and positioned at the opposite end of the bus from the main circuit breaker, is a requirement of *Rule 64–058 (4)*. See Figures 26-7 and 26-8 for a representation of the connection methods discussed.

An AC photovoltaic module is essentially a normal DC module with the micro-inverter installed at the factory. The DC conductors are covered and inaccessible on an AC module. The absence of field-installed DC string conductors is a big advantage with both micro-inverters and AC modules. Hazards associated with the DC string conductors are minimized with the use of micro-inverters and eliminated with AC modules. *CEC* requirements for AC circuits connecting micro-inverters/AC modules to service panels are similar to normal branch-circuit rules.

REVIEW

Refer to the *CEC* when necessary. Where applicable, responses should be written in complete sentences.

1. ______________________________ contains most of the requirements for installation of photovoltaic systems.
2. *CEC Rules* ____________________ through ____________________ will apply to PV installations except as modified.
3. Name four electrical hazards associated with photovoltaic systems. ____________________
4. Name five components of a utility-interactive photovoltaic system. ____________________
5. Which conductor types are permitted to be installed exposed for array circuits? ____________________
6. To size conductors and overcurrent devices for source circuits, you must multiply the module string short-circuit current by ____________________.
7. When is a metallic raceway required for photovoltaic source circuits? ____________________
8. Photovoltaic modules that also serve as an outer protective finish for a building are known as ____________________
9. Grounding electrode system requirements for a utility-interactive photovoltaic system are found in which section of the *CEC*? ____________________
10. The two methods permitted for connection of a utility-interactive photovoltaic system to a service are known as ____________________ and ____________________.

Appendix A

Specifications for Electrical Work—Single-Family Dwelling

1. ***General*****:** The "General Clause and Conditions" shall be and are hereby made a part of this division.
2. ***Scope*****:** The electrical contractor shall furnish and install a complete electrical system as shown on the drawings and/or in the specifications. Where there is no mention of the responsible party to furnish, install, or wire for a specific item on the electrical drawings, the electrical contractor will be responsible completely for all purchases and labour for a complete operating system for this item.
3. ***Workmanship*****:** All work shall be executed in a neat and workmanlike manner. All exposed conduits shall be routed parallel or perpendicular to walls and structural members. Junction boxes shall be securely fastened, set true and plumb, and set flush with the finished surface when the wiring method is concealed.
4. ***Location of Outlets*****:** The electrical contractor shall verify location, heights, outlet and switch arrangements, and equipment prior to rough-in. No additions to the contract sum will be permitted for outlets in wrong locations, in conflict with other work, and so on. The owner reserves the right to relocate any device up to three metres prior to rough-in, without any charge by the electrical contractor.
5. ***Codes*****:** The electrical installation is to be in accordance with the latest edition of the *Canadian Electrical Code (CEC), Part I,* all local and provincial electrical codes, and the utility company's requirements.
6. ***Materials*****:** All materials shall be new and shall be listed and bear the appropriate label of the Canadian Standards Association (CSA) or another nationally recognized testing laboratory for the specific purpose. The material shall be of the size and type specified on the drawings and/or in the specifications.
7. ***Wiring Method*****:** Wiring, unless otherwise specified, shall be non-metallic-sheathed cable (NMSC), armoured cable, or EMT adequately sized and installed according to the *CEC* and local ordinances.

8. ***Permits and Inspection Fees*:** The electrical contractor shall pay for all permit fees, plan review fees, licence fees, inspection fees, and taxes applicable to the electrical installation, and these amounts shall be included in the base bid as part of this contract.

9. ***Temporary Wiring*:** The electrical contractor shall furnish and install all temporary wiring for handheld tools and construction lighting according to the *CEC* and *National Building Code of Canada* standards and include all costs in the base bid.

10. ***Number of Outlets Per Circuit*:** In general, not more than 12 lighting and/or receptacle outlets shall be connected to any one branch circuit. Exceptions may be made for low-current-consuming outlets.

11. ***Conductors*:** General lighting and power branch circuits shall be #14 AWG copper protected by 15-ampere overcurrent devices. All other circuits: wire and over-current device as required by the *CEC*. All conductors shall be part of an approved cable assembly or, if installed in raceway, Type R90 XLPE without jacket or RW90 XLPE, unless specified otherwise.

12. ***Load Balancing*:** The electrical contractor shall connect all loads, branch circuits, and feeders per Panel Schedule, but shall verify and modify these connections as required to balance connected and computed loads to within 10% variation.

13. ***Spare Conduits*:** The electrical contractor shall furnish and install two empty 27-mm EMT conduits between the workshop and the attic for future use.

14. ***Guarantee of Installation*:** The electrical contractor shall guarantee all work and materials for a period of one full year after final acceptance by the architect/engineer and owner.

15. ***Appliance Connections*:** The electrical contractor shall furnish all wiring materials and make all final electrical connections for all permanently installed appliances such as, but not limited to, furnace, water heater, water pump, built-in ovens and ranges, food waste disposer, dishwasher, central vacuum, food processor power unit, and clothes dryer.

    These appliances are to be furnished by the owner.

16. ***Chimes*:** Furnish and install two two-tone door chimes where indicated on the plans, complete with two pushbuttons and suitable chime transformer. Allow $100 for above items. Chimes and buttons are to be selected by the owner.

17. ***Dimmers*:** Furnish and install dimmer switches where indicated.

18. ***Exhaust Fans*:** Furnish, install, and provide connections for all exhaust fans indicated on the plans including, but not limited to, ducts, louvres, trims, speed controls, and lamps. Included are recreation room, laundry, rear-entry powder room, range hood, and bedroom hall ceiling fan. Allow a sum of $2000 in the base bid for this. This allowance does not include the two bathroom heat/vent/light/night-light fixtures.

19. ***Fixtures*:** A fixture allowance of $2750 shall be included in the electrical contractor's bid. This allowance shall include the furnishing and installation of all surface, recessed, track, strip, pendant, and/or hanging fixtures, complete with lamps. This allowance includes the three bathroom medicine cabinets with lights.

    This allowance does not include the two bathroom ceiling heat/vent/light/night-light fixtures.

    Labour for installation of the fixtures shall be included in the base bid.

20. ***Heat/Vent/Light/Night-Light Ceiling Fixtures*:** Furnish and install two heat/vent/light/night-light units where indicated on the plans, complete with switch assembly, ducts, and louvres required to perform the heating, venting, and lighting operations as recommended by the manufacturer.

21. ***Plug-In Strip*:** Where noted in the workshop, furnish and install a multioutlet assembly with outlets 1.2 m from floor. Total outlets: 6.

22. ***Switches, Receptacles, and Faceplates*:** All flush switches shall be of the quiet AC-rated toggle type. They shall be mounted 1300 mm to centre above the finished floor unless otherwise noted.

    Receptacle outlets shall be mounted 300 mm to centre above the finished floor unless otherwise noted. All convenience receptacles shall be of the grounding type. Furnish and install Ground-Fault Circuit Interrupter (GFCI) receptacles where indicated, to provide ground-fault circuit protection as required by the *CEC*. All branch circuits except bathroom and washroom receptacles, refrigerator, kitchen counter work area, island receptacle, and water pump shall be protected by a combination-type arc-fault circuit interrupter. All wiring devices are to be provided with ivory handles or faces and shall be trimmed with ivory faceplates except in the kitchen, where chrome-plated steel faceplates shall be used.

Receptacle outlets, where indicated, shall be split-switched. All receptacles within the dwelling shall be tamper-resistant (TR), with the exception of microwave, refrigerator, freezer, and kitchen counter receptacles, as per *Rule 26–712(g)*.

23. ***Television Outlets***: Furnish and install a 4 × 4 single-gang raised plaster cover at each television outlet where noted on the plans. Provide a vapour boot in all insulated walls. Mount at the same height as receptacle outlets. Furnish and install RG-6 coaxial cable to each television outlet from a point in the workshop near the main service entrance switch. Allow 2 m of cable in workshop. Furnish and install television plug-in jacks at each location. Faceplates are to match other faceplates in the home. All remaining work is to be done by others.

24. ***Telephones***: Furnish and install a 3-inch deep device box or 4-inch square box, 1½-inches deep with suitable single-gang raised plaster cover, at each telephone location, as indicated on the plans.

    Furnish and install four-conductor #22 AWG copper telephone cable or Cat 6 cable to each designated telephone location; terminate in proper modular jack, complete with faceplates. Installation shall be according to any and all applicable *CEC* and local code regulations.

25. ***Overhead Garage Door Opener***: Receptacle outlet mounted 3.2 m from inside edge of overhead door opening. The overhead garage door opener to be purchased and installed by a specially trained overhead door installer

26. ***Service Entrance***: Furnish and install one 30-circuit, 200-ampere, 120/240-volt single-phase, three-wire combination panel (rated for continuous operation at 80%), complete with a 200-ampere main breaker in the workshop where indicated on the plans. Branch-circuit protection in the panel is to incorporate circuit breakers. The panel is to have a 10 000-ampere interrupting rating.

27. Service entrance underground consumer's service conductors are to be furnished and installed by the utility. The meter equipment model number is to be furnished by the utility and installed by the electrical contractor where indicated on plans. The electrical contractor is to furnish and install all panels, conduits, fittings, conductors, and other materials required to complete the service entrance installation from the demarcation point of the utility's equipment to and including the main panel.

28. Service entrance conductors supplied by the electrical contractor shall be three 3/0 RW75 XLPE (600 V with jacket) or equivalent for the phase conductors and neutral conductor. Install 53-mm trade-size PVC conduit from main Panel A to the meter base.

29. Bond and ground service entrance equipment in accordance with the *CEC* and local and utility code requirements. Install #6 AWG copper system grounding conductor.

30. ***Subpanel***: Furnish and install one 24-circuit, 120/240-volt, single-phase, three-wire load centre in the recreation room. The load centre is to have 100-ampere mains. Feed the load centre with three #3 RW75 XLPE (600 V with jacket) conductors or equivalent protected by a 100-ampere, two-pole overcurrent device in the main panel. Install conductors in 27-mm EMT conduit buried under the basement floor slab. Include a #6 AWG copper bonding conductor with the panel feeds, as per *Table 16A*. Branch-circuit protection in this panel is to incorporate circuit breakers.

31. ***Circuit Identification***: All panel boards shall be furnished with typed-card directories with proper designation of the branch-circuit feeder loads and equipment served. The directories

shall be located in the panel in a holder for clear viewing. Test charts for GFCI and Arc Fault Circuit Interrupter (AFCI) breakers will be located on the panel enclosure.

32. The electrical contractor shall seal and weatherproof all penetrations through foundations, exterior walls, and roofs.
33. Upon completion of the installation, the electrical contractor shall review and check the entire installation, clean equipment and devices, and remove surplus materials and rubbish from the owner's property, leaving the workplace neat and clean and in complete working condition. The electrical contractor shall be responsible for the removal of any cartons, debris, and rubbish for equipment installed by the electrical contractor, including equipment furnished by the owner or others and removed from the carton by the electrical contractor.
34. ***Special-Purpose Outlets***: The electrical contractor shall install, provide, and connect all wiring for all special-purpose outlets (see Table A-1). Upon completion of the job, all

TABLE A-1

Schedule of special-purpose outlets.

SYMBOL	DESCRIPTION	VOLTS	HORSE-POWER	APPLIANCE AMPERE RATING	TOTAL APPLIANCE WATTAGE RATING (OR VA)	CIRCUIT AMPERE RATING	POLES	WIRE SIZE NMD 90	CIRCUIT NUMBER	COMMENTS
A	Hydromassage tub, master bedroom	120	½	10	1200	15	1	14	A9	Connect to Class A GFCI. Separate circuit.
B	Water pump	240	1	8	1920	20	2	12	A5/A7	Run circuit to disconnect switch on wall adjacent to pump; protect with Fusetron dual-element time-delay fuses sized at 125% of motor's FLA.
C	Water heater: top element 2000 W, bottom element 3000 W.	240	-0-	8.33 12.50	2000 3000	20	2	12	A6/A8	Connected for limited demand.
D	Dryer	120/240	120 V 1/6 Motor Only	23.75	5700 Total	30	2	10	B1/B3	Provide flush-mounted 30-A dryer receptacle. 14–30R.
E	Overhead garage door opener	120	1/4	5.8	696	15	1	14	B22	Unit comes with 3 W cord. Unit has integral protection.
F	Wall-mounted oven	120/240	-0-	27.5	6600	40	2	8	B6/B8	
G	Countertop range	120/240	-0-	31	7450	40	2	8	B2/B4	
H	Food waste disposer	120	1/3	7.2	864	15	1	14	B19	Controlled by single-pole switch on wall.

(*continues*)

TABLE A-1 *(Continued)*

SYMBOL	DESCRIPTION	VOLTS	HORSE-POWER	APPLIANCE AMPERE RATING	TOTAL APPLIANCE WATTAGE RATING (OR VA)	CIRCUIT AMPERE RATING	POLES	WIRE SIZE NMD 90	CIRCUIT NUMBER	COMMENTS
▲ I	Dishwasher	120	1/4 Motor Only	Motor Htr. Total	5.80 696 6.25 750 12.05 1446	15	1	14	B5	Motor and heater are not on simultaneously.
▲ J	Heat/vent/night-light/light master bedroom bath		120	14.9	1788	20	1	12	A13	Minimum circuit ampacity 18.6 A
▲ K	Heat/vent/night-light/light front bedroom bath		120	14.9	1788	20	1	12	A11	Minimum circuit ampacity 18.6 A
▲ L	Attic exhaust fan	120	1/4	5.8	696	15	1	14	A10	Run circuit to 4 inch. square box. Locate near fan in attic. Unit has integral protection.
▲ M	Electric furnace	240	1/3 Motor	Motor 3.5 Htr. 50.7 Total 54.2	13 000	70	2	4	A1/A3	The overcurrent device shall not be less than 125% of the total load of the heaters and motor. 54.2 × 1.25= 67.75 *(Sections 62–114(7)(a)(b); 62–114(8).*
▲ N	Air conditioner	240	-0-	30	7200	40	2	8	A2/A4	Compressor rated load amperes 27.8. Compressor locked rotor amperes 135.0. Condenser fan full load amperes 2.2. Condenser locked rotor amperes 4.5. Branch-circuit O.C. device 40-A 40.0. Minimum circuit ampacity 37.5.
▲ O	Freezer	120	1/4	5.8	696	15	1	14	A12	Install single receptacle outlet. Do not provide GFCI protection.
▲ P	Microwave oven outlet	120	-0-	12	1440	15	1	14	B17	
▲ Q	Central vac. power unit	125	2	12	1492	15	1	14	B24	Receptacle in garage.

fixtures and appliances shall be operating properly. See plans and other sections of the specifications for information as to who is to furnish the fixtures and appliances.

35. ***Alternative Bid Low-Voltage, Remote-Control System*****:** The electrical contractor shall submit an alternative bid on the following:

 Furnish and install a complete extra-low-voltage, remote-control system to accomplish the same results as would be obtained with the conventional switching arrangement as indicated on the electrical plans.

 Furnish and install one 12-position master selector switch in the master bedroom or as directed by the architect or owner. Outlets to be controlled by this switch are to be selected by the owner.

 Furnish and install two motorized 25-circuit masters. These motor-operated controls shall be run from the front hall and master bedroom, or as indicated by the architect or owner. Connect the motor-operated master control so that each and every switch-controlled lighting outlet and switch-controlled receptacle outlet can be turned off or on from the two control stations.

 All extra-low-voltage wiring is to conform to the *CEC*.

Appendix B

Jack Pin Designations and Colour Codes

Standard 4-Pair Wiring Colour Codes

Pair 1	T R	White/Blue Blue/White
Pair 2	T R	White/Orange Orange/White
Pair 3	T R	White/Green Green/White
Pair 4	T R	White/Brown Brown/White

Note: For 6-wire jacks use pair 1, 2 and 3 colour codes.
For 4-wire jacks use pair 1 and 2 colour codes.

Pin #	***Jack Type***		
	8P8C/ 8P8C Keyed	6P6C	MMJ
1	Blue (L)	White (W)	Orange (O)
2	Orange (O)	Black (B)	Green (G)
3	Black (B)	Red (R)	Red (R)
4	Red (R)	Green (G)	Yellow (Y)
5	Green (G)	Yellow (Y)	Black (B)
6	Yellow (Y)	Blue (L)	Brown (N)
7	Brown (N)	—	—
8	White (W)	—	—

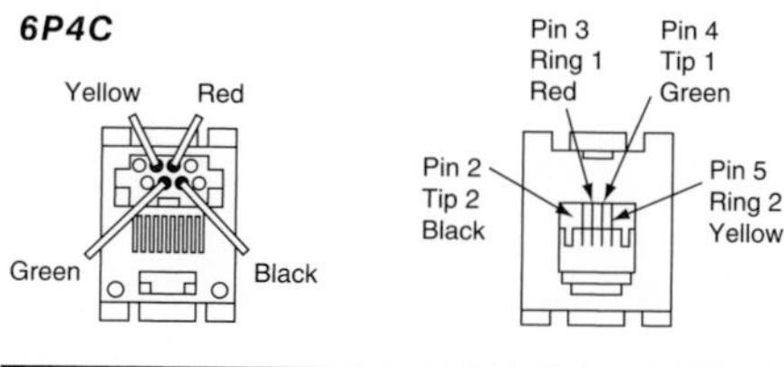

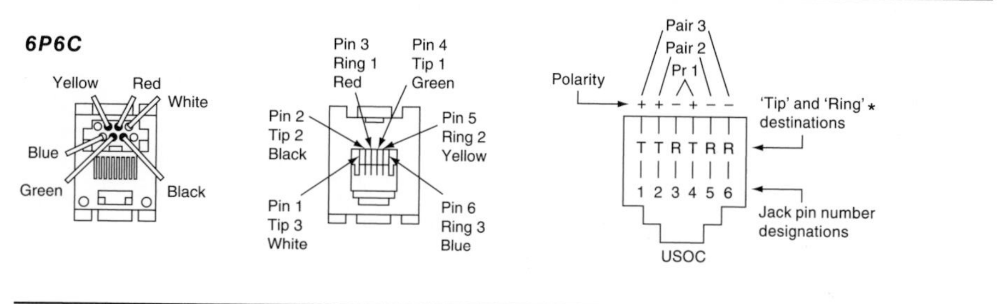

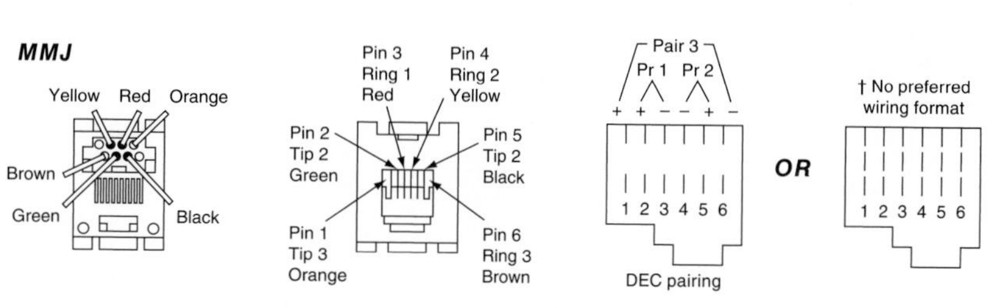

**Some equipment standards may vary from the standards shown here.*

FIGURE B-1 Jack pin designations and colour codes for communication jacks and cables.

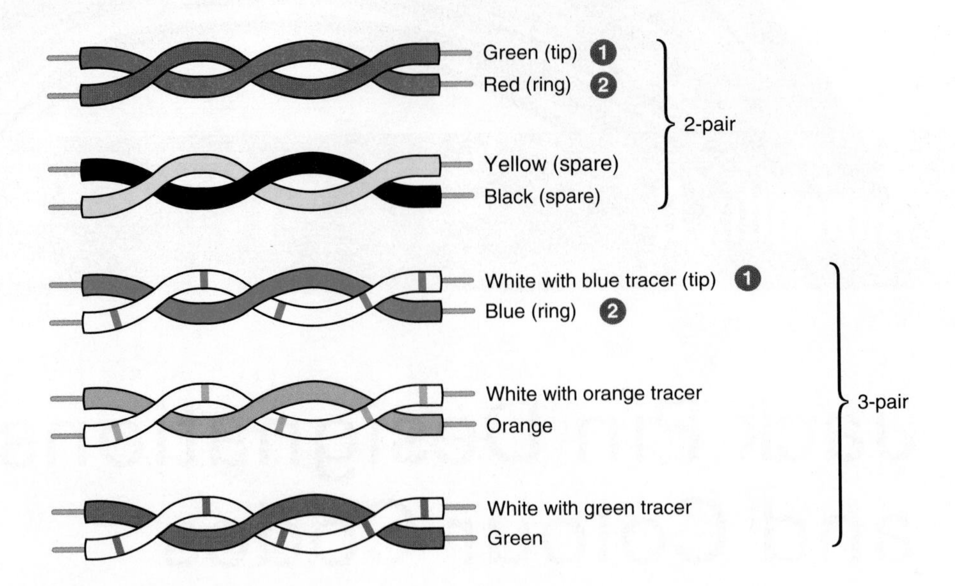

1. *Tip* is the conductor that is connected to the telephone company's positive terminal. It is similar to the neutral conductor of a residential wiring circuit.

2. *Ring* is the conductor that is connected to the telephone company's negative terminal. It is similar to the "hot" ungrounded conductor of a residential wiring circuit.

FIGURE B-2 Colour codes for communication jacks and cables.

Appendix C

EIA/TIA 568A and 568B: Standards for Cabling Used in Telecommunications Applications

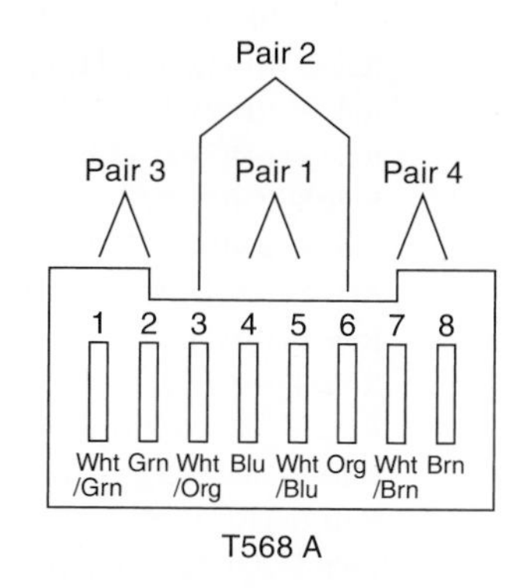

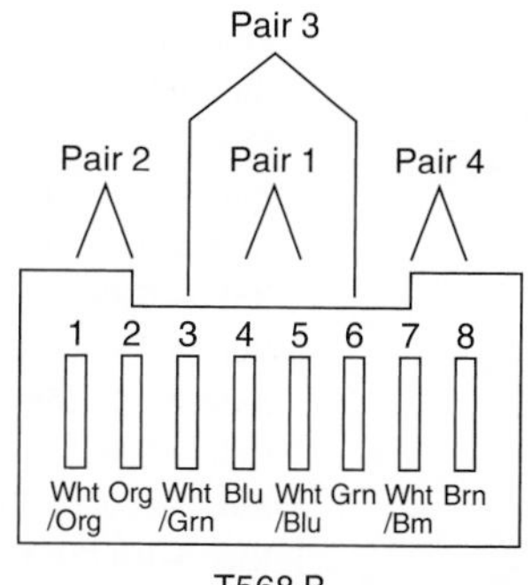

Wht /Grn	Solid white with green stripe
Wht /Blu	Solid white with blue stripe
Grn	Solid green
Blu	Solid blue

Wht /Org	Solid white with orange stripe
Wht /Brn	Solid white with brown stripe
Org	Solid orange
Brn	Solid brown

Standard Colour Connections for Categories 5 and 6

FIGURE C-1 EIA/TIA 568A and 568B

Code Index

S

T

Subject Index

D

E

J

K

L

M

N

O

P